1st International Conference on Frontiers in Engineering Science and Technology (ICFEST 2020)

IOP Conference Series: Materials Science and Engineering
Volume 1065

Mangalore, India
18-19 December 2020

Editors:

R. Badiger
Manjaiah
M. S. Srinath
K. Satyanarayana

ISBN: 978-1-7138-4576-8
ISSN: 1757-8981

Printed from e-media with permission by:

Curran Associates, Inc.
57 Morehouse Lane
Red Hook, NY 12571

Some format issues inherent in the e-media version may also appear in this print version.

Institute of Physics
Dirac House, Temple Back
Bristol BS1 6BE UK

Phone: 44 1 17 929 7481
Fax: 44 1 17 920 0979

techtracking@iop.org

Additional copies of this publication are available from:

Curran Associates, Inc.
57 Morehouse Lane
Red Hook, NY 12571 USA
Phone: 845-758-0400
Fax: 845-758-2633
Email: curran@proceedings.com
Web: www.proceedings.com

TABLE OF CONTENTS

IOP Conf. Series: Materials Science and Engineering 1065 (2021) 011001 doi:10.1088/1757-899X/1065/1/011001

International Conference on Frontiers in Engineering Science and Technology (ICFEST-2020)
Yenepoya Institute of Technology, Moodbidri, Mangalore, India.
18th - 19th December, 2020

Yenepoya Institute of Technology, Moodbidri conducted its first International conference during 18-19 December 2020 and is pleased to publish the selected papers in IOP Conference Series: Materials Science and Engineering. The conference was interdisciplinary and covered several scientific thematic areas from Mechanical, Electronics and communication, Electrical and Computer Engineering.

Virtual Disclaimer for ICFEST-2020: Due to the uncertainty associated with COVID-19 pandemic and travel restrictions across the globe and considering the safety of the participants it was decided to conduct the conference in virtual mode using the Google meet platform. The conference was not postponed to a later date, as the scientific community would have been facing a delay in publishing their research work.

The organizers were located at Yenepoya Institute of Technology; Moodbidri and the participants across the globe presented their research works through oral presentation. Each registered participant was allotted 12 minutes including questionnaire. The conference was conducted smoothly without any major bottlenecks and the organizers would like to express their gratitude to all the people associated especially the invited keynote speakers, session chairs and the participants who supported this conference. Also, the thanks go to the IOP Science for agreeing to publish the selected papers in its Conference Series: Materials Science and Engineering journal.

Day 1 (Friday, 18-12-2020):

The conference was inaugurated by the Chief Guest Dr. S A Khan, Professor, International Islamic University, Malaysia; Mr. Yenepoya Abdulla Javeed, Director (Operations), Yenepoya Group; Dr. Rajesh G D'Souza, Principal; Dr. N Sathisha, HOD Mechanical Engineering and Convener, ICFEST-2020. After the inaugural function keynote speech was delivered by Dr. S A Khan on **Recent Advances in Aerospace Engineering**.

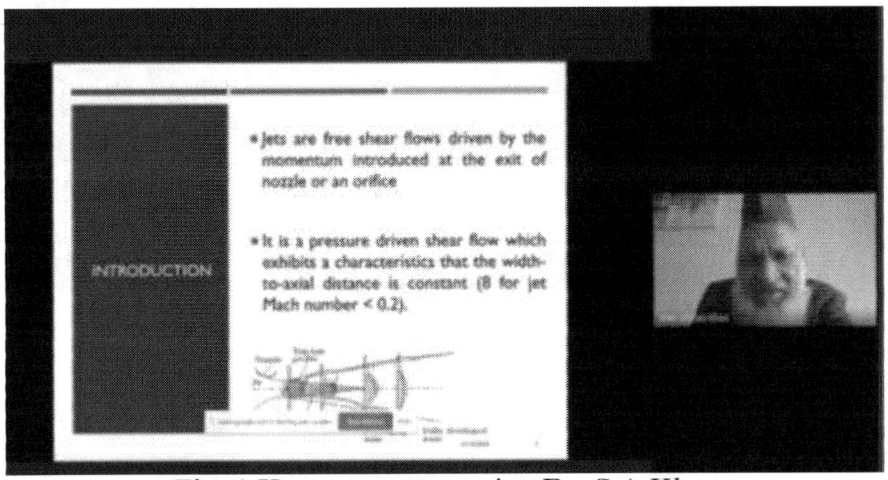

Fig. 1 Keynote presentation Dr. S A Khan

ICFEST-2020 received a total of 92 papers; however, after a preliminary review process only 74 papers were selected for presentation. Three parallel sessions were conducted from 11 AM to 5 PM in

different streams such as Design/Materials Characterization, Thermal Science, Production and Industrial Management, Computer and Wireless Sensor Networks, Artificial Intelligence, VLSI and Embedded Systems and Renewable Energy Sources. Each session was chaired by the experts from several reputed institution across the country and abroad.
Dr. Nusrathulla, Manager (Planning and Operations), Nardoni Institute of Advanced NDT, Muscat in his invited talk spoke about the importance of **Inevitable Renewable Sources of Energy**. Afternoon session was chaired by Dr. Manjaiah M, Asst. Professor, Department of Mechanical Engineering, NIT Warangal. His topic of presentation for the invited talk was **Evolution in Digital Manufacturing.** Dr. Suresha D, Assoc. Professor, Department of Computer Science and Engineering was session chair for computer and information science stream while, Dr. Subramanya Bhat, Assoc. Professor, Department of Electronics and Communication Engineering, NMAMIT, Nitte chaired the session for Electrical and Electronics stream. At the end the of the Day-1 total of 46 papers were presented by the registered authors from various institution across the globe.

Day-2 (Saturday, 19-12-2020):
On Day-2 two sessions were conducted from 10 AM to 5 PM. Each session was chaired by the experts from various reputed institution across the globe. Morning session started with the invited talk by session chair Dr. Ahammed Saleel C, Associate Professor, Department of Mechanical Engineering, King Khalid University, Saudi Arabia. The topic of presentation was **Use of coconut oil as an organic phase change material-scope and applications**. The afternoon session started at 2 PM with the invited talk by Dr. S Somashekhar Hiremath, Associate Professor, Department of Mechanical Engineering, IIT Madras. In his talk Dr. S Somashekhar Hiremath addressed on the issues in **Micromachining: Present and Future Challenges.** Total of 28 papers were presented by the registered authors from various institutions across the globe.

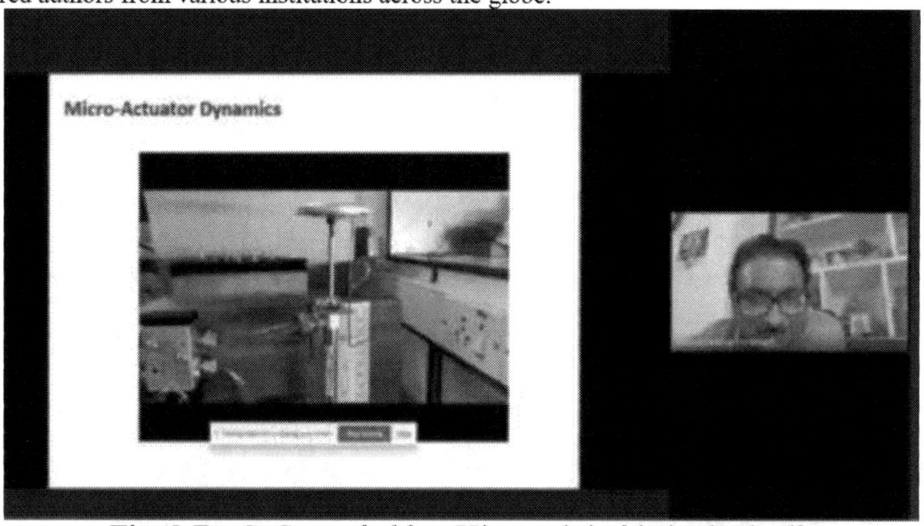

Fig. 2 Dr. S. Somashekhar Hiremath in his invited talk

During the valedictory session Dr. Rajesh D'Souza, Principal, Yenepoya Institute of Technology, expressed the need of research in current scenario with the innovative areas of Engineering. He expressed his satisfaction about quality of paper presented in the conference.
Out of total 92 registered papers, 74 papers were presented. 50 papers will be published in the IOP Conference Materials Science and Engineering Journal. The program came to an end with concluding remarks by Dr. N Sathisha, Convener and HoD of Mechanical Engineering followed by vote of thanks.

IOP Conf. Series: Materials Science and Engineering 1065 (2021) 011001 doi:10.1088/1757-899X/1065/1/011001

IOP Conf. Series: Materials Science and Engineering 1065 (2021) 011001 doi:10.1088/1757-899X/1065/1/011001

Conference Themes

Design/Material characterization

CAD / CAM / CAE, Finite Element Analysis, Material Science & Metallurgy, Composite Materials, Product Design & Development, Nano Materials, Experimental Stress Analysis.

Thermal science and engineering

Energy Conservation and Management, IC Engines, Computational Fluid Dynamics, Nano Fluids, Refrigeration and Air-conditioning, Green Energy Technologies.

Production and Industrial management

Non-conventional Machining, NDT, Industrial Automation, Robotics, Rapid Prototyping and Processing Techniques.

Computer Science

Information and communication technology, Data mining and predictive modeling, Computer vision and Pattern Recognition, Forensics computing and security, Computer Networks, Wireless Sensor Networks, Wireless Sensor Networks, Artificial Intelligence, 5G Networks, Cyber Security, High Performance and Parallel Computing.

Electrical and Electronics

Multimedia Processing, Speech/Signal/Image Processing, VLSI & Embedded System Design, Wireless Communication, MEMS and Nanotechnology, Power Quality Improvement Techniques, Hybrid Vehicles

IOP Conf. Series: Materials Science and Engineering 1065 (2021) 011002 doi:10.1088/1757-899X/1065/1/011002

Peer review declaration

All papers published in this volume of *IOP Conference Series: Materials Science and Engineering* have been peer reviewed through processes administered by the Editors. Reviews were conducted by expert referees to the professional and scientific standards expected of a proceedings journal published by IOP Publishing.

- **Type of peer review:** Double-blind
- **Describe criteria used by Reviewers when accepting/declining papers. Was there the opportunity to resubmit articles after revisions?**
  A basic plagiarism check was done as a primary filtration and the papers with less than 20% plagiarism were sent to the Reviewers and furthermore, they were suggested and briefed out to keep a check on novelty and continuity of flow along with the clarity of thought and correctness concerning the data. Upon satisfying the above-mentioned criteria, the papers were submitted for final publication with IOP. The papers, that failed to meet the proposed criteria, were sent to the respective authors along with comments from reviewers and a second chance was given to resubmit their work addressing the comments.
- **Conference submission management system:** Manual
- **Number of submissions received:** 92
- **Number of submissions sent for review:** 92
- **Number of submissions accepted:** 50

(NOTE: Out of 92 submissions, 74 papers were shortlisted for ICFEST-2020, out of which only 50 papers were considered for publication with IOP, and the remaining 24 papers were considered under the category of the presentation without publication)

- **Acceptance Rate (Number of Submissions Accepted / Number of Submissions Received × 100:** 54.34%
- **Average number of reviews per paper:** 2
- **Total number of reviewers involved:** 35
- **Any additional info on review process (i.e., plagiarism check system):**
  Every submitted paper has gone through a plagiarism check using the software **Turnitin** and every key factor like novelty, continuity of flow, clarity, and correctness of thought was taken into consideration and based upon these; a strict acceptance process was adopted. Papers meeting the high proposed standards were accepted for publication with IOP.
- **Contact person for queries:**

Mr. Anfas Mukram T,
Assistant Professor,
Department of Mechanical Engineering,
Yenepoya Institute of Technology,
Coordinator, ICFEST-2020.
anfasmukram@yit.edu.in
anfasmukramt@gmail.com

Mechanical properties of natural fiber reinforced polylactide composites: A review

Santosh kumar[1], M S Shamprasad[2], Y S Varadarajan[3]

[1]Assistant Professor, Department of Industrial and Production Engineering, The National Institute of Engineering, Mysuru-570008, Karnataka-India; santoshkumar.ip@nie.ac.in; https://orcid.org/0000-0002-6129-4466

[2] Professor, Department of Industrial and Production Engineering, The National Institute of Engineering, Mysuru-570008, Karnataka, India; shamprasad.nie@gmail.com

[3]Professor, Department of Industrial and Production Engineering, The National Institute of Engineering, Mysuru-570008, Karnataka-India; varadarajan.ys@gmail.com; https://orcid.org/0000-0001-8554-213X

Corresponding Author: santoshkumar.ip@nie.ac.in

Abstract. Nowadays, there is an augmented concern for the development of not only well-designed but also fully sustainable materials. The extraction of renewable resources and disposal processes that would not harm our eco-system is associated with material sustainability. In this respect, biodegradable polymers and natural fibers are suitable alternatives to many non-degradable composites which one lightweight. Many researchers are trying to fabricate composites by adding these two through various means and characterizing their mechanical properties to find the suitability of these for different applications. The present article reviews the literature on the fabrication of polylactide composite reinforced with natural fibers, both of which are biodegradable fully; further, the article reviews the methods used by the researcher for characterizing mechanical properties of these composites.

1. Introduction

Nowadays, biodegradable polymers are gaining popularity because of awareness of environmental pollution caused by conventional polymers. The development of biodegradable polymers promotes the use of alternative materials that are environmentally friendly and use them in composite making for various engineering applications. Agricultural crop residues such as rice husk, coconut, flax, oil palm, rice straw, pineapple leaf, banana, jute, bamboo, and sugar palm are produced in very large quantities across the globe[1,2]. They are low cost, easily degradable, and available in abundance. Among these huge quantities of residues, only a little quantity of these are used as domestic fuel or fertilizer, and the maximum portion of these residues are burnt in fields, resulting in Air pollution[3]. These can be overcome by utilizing suitable agricultural residues as reinforcement in the fabrication of natural fiber reinforced polymer composites[4–8]

Many researchers have successfully used these natural fibers in thermoplastics as reinforcement; the advantage being that these composites are recyclable. Most polymers themself do not possess load-bearing capacity as they don't possess the required strength, stiffness, stability. But fibers do have good stiffness and strength but difficult to be used in load-bearing applications because their structure is typically fibrous like say in order to exploit the advantages of these to compensate for the disadvantages these materials are put together in the form of composites[9].

However, research on biodegradable polymers as composite matrices are limited in comparison to conventional polymers (both thermoplastic and thermoset) since they are not readily available and are priced more [10]. The existing literature generally dealt with the study on the effect of adding the natural fiber in conventional polymers on mechanical properties only; the influence of adding natural fibers in the biodegradable matrix-like PLA is hardly dealt with. Adding natural fiber with PLA results in completely biodegradable composites which can be treated to be sustainable development, taking into consideration, both economic and ecological aspects [11]. The classifications of Natural fiber-reinforced bio-composites are shown in figure 1.

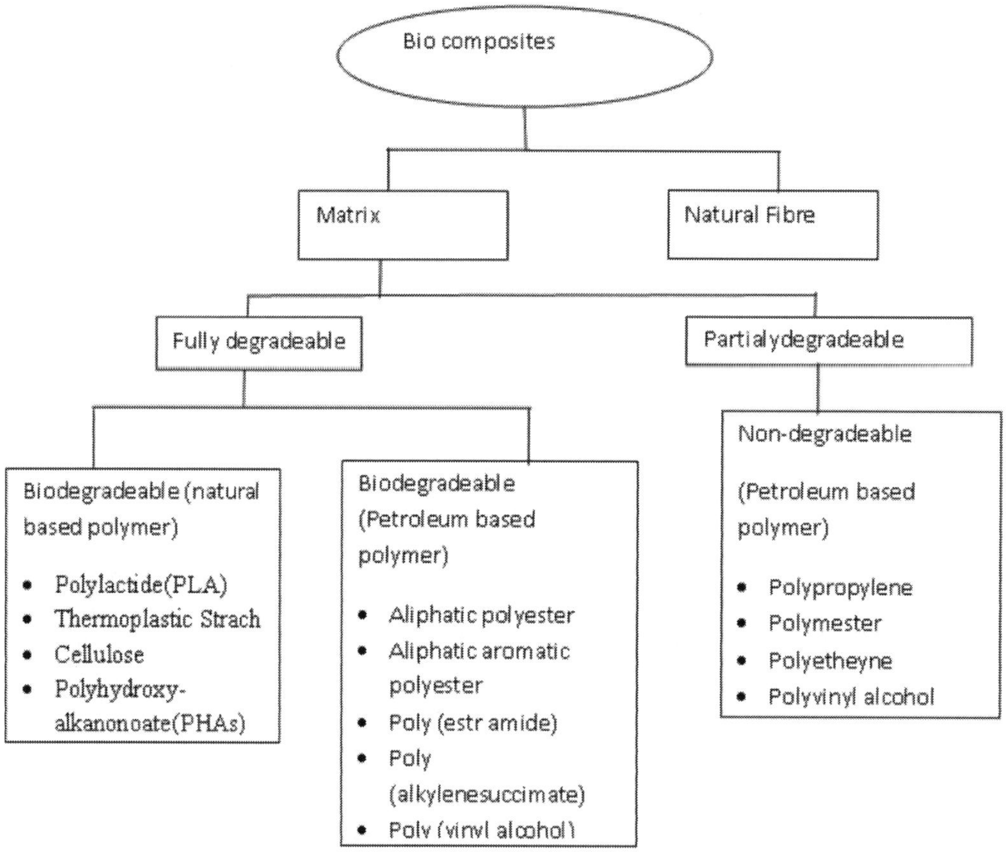

Figure 1. Classifications of bio-composites (adapted from Mohanty et al. [8]).

1.1 Polylactide (PLA)

Polylactide is a biodegradable thermoplastic aliphatic polymer based on lactic acid which can be derived from100% renewable resources like corn starch, sugar cane, cassava roots, or chips. The molecular Structure of polylactide is shown in figure 2.

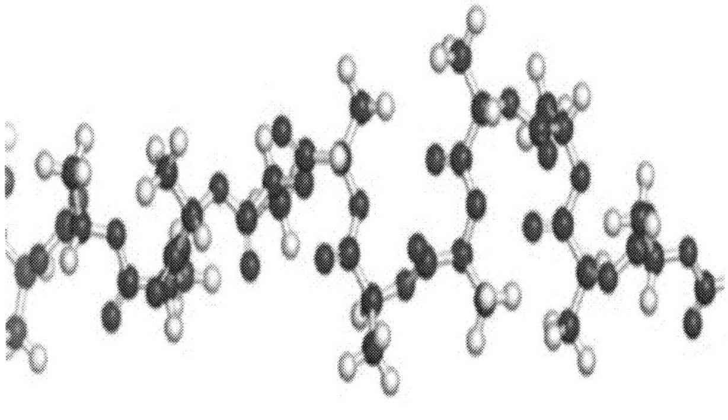

Figure 2. Molecular Structure of polylactide
(Courtesy: Omnexus, the material selection platform)

Polylactide is a biodegradable polymer and has similar characteristics as polystyrene (PS), polyethylene (PE), or polypropylene (PP). It can be manufactured by the same equipment used to produce petrochemical based industry plastic[12]. This method of fabrication is relatively cost-efficient. Among a variety of bioplastics, Polylactide has a second huge quantity of bioplastic production. PLA can be soluble in hot benzene, chlorinated solvents, dioxane, and tetrahydron[13,14]. The properties of polylactide are presented in table 1.

Table 1. Polylactide's properties.

Property	**Value**
Melting temperature	157^0 C to 170^0 C
Injection molding temperature	178^0 C to 240^0 C
Heat deflection temperature	49^0 C to 52^0 C at 4.6 MPa
Shrink rate	0.37 to 0.41%
Flexural strength	48 to 110 MPa
Tensile Strength	61 to 66 MPa

2. Fabrication of fiber-reinforced plastic composites (FRPCs)

FRPCs are mainly fabricated using the production process of extrusion, injection, thermoforming process, or by compression molding methods, and the details are given below.

2.1 Extrusion method: Two types of extrusion methods available; single screw, and twin screw. The schematic set-up of the Polymer Extrusion device is shown in figure 3.

For pure polymers, single screw extrusion is mostly used. Twin-screw extrusion is used if it is made of distinct materials. As part of a large extrusion system, single screw extruders can be used or used as a stand-alone extrusion machine. Twin-screw extruders are suitable for mixing, compounding, and processing viscous materials. Twin-screw extrusion has consistency in production and control of product quality as the process ensures uniform mixing.

In an extruder, a fiber and plastic mixture is fed into the hopper, then they pass through the heating barrel pushed by the rotating screw; during heating, the barrel softens and melts the plastic; the screw supplies the mixture continuously. Melted plastic goes through the die, take the intended shape. The product is then cooled down and drawn out in the form of a pellet, thin sheets, or granules for further molding. The process of extrusion is suitable to produce composite filaments or pellets for further processing[15].

2.2 Injection Molding: In injection molding, the schematic set-up of which is shown in figure 4, the molding compound is poured in a hopper and the same is fed to an extruder. The plastic is pushed by the extruder screw into the heating chamber in which the molding compound is heated, changing it into a liquid state; then, at high pressure, it is injected into the mold of the desired shape. Once the mold is filled, the part is allowed to cool; the mold opens and the final object comes out. Injection molding is a suitable manufacturing process for mass production in large volumes, but it is confined to short

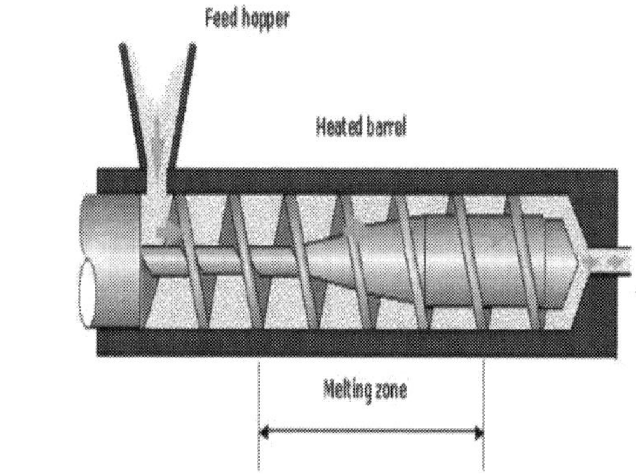

Figure 3. Schematic set-up of polymer extrusion (Courtesy: www.substech.com)

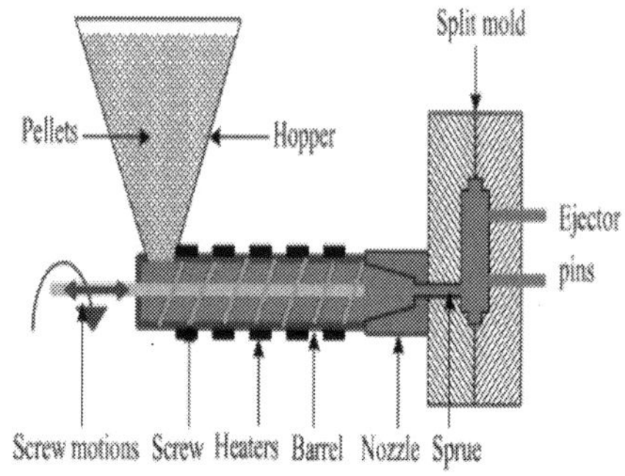

Figure 4. Schematic set-up of Injection molding (Courtesy: www.substech.com)

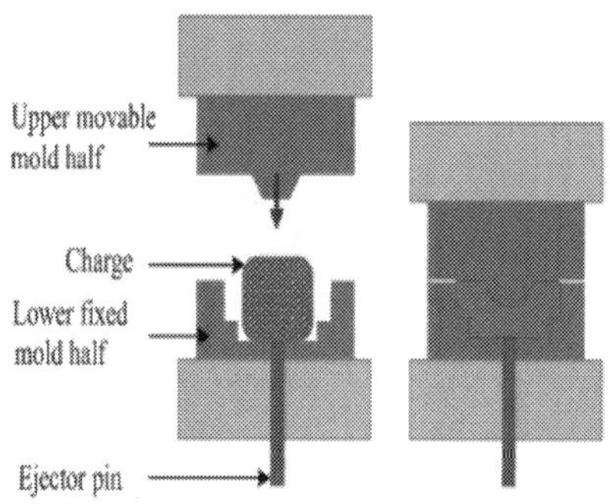

Figure 5. Schematic set-up of compression molding (Courtesy: www.substech.com)

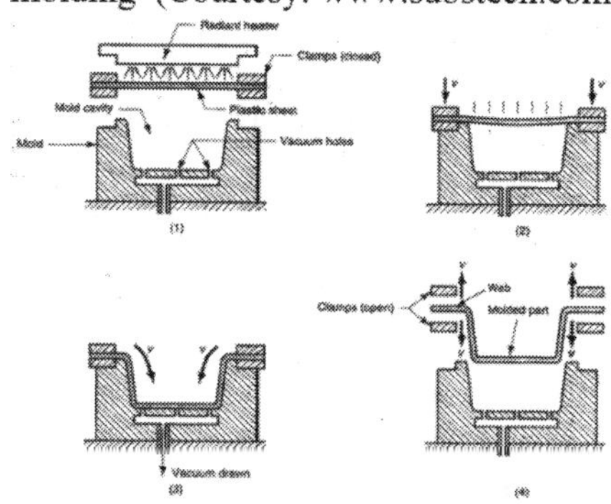

Figure 6. Schematic set-up of thermoforming (adapted from engineering clicks)

fibers[16,17].

2.3 Compression molding: in this process pressure and heat are applied in molding the material to the desired shape, the schematic set-up of which is shown in figure 5. The molds are heated to a preset heat-pressure cycle; a preform/charge is placed in the cavity of the matched mold in the open position; then by bringing two halves of the mold together, the pressure is applied in making the resin fill the mold cavity; heated under a pressure, mold that cures the resin, the composite component is then ejected from the mold[18].

2.4 Thermoforming: Thermoforming is the process where the thermoplastic sheet is heated and deforms into the desired shape, the schematic set-up of which is shown in figure 6.

Process:

- Heating plastic sheet to the temperature wherein softens
- Forming or converting soft sheet into desired die shape by either air pressure, vacuum, or mechanical power
- Cooling the finished part and trimming excess plastic[18].

Eco-composites are the one which has environmental and ecological advantages in comparison with traditional composites. In eco-composites, natural fibers are reinforced in the biodegradable polymer matrix. A variety of biodegradable polymers are available these days[19]. The objective of the present article is to review the literature on natural fiber reinforced polylactide polymers and their properties.

3. Mechanical properties of FRPCs

The properties of natural fiber reinforced polylactide composites(NFRPCs) are depending on several parameters such as types of fiber used, fiber orientation, fiber volume fraction, fabrication process, processing of polylactide polymer, fiber structure, source of fiber, moisture conditions, fiber aspect ratio, treatment of fiber and bonding.

The mechanical properties of the polylactide matrix can be improved by adding fibers since fibers have higher strength and stiffness values. The mechanical properties of a few selected fibers are mentioned in Table 2.

Several researchers experimented on the mechanical properties of FRPCs. Yu et al., carried out experiments on PLA/ ramie fiber and PLA/jute fiber composites [21], Oksaman et al., on PLA/ flax fiber composites [22], Plackett et., on PLA/ jute fiber composites [23], Hu and Lim on PLA/ Hemp fiber composites[24], Rajesh Gunti et al, on PLA /jute fiber, PLA /sisal fiber and PLA / elephant grass fiber composites[25], Bax and Mussig reviewed the impact and tensile properties of PLA/ Cordenka and PLA/ flax composites[26], Ochi, 2008 worked on PLA/ kenaf fiber composites[27], Nam et al., on Coir fiber reinforced PLA composites[29].

Table 2. Properties of Natural fibers selected[20].

Fibers	Specific mass in g/cm^3	Tensile Strength in MPa	Modulus of elasticity in GPa	Elongation at Break Percentage in %
Ramie	1.5	500	44	2
Jute	1.46	400 to 800	10 to 30	1.8
Flax	1.4	88 to 1500	60 to 80	1.2 to 1.6
Hemp	1.48	550 to 900	70	1.6
Sisal	1.33	600 to 700	38	2.3
Elephant grass	0.8175	185 to 327	7.4	3.23
Kenaf (Bast)	1.2	295	-	2.7 to 6.9
Coir	1.25	220	6	15 to 25

3.1 *Tensile strength of fiber-reinforced PLA composites*

The tensile strength of pure PLA is less compared to PLA composites as shown in figure 7 [21]. With the inclusion of either ramie fiber or jute fiber in different weight percentages to the PLA polymer, an increment in tensile strength was observed. But the tensile strength of composites decreased less than that of pure PLA when the addition of fiber is more than 30wt percent. This is due to poor fiber dissipation in the PLA polymer. The tensile strength of ramie fiber reinforced PLA composite is higher than that of jute fiber-reinforced composite since the strength of ramie fiber is more than jute fiber[21]. These results are in agreement with the findings of Rajesh et. al[25].

Further, efforts were made by Plackett et. al[23]. to find the suitable heating temperature of molding of jute fiber-reinforced composites for getting maximum tensile strength. Under a heating temperature of 2100C in a hot press, the tensile strength was increased more than two times that of the polylactide. When the temperature in the heating stage was increased to 2200C, which is far above the recommended processing temperature for the polylactide, a statistically equivalent result was obtained.

IOP Conf. Series: Materials Science and Engineering 1065 (2021) 012001 doi:10.1088/1757-899X/1065/1/012001

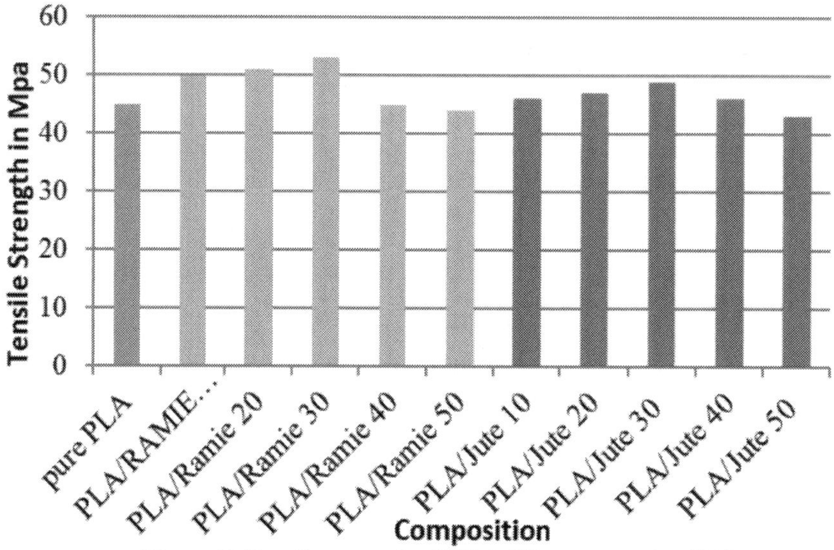

Figure 7. Tensile strength of PLA /fiber composites[21].

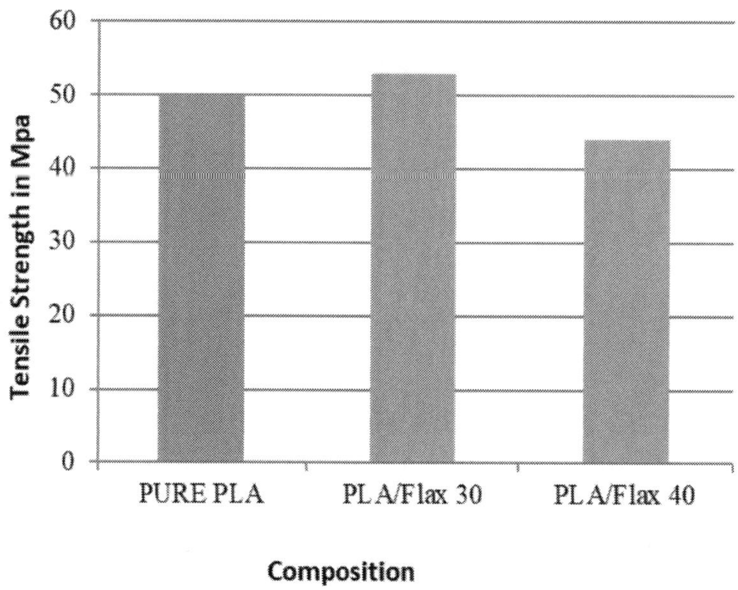

Figure 8. Tensile strength of PLA/flax composites[22].

Figure 8 shows the tensile strength of pure PLA which is 50 MPa as found by Oksman et al[22]. The inclusion of flax fibers of 30weight percent improved the tensile strength marginally; thereafter strength decreased. This was attributed to the poor adhesion of flax fiber with PLA polymer. For composites of PLA/ 40weight percent flax fiber treated with triacetin, the tensile strength was reduced with the inclusion of triacetin content from 5 to 15weight percent. This was anticipated due to the softening effect. The greater triacetin inclusion (15percent) showed an adverse effect in flax fiber reinforced PLA composites[22,26].

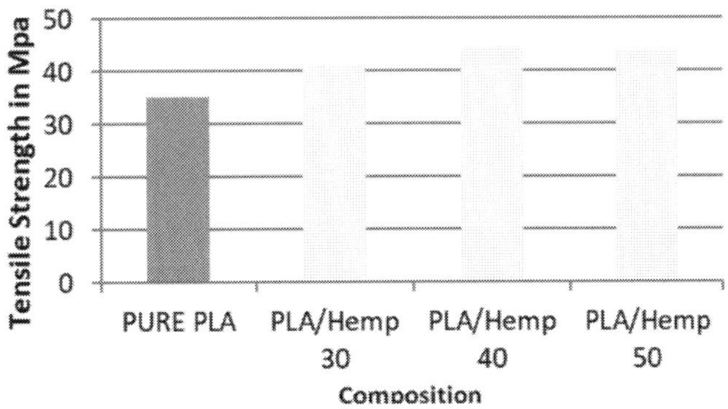

Figure 9. Tensile strength of PLA/Hemp fiber composites[24].

Figure 9 shows the tensile strength of pure PLA is lower than Hemp fiber reinforced PLA composites. The addition of hemp fiber to PLA helps in transferring the load from the matrix to the fiber resulting in increased tensile strength. When the volume percent of hemp fiber in PLA polymer composite is 40 percent, higher tensile strength was observed. But in PLA with 50 volume percent hemp fiber, the tensile strength slightly decreased.

PLA with 40 volume percent hemp fiber treated by 6percent alkali, showed a greater tensile strength of 54.6MPa compared to PLA with 50 volume percent hemp fiber treated with the same percent of alkali. This is possibly due to inadequate fiber wetting as the fiber volume fraction increases[24].

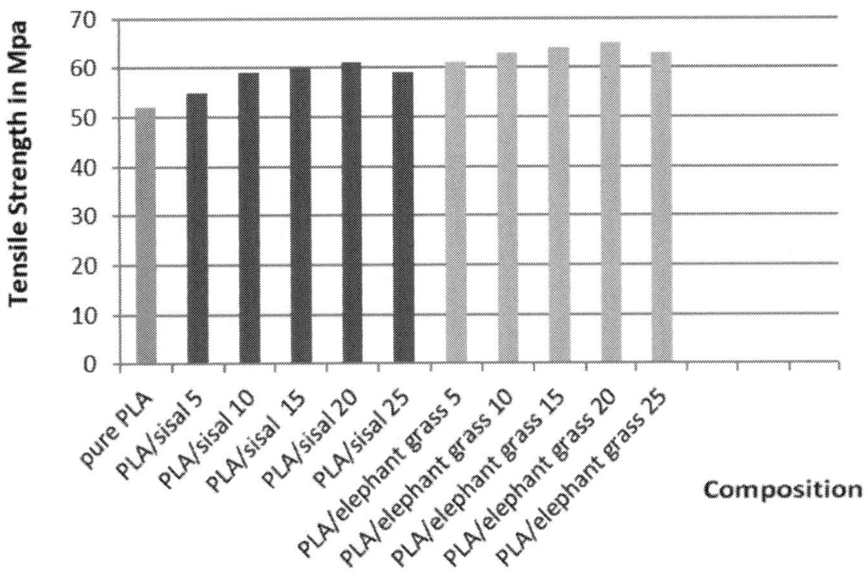

Figure 10. Tensile strength of PLA /fiber composites[25].

Figure 10 shows the tensile strength of PLA/ sisal fiber composites and PLA/ elephant grass fiber composites. It was observed that the strength value of both the composites is more than pure PLA. The tensile strength showed an increasing trend up to 20 percentage by wt of fiber addition; thereafter a declining trend was observed. The PLA/elephant grass composite showed superior strength than that of PLA / jute fiber and PLA/ sisal fiber composites.

Treating all the types of fibers mentioned above with 5 percent alkaline and reinforcing the same with PLA increased the tensile strength further. The PLA/ 5 percent alkaline treated elephant grass fiber composite showed a superior tensile strength than other types of composites[25].

Treating the fiber with alkaline would interrupt the existing hydrogen bonding in the structure, resulting in roughening the surface; further, a small amount of lignin, wax, and oils covering the outer surface of the fiber is removed; all these put together results in enhanced interlocking between the matrix and the fiber and effective stress transfer. As a consequence, treated fiber-reinforced composites showed improved strength[30].

Bleaching with H2O2 removed the lignin without fiber oxidative damage. This treatment ensured compact, clean, and smooth fiber surfaces. Increased NaOH concentration during treatment resulted in the excessive deletion of non-cellulosic material, thus weakening the fiber. This resulted in decreased fiber strength. Further, it was suggested that the lignin amount was not significantly affected by strong NaOH concentrations, but it could contribute to the degradation of the extracted fibers.

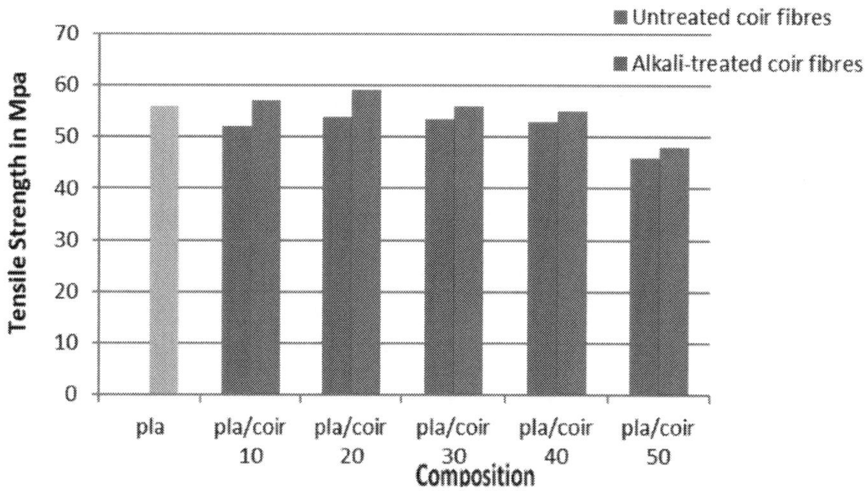

Figure 11. Tensile strength of untreated and 5percent alkali (NaOH solution) treated coir/PLA composites[31].

Nam T H et. al[31]. Experimented assess the influence of adding coir fibers to PLA matrix. Increase presence of fiber improves the tensile strength up to 20 wt percent of fiber content; further addition of fiber did not improve the strength. 50 wt percent of coir presence deteriorated the situation. From figure 11 it is to be noted that the strength values of all the composite with different wt percent of fiber is less than pure PLA.

In order to improve the strength of composites in comparison with pure PLA authors treated the coir fiber with 5 percent alkali (NaOH solution) and then reinforced the same with PLA matrix. A marginal improvement in strength was observed over pure PLA up to 20 wt percent of fiber content, further, addition resulted in decreasing the strength of composite and presence of 50 percent by weight of coir fiber saw a steep decrease in strength

Using kenaf fiber bundles, Kenaf / PLA biodegradable composites have been produced. Figure 12 reveals that tensile strengths increased linearly with an increase in fiber reinforcement up to 50 percent by weight of fiber content[27]. The tensile strength of PLA with 70% by wt of fiber content was observed to be 223 MPa. The tensile strength of kenaf fiber cultivated at the mean temperature of 30^{0}C was higher than that of fiber cultivated at the mean temperature of 22^{0}C. The kenaf cultivated at a mean temperature of 22^{0}C grew up to about 2000 mm, while that cultivated at a mean temperature of 30^{0}C by about 3650 mm; this means that the kenaf fibers which grew over longer length showed improved strength.

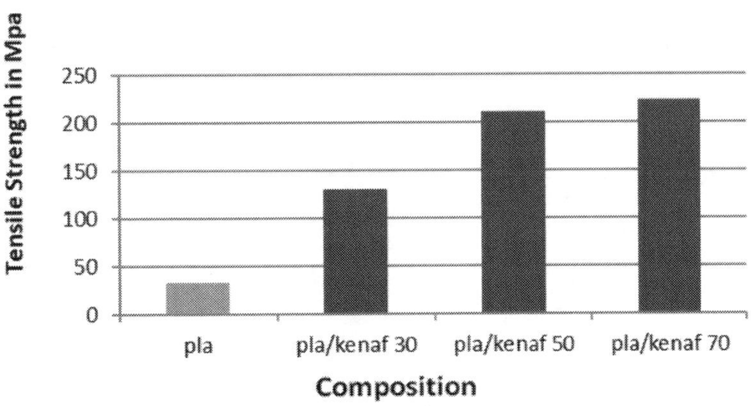

Figure 12. Tensile strength of PLA/ kenaf fiber composites[27].

The addition of fiber in PLA showed a definite increase in tensile strength irrespective of the type of fiber used. The increase in strength was dependent on the type of fiber used as a reinforcement. The addition of elephant grass fiber improved the tensile strength of composite compare to all other types of fibers. Reinforcing PLA with kenaf fiber increased the strength substantially compared to all other types of fiber addition. The presence of 70 percent by wt of kenaf fiber in the composite increases the tensile strength up to 225Mpa which is more than the strength of many of the metals and their alloys.

3.2 Flexural strength of fiber-reinforced PLA composites

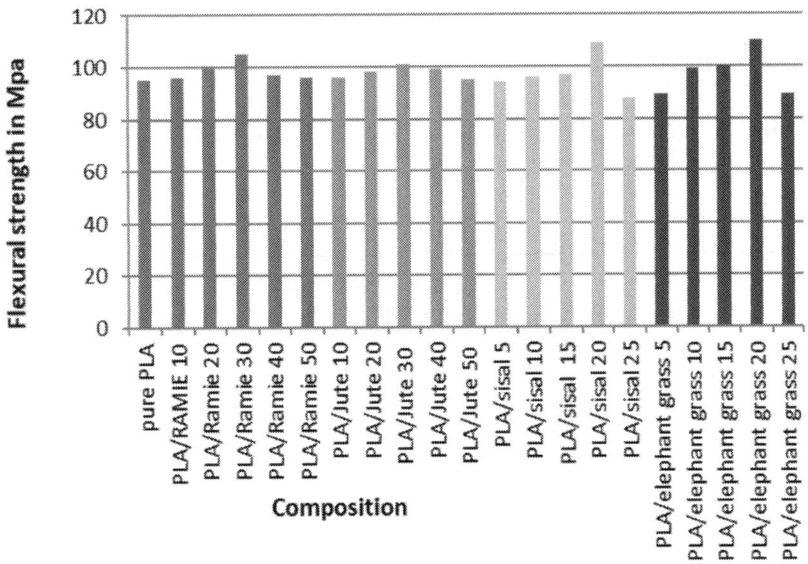

Figure 13.flexural strength of PLA/fiber[25].

Figure 13 shows the flexural strength of pure PLA and PLA-based composites. From the results, it is observed that the flexural strength of PLA/ ramie fiber and PLA/ jute fiber composites is marginally high with pure PLA up to 30 wt percent of fiber loading; the strength falls with further increase in fiber content; in many cases, the strength decreased lower than that of pure PLA. This is attributed to the poor dispersion of fiber in PLA polymer[25,30].

The flexural strength of PLA/ sisal and PLA/ elephant grass reinforced composites increased with the fiber loading initially and then decreased when the fiber loading is more than 20 wt percentage [30]. Flexural strength was found to be maximum for PLA/ elephant grass composites with 20 wt percentage reinforcement, both for treated and untreated fibers. The flexural strength of the PLA/ sisal and PLA/ elephant grass reinforced composites was higher in comparison with other types of composites.

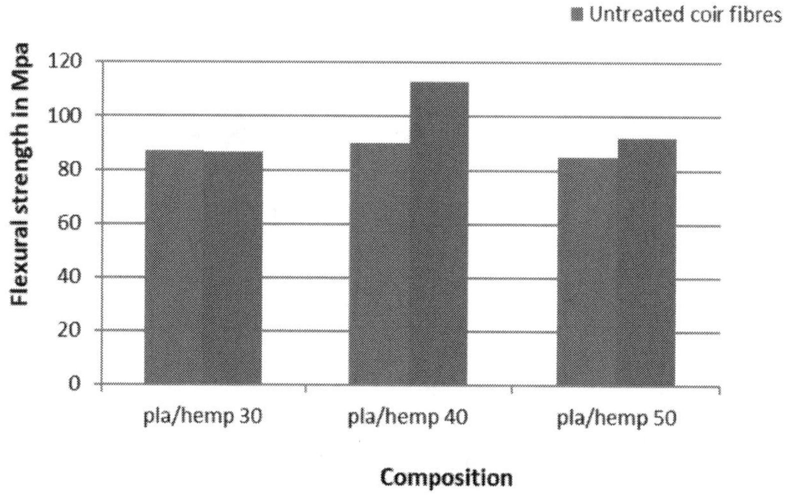

Figure 14. flexural strength of PLA/hemp fiber composites[24].

The flexural strength of PLA/ hemp fiber composites at different fiber volume fractions is shown in Figure14 [24]. Flexural strength for PLA/untreated hemp fiber-reinforced composite does not improve with increasing fiber loading. This is due to the lack of fiber surface wetting by the PLA, resulting in poor bonding between the fiber and the matrix.

According to the rule of mixtures, the strength and modulus should increase with the increase of fiber volume fraction but results showed that this tendency is not followed. Flexural strength increased noticeably with alkali treating of fiber in composites. The maximum flexural strength was found to be 112.7MPa obtained for alkali-treated hemp fiber/PLA composite with 40 percent by wt. of fiber presence; this value was comparably larger than that for PLA with 30wt percent of chopped glass fiber composite and much larger than that of polypropylene reinforced with 30 wt percent of chopped glass fiber composite.

Figur15 shows that flexural strength rises directly with increasing fiber loading in PLA/kenaf fiber composites up to 50 percent by weight [27]. With further increase in fiber content to 70 percent by weight showed only a marginal improvement in strength; anyway, the strength was found to be 254MPa for this composite which is much higher compared to any other natural fiber reinforced PLA composites.

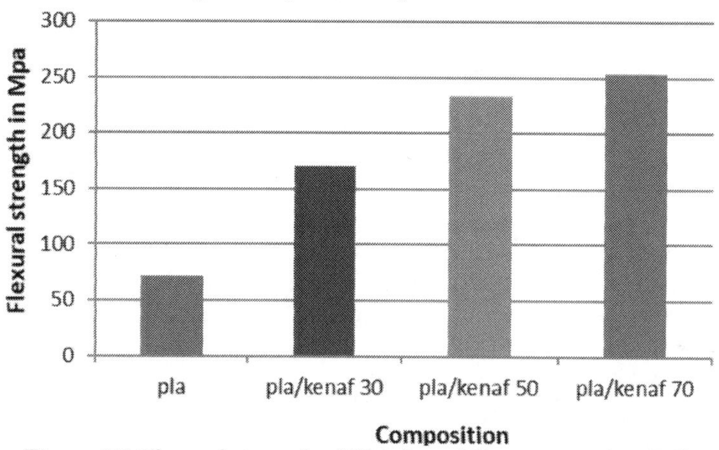

Figure 15. Flexural strength of PLA/kenaf fiber composites [27].

4. Conclusions

Considerable research is being carried out on FRPCs over the last few years because of their excellent properties. PLA is a fully biodegradable thermoplastic polymer, which has certain excellent properties. Several different types of fibers are employed as reinforcement with PLA to fabricate a variety of composites. Kenaf fibers are very strong amongst all the natural fibers; thus kenaf fiber reinforced PLA polymer composites showed the best mechanical properties.

There are certain drawbacks observed in natural fibers while used in PLA for making composites like poor interfacial bonding with polymer, absorbing moisture, low processing temperature of about 2000C, and inadequate dimensional stability. Surface treatment of fiber using NaOH removed the impurities from the surface of fiber making the surface rougher which helps in getting better mechanical interlocking with the matrix resulting in improvement in properties, but there is an increase in the final product cost.

FRPCs have a bright future, as PLA is a fully biodegradable thermoplastic polymer. Future researchers could look for enhancing the mechanical properties of FRPCs. Also in view of the environmental consideration, the emphasis should be on replacing synthetic fibers with natural fibers in PLA composites.

The objective of the present article is to throw light on whether PLA can be used as a matrix in which bio-fibers can be reinforced to fabricate the composite structure. Many of the works as briefed above suggest that composites with renewable fiber reinforcement show qualities comparable to those of synthetic fiber

References

[1] Li Z, Zhou X and Pei C 2011 Effect of sisal fiber surface treatment on properties of sisal fiber reinforced polylactide composites Int. J. Polym. Sci.
[2] https://bioplasticsnews.com/polylactic-acid-or-polylactide-pla/
[3] Bajpai P K, Singh I and Madaan J 2014 Development and characterization of PLA based green composites: A review J. Thermoplast. Compos. Mater.27 52–81
[4] Mohammed L, Ansari M N M, Pua G, Jawaid M and Islam M S 2015 A review on natural fiber reinforced polymer composite and its applications Int. J. Polym. Sci.2015
[5] Nabi Saheb D and Jog J P 1999 Natural fiber polymer composites: A review Adv. Polym. Technol.18 351–63
[6] Garlotta D 2019 A Literature Review of Poly (Lactic Acid) A Literature Review of Poly (Lactic Acid) J. Polym. Environ.9 63–84
[7] Sahari J and Sapuan S M 2012 Natural fibre reinforced biodegradable polymer composites Rev. Adv. Mater. Sci.30 166–74
[8] Amar K. Mohanty, Manjusri Misra L T D 2005 Natural Fibers, Biopolymers, and Biocomposites (CRC Press, Boca Raton)
[9] Medina L A and Dzalto J 2017 Natural fibers Compr. Compos. Mater. II 269–94
[10] Islam M S, Pickering K L and Foreman N J 2010 Influence of alkali treatment on the interfacial and physico-mechanical properties of industrial hemp fibre reinforced polylactic acid composites Compos. Part A Appl. Sci. Manuf.41 596–603
[11] Kumar A A, Karthick K and Arumugam K P 2011 Biodegradable Polymers and Its Applications 1 1–4
[12] Anon 2015 Everything You Need To Know About Polylactic Acid (PLA)
[13] Santos Rosa D dos and Maria D 2013 Biocomposites: Influence of Matrix Nature and Additives on the Properties and Biodegradation Behaviour Biodegrad. - Eng. Technol.
[14] Lu D R, Xiao C M and Xu S J 2009 Starch-based completely biodegradable polymer materials Express Polym. Lett.3 366–75
[15] Shubhra Q T H, Alam A K M M and Quaiyyum M A 2013 Mechanical properties of polypropylene composites: A review J. Thermoplast. Compos. Mater.26 362–91
[16] Morales G, Barrena M I, Salazar J M G de, Merino C and Rodríguez D 2010 Conductive CNF-reinforced hybrid composites by injection moulding Compos. Struct.92 1416–22

[17] Huda M S, Drzal L T, Mohanty A K, and Misra M 2006 Chopped glass and recycled newspaper as reinforcement fibers in injection molded poly(lactic acid) (PLA) composites: A comparative study Compos. Sci. Technol.66 1813–24
[18] http:// www.engineeringclicks.com
[19] G. Bogoeva-Gaceva, 1 M. Avella, 2 M. Malinconico, 2 A. Buzarovska, 3 A. Grozdanov, 3 G. Gentile 4 M.E. Errico4 2008 Natural Fiber Eco-Composites G. Polym. Polym. Compos.16 101–13
[20] Jawaid M and Abdul Khalil H P S 2011 Cellulosic/synthetic fibre reinforced polymer hybrid composites: A review Carbohydr. Polym.86 1–18
[21] Yu T, Li Y and Ren J 2009 Preparation and properties of short natural fiber reinforced poly(lactic acid) composites Trans. Nonferrous Met. Soc. China 651–5
[22] Oksman K, Skrifvars M and Selin J F 2003 Natural fibres as reinforcement in polylactic acid (PLA) composite Compos. Sci. Technol.63 1317–24
[23] Plackett D, Andersen T L, Pedersen W B and Nielsen L 2003 Biodegradable composites based on L-polylactide and jute fibres Compos. Sci. Technol.63 1287–96
[24] Hu R and Lim J K 2007 Fabrication and mechanical properties of completely biodegradable hemp fiber reinforced polylactic acid composites J. Compos. Mater.41 1655–69
[25] Rajesh Gunti, Ratna Prasad A V, Gupta A V S S K S 2016 Mechanical and Degradation Properties of Natural Fiber Reinforced PLA Composites: Jute, Sisal, and Elephant Grass wileyonlinelibrary.com 2016
[26] Bax B and Müssig J 2008 Impact and tensile properties of PLA/Cordenka and PLA/flax composites Compos. Sci. Technol.68 1601–7
[27] Ochi S 2008 Mechanical properties of kenaf fibers and kenaf/PLA composites Mech. Mater.40 446–52
[28] Huda M S, Drzal L T, Mohanty A K and Misra M 2008 Effect of fiber surface treatments on the properties of laminated biocomposites from poly(lactic acid) (PLA) and kenaf fibers Compos. Sci. Technol.68 424–32
[29] Dong Y, Ghataura A, Takagi H, Haroosh H J, Nakagaito A N and Lau K T 2014 Polylactic acid (PLA) biocomposites reinforced with coir fibres: Evaluation of mechanical performance and multifunctional properties Compos. Part A Appl. Sci. Manuf.63 76–84
[30] Tao Yu, Jie Ren, Shumao Li, Hua Yuan and Yan Li 2010 Effect of fiber surface treatments on the properties of poly(lactic acid)/ramie composites Composites: Part A 41 499–505
[31] Nam T H, Ogihara S and Kobayashi S 2012 Interfacial, mechanical and thermal properties of coir fiber-reinforced poly(lactic acid) biodegradable composites Adv. Compos. Mater.21 103–22
[32] http://www.substech.com
[33] Vijay kumar, Thakur micael R and Kessler 2016 Green Biorenewable Biocomposites: From Knowledge to Industrial Applications, CRC press2016 568.
[34] Sawpan M A, Pickering K L and Fernyhough A 2011 Effect of fibre treatments on interfacial shear strength of hemp fibre reinforced polylactide and unsaturated polyester composites Compos. Part A Appl. Sci. Manuf.42 1189–96.

Application of value stream mapping using simulation tool in manufacturing assemble line: A case study

Latha B.M[1*], Raghavendra N V[2] and Jangam Ramesh[3]

[1]Assistant Professor, Department of Industrial and Production Engineering, The National Institute of Engineering, Mysuru, Karnataka, India,

[2] Professor, Department of Mechanical engineering, The National Institute of Engineering, Mysuru, Karnataka, India.

[3] Research scholar,Department of Industrial and Production Engineering, The National Institute of Engineering, Mysuru, Karnataka, India,

Corresponding Author: lathabm@nie.ac.in

Abstract. Value stream mapping is one of the most convenient graphic tools in the manufacturing industry to eliminate wastes. This study has been carried out for the heat sink part assembly line of a manufacturing unit. The basic objective was to employ computational tools for carrying out value stream mapping and to make improvements. In the first stage, value stream mapping was done for the existing assemble line with details such as cycle time, setup time, number of employees, etc., to identify bottlenecks. Future value stream mapping was carried out using a software tool. Simulation is employed for identifying the real-time production bottlenecks; this in turn leads to improvement in productivity. This proposed method can be implemented for any manufacturing industry to eliminate non-value-added activities (NVA). The study demonstrates that the use of appropriate computational tools ensure standard methodology for a complex problem as well as produces quick results.

1. Introduction

To be competitive in the national and global markets, manufacturing organizations have to adopt and implement better strategies to be successful. The current industrial revolution is moving ahead with rapid progress in technology transfer along with better practices to eliminate non-value added activities in the manufacturing sector. It is the larger consumer of natural resources, this will lead to a severe impact on the industrial sector. It poses a severe effect on the growth of developing nations such as India, China, Brazil, and the like. In the Indian economy, the manufacturing sector is one of the main contributorsto economic development. But Indian manufacturing firms are not coping up with the global market in competition and growth due to obsolete technology and financial constraints. This can be overcome by less investment, better management strategies, computational tools, and techniques in the manufacturing sector. Therefore, there is a need to fill the gap in the manufacturing sector through computational tools and lean manufacturing principles with less investment for

IOP Conf. Series: Materials Science and Engineering 1065 (2021) 012002 doi:10.1088/1757-899X/1065/1/012002

enhancing economic performance and competitiveness. A lean manufacturing system is an effective technique in identifying and eliminating different wastes for improving productivity by reducing lead time in production.

Manufacturing layout plays a key role, it incorporated with numerous work stations and workflow path; to carry a raw material till a finished good is produced from it. Each operation is performed as per schedule and sequencing to meet production volume; each entity is interdependent on the other for completing a required task in a given time. Shortage of inventory, unavailability of machine tools, and breakdown of machines causes a delay in meeting the production rate. Improper layout design, unidentified items, and unnecessary flow of material create fatigue to the operator. This eventually results in a slower production rate. VSM analyzes the manufacturing processes from the beginning to the completion of tasks with the help of available facility, the flow of material, people, products, and information. A visual representation map can be drawn based on this. Value stream mapping (VSM) is an efficient tool to make the entire stream valuable by eliminating the non-value added activities.It describes the value map which is a collection of value-added and non-value-added activities.This map comprises every single process involved in the material and information flow. Based on the present data, a future state map is drawn to show how things would work to attain competitive advantage for a firm. Due to severe competition, firms are working for achieving minimal lead time, lower production costs, and greater customer satisfaction. In the present work, lean techniques along with computational tools are successfully used to reduce lead time in a heat sink part assembly line. Tecnomatix Plant Simulation software is used in the present study to capture non-value added activities in the manufacturing unit.

2. Literature survey

Lean manufacturing, pioneered by the Toyota Motor company, has become very popular among manufacturing organizations around the world, to reduce waste. Value stream mapping (VSM) is a lean management method for analyzing the current state and designing a future state for the series of events that take a product or service from its beginning through to the customer. Several research studies have been carried out to analyze the effect of VSM on productivity. Life usage monitoring system (LUMS), which is one of the best methods to identify the component wear out, fatigue failure, and damages of aerospace components [1]. LUMS method uses lean principles to make the process scheduled, process planning and controlling. It was concluded that the LUMS method can be used to find failure mechanisms in engine parts (Blades, Discs). Lean startup methodology (LSM)focuses on failures in the business area and agility (market response) [2]. It also helps an entrepreneur in identifying the different causes of business failure. It interpreted that LMS gives a clear view of the improvement in their business.

Manufacturing strategies have explored procedure on knowledge creation and lean performance[3]. This method helps the management in creating knowledge on lean indicators in explicit and tacit ways. It played a major role in lean practice and it has been used as a prerequisite for lean adoption. Further,it suggested that the top management should have more awareness of knowledge creation and its performance to improve the performance level. A manufacturing sector has chosen to discuss and assess the work in the process in the manufacturing line with the help of a mapping tool[4]. Itwas found that the mapping tool plays a major role in identifying, analyzing, and evaluate certain non-value added activities in a flow line and hence improves productivity.

VSM helps in reducing different non-value-added activities like process variation, product defect identification, machine variability, etc [5]. It represents the whole process of the assembly line bygraphical methodwhereas VSMII shows the process mapping for a product family i.e. group of product which has gone through a similar process and with the help of variability index, processing time can be calculated.

VSM tool found reduces the wastes and increases the efficiency of the production line[6].They interpreted 5S, kaizen, andkanban are the most important tools in a manufacturing firm to eliminate wastes. Companies require improvement in various factors such as production lead time, customer service, etc to survive in the competitive world [7]. This paper details how to draw the value stream mapping by collecting relevant data about the process flow and identify the different non-value-added activities and suggest the future value stream map to eliminate different wastes and putting significant effort in improving the production efficiency. VSM can be implemented for both manufacturing and service sectors [8]. The tool explores different Muda, which is a serious issue in the organization which affects the production level tremendously. By representing the future mapping graphically, improvements in lead time, overproductions can be done effectively. Therefore, the authors declared that value stream mapping is the best tool that eliminates different wastes. Value stream mapping is the fundamental tool of lean practice which has been proved efficient in reducing the different mudas. Finally, it was proposed that value stream mapping is one of the best tools of the Toyota production system which can be implemented successfully for the organization to reduce or eliminate the wastes.

A series of different scenarios of the present and future map and to detectvarious wastes that can be eliminated and the flow line can be made smoother. Interestingly, ithas shown that VSM can help in the ergonomic design of work stations and make the production line streamlined [9,10,11,]. Hence, productivity also can be improved. An application of value stream mapping in the manufacturing sector and investigated the main difficulties and limitations in drawing the current value map [12]. Finally, it tried to explain that rather than benefits in the implementation of value stream mapping tools try to get an exposure on basic knowledge in drawing in the value map and also its limitation

Further, computational tools are very important for lean manufacturing systems. The combined effect of lean practice with simulation software makes the system feasible. Many research works discussed various simulation software such as simul8, ARENA, Discrete event simulation (DES), Tecnomatix simulation (TX) have been integrated with lean with to improve productivity and material flow can be optimized by implementing optimized simulation techniques[14]. The evaluation of present and future value maps based on discrete simulation process and control feedback also be an essential element to help in preventing deviations in the production line. Finally, it interpreted that the simulation software is particularly suitable for medium and high volume production industry. An earthmoving bucket manufacturing industry for the case study [13]. Collected data such as set up time, work in the process as (WIP), cycle time at the individual workstation, and identified the wastes. Value stream mapping for the present process can be drawn and by inputting the value to the arena software for model creation, the simulation will be done and wastes are identified virtually, analyze the process.

As can be observed, frameworks to extend VSM to assess shop floor performance have been proposed by many. However, there is a need to incorporate better strategies for enhancing productivity with optimal utilization of resources. A comprehensive application of VSM through simulation tool highlights the different scenarios for optimizing the layout before implementing the proposed plans and not only serve as a tool to identify and eliminate waste in any activity that does not add value to the final product in the assembly process. With this objective, a case study has been carried out in a heat sink part manufacturing assembly line to capture the benefits of VSM along with simulation tools. Simulation technique along with lean manufacturing principles in the manufacturing process gives valuable insights to compare the present value stream mapping (PVSM) to identify the value-added and non-value-added activity and propose future value stream mapping (FVSM) to eliminate the wastes

3. Methodology

The VSM methodology and Tecnomatix plant simulation software are adopted in the present study. The systems approach will help to provide a holistic view during the project. A combination of the case study and action research has been used as the research method. Initially, the researchwill start with a case study to build a thorough understanding of the organization and its processes. After this has been achieved the focus will be altered to improving the process while observing it. This means that the research method will be changed into action research. The important steps to getting started with the value stream mapping process which has been followed in this work are shown in Figure.

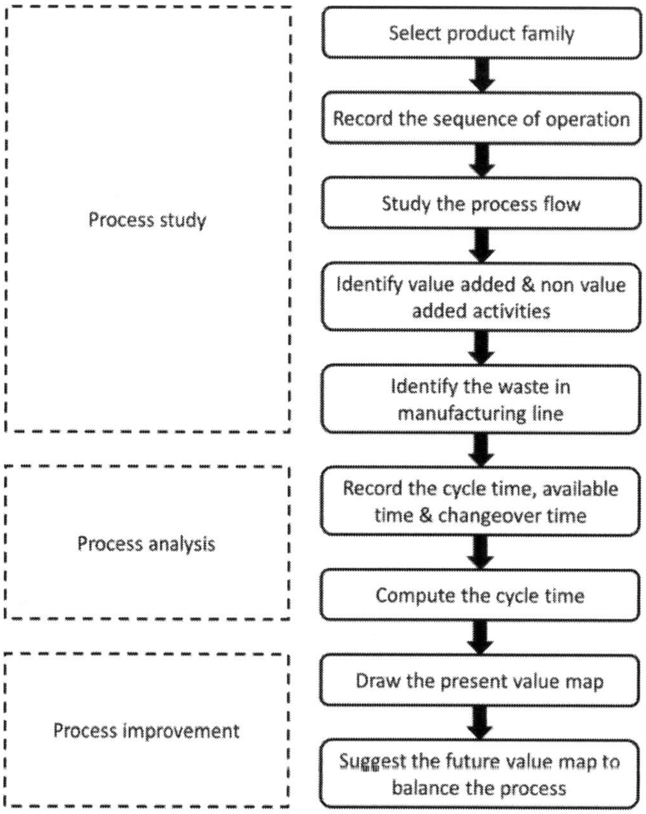

Figure.1 VSM Methodology

The following wastages were identified in the current of the heat sink part assembly line of a manufacturing unit

- It has bottlenecks such as stage inspection in the manufacturing line which leads to waiting time and delays in production order.
- There will be waiting time in the final (workstation) inspection stage.
- Excess stock in the inventory

The well-reputed manufacturing industry was selected for the case study. The first step is to visit the company assembly line to get a clear idea about the manufacturing process in the assembly line. The main objective of the VSM tool is to expose and reduce the time and price which is not adding any value to the product as much as possible. By implementing VSM with the Kanban tool, the company can reduce its WIP by converting the push systems into a pullsystem. Current data was collected and identify the different non-value added activities (NVA) such as excessive inventory level, huge production lead time, overproduction, rework, scrap, set up time, mismanagement of resources, errors in production layout, etc. VSM is the only process that includes both flows in current and future state maps and helps establish a relation between them. The initial stages of VSM consist of mapping the current state of the manufacturing plant. The current state is studied to identify wastes and bottleneck processes present in the process. Appropriate solutions are proposed to eliminate wastes, based on which a future state map is created[14].

It is very essential to understand the flow of material and information along with the machine stations involved with the process. Data collection should start with the "end customer" and walk the process backward. Information such as non-value-added time (time which does not contribute to final product), changeover time (time to change tool or machine time, etc.), amount of WIP(Work-In-Process), a number of operators at each station, standard set up for each operation, inspection or testing time and the like are collected for evaluating waste. It is also beneficial to calculate Takt time for comparison to the actual process time once the VSM is complete.

It has been observed from the literature there are various wastes identified in the manufacturing firm that reduce productivity like over-processing, unwanted or excess inventory, defects (scrap/rework), waiting (unnecessary delay), unnecessary motion or movement, and the like.

A simulation study is not only restricted to optimize the process, rather it gives an immense idea to the manager of the company a clear knowledge of lean implementation and also LM elements effects on the performance of the organization. This software can experiment on the various process parameters in the organization such as the design of the layout, allocation of resources, and the like, which helps in minimizing the production throughput and cost.

The study of the current layout is carried out as follows: In the present scenario, there are 7 workstations in the heat sink part assembly line with one manpower at each work station.Therewas a visual inspection between the station 3rd and 4th between which consume 12 seconds for inspection of each part which leads to waiting time in the line. There is a need to improve productivity by eliminating visual or stage inspection; because the number of defects identified from the production histories is 4 out of 1000 parts produced in the series if workstations kept before the visual inspection, it leads to an increase in production time, waiting time, blocked time for next operation and interruption for smooth production.It was observed that at end of the assembly line there is a final inspection that will cover the operation performed at the 4th work station and it consuming more time. Therefore, there is a need to improve the typical manufacturing assembly line with simulation techniques and with different scenarios to enhance its performance. Hence a proposed model is developed to eliminate the state inspection in the line and reduce the workforce to30% with the same production order (output).

The current value map is used for developing the model with process parameters captured from the assembly line for each work station. Information such as setup time, number ofoperators, machine arrangement, cycle time, and change over time were collected in the workstation are noted in table 1.

Table1: Details of operations carried out in current value map

Sl.No	No of operators at each work station	Operation involved	Cycle time(sec)	Changeover time(sec)
1	1	Bolt pressing	10.28	0.93
2	1	Diode pressing	10.28	0.28
3	1	Diode Testing	12	0.22
4	1	Visual inspection	12	2
5	1	Riveting & flaring	16.58	6.58
6	1	Crimping &welding	14.40	4.40
7	1	Final inspection	23.07	4.2

A snapshot of the simulation model for the current VSM is shown in figure 2. The entities of different machines were identified from the manager class library and are built-inthe "Simulation world".The existing plant was abstracted by placing the different workstations as per the distance allocated and labor was imported to each workstation as per thetable1. For each workstation, the details of setup time, processing time, recovery time,etc, were feed in the associated box. The sequence of operations for processing heat sink part assembly is established in the material flow line as shown in figure 3. Some of the bottlenecks observed in the manufacturing line are visual or stage inspection between the stations, unnecessary labor allocated instage inspection which leads to labor cost, and also waiting time inthe final inspection. With the help of simulation software build an efficient layout and set an optimal number of the operator.

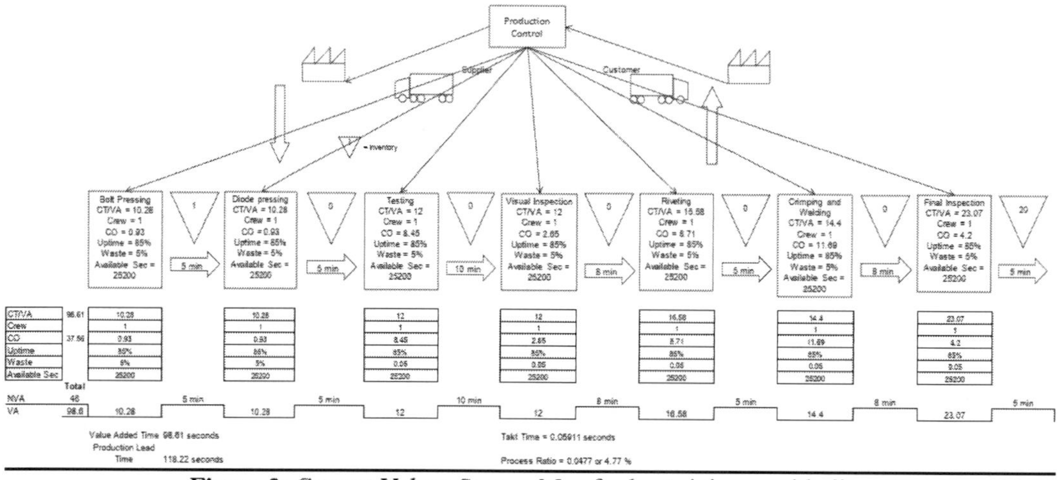

Figure 2: Current Values Stream Map for heat sink assembly line

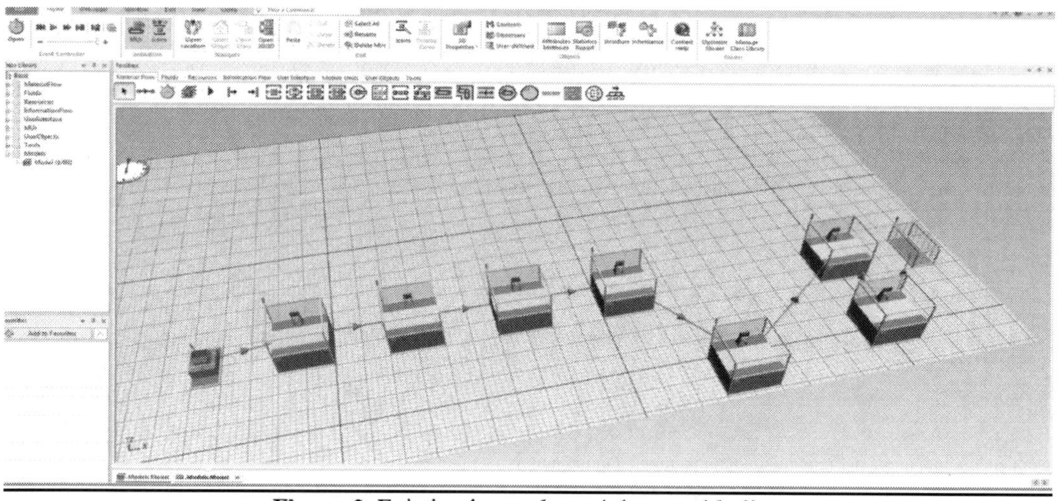

Figure 3: Existing layout heat sink assembly line

The current value map revealed many non-value-added wastes in the assembly line. To reduce this waste and increase productivity; different scenarios are analyzed and it has been proposed to eliminate stage inspection to reduce the cycle time and also suggested implementing them with the automated inspection at the final inspection stage and the results were quite encouraging. Tecnomatix Plant Simulation (TX) software is proved as an improvement tool in yielding the results in seconds with minimum cost to real-life implementation and testing. It explores the dynamic flow effects hidden in the static value chain. The goal is not only to eliminate such waste but to ensure the manufacture of productsas per the customer's demand or requirement as quickly as possible and at the lowest cost. The proposed plant layout was abstracted by placing the different workstations to optimize the process parameters and the layout for enhancing productivity as per table 2. The proposed layout can be visualized by future value stream mapping as shown in figure 4 and simulating the proposed layout to estimate throughput as shown in figure 5.

Table 2: Details of operations carried out in future value map

Sl.no	No of persons	Operation involved	Cycle time (sec)	Changeover time
1	1	Bolt pressing	10	0.65
2	1	Diode pressing	12	2
3	1	Diode Testing	11	0.78
4	1	Riveting & flaring	12.88	0.12
5	1	Crimping &welding	13	0.91
6	1	Final inspection	24	0.93

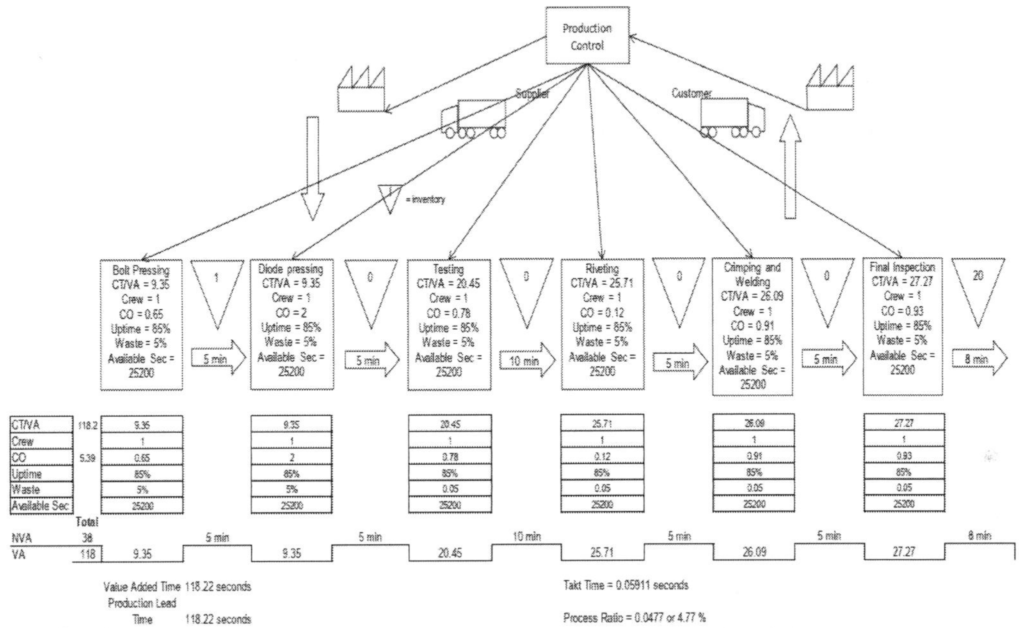

Figure 4: Future value map for heat sink assembly line

The following figure 5 shows the simulation model future value map

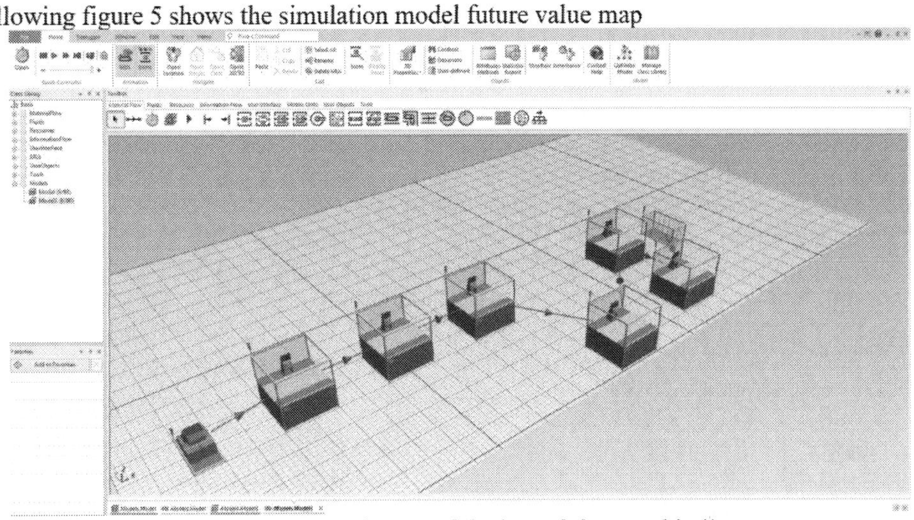

Figure 5: Future layout of the heat sink assembly line

4. Results and Discussion

Most of the literature survey focussed on LM elements. From the literature survey, it has been found that simulation tools can be utilized along with value stream mapping. This paper attempted that integrating a software simulation tool with conventional VSM can overcome the various non-value added activities in a manufacturing assembly line of a selected organization. These simulation model proved for the shop floor managers and engineers practically observe and feel how the future process (or) layout before implementing the actual design of lean manufacturing system (LMS) it makes the production process more flexible. The transition from the conventional method to the digital method using Tecnomatix plant simulation tools gives concrete results for the organization .by comparing the present and future data; it has been clear that cycle time has been reduced from 98.61 seconds to 82.88 seconds and workforce also reduced from 7 to 6.

5. Conclusion

The need for the present study was to develop a value stream map for the heat sink part assembly line to identify and eliminate waste which is any activity that does not add value to the final product in the assembly process. It is also aimed at reducing lead time and increasing the throughput rate of parts. VSM is a continuous improvement process; we must keep on changing the future state into the current state that will not end during our life. VSM has been proven to be a greatly useful tool to eliminate some waste in a cycle and find there is more waste for us to eliminate in the next cycle, during which lean becomes a habit or culture. The technique of lean tool can be applied to every situation in a company by finding out what customer wants and eliminating waste. The idea is to create a culture in which people at various levels of an organization are continuously improving their products every day & in every way.Identified the bottlenecks such as cycle time, unnecessary inventory, stage inspection between the stations, and downtime & layout design. Hence, proposed a new method with the help of a future value map with the support of simulation software to get quick results. In the proposed layout the workforce is reduced from 7 to 6 numbers by redesigning the layout.Cycle time is reduced from 98.61 seconds to 82.88 seconds and hence the productivity will be increased by 15%. There is a need to set up automated inspection (buffer) to avoid piling up inventory at the final inspection.

References

[1] Shanmuganathan, V. K., Haran, A. P., & Gayathri, N. (2015). Condition monitoring maintenance of aero-engines through LUMS - A method for the implementation of Lean tools. Measurement: Journal of the International Measurement Confederation, 73, 226–230. https://doi.org/10.1016/j.measurement.2015.05.017

[2] Nirwan, M. D., & Dhewanto, W. (2015). Barriers in Implementing the Lean Startup Methodology in Indonesia – Case Study of B2B Startup. Procedia - Social and Behavioral Sciences, 169(August 014), 23–30.https://doi.org/10.1016/j.sbspro.2015.01.282

[3] Lianying Zhang and Zhen Zhang. (2008). Project Portfolio Control and Portfolio. Project Management Journal, 39(1999), 28–42. https://doi.org/10.1002/pmj

[4] Gunaki, P., & Teli, S. N. (2015). Productivity Improvement by Value Stream Mapping in Die Casting Industry. Journal of Emerging Technologies and Innovative Research, 2(6), 2049–2064.

[5] Deif, A. (2012). Assessing lean systems using variability mapping. Procedia CIRP, 3(1), 2–7. https://doi.org/10.1016/j.procir.2012.07.002

[6] Venkataraman, K., Ramnath, B. V., Kumar, V. M., & Elanchezhian, C. (2014). Application of Value Stream Mapping for Reduction of Cycle Time in a Machining Process. Procedia Materials Science, 6(Icmpc), 1187–1196. https://doi.org/10.1016/j.mspro.2014.07.192

[7] Saraswat, P., Kumar, D., & Kumar Sain, M. (2015). Reduction of Work in Process Inventory and Production Lead Time in a Bearing Industry Using Value Stream Mapping Tool. International Journal of Managing Value and Supply Chains, 6(2), 27–35. https://doi.org/10.5121/ijmvsc.2015.6203

[8] Jadhav, J. R., Mantha, S. S., & Rane, S. B. (2015). Analysis of interactions among the barriers to JIT production: Interpretive structural modelling approach. Journal of Industrial Engineering International, 11(3), 331–352. https://doi.org/10.1007/s40092-014-0092-4

[9] Kapoun, E., & Börjesson, T. (2015). Identifying Organizational Improvement Potential using Value Stream Mapping and Simulation A Case Study at Swedish Match Göteborg Department of Product and Production Development.

[10] Maheshwari, A. (2017). Analysis of Industrial process Implementing VSM with software simulation based approach.5(3).

[11] Sarhan, S., & Fox, A. (2012). Trends and challenges to the development of a lean culture among uk construction organisations. IGLC 2012 - 20th Conference of the International Group for Lean Construction, July. https://doi.org/10.13140/2.1.2520.0323

[12] Deshkar, A., Kamle, S., Giri, J., & Korde, V. (2018). Design and evaluation of a Lean Manufacturing framework using Value Stream Mapping (VSM) for a plastic bag manufacturing unit. Materials Today: Proceedings, 5(2), 7668–7677. https://doi.org/10.1016/j.matpr.2017.11.442

[13] Trebuna, P., Pekarcikova, M., & Edl, M. (2019). Digital value stream mapping using the tecnomatix plant simulation software. International Journal of Simulation Modelling, 18(1), 19–32. https://doi.org/10.2507/IJSIMM18(1)455

Optimization of process parameters in Electro Chemical Discharge Machining of silica glass through Analysis of Means

Sadashiv Bellubbi[1,4,*], Maheshwar A Hipparagi[2], Ravindra Naik[3,4], Sathisha N[3]

[1]Dept. of Mechanical Engineering, Alva's Institute of Engineering and Technology, Moodabidri, Karnataka, India.

[2]Dept. of Mechanical Engineering, Hirasugar Institute of Technology, Nidasoshi, Karnataka, India.

[3]Dept. of Mechanical Engineering, Yenepoya Institute of Technolgy, Moodabidri, Karnataka, India.

[4]Research scholar, Visvesvaraya Technological University, Belagavi, Karnataka, India

*Corresponding author: bellubbisadashiv@gmail.com

Abstract. In recent years, the Electro Chemical Discharge Machining (ECDM) process has advanced as a substitute for machining of conductive and non-conductive materials. Optimization of the process parameters plays a crucial role in improving the quality of the processed components. The machining of microchannel in silica glass has been successfully attempted in present work by considering Analysis of Means (ANOM) of ECDM process parameters. The experiments were planned using Taguchi's L_9 orthogonal array with applied voltage, stand-off distance (SOD), electrolyte concentration (EC) and pulse on time (TON) as process parameters. The Material Removal Rate (MRR), Machining Depth (MD) and Overcut (OC) were observed as output characteristics. The ANOM study revealed that the error between experimental and predicted values of MRR, MD and OC are 4.22%, 1.92% and 3.46% respectively. Stand-off distance was found to be the most significant factor followed by pulse on time, applied voltage and electrolyte concentration from ANOVA. Confirmation test was performed by setting the optimum combination of process parameters and the predicted values from regression analysis exhibited a good agreement with experimental results.

Key word: MRR, ECDM, silica glass, overcut, microchannel, Machining depth, SOD, ANOVA, analysis of mean (ANOM).

1. Introduction

ECDM is a nontraditional machining process which includes a combination of different physical and chemical phenomena such as electrical discharge and chemical etching to remove the materials from the workpiece [1]. During ECDM material removal takes place when the applied voltage increases beyond a critical value, a high current density at the tip of cathode tool causes sparking which is attributed to the difference between the surface areas of electrodes [2]. ECDM has been widely employed for processing various nonconductive materials such as quartz, glass and ceramics. However, the process is popularly used for machining of variety of glasses including borosilicate glass, sodalime glass, pyrex glass and silica glass. Glasses have enormous applications in several sectors as micro fluidics, MEMS- microsensors, micropumps, medical micro devices [3] due to its distinctive properties such as transparency, chemical resistance, biocompatibility and low thermal and electrical conductivity [4]. However, it has been reported by many researchers that producing microholes and microchannel in glass material is very difficult owing to its brittle nature through conventional processes [5]. During the micromachining of borosilicate glass Paul and Hiremath [2] experimented with various tool tip geometry and reported that tapered tool tips exhibit better machining characteristics as the conical profile of tool tip facilitates the electrolyte contact in the machining zone and concentrates sparking action at the tool tip.

Crichton et al. [6] investigated the mechanism of spark generation in electrolyte and stated that incident of sparks is mainly due to sudden transient and noisy discharge between two electrolytes. Authors also mentioned that sparks of shorter duration favored maximize sparking which consequently resulted in increased MRR. Yang et al. [7] studied the effect of surface roughness on gas film formation of various tool materials and stated that tungsten carbide in comparison with tungsten and stainless steel exhibits the highest machining stability whereas it experiences the highest wear under gravity-feed processing. Bellubbi et al. [8] produced microholes in borosilicate glass using ECDM process and stated that applied voltage was the most dominant factor which affected the MRR followed by electrolyte concentration and machining time. Doloi et al. [8] used Taguchi method [9-11] for the optimization of different process parameters in ECDM of holes in ZrO_2 ceramic components and mentioned that voltage, electrolyte concentration and gap between the electrodes are the foremost parameters affecting MRR and radial overcut. Most of the researchers adopted Taguchi orthogonal array experimental design to reduce the experimental trials during machining of various engineering materials, which reduce machining cost and time [12-14], also this taguchi method provides simple design approach to optimize the response characteristics [15-17]. Further, Several researchers emphasized on optimizing the process parameters of ECDM for different work materials and profile geometry. The literature review reveals abundant research on both experimental and theoretical analysis of the mechanism involved in ECDM. However, it is found that very limited efforts have been published in the past to investigate the effect of process parameters in ECDM of silica glass.

Silica glass is widely used in production of mercury discharge lamps and optical laser corner reflectors. Since the use of traditional machining process to machine silica glass is very difficult, non-conventional machining of the same is required. In the past several researchers have investigated on machining of different profiles in silica glass. These works emphasize studies on effect of process parameters; MRR, width over-cut, surface roughness, electrode wear rate etc. [18-19]. However, from the literature review it is noticed that the published works lack in studies of optimization of process parameters and therefore there is a need to address the optimization of process parameters in ECDM of silica glass.

In the present work microchannel of 250μm were machined successfully in silica glass using ECDM process, followed by optimization of process parameters involved. Applied voltage, electrolyte concentration, pulse-on time and stand-off distance were considered as machining parameters. The material removal rate (MRR), machining depth (MD) and overcut (OC) were considered as response characteristics. The optimization of process parameters has been carried out by considering L_9 orthogonal array through analysis of means (ANOM).

2. Experimental Details

In the present work, machining of microchannel in silica glass has been carried out with ECDM and the effect of process parameters on MRR, MD and OC has been investigated. Silica glass (Na_2SiO_3+SiO_2) with dimension 76×26×5mm^3 was considered as work material whereas, tungsten carbide wire with 250μm diameter was used as tool for the study. Figure 1 show the schematic of the experimental set up used in the present work. Experiments were planned according to Taguchi's L_9 orthogonal array by considering applied voltage, stand-off distance, electrolyte concentration and pulse-on time as process parameters. The control factors and their levels are given in Table 1. Inter-electrode gap of 40mm and machining time of 20min were maintained constant for all the experiments. NaOH solution was used as electrolyte prepared by blending 10g, 17g and 24g of NaOH pellets in 100ml of distilled water to get the concentration of 10wt.%, 17wt.% and 24wt.% respectively. Stand-off distance was controlled by inserting slip gauges between the tool-electrode and the workpiece.

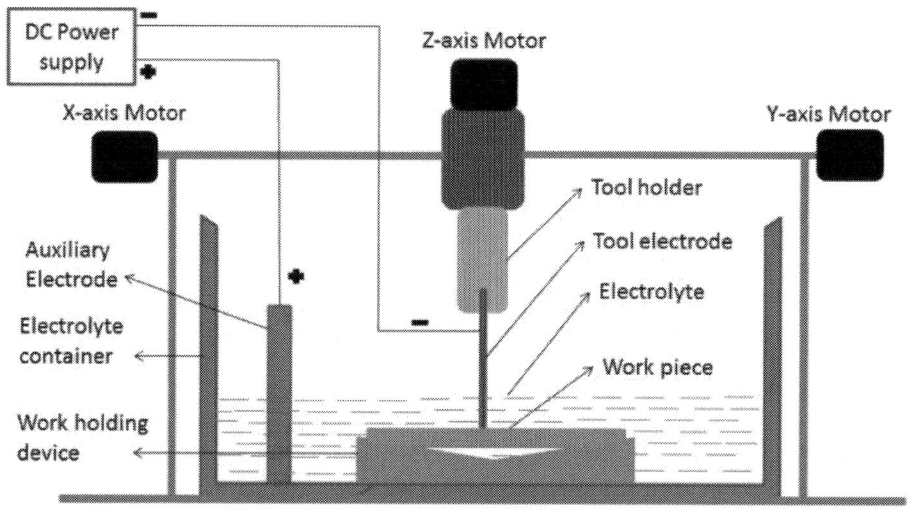

Figure 1. Schematic view of ECDM Process

Table 1. Control factors and their levels

Parameter	Symbol	Factor	Level 1	Level 2	Level 3
Applied Voltage (V)	V	A	45	50	55
Stand-off Distance (mm)	SOD	B	0.5	1.0	1.5
Electrolyte Concentration (wt%)	EC	C	10	17	24
Pulse on time (μs)	T_{on}	D	40	50	60

2.1. *Evaluation of output characteristics*

Some of the MRR was determined using equation (1) by weighing the specimens before and after the machining of microchannels [19].

$$MRR = \frac{Wt_b - Wt_a}{T} \text{ mg/hr} \quad \ldots\ldots\ldots (1)$$

Where: Wt_b - Weight of work-piece before machining in mg.
Wt_a - Weight of work-piece after machining in mg.
T - Machining time in hr.

The over cut and machining depth were measured at different magnifications with Leica Microscope. The analysis of mean (ANOM) technique was used to obtain optimal combination of process parameters with an objective of minimum overcut and maximum MRR and MD. The analysis of variance (ANOVA) has been performed to determine the % contribution of process parameters. Table 2 depicts the results obtained after performing the experiments according to L_9 OA. Micrographic observation of machined surfaces was carried out with Scanning Electron Microscopy (Model: JEO JSM–638OLA from JEOL, USA) operated at 30kV.

3. Results and Discussion

3.1. *Analysis of means*

The analysis of means (ANOM) on signal to noise ratio (SNR) was achieved for determining the optimal setting parameters [20]. In this study, Taguchi's design is proposed for optimizing the multiple objecting responses i.e. MRR, MD and OC. Here, MRR and MD are to be maximized and OC is to be minimized. Hence larger the better characteristic for MRR and MD, smaller the better characteristic for OC have been selected using equations (2)-(4). The SN ratios associated with the responses are given as follows [21-23].

$$SNR_{MRR} = -10\log \frac{1}{MRR^2} \quad dB \; ..... (2)$$

$$SNR_{MD} = -10\log \frac{1}{MD^2} \quad dB \; ...... (3)$$

$$SNR_{OC} = -10\log OC^2 \quad dB \; ...... (4)$$

The output data and signal to noise ratio for MRR, MD, OC are given in Table 2. Averaging of signal to noise ratio of material removal rate, machining depth and overcut was determined using weight function method as shown in equation (5). The weight functions assumed were 0.4, 0.3 and 0.3 for MRR, MD and OC respectively based upon priority of response characteristics.

$$SNR_{AVG} = SNR_{MRR} \times W_1 + SNR_{MD} \times W_2 + SNR_{OC} \times W_3 \quad dB \; .... (5)$$

Where SNR_{AVG} is the average signal to noise ratio in dB, SNR_{MRR} is the signal to noise ratio for material removal rate in dB, SNR_{MD} is the signal to noise ratio for machining depth in dB and SNR_{OC} is the signal to noise ratio for overcut in dB.

The computed SNR_{AVG} were ranked based on the larger-the-better criteria. Hence, 4th experimental trial is considered as best run out of all 9 trials and ranked first as given in Table 2. Further, ANOM mean SNR values were calculated to identify optimal levels of each parameter presented in Table 3 and A_3-B_1-C_2-D_3 combination was considered as optimal setting parameters while, Table 4 presents ANOVA for ANOM Signal to Noise Ratio. Table 4 shows that factor stand-off distance is the most influential factor followed by factors pulse on time, applied voltage and electrolyte concentration based on ANOVA by considering 95% confidence level. Furthermore, factor SOD exhibits highest contribution of 64.785% whereas, pulse on time shows 26.557% contribution. However, it was observed that factors applied voltage and electrolyte concentration influence to a minimum level of 5.995% and 2.663% respectively.

Table 2. ANOM data analysis

Run No	A	B	C	D	MRR (mg/hr)	MD (µm)	OC (µm)	SNR_{MRR} dB	SNR_{MD} dB	SNR_{OC} dB	SNR_{AVG} dB	Rank
1	45	0.5	10	40	73.51	270.21	122.73	37.3269	48.6340	-41.7790	16.9873	3
2	45	1	17	50	65.73	263.92	152.39	36.3553	48.4294	-43.6591	15.9732	6
3	45	1.5	24	60	76.83	307.37	241.23	37.7106	49.7532	-47.6486	15.7156	7
4	**50**	**0.5**	**17**	**60**	98.22	392.88	227.494	39.8439	51.8852	-47.1393	**17.3613**	**1**
5	50	1	24	40	87.32	349.23	248.376	38.8223	50.8622	-47.9022	16.4169	5
6	50	1.5	10	50	56.87	220.48	246.797	35.0976	46.8674	-47.8468	13.7452	9
7	55	0.5	24	50	96.223	312.34	232.34	39.6655	49.8926	-47.3225	16.6372	4
8	55	1	10	60	95.432	381.92	236.81	39.5938	51.6394	-47.4880	17.0830	2
9	55	1.5	17	40	82.591	318.25	282.91	38.3386	50.0554	-49.0329	15.6422	8

Table 3. Means of SNR_{AVG}

Symbol	Parameters	Level 1	Level 2	Level 3	Optimal level
A	**Applied Voltage (V)**	16.2254	15.8412	16.4541	A_3
B	**Stand-off Distance (mm)**	16.9953	16.491	15.0343	B_1
C	**Electrolyte Concentration (wt.%)**	15.9385	16.3256	16.2566	C_2
D	**Pulse on time (µs)**	16.3488	15.4519	16.7200	D_3

Table 4. ANOVA for ANOM Signal to Noise Ratio

Source	Sum of squares	DOF	Mean squares	%P (Percentage contribution)
Applied Voltage (V)	0.5757	2	0.2879	5.995
Stand-off Distance (mm)	6.2215	2	3.1107	64.785
Electrolyte Concentration (wt%)	0.2557	2	0.1279	2.663
Pulse on time (μs)	2.5503	2	1.2752	26.557

3.2. *Regression model*

From the experimental data, regression model has been obtained by conducting analysis with MINITAB 17 software. The equations (6)-(8) represent the predicted values of MRR, MD and OC in terms of applied voltage (A), stand-off distance (B), electrolyte concentration (C) and pulse on time (D).

$$\text{Material Removal Rate} = -34.9 + 1.939\,A - 17.22\,B + 0.823\,C + 0.451\,D \quad \ldots\ldots (6)$$

$$\text{Machining Depth} = -88 + 5.70\,A - 43.1\,B + 2.29\,C + 2.41\,D \quad \ldots\ldots (7)$$

$$\text{Over cut} = -324 + 7.86\,A + 62.8\,B + 2.75\,C + 0.86\,D \quad \ldots\ldots (8)$$

3.3. *Confirmation test*

Confirmation test was conducted for the A_3-B_1-C_2-D_3 combination of machining parameters obtained from ANOM and experimental responses are presented in Table 5. Predicted values for optimal combination were calculated by using regression equations (6)-(8). From table 5, showed that the error between experimental and predicted values of MRR, MD and OC are 4.22%, 1.92% and 3.46% respectively.

Table 5. ANOM confirmation test results

Technique	Response	Optimal Parameters	Predicted	Experimental	% Error
ANOM	MRR(mg/hr)	$A_3B_1C_2D_3$	104.186	99.79	4.22
	Machining depth (μm)	(55V, 0.5mm,	387.48	380.02	1.92
	Overcut (μm)	17wt.%, 60μs)	238.05	229.81	3.46

Figure 2 shows the machined sample and SEM morphology of machined surface obtained after confirmation test with optimal setting parameters (A_3-B_1-C_2-D_3). It is observed that heat affected zone occurred along the microchannel due to side spark with increase in voltage. Electrochemical discharge machining (ECDM) is a promising operation for micromachining of glass. Though it avoids surface cracking, it repeatedly produces surface damage due to its tendency to form subsurface cracks.

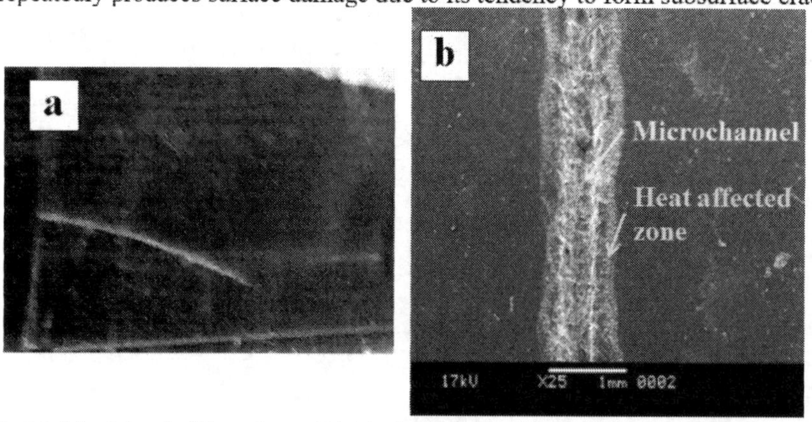

Figure 2. (a) Machined silica glass. (b) Typical SEM micrograph of machined surface from confirmation test.

4. Conclusions

The machining of microchannel in silica glass attempted successfully in this work by considering ANOM of ECDM process parameters. Following conclusions were drawn out.

1. The recommended optimum setting of machining parameters through ANOM for obtaining maximum MRR, Maximum MD and minimum OC was found to be A_3-B_1-C_2-D_3 (55V, 0.5mm, 17wt%, 60 µs)
2. The ANOM confirmation test result showed that the error between experimental and predicted values of MRR, MD and OC are 4.22%, 1.92% and 3.46% respectively.
3. The analysis of variance resulted that the stand-off distance has major influence on MRR (mg/hr), MD (µm) and OC (µm).

References

[1] Mehrabi, F., et al. "Application of electrolyte injection to the electro-chemical discharge machining (ECDM) on the optical glass." Journal of Materials Processing Technology 255 (2018): 665-672.Another reference

[2] Lijo Paul and Somashekhar S Hiremath. Evaluation of process parameter in ECDM using grey relation analysis. Int. Con. On Advance in Manufacturing and Materials Engineering 2014; 2273–2282.

[3] Bassyouni, Zahraa, and Jana D Abou Ziki. "The Capabilities of Spark-Assisted Chemical Engraving: A Review." Journal of Manufacturing and Materials Processing 4.4 (2020): 99.

[4] Madhavi, J. Bindu, and Somashekhar S. Hiremath. "Investigation on machining of holes and channels on borosilicate and sodalime glass using µ-ECDM setup." Procedia Technology 25 (2016): 1257-1264.

[5] N. Sathisha, Hiremath S., J. Shivakumar, Int. J. Recent Adv. Mech. Eng. 3 (2) (2014) 69–81.

[6] Crichton IM, McGeough JA, Munro W and White C. Studies of the discharge mechanisms in electrochemical arc machining. J of Appl. Electrochemistry 1985; 15: 113-119.

[7] Cheng-Kuang Yang, Chih-Ping Cheng, A. Chao-Chuang Mai, Cheng Wang, Jung- Chou Hung, Biing-Hwa Yan, Int. J. Mach. Tools Manuf. 50 (2010) 1088–1096.

[8] Bellubbi, S., Naik, R., & Sathisha, N. (2020). An experimental study of process parameters on material removal rate in ECDM process. Materials Today: Proceedings.

[9] Doloi, B., B. Bhattacharyya, and S. K. Sorkhel. "Electrochemical discharge machining of non-conducting ceramics." Defence Science Journal 49.4 (1999): 331.

[10] Naik, Gajanan M., et al. "An investigation on effects of wire-EDT machining parameters on surface roughness of INCONEL 718." Materials Today: Proceedings (2020).

[11] S. Bellubbi and N. Sathisha. "Reduction in through put time of drum shell manufacturing by single-V welding configuration." MS&E 376.1 (2018): 012104.

[12] Badiger, Ravindra I., S. Narendranath, and M. S. Srinath. "Optimization of process parameters by Taguchi grey relational analysis in joining Inconel-625 through microwave hybrid heating." Metallography, Microstructure, and Analysis 8.1 (2019): 92-108.

[13] Badiger, Ravindra I., S. Narendranath, and M. S. Srinath. "Optimization of parameters influencing tensile strength of Inconel-625 welded joints developed through microwave hybrid heating." Materials Today: Proceedings 5.2 (2018): 7659-7667.

[14] Gajanan M. Naik, S. Narendranath, A parametric optimization of wire-ED turning process parameters on material removal rate of INCONEL 718, Journal of Mechanical Engineering and Biomechanics, Volume 2, Issue 2, Page 8-14, 2017 ISSN- 2456-219X.

[15] Shetty, Abhishek S., K. S. Akshar, B. Y. Prashanth, and Gajanan M. Naik. "Optimization of machining parameters on MRR for EN19 & EN31 steel using Taguchi method." Int. J. of Emerg. Res. Mngmt 6 (2017): 2278-9359.

[16] Vishwas, C. J., L. V. Girish, Gajanan M. Naik, B. Sachin, Abhinaba Roy, B. Y. Prashanth, and RavindraIshwar Badiger. "Effect of Machining Parameters on Surface integrity during Dry Turning of AISI 410 martensitic stainless steel." In IOP Conference Series: Materials Science and

Engineering, vol. 376, no. 1, p. 012127. IOP Publishing, 2018.
[17] Sadashiv Bellubbi, Vijeath A, Mithesh Gowda J R, Karthik Prabhu. Experimental Investigation of Process Parameters on Machining Force, MRR and Power in Turning of AISI 316 Steel. International Journal of Computer Aided Manufacturing. 2019; 5(1): 18–25p.
[18] Mallick, B., Sarkar, B.R., Doloi, B. and Bhattacharyya, B. (2017) 'Analysis on electrochemical discharge machining during micro-channel cutting on glass', Int. J. Precision Technology, Vol. 7, No. 1, pp.32–50.
[19] Nagaraj, Y., Jagannatha, N., Sathisha, N. et al. Prediction of Material Removal Rate and Surface Roughness in Hot Air Assisted Hybrid Machining on Soda-Lime-Silica Glass using Regression Analysis and Artificial Neural Network. Silicon (2020).
[20] Manjaiah, M., Narendranath, S., Basavarajappa, S. and Gaitonde, V.N., 2014. Wire electric discharge machining characteristics of titanium nickel shape memory alloy. Transactions of Nonferrous Metals Society of China, 24(10), pp.3201-3209.
[21] M. Durairaj, D. Sudharsun, N. Swamynathan, Analysis of Process Parameters in Wire EDM with Stainless Steel using Single Objective Taguchi Method and Multi Objective Grey Relational Grade, Procedia Engineering 64 (2013) 868 – 877.
[22] Amitesh Goswami, Jatinder Kumar, Optimization in wire-cut EDM of Nimonic-80A using Taguchi's approach and utility concept, Engineering Science and Technology, an International Journal 17 (2014) 236-246.
[23] Jagannatha, N., S. S. Hiremath, and K. Sadashivappa. "Analysis and parametric optimization of abrasive hot air jet machining for glass using Taguchi method and utility concept." International Journal of Mechanical and materials engineering 7.1 (2012): 9-15.

IOP Conf. Series: Materials Science and Engineering 1065 (2021) 012004 doi:10.1088/1757-899X/1065/1/012004

Experimental investigation on characterization of microwave welded Al6061-Gr-SiC hybrid metal matrix composites

Zaheerabbas B Kandagal[1], V G Akkimardi[2] and A N Sonnad[3]

[1]Assistant Professor, Mechanical Department, AGMR College of Eng. and Technology, Varur, Karnataka, India

[2]Associate Professor, Industrial and Production Engineering Department, Basaveshwar Engineering College (A), Karnataka, India

[3]Assistant Professor, Mechanical Department, Basaveshwar Engineering College (A), Karnataka, India

Abstract – In the present work, experimental microwave welding of Al6061-6%Gr-SiC hybrid metal matrix composites (MMC) in bulk form has been carried out. The principle of microwave hybrid heating was applied with domestic microwave having the frequency of 2.45GHz and 900W power. The interface powder Al6061 and susceptor as charcoal powder were used for the welding process. The stir casting process was used to prepare Al6061-6%Gr-SiC hybrid MMCs by varying SiC content by 3%, 6% and 9% and keeping 6% graphite constant by wt%. The composite weld joints were characterized by scanning electron microscope (SEM) equipped with energy dispersive spectrum (EDS) and mechanically through tensile test, Rockwell hardness test and impact test. The microstructure study showed the good metallurgical bonding between the interface and base composites without any visible cracks. The SEM-EDS study revealed the formation of various carbides and inter metallics at the weld interface. The average ultimate tensile strength of the joint was found to be 135MPa with 4.3% elongation. The average Rockwell hardness at the weld joint observed was 61.5HRB. The average impact strength of welded composite was 1.62J/mm^2.

Keywords: Al6061-Gr-SiC MMCs, Microwave, Welding, Microstructure, Tensile strength, Hardness, Impact strength, Susceptor.

1. Introduction

Metal matrix composites are fast replacing conventional metallic alloys in aerospace, automobile, defence, marine and sports industries due to their excellent combination of properties. Among the alloys, aluminium alloys are the most utilized metallic alloys as matrix material both in research and industry. The aluminium alloys are used in light weight applications due to their properties such as low density and high strength to weight ratio [1]. The major advantages of aluminium based MMCs are the reduction in overall weight, lower fuel consumption and emissions in automobiles. The hybrid composites possess better microstructural and mechanical properties than the single reinforced composites. Aluminium matrix composites reinforced with silicon carbide (SiC) and graphite (Gr) have attracted a considerable attention due to their better mechanical properties and excellent wear resistance. Hence proper volumetric or weight shares of graphite and SiC reinforcements is required in order to get optimum mechanical and tribological properties [2]. Some of the applications of aluminium based composites in automobile industry are pistons, connecting rods, engine cylinders, brake systems and many more. These MMCs are also finding applications in aerospace and space industries [3]. The growth in the field of engineering and technologies led the industries to find new and improved methods that can be used for processing of wide range of materials like metals, non-metals, ceramics and composites. The developed new methods should have the characteristic of lower energy consumption and higher production rates.

The welding process plays a very significant role in the manufacturing industries. The reasons for welding are: 1) achieve large size of components and geometric complexity which is difficult to obtain by casting, forging or other methods, 2) adding material where it is required and 3) material optimization and many more. The commonly used welding processes are soldering, brazing, friction stir welding, electron beam welding, laser welding etc. But these welding techniques have their own disadvantages like higher power and processing time, costlier set up, operator safety and poor characteristics of the joint. The welding of composites is complex process due to the possible chemical reaction between matrix and reinforcement at higher temperatures, low weld efficiency and settlement of impurities. So, for the industries a more versatile, faster and cleaner process is required that not only improves the production rate but also the microstructural and mechanical properties. The microwave welding of materials is not only an alternative to high energy consumption heating techniques, but also it can meet the needs of the modern industrial applications [4].

In the past, the microwaves were used widely in the field of food processing. Later further research revealed the use of microwave energy in material processing. It is not only a green manufacturing process, but also significantly faster at relatively low investment with selective and volumetric heating capability and improved microstructure. The frequencies of microwaves vary from 300MHz to 300GHz with wavelengths from 1mm to 1m. The microwave energy is produced by the interaction of electric field and magnetic field. The interaction of electromagnetic waves with molecules produces volumetric heating and the energy is directly transferred to the material. In conventional method, the heat is generated from the external sources but in microwave heating, it is generated internally within the material and gets transmitted outward [5].

The welding of metallic materials in bulk form using microwave energy is challenging as the most of the metals reflect microwaves at room temperature. The unique characteristics led many researchers to carry out microwave welding of bulk metals such as stainless steel, aluminium, copper, iron, tungsten etc. with significant importance. M. S. Srinath et al. [6] have reported the welding of dissimilar materials such as mild steel (MS) with stainless steel (SS) through microwave energy technique which resulted in good metallurgical bond and lesser defects. Gupta and Kumar [7] have studied the microwave welding of stainless-steel using nickel metallic powder as sandwich layer. The joint showed no visible cracks and strong interfacial bond with uniform microstructure was observed. Badiger et al. [8] have joined Inconel-625 alloy using microwave energy technique and studied the microstructural and mechanical properties of the joint. Bansal et al. [9] have microwave welded stainless steel (SS-316) using nickel-based powder as interfacial material and studied joint microstructural and mechanical properties. Further many other researchers were encouraged to weld bulk metallic materials using microwave energy and reported better microstructure with lesser defects [10] [11] [12].

The available literature showed that the microwave energy is used successfully to weld bulk metals with very fewer reports on the microwave welding of hybrid MMCs. The aim of the present investigation is to weld Al6061-6%Gr-SiC hybrid MMCs by using the principle of microwave hybrid heating and study their microstructural and mechanical properties.

2 Experimentation

The welding of MMCs using microwave energy is very challenging as they do not allow the microwaves to penetrate inside the bulk composites and get reflected at room temperature. This can be overcome by using the principle of microwave hybrid heating with proper susceptor medium. The susceptor transfers the heat to the bulk composites by absorbing microwave energy and a number of trials were carried out. Al6061 based powder was used as filler material between the welding surfaces. The present study showed the successful welding of Al6061-6%Gr-SiC hybrid composites using microwave energy technique. The experiments were conducted using the domestic multimode microwave oven (Make: Samsung) with 2.45GHz frequency and 900W power. The mechanical properties of the joints were studied by performing tensile, Rockwell hardness and Charpy impact tests. All the specimens were prepared as per ASTM standards for the experiment.

2.1 Material selection

The stir casting method was used for the preparation of Al6061-6%Gr-SiC hybrid MMCs. In the preparation of hybrid MMCs, the commercially available Al6061 alloy billets as matrix material, the reinforcements SiC with a particle size of 40μm and graphite with particle size of 100μm were used. The reinforcement SiC was varied with 3%, 6%, 9% and graphite at 6% constant by weight percentage. This produced a very hard and strong composite. It has characteristics like low density, high strength and thermal conductivity etc. [13]. The chemical composition of Al6061 alloy and Al6061 interface powder are given in table 1. The physical properties of Al6061 alloy, reinforcements SiC and graphite are given in table 2. In the present experiment, charcoal powder as susceptor and commercially available Al6061 based powder with 97.9% purity having average particle size of 30μm as an interfacial material was used for microwave welding. During microwave hybrid heating (MHH), the conventional mode of heat transfer was provided by the charcoal powder since it is a good absorber of microwave energy.

Table 1. Chemical composition of Al6061 alloy [14].

Element	Al6061 alloy (%)	Al6061 Powder (%)
Mg	0.9	0.85
Si	0.6	0.39
Fe	0.5	0.35
Cu	0.32	0.28
Zn	0.2	0.15
Ti	0.12	0.10
Mn	0.12	0.12
Cr	0.3	0.28
Al	Balance	Balance

Table 2. Physical properties of Al6061 alloy, SiC and Graphite [15].

Physical properties	Al6061 alloy	SiC	Graphite
Tensile strength MPa	115 (T)	3900 (C)	20-200 (T)
Melting point °C	650	2200-2700	3915
Hardness (HB500)	30	2800	1.7mohs scale
Density g/cm^3	2.7	3.1	2.09
Poisson's Ratio	0.33	0.14	0.14

2.2 Microwave Welding

Prior to the welding, Al6061-6%Gr-SiC hybrid MMC specimens were polished mechanically with emery paper of different grades and cleaned with acetone to remove dust and impurities. The welding thickness of approximately 1mm to 2mm was maintained with butt joint configuration. The interfacial powder Al6061 was mixed with epoxy resin and hardener to prepare paste like slurry. The paste was applied uniformly between the composite specimen interfaces. The epoxy resin binds the interface material with specimens. Then the graphite sheets of approximately 0.5mm thickness were used to cover the specimens. These graphite sheets avoid the intermixing between the charcoal susceptor with specimens and prevents the specimens from direct exposure to microwaves. The initial coupling of microwaves with composite specimens was provided by charcoal powder and poured near the weld region. The microwave oven was switched on with required power and exposure time after the covered specimen kept inside. As the microwave exposure started, the charcoal powder began to couple with microwaves and increases the temperature rapidly at the weld region. The figure 1 shows the schematic diagram of experimental microwave welding of Al6061-6%Gr-SiC hybrid MMCs. The table 3 gives the microwave processing parameters used in the experiment.

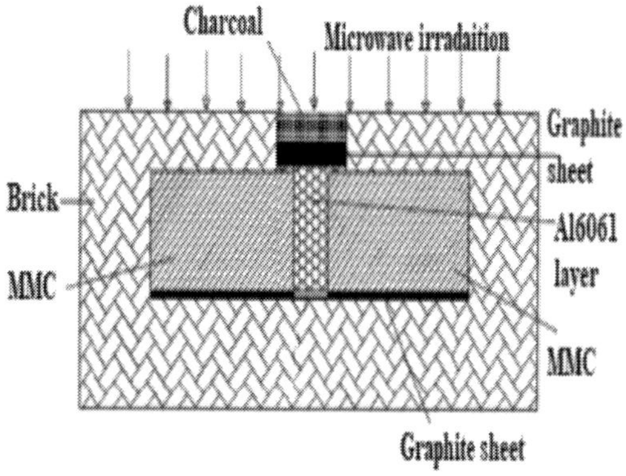

Figure 1. Microwave welding of Al6061-6%Gr-SiC MMCs.

Table 3. Microwave processing parameters.

Microwave frequency	Multimode -2.45 GHz
Exposure power	900 W
Base Material	Al6061-6%Gr-SiC MMC
Interfacial material	Al6061 based powder
Susceptor material	Charcoal powder
Exposure time	12mins (720s)

The graphite sheets transfer this heat by conventional mode of heat transfer to the interface layer Al6061 powder. The epoxy resin got evaporated at higher temperature without affecting the weld strength. During this period, microwaves were absorbed by Al6061 powder particles directly and wetting takes place at the joint region. The molten particles of Al6061 fuse with the melted thin interface layers of composite specimens. The microwave welding of composite specimens was carried out with exposure time of 720secs (12mins). Before conducting the final experiment, the trails were carried out and the exposure time was determined. After the process, the entire specimen assembly was taken out from the oven and kept for cooling at atmospheric conditions. On cooling, the homogenous and dense joint was formed between specimens with complete melting of the interfacial layer.

2.3 Characterization of joints

The microstructure study of the joints was carried out through scanning electron microscopy (SEM) equipped with energy dispersive spectrum (EDS) (Make: JEOL JSM, Model: JSM-IT500LA, Japan). Prior to the SEM-EDS characterization test, the specimens were polished mechanically with series of abrasive papers of different grades ranging from 200 to 2000 grades and finally polished by velvet cloth. The Keller's' reagent was used to etch the polished specimen to reveal the microstructure. For the mechanical property evaluation, the ultimate tensile strength and percentage elongation were determined using the universal testing machine at room temperature, the hardness measurements were carried out using Rockwell hardness tester and the Charpy impact test was used to determine the impact strength.

3 Results and discussion

In the present work, Al6061-6%Gr-SiC hybrid composites were welded successfully by exposing them to microwave radiations for 720s (12mins) at 2.45 GHz frequency and power of 900W in a multimode domestic microwave oven. The characterization results were discussed with appropriate illustrations in the following paragraphs.

3.1 SEM-EDS observations

The figures 2-5 illustrates the SEM microstructure of hybrid composite joints. The images show the good metallurgical bond between interface material and hybrid composites with complete melting of Al6061 based powder particles.

All hybrid composite joints are dense and homogenous as illustrated in SEM images. The uniform heating has resulted in homogenous joint. The well bonded microstructure between interface layer and bulk composite results in more efficient joint. The weld zone contains more Al6061 powder particles than the reinforcements SiC and graphite. The SEM images of composite joints with different content of reinforcements show fewer voids and absence of cracks. In figure 4, it is observed that Al_2O_3 and graphite particles have tendency to segregate and cluster at inter-dendritic regions. The images also show the unreacted carbons particles in the joint area [8-9][16].

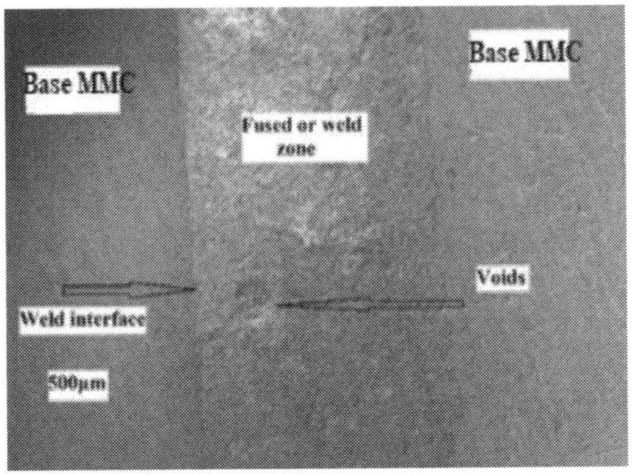

Figure 2. SEM of Al6061-Gr-SiC weld.

Figure 4. SEM of Al6061-6%Gr-6%SiC weld.

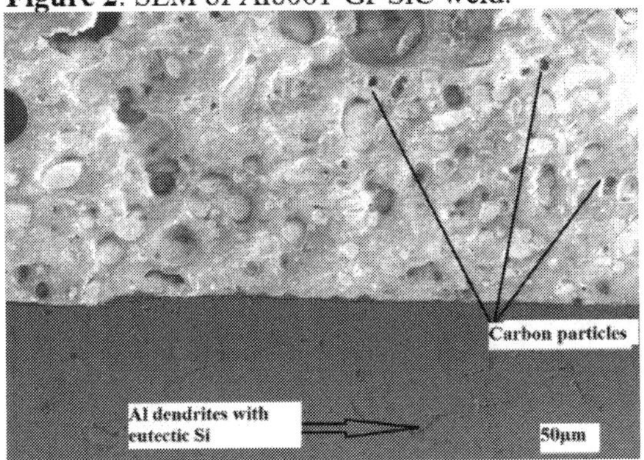

Figure 3. SEM of Al6061-6%Gr-3%SiC weld.

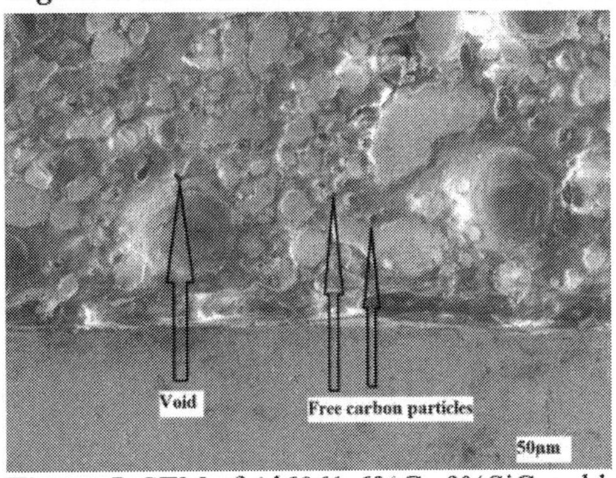

Figure 5. SEM of Al6061-6%Gr-9%SiC weld.

The EDS analysis of Al6061-6%Gr-SiC composite joints is carried out at different positions and are shown in figures 6-8. The EDS figures indicate the aluminium as major constituent element with other elements such as oxygen, carbon and sulphur (S). The presence of carbon is appeared due to diffusion from graphite sheet or decomposition of SiC reinforcement at higher temperature. The carbon and oxygen elements in EDS analysis confirms the presence of reaction products aluminium carbide and aluminium oxide. The aluminium oxide (Al_2O_3) is formed during slow cooling of weld bead at atmospheric conditions due to the reaction between aluminium and oxygen. The molten aluminium reacted with carbon to form aluminium carbide (Al_4C_3) phase at higher temperature. Many researchers have observed that the formation of aluminium carbide is unavoidable in welding of Al-based composites using different methods [17-20]

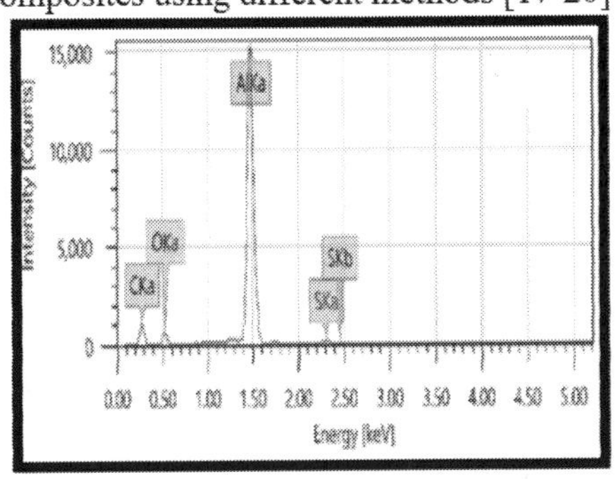

Figure 6. EDS of Al6061-6%Gr-3%SiC weld.

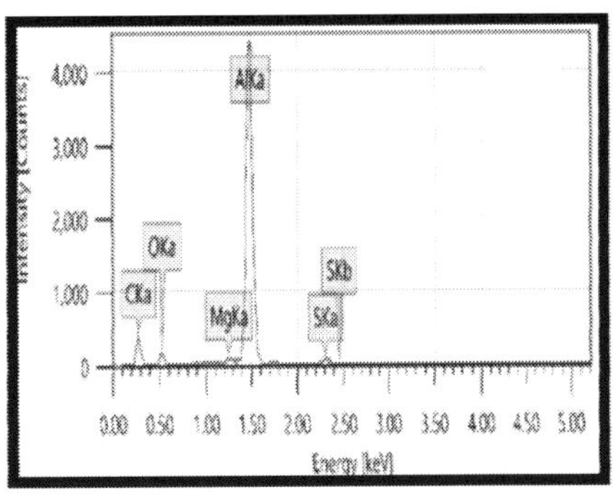

Figure 7. EDS of Al6061-6%Gr-6%SiC weld.

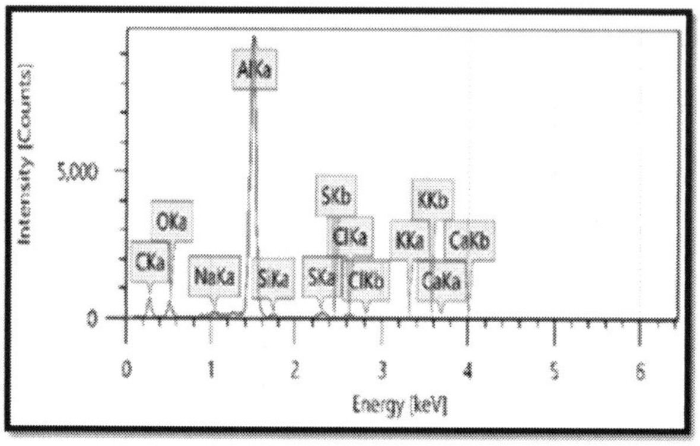

Figure 8. EDS of Al6061-6%Gr-9%SiC MMC weld.

3.2 Observations from tensile test

The universal testing machine (Make: Testwell Instruments, Model: TUTE60T, India) of capacity of 60 tonnes was used to evaluate the tensile properties of microwave welded hybrid MMCs. The specimens were prepared as per ASTM E-8 standards having the gauge length of 50mm and width of 12.5mm as shown in figure 9. The microwave welded tensile specimens are shown in figure 10. The table 4 summarizes the recorded ultimate tensile strength and percentage of elongation of Al6061-6%Gr-SiC hybrid composites.

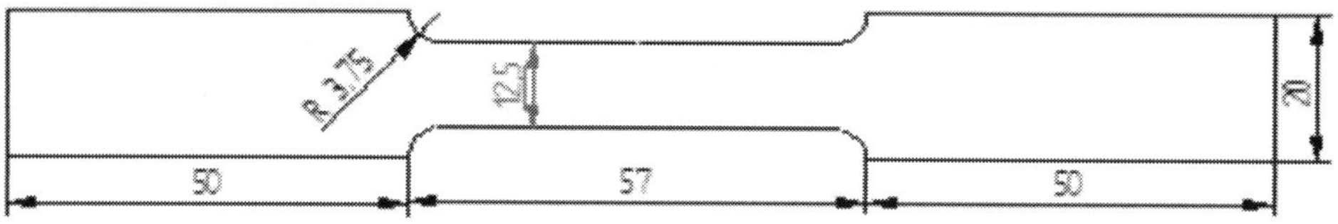

Figure 9. Tensile test specimen of composite joint.

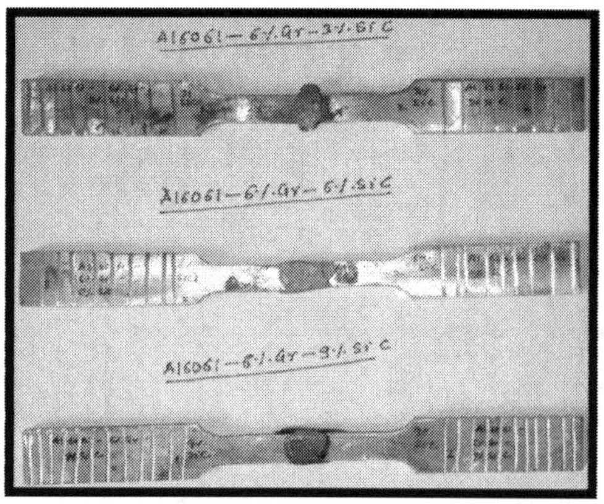

Table 4. Tensile test results.

Composites	UTS (MPa)	% Elong.
Al6061	85	2.8
Al6061-6%Gr-3%SiC	115	6.8
Al6061-6%Gr-6%SiC	145	4.4
Al6061-6%Gr-9%SiC	195	3.2

Figure 10. Microwave welded tensile specimen.

From the table 4, the ultimate tensile strength of Al6061 alloy joint was found to be 85MPa with 2.8% elongation. For the Al6061-6%Gr-3%SiC MMC joint, the ultimate tensile strength was found to be 115MPa with 6.8% elongation. For the Al6061-6%Gr-6%SiC MMC joint, the ultimate tensile strength was recorded as 145MPa with 4.4% elongation and for Al6061-6%Gr-9%SiC MMC, the ultimate tensile strength was found to be 195MPa with 3.4% elongation. The average ultimate tensile strength of microwave welded hybrid MMC joint was 135 MPa with average elongation 4.3%. From the test results, it is observed that as the content of SiC is increased, the ultimate tensile strength also increased for the welded joints. This is due to the strong metallurgical bond between bulk composites and interfacial Al6061 layer along with plastic deformation resistance of harder SiC particles. The elongated grains and few semi-molten Al6061 particles as observed in SEM images also enhanced the strength. The reduction in elongation or ductility is due to the brittleness of SiC particles and presence of aluminium carbide phases in the weld region. Similar observations were made by other researchers [18-21]. The figure 11 indicates the variation of ultimate tensile strength for welded composites as the content of SiC is increased.

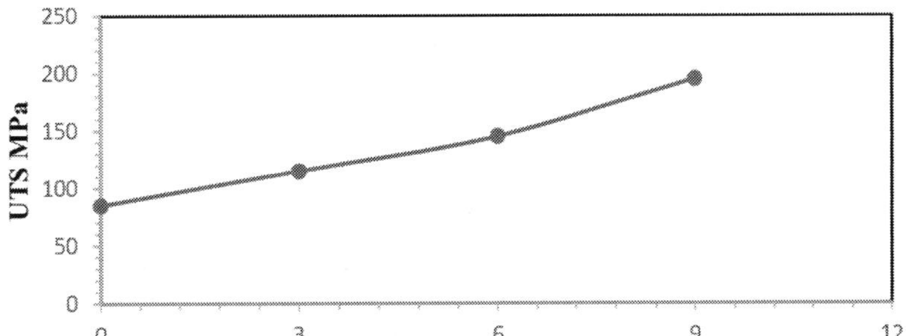

Figure 11. Ultimate tensile strength of welded Al6061-6%Gr-SiC MMCs.

3.3 Observations from Rockwell hardness test

The Rockwell hardness test (TRAB-2010-263, Testwell Instruments, India) was carried out on welded hybrid MMCs using steel ball indenter of diameter (1/16) inch and load of 100 kgf. The specimens were prepared as per ASTM E28 standards as shown in figure 12. The hardness was measured on the base composite, weld interface and weld zone. For the analysis, the average hardness value was calculated. The table 5 summarizes the hardness test results.

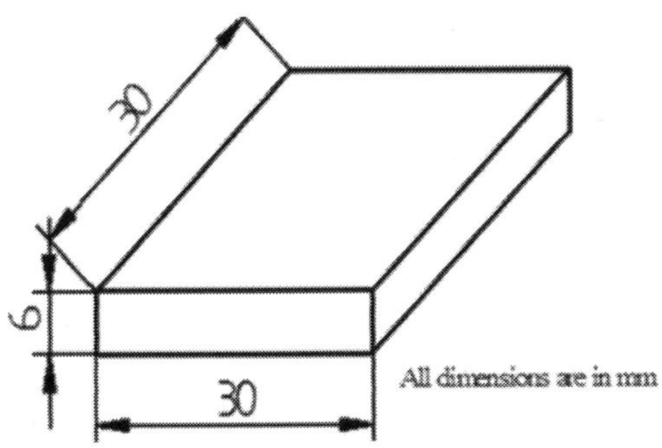

Figure 12. Hardness specimen of composite joint

Table 5. Hardness test results.

Composites	Rockwell Hardness HRB
Al6061	56
Al6061+6%Gr+3%SiC	59
Al6061+6%Gr+6%SiC	64
Al6061+6%Gr+9%SiC	67

From the table 5, it is observed that the hardness value increases as the content of SiC increases for the welded composites. The average hardness value on the base composite was found to be 55HRB and on the weld interface 61.50HRB. The higher hardness on the weld interface is due to the presence of aluminium oxides, aluminium carbides and free carbon which make it harder than the base composite. It is observed that the hardness value decreased while traversing from weld interface to the weld zone, due to lesser harder particles in the weld zone [7-8] [22-23] [25] The variation of Rockwell hardness for the welded hybrid MMC joints with increase of SiC content is shown in figure 13.

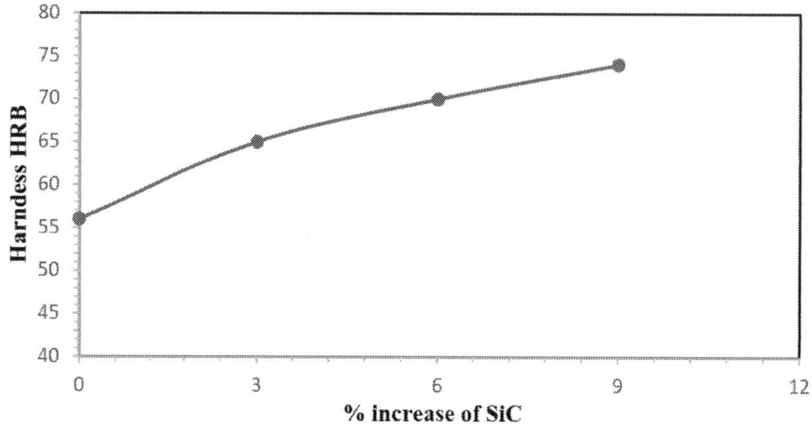

Figure 13. Hardness of welded Al6061-Gr-SiC MMC joints.

3.4 Observations from Charpy impact test

The Charpy impact test (Make: Testwell Instruments, India) was carried out on microwave welded Al6061-6%Gr-SiC composites. The specimens were prepared as per ASTM E23 standards with 45^0 V-notch as shown in figure 14. The impact test was carried out on two specimens and the average value was taken for the study. The table 6 represents the experimental impact test results of welded composites.

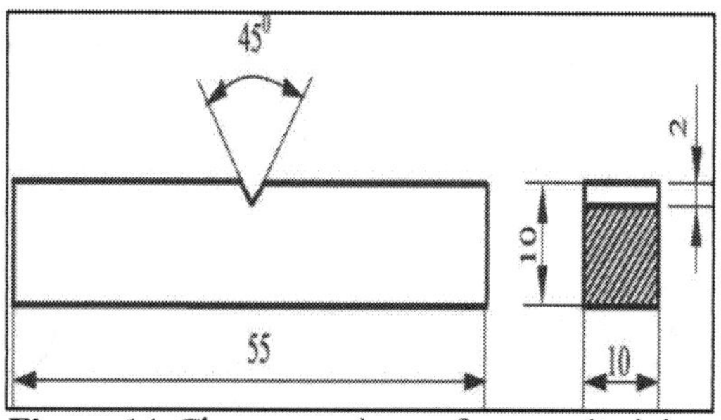

Figure 14. Charpy specimen of composite joint

Table 6. Impact test results.

Composites	Impact strength (J/mm^2)
Al6061	1.44
Al6061+6%Gr+3%SiC	1.60
Al6061+6%Gr+6%SiC	1.70
Al6061+6%Gr+9%SiC	1.75

From the table 6, the average impact strength of welded composites was found to be $1.53J/mm^2$. It is observed that the impact strength of welded Al60061-6%Gr-SiC MMCs was increased as the content of SiC increased. The more load carrying capacity of SiC particles, good fusion of interface Al6061 layer with composites and good metallurgical bond enhanced the overall impact strength of weld joint [8-9]. The combination of hard SiC particles and soft aluminium matrix improves the impact property of composites up to 6% SiC. For the Al6061-6%-9%SiC MMC joint, there was no drastic increase in the impact strength due to more porosity [22]. The figure 15 shows the variation of impact strength for welded Al60061-6%Gr-SiC MMCs with increase of SiC content [24].

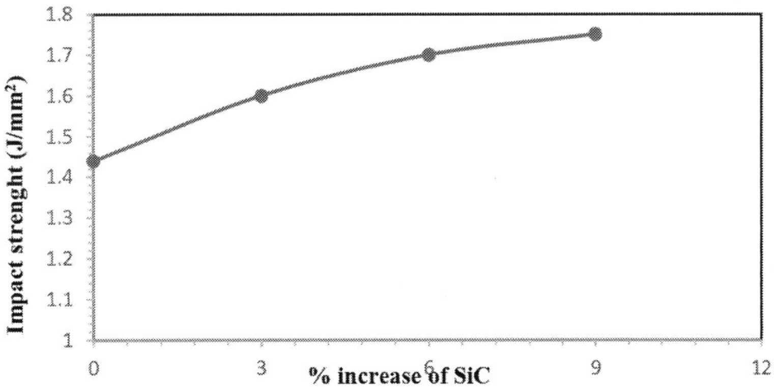

Figure 15. Impact strength of welded Al6061-6%Gr-SiC MMCs.

4. Conclusions

The microwave welding of bulk composites is challenging as microwaves get reflected at room temperature. The present investigation shows that the microwave welding of composites can be an alternative joining method which is fast and eco-friendly. The following are the conclusions drawn from the present investigation.

1. The microwave welding of Al6061-6%Gr-SiC hybrid composites varying SiC by 3%, 6% and 9% with keeping graphite constant at 6% was carried out successfully using Al6061 powder as interface material.
2. The SEM images of composite joints showed homogenous and dense weld joint without any visible cracks due to the complete melting of interface material. The increase in porosity of the joint was observed with increase of SiC content from 6 to 9%.
3. The EDS analysis of joints indicate the presence of aluminium, carbon, oxygen and other elements. The oxygen and carbon react with aluminium at higher temperature to form aluminium oxide and aluminium carbide phases respectively. The images showed the presence of unreacted carbon in the joint area.
4. The average ultimate tensile strength of microwave welded Al6061-6%Gr-SiC hybrid composites was found to be 135MPa with 4.3% elongation. For the welded composites, the tensile strength increases as percentage of SiC increases due to the hardness of SiC particles and the ductility of weld zone.
5. The average Rockwell hardness of weld joint was found to be 55HRB. The presence of harder aluminium carbide, aluminium oxide and SiC particles at the weld interface made it harder than the base composites.
6. The average impact strength of microwave weld was found to be 1.53J/mm^2. The impact strength is improved as the content of SiC increased.

The quality of weld joints can be improved further by investigating the microwave hybrid heating process parameters.

Acknowledgements

The authors are very grateful to for Sophisticated Analytical Instrument Facility laboratory (SAIF), Karnataka University, Dharwad, Karnataka for kind permission to perform SEM-EDS testing in the centre.

References

[1] Khatkar S K, Suri N M, Kant S and Pankaj 2018 *Rev. Adv. Mater. Sci* A review on mechanical and tribological properties of graphite reinforced self lubricating hybrid metal matrix composites **56** 1-20

[2] Karuppusamy T, Velmurugan, Saran S and Babu S K 2017 *Journal of advanced research in dynamical and control systems* Investigation on the microstructure and wear characteristics of heat treated hybrid aluminium matrix **9** 1895–12

[3] Blaža Stojanović and Lozica Ivanović 2015 *Tehnički vjesnik* Application of aluminium hybrid composites in automotive industry **22** 247-51

[4] Robert W and Messler Jr January 2004 *Journal of thermoplastic composite materials Sage Publications* Joining composite materials and structures: Some thought-provoking possibilities **17**

[5] Bansal A, Sharma A K, Kumar P and Das S May 2014 *Materials Characterization* Characterization of bulk stainless steel joints developed through microwave hybrid heating **91** 34-41

[6] Srinath M S, Sharma A K and Pradeep K 2011 *Journal of Manufacturing Processes* Investigation on microstructural and mechanical properties of microwave processed dissimilar joints **13** 141–46.

[7] Gupta P and Kumar S 2014 *Material and manufacturing process* Investigation of stainless steel joint fabricated through microwave energy **29** 910-15

[8] Badiger R, Narendranath S and Srinath M S 2015 *Journal of Manufacturing Processes* Joining of Inconel-625 alloy through microwave hybrid heating and its characterization **18** 117-23

[9] Bansal A, Sharma A K and Das S 2013 *Sadhana* Metallurgical and mechanical characterization of mild steel-mild steel joint formed by microwave hybrid heating process **38** 679–86

[10] Srinath M S, Sharma A K and Kumar P 2011 *Mater. Des* A new approach to joining of bulk copper using microwave energy **32** 2685–94

[11] Shula M and Ghosh S 2018 *Material Today: Proceedings* Analysis of interfacial joint strength of microwave and conventionally brazed alumina ceramic joints using scratch test method **5** 2010-15

[12] Singh S, Gupta D and Jain V 2015 *Proc. Inst. Mech. Eng. Part B J. Eng. Manuf.* Recent applications of microwaves in materials joining and surface coatings 1–15

[13] Jayashree P K, Gowrishankar M C, Kinia A, Sharma S and Raviraj S August 2013 *International Journal of Current Engineering and Technology ISSN 2277* Review on effect of silicon carbide (SiC) on stir cast aluminium metal matrix composites **3**

[14] Rayjadhav S B and Naik V R *ISBN* Characterization of developed al 6061-si metal matrix composites produced by the stir casting method 978-81-932074-5-1

[15] Bhandare R G and Parshuram M S April 2014 *International Journal of Current Engineering and Technology ISSN 2277 – 4106* Preparation of aluminium matrix composite by using stir casting method & it's characterization Special Issue-**3**

[16] Sudindra S and Anil Kumar C *International Journal of Metallurgical & Materials Science and Engineering* Aug 2013 Studies on Al6061/Al2O3 and graphite hybrid metal matrix composites **3** Issue 3 35-42

[17] Beihai Ma, Jie Wang, Tae H. Lee, Stephen E. Dorris, Jianguo Wen and Uthamalingam Balachandran Journal *of Materials Science* April 2018, Microstructural characterization of Al_4C_3 in Aluminum-Graphite composite prepared by electron beam melting

[18] Omar S Salih, Hengan Ou, W Sun and D G McCartney 2015 *Materials and Design* A review of friction stir welding of aluminium matrix composites 86 61-71

[19] Adel M H, Mohammed A, Qasim T and Ahmed Ghaithan *Materials and Manufacturing Processes* 2012 Effect of processing parameters on friction stir welded aluminium matrix composites wear behavior **27** 1419–23

[20] Shivinder Singh, Belokar R M and Suri N M *International Journal of Scientific Research and Review* February 2019 XRD characterization of aluminium joint processed by microwave hybrid heating **07** Issue 02 15-21

[21] Mohamed S S, Sayed A Abdallah and Hamdan K H Alazem *International Journal of Engineering and Applied Sciences* March 2018 An investigation on the mechanical and physical properties of Al6061/SiCp/Gr metal matrix composites **05** Issue-3

[22] Sujan D, Yeo C W, Rahman M E, Reddy M M, Maleque M A and Ali M Y *Advanced Materials Research* 2012 Aluminum-silicon carbide composites for enhanced physio-mechanical properties **576** 370-73

[23] Manish M, Nagendra Kumar M and Vivek B *Evergreen Joint Journal of Novel Carbon Resource Sciences and Green Asia Strategy* September 2019 Effect of SiC reinforced particle parameters in the development of aluminium based metal matrix composite **6** Issue 3 200-06

[24] Manoj S, Deepak D, Lakhvir Singh and Vikas Chawla *Journal of Minerals and Materials Characterization & Engineering* 2009 Development of aluminium based silicon carbide particulate metal matrix composite **8** No 6 455-67

[25] Ravindra I Badiger, Narendranath S, Srinath M S and Ajit M Hebbale *Trans Indian Inst Met* 2019 Effect of power input on metallurgical and mechanical characteristics of Inconel-625 welded joints processed through microwave hybrid heating

A Comparative Analysis of Biodiesel Properties Derived from Meat Stall Wastes through Optimized Parameters

R Ravikumar [1], K Kiran[1], M Harish Kumar[2], Gurumoorthy S Hebbar[3], A S Divakara Shetty[4]

[1,2,3]Department of Mechanical and Automobile Engineering, SoET, CHRIST (Deemed to be University), Bangalore, India.

[4]Engineering Department, University of Technology and Applied Sciences, Al Dhahirha Governorate, Sultanate of Oman.

r.ravikumar64@gmail.com

Abstract. Biodiesel is considered as alternative green fuels that can be used in Internal Combustion engines as a replacement fuel for conventional diesel. Biodiesel is extracted from vegetable and animal sources which are rich in triglycerides. In this work, an attempt has been made to obtain and characterize the biodiesel from animal wastes such as chicken skin and pig tallow which are available in abundance and at an economical cost within the authors' geographical location. Initially, the feedstock is decontaminated and subjected to conventional heating to convert it into fatty oil. Heating is carried out at different temperatures and for varying time to find out the optimal combination of time and temperature, which would result in maximum fat yield. The fatty oil is then subjected to the trans-esterification process with methyl alcohol in the presence of a catalyst to extract crude biodiesel. A de-canter funnel is used to separate the glycerine and biodiesel from the crude extract. The extracted biodiesel is mixed in different volume percentages with conventional diesel, and various thermochemical properties were evaluated as per ASTM standards. The test result indicated that the properties of the biodiesel blends were well within the limits as prescribed by ASTM standards.

1. Introduction

The last few decades have witnessed a huge transformation in the life of a common person due to the revolution in the growth of technology, in all sectors of industries. The lifestyle changes, particularly in developing countries, resulted in higher demands for industries, and higher energy requirements. The growing energy requirement of these industries is usually satisfied using petroleum distillates [1]. As per a study, it is said that the growth of a country through social and economic status is highly dependent on the availability and optimum utilization of its energy resources [2]. In general, any energy requirement of the industries in any country is largely fulfilled by fossil fuels, which include the burning of coal, petroleum products, and natural gas [3, 4].

Fossil fuels are categorized under conventional or non-renewable energy sources as they require more than thousands of years to be restored. From an economic point of view, these conventional energy sources are becoming expensive due to limited availability and high demands, which will ultimately result in increased cost of production in industries [5, 6]. Also, the burning of these fossil fuels releases carbon dioxide into the atmosphere to cause environmental pollution, the greenhouse effect, and one of the reasons for global warming. These disadvantages of conventional fuels have resulted in the discovery and use of non-conventional energy sources such as solar energy, hydro energy, wind, and tidal sources, which are considered green energy sources as they do not cause much environmental pollution and also they are renewable energy sources [7, 8]. Biodiesel, which is an alternative energy source similar to conventional petrochemical diesel and unlike petrochemical diesel, which can be extracted directly from vegetable oil and animal fats. Unlike conventional petrochemical diesel, biodiesel is environmentally friendly fuel as it is low in toxic, biodegradable, emits less sulfur when burned. Biodiesel can be extracted from both edible and non-edible oils such as palm, peanut, sunflower, jatropha, castor oil, cottonseed oil, etc. [9, 10].

The disadvantage of utilizing cooking oils for biodiesel production may directly affect the food chain, influence the scarcity of cooking oil, and contribute to economic inflation [11]. Besides, using vegetable oils as the source of biodiesel production may pose some challenges such as procurement, cost, transportation, and storage of feedstock, which inhibits the use of fresh vegetable oils [12, 13]. In recent years, animal waste fats, which are composed of "triglycerides" have been tried and used for biodiesel production. Animal fats are generally composed of triglycerides, some form of proteins, water, different minerals, and unwanted impurities. Compared to edible oils, animal waste fats are available in abundance and usually generated from animal slaughterhouses, restaurant wastes, veterinary services, food farming industries, and so on. Animal fats from these sources are not suitable for human consumption, and hence they are either used as low-cost feeding for animals or disposed of as a waste product [14, 15]. The biodiesel extracted from vegetable oils, plant-based sources, and animal fats can be readily used in the existing diesel engines without much engine modifications as it is primarily composed of fatty acid methyl ester obtained from "Trans-esterification" triglycerides of the feedstock [16, 17].

The resulting biodiesel is similar to the petrochemical diesel, and both can be blended in any proportions without many difficulties. Given these characteristics and attractive properties of biodiesel, the extraction of biodiesel from different renewable sources has become a topic of interest [18]. Many experiments and researches are going on in the field of production of biodiesel from animal fats due to the availability of animal feedstock in abundance. Animal fats such as chicken wastes, mutton fats, beef, and pork tallow, duck tallow, pork skin, and pork lard are a few of the many available animal-based feedstocks, which are used for biodiesel production [19]. They produced biodiesel and verified the quality of it against the petrochemical diesel as per the standards of the National Agency of Petroleum. A blend of beef tallow and soybean oil has been used for biodiesel production using sulfonated polystyrene compounds as catalysts [20]. The production of beef tallow into biodiesel and its comparison with the biodiesel produced from vegetable oils is studied and reported by Nelson et al., the study was conducted based on the parameters such as availability of raw materials, the efficiency of biodiesel produced, and the economy of conversion [21].

Similarly, a study conducted by Chakraborthy et al. [22] reports the conversion of goat tallow into biodiesel using infrared radiation assisted reactor, and they reported a substantial reduction in conversion time compared to the conventional process. Some studies also report the use of fish oil for biodiesel production and they compared the properties of fish oil biodiesel with cooking oil biodiesel. Beef tallow, chicken skin, and pork lad were used for the production of biodiesel, and different purification methods [23] and they concluded that animal fats could be considered as a potential source for biodiesel production. From the review of literature, it can be summarized that in today's energy requirements, biodiesel can be one of the promising alternatives to conventional energy sources. The biodiesel extracted from edible oil and other plant-based feedstock is commercially produced and used as a blend with petrochemical diesel. But, the plant-based feedstock for biodiesel has its disadvantages, such as causing inflation and other economic concerns [24]. Hence it is required to look at the alternative source of feedstock for biodiesel production. Animal-based sources are one of the alternatives that have the potential to be used as feedstock for biodiesel production as they are cheap and available in abundance. The research is undertaken by many researchers worldwide to analyze and characterize different animal-based feedstock for biodiesel production. But, as the geographic location varies, the availability of feedstock sources also differs [25]. As per the geographic location and culture, the food habits of the population varies, and similarly, the animal waste generated also changes. Because of these reasons, the current work is undertaken to analyze and characterize the animal feedstock available in the authors' geographic location [26]. The non-biodegradable waste such as waste plastics are also can consider and processed into biodiesel by adopting the pyrolysis process, and obtained fuels are characterized and tested in a diesel engine [27]. In this current work waste chicken skin, and pig tallow are considered as a potential source of feedstock for biodiesel production. The economical and widely used extraction method, i.e. Trans-esterification process is used for the conversion of animal fats into biodiesel. The extracted biodiesel is then subjected to different tests to find its thermochemical properties and to test its compatibility with conventional petrochemical diesel.

2. Materials and Methods

2.1. Materials

The present study is carried out to extract biodiesel from animal wastes which are geographically available in abundance from an economic point of view. For this purpose, waste chicken skin and pig tallow were selected as feedstock for the biodiesel production process. Chicken skin and pig tallow were procured from a local slaughterhouse. From the initial study, it was found within the Author's geographical location; a local slaughterhouse can produce around 20-25 kg of chicken skin and pig tallow per day, which is generally considered as a waste product.

2.2. Experimental Method

The biodiesel extraction from chicken skin and pig tallow involves a series of different procedures that are carried out sequentially on the feedstock. The feedstock gathered from the source is initially cleaned with water to remove the contaminants from it. Each 1 kg of feedstock is agitated with 5L of filtered water for 10 minutes. This washing cycle is repeated 3-4 times until all the dust, and other impurities are washed out thoroughly from the feedstock.

2.3. Fat Extraction by Conventional Heating

The separation of "fat" from the feedstock is carried out by following the conventional heating method. Both the feedstock, chicken skin, and pig tallow are subjected to the heating process separately to extract the fat. One kilogram of feedstock (chicken skin and pig tallow) and three liters of water were added in a heating pot and kept on a conventional heater. The mixture of feedstock and water is then heated to a temperature above 50°C for a specific duration of time until the fat tissue of the feedstock breakdown and oil is separated. The heating was carried out at different temperatures and for varying durations of time and the same is shown in table 1.

Table 1. Parameters of Fat Extraction by Heating

Temperature, °C	60	70	80	90	100
Time, Minutes	20	25	30	35	40

2.4. Purification of Extracted Fat

At specific temperature and heating duration, the fat breaks down from the feedstock, and the "fat" separate from the fat tissues inside the pot. Then the pot is rested for 24 hours until all the water, and fat oil is separated. The oil is collected out and passed through filter papers to further purify it from the external wastes such as skin and fat tissue. The purified oil is again rested for 24 hours to separate any remaining water content, and pure oil is collected out.

2.5. Conversion of fat into biodiesel by the trans-esterification process

Trans-esterification is a chemical process where the triglycerides are reacted with short-chain alcohols such as ethanol and methanol using an acid or base catalyst to exchange the organic group of the triglycerides with the organic group of methanol or ethanol. In the present study, this trans-esterification method was used to convert the triglycerides extracted from chicken skin and pig tallow into biodiesel.

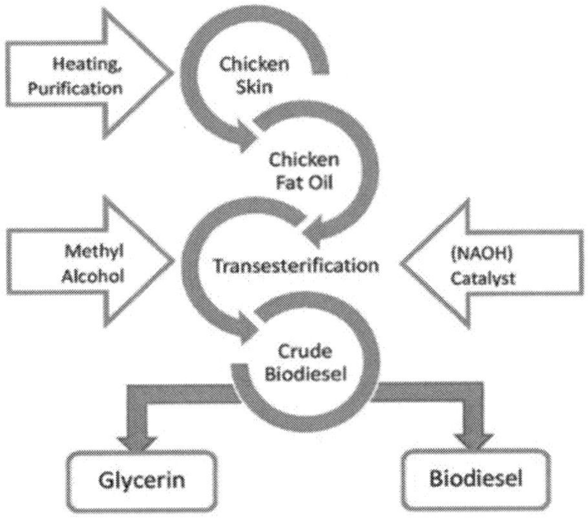

Figure 1. Trans-esterification of chicken skin into biodiesel

Trans-esterification of chicken skin into biodiesel: The chicken skin feedstock is cleaned and heated to get triglycerides in the form of the fat oil as per the procedure explained earlier. The separated oil from the chicken skin is passed through the filter paper to purify it from remains of skin and other impurities further. The purified oil is then subjected to the trans-esterification process to convert it into biodiesel, as shown in figure 1. A measured quantity of chicken fat oil (500ml) is transferred into a three-neck flask and heated gradually until the temperature of the oil reaches to 80°C. The temperature is monitored using a thermometer dipped inside the oil through the three-neck flask. When the oil temperature reaches 80°C, 125ml of methane alcohol and 3gm of sodium hydroxide (NaOH) is added gradually to the oil, and the mixture is continuously stirred. NaOH acts as a catalyst, and the methyl alcohol acts as a driver to start the trans-esterification reaction. A condensing system is used while heating the mixture to prevent any evaporation of methyl alcohol out of the trans-esterification setup.

The system is heated and continuously stirred for 2 hours after adding the methyl alcohol and NaOH. During this period the solution temperature is monitored continuously and maintained at 80°C. After the reaction time is complete, it can be observed that the chicken fat oil is converted into crude biodiesel which is again a mixture of biodiesel and glycerine. The crude biodiesel is then transferred into a de-canter funnel for the separation process. The solution is allowed to rest for about 5 hours in the de-canter funnel at room temperature. Gradually the biodiesel, NaOH, and glycerine are separated in the de-canter funnel. The glycerine is collected at the bottom of the de-canter funnel whereas the pure biodiesel settles above it. The glycerine is then removed from the funnel, and pure biodiesel is collected out for the further process.

Trans-esterification of Pig Tallow into biodiesel: Similar to the chicken skin trans-esterification process, the feedstock in the form of pig tallow is cleaned and heated to get triglyceride fat oil as per the procedure explained earlier. The separated oil from the pig tallow is purified by passing it through the filter papers. The purified oil is then subjected to the trans-esterification process to convert it into biodiesel, as shown in figure 2.

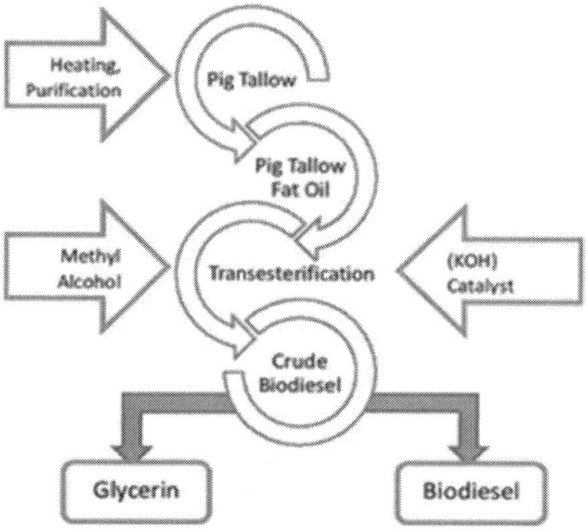

Figure 2. Trans-esterification of pig tallow into biodiesel.

As explained earlier, the fat oil from pig tallow (500ml) is transferred into a three-neck flask and heated up to 80°C. A solution of Potassium Hydroxide (3gm) and methyl alcohol (125ml) is mixed with the heated fat oil and continuously stirred in the three-neck flask. The process is continued for 2 hours, and the temperature is maintained at 80°C. After the reaction period is complete, the crude biodiesel is transferred to the de-canter funnel for the separation process and allowed to rest for 5 hours. The glycerine is settled at the bottom of the funnel and biodiesel gets collected at the top. The biodiesel is collected out from the de-canter funnel for further processing.

3. Results and Discussion

3.1. Fat Extraction from the Feedstock

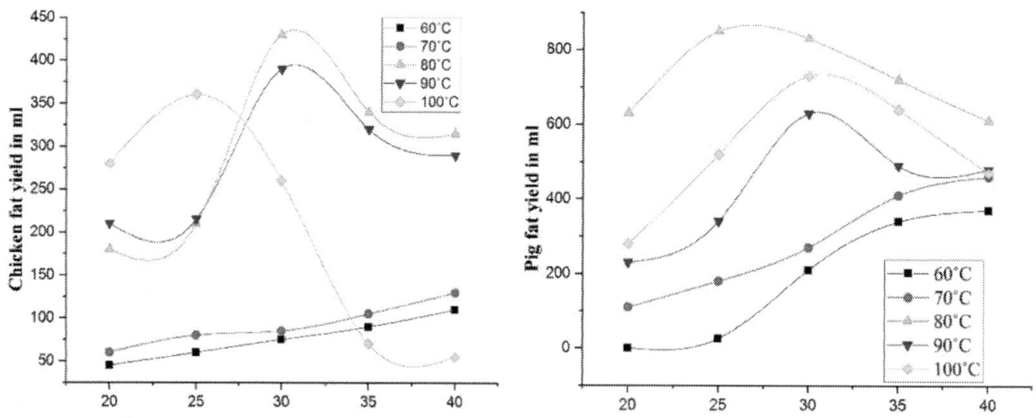

Figure 3. Graph showing the fat oil yield from heating (a) Chicken Skin and (b) Pig Tallow for different heating durations at different temperatures

The initial stage of biodiesel extraction from animal waste is to convert animal waste into triglycerides, i.e. in common words "fat." This is achieved through the heating process. The fat oil extraction from the chicken skin and pig tallow are explained in detail earlier. The heating process is carried out for different durations of time at different temperatures to find out the optimum time and temperature that would yield the maximum fat oil from the feedstock, and these parameters are shown in Table 1.

As per the selected heating parameters, the experiments were conducted by individually heating the chicken skin and pig tallow for different durations at different temperatures. The full factorial experimental design was framed by changing one variable at a time for each trial of the experiment. A total of 25 experiments were conducted twice as per the experimental design, and the oil yield is measured after each experiment trial. The average results of the two trials are considered as the final fat oil yield, and the results are noted down.

It was found from the experiments that the oil yield varies with change in the heating time and temperature. Figures 3(a) and 3(b) indicate the graphs of the fat oil yield from heating chicken skin and pig tallow for different heating durations at different temperatures respectively. The optimum combination of heating temperature and heating time which would give maximum fat yield from chicken skin and pig tallow are indicated in Table 2.

Table 2. The Optimum Combination of Temperature and Time for Maximum Fat Yield.

	Chicken Skin (per kg)	Pig Tallow (per kg)
Optimum Temperature, °C	80	80
Optimum Time, minutes	30	25
Fat yield, ml	460	890

3.2. Biodiesel Blends

Table 3. Biodiesel Blends and Their Nomenclatures

Biodiesel Nomenclature	Blend (% by Volume)		
	Chicken Skin, %	Pig Tallow, %	Diesel, %
CP0D100	00	00	100
C100D0	100	00	00
C10D90	10	00	90
C20D80	20	00	80
C30D70	30	00	70
C40D60	40	00	60
C50D50	50	00	50
P100D0	00	100	00
P10D90	00	10	90
P20D80	00	20	80
P30D70	00	30	70
P40D60	00	40	60
P50D50	00	50	50
C05P05D90	05	05	90
C10P10D80	10	10	80
C20P20D60	20	20	60

The trans-esterification process is used for the conversion of triglycerides into biodiesel, and the process of trans-esterification for chicken skin and pig tallow is explained in other sections. 500ml of triglycerides obtained from chicken skin and pig tallow are individually subjected to the trans-esterification process by adding measured quantities of methyl alcohol and a suitable catalyst. The obtained crude biodiesel after the trans-esterification process contained both glycerine and biodiesel and they were separated by using the de-canter funnel. It was found that 500ml of chicken skin and pig tallow have yielded 365ml and 475ml biodiesel respectively. The obtained biodiesel from chicken skin and pig tallow are blended individually and in combination with the petrochemical diesel in different volume fractions. Thus obtained blended biodiesel is characterized to find out various fuel compatibility properties. Table 3 indicates the different biodiesel blends and their nomenclatures.

3.3. Properties of Chicken Biodiesel and its Blends.

Different thermochemical tests were carried out on the biodiesel blends to find out their thermo-chemical properties. Properties such as flash point, fire point, kinematic viscosity, density, calorific values, and carbon residue (content) were evaluated for all biodiesel blends as per ASTM standards. The variation of these thermo-chemical properties with change blending volume is discussed in the following sections.

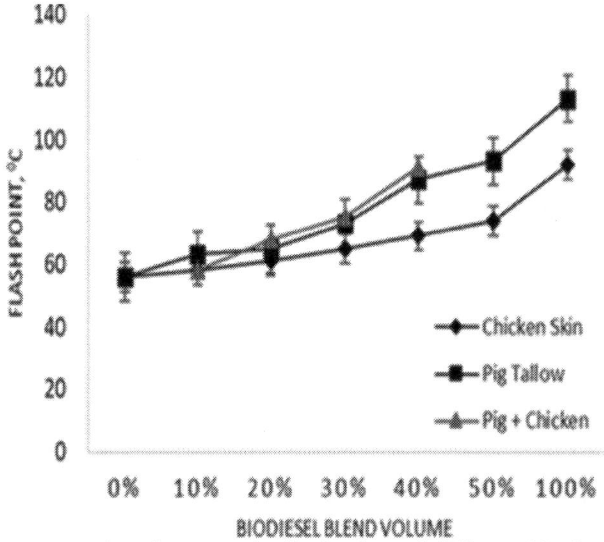

Figure 4. Comparison between the flashpoints of different biodiesel blends

Flashpoint or flash point temperature of the biodiesel blends are evaluated as per the ASTM D93 standards. In general, flash point temperature has defined a measure of biodiesel tendency to form a flammable mixture with air under laboratory conditions, and it is one of a property to determine the flammability hazard of the biodiesel. Figure 4 shows a graph of the variation in flashpoints of various biodiesel blends. The comparison between the blending volume of chicken skin and pig tallow based biodiesel can also be observed from the graph. The pure petrochemical diesel has a flashpoint of 56°C, and it can be observed that the flashpoint temperature is increasing gradually with an increase in blending volume, and the pure biodiesel exhibits the highest flash point temperature. It can be observed from the graph that the biodiesel extracted from pig tallow is more flammable compared to the biodiesel extracted from the chicken skin. And, the flashpoint temperature further increases if both the extracted biodiesels are blended in equal quantities with the petrochemical diesel.

Figure 5 depicts the kinematic viscosity of different biodiesel blends and their comparison with one another. The kinematic viscosity of the biodiesel and its blends are estimated as per the ASTM D445 standards. As any fuel used in an engine acts as a lubricant also, it becomes essential to evaluate its kinematic viscosity. And having the right values of these parameters ensures that the biodiesel's storage, handling, and operational conditions. Thus, it becomes essential to evaluate the kinematic viscosity of all the biodiesel blends. The kinematic viscosity of all the biodiesel combinations are evaluated, and the results are shown graphically in Figure 5. It can be observed that the kinematic viscosity of petrochemical diesel is lowest compared to the biodiesel blends and the kinematic viscosity was found to be highest for pure extracted biodiesel without any mix. Compared to the blends based on chicken skin the pig tallow based biodiesels have exhibited higher values of kinematic viscosity and the hybrid biodiesel exhibited still higher level of kinematic viscosity values.

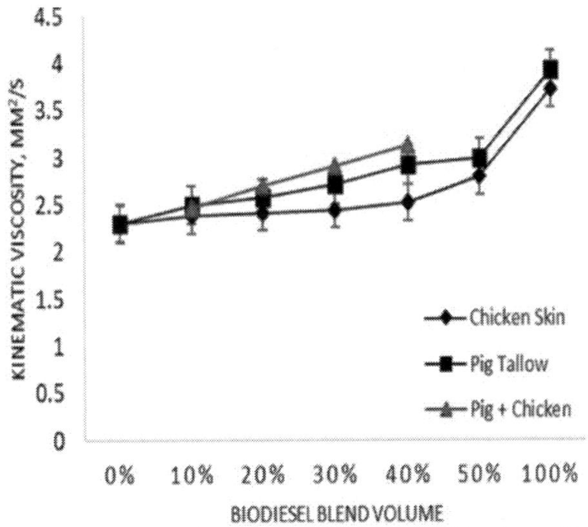

Figure 5. Kinematic viscosity of biodiesel blends at 40°C

The calorific value in kJ/kg of the biodiesel blends was evaluated using a standard bomb calorimeter by following the standard test methods. The obtained values are graphically tabulated as shown in Figure 6. The graphs indicate that the pure petrochemical diesel is having more calorific value and the calorific value reduced with the addition of biodiesel to it. The blending percentage is inversely proportional to the calorific value, and it may be attributed to the higher oxygen content of the biodiesel than the petrochemical diesel. Compared to pig tallow based biodiesel blends, the chicken skin based biodiesel was found to be having higher calorific values. The hybrid blends which contained both the chicken fat and pig tallow based biodiesel exhibited the lowest range of calorific values compared to their counterparts due to its higher oxygen contents.

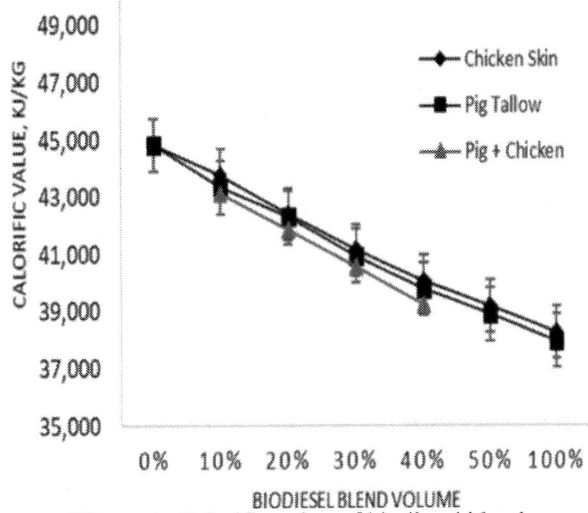

Figure 6. Calorific value of biodiesel blends

The carbon content or carbon residue value of any fuel indicates an approximation of its use and its inclination to form carbon residue inside the burning chambers of the engine and the carbon residue value of biodiesel indicates its tendency to form carbon residue inside the combustion chamber over its continued use as a fuel. The carbon residue value of the biodiesel blends was estimated by following the standard test method as prescribed by the ASTM D524 – 15 standards. All the biodiesel blends are subjected to a standard test to evaluate their carbon content, and the values obtained are graphically compared through Figure 7. It can be observed that the biodiesel and its blends have lower carbon content compared to the petrochemical diesel. As per ASTM D524 – 15, the maximum carbon content for any biodiesel is less than 0.05 or 5%, and it can be observed that almost all the blends fall under the standard prescribed limits.

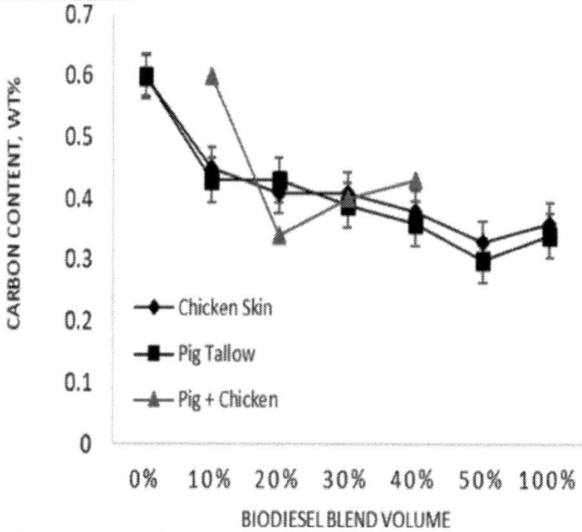

Figure 7. Carbon Content (wt.%) of biodiesel blends

4. Conclusion

The research paper presents experimental procedures followed to extract biodiesel from animal wastes such as chicken skin and pig tallow, which are available in abundance and at economical cost. The extracted biodiesel from chicken skin and pig tallow are blended with petrochemical diesel in different volume percentages. The investigation of the thermochemical properties of all the biodiesel blends is carried out, and the conclusions are as follows:

Chicken skin and pig tallow – waste products from animal slaughterhouses can be successfully used for biodiesel extraction. The heating process was used to convert the feedstock into triglycerides, and per kg of each type of feedstock has yielded around 460 ml and 890 ml of fat oil from chicken skin and pig tallow respectively. The optimum parameters that would yield maximum fat yield were found to be 80°C and 30 minutes for chicken skin, 80°C, and 25 minutes for pig tallow. The trans-esterification process was carried out to convert the extracted triglycerides into crude biodiesel. The process was carried out at 80°C for 2 hours by mixing a measured quantity of methyl alcohol and a catalyst. It was found that 500ml of chicken skin and pig tallow have yielded 365ml and 475ml biodiesel, respectively.

The different volume percentage of extracted biodiesel was mixed with petrochemical diesel to form various blends. Thermo-chemical properties such as flash point, kinematic viscosity, calorific value, and carbon content were evaluated by following standard test methods as specified by ASTM. It was found that the test results of almost all blends were within limits prescribed by ASTM D 6751 and hence can be concluded that they may be used as a compatible replacement of conventional diesel.

The fire point evaluation of different biodiesel blends was carried out, and it was found that, compared to conventional diesel, the blends exhibited higher fire point temperature and pure biodiesel extract exhibited the highest fire temperature. The blending volume is found to be directly proportional to the fire point temperature, i.e. as a larger volume of biodiesel lesser will be the flammability of the blend. Similar results were obtained concerning kinematic viscosity of the biodiesel blends. As the volume of mixing increased, the kinematic viscosity of the blend is found to be reducing, and comparatively pure biodiesel extract was found to be having the lowest value of kinematic viscosity.

The carbon residue value of the biodiesel is one of the criteria to predict the carbon deposits in the combustion chamber. Compared to conventional diesel, the biodiesel blends were found to be having lesser carbon content, and the pure biodiesel is the one with the lowest one. As the biodiesel from animal extract has more oxygen content, their combustion will result in lesser carbon emission, and hence the biodiesel and its blends can be considered as green fuels. The calorific value of the biodiesel was found to be lower compared to petrochemical diesel and increasing the blending volume has resulted in reducing the calorific value of the biodiesel. The 10% blended biodiesel was found to be having nearly equal calorific value as that of conventional diesel.

References

[1] Devaraj Kubendran, Abdur Rawoof Salma Aathika, Thanarasu Amudha, Devaraj Thiruselvi, Mani Yuvarani and Subramanian Sivanesan 2017 Utilization of leather fleshing waste as a feedstock for sustainable biodiesel production. Energy Sources, Part A: Recovery, Utilization, and Environmental Effects, 1-7, http://dx.doi.org/10.1080/ 15567036.2017. 1349218.

[2] Vishal Saxena, Niraj Kumar and Vinod Kumar Saxena 2017 A comprehensive review on combustion and stability aspects of metal nanoparticles and its additive effect on diesel and biodiesel fuelled C.I. engine, Renewable and Sustainable Energy Reviews, 563–588, http://dx.doi.org/10.1016/j.rser. 2016.11.067.

[3] Ayhan Demirbas and M Fatih Demirbas 2010 Importance of algae oil as a source of biodiesel. Energy Conversion and Management, 163-170, doi:10.1016/j.enconman.2010.06.055.

[4] Qing Li, Longyu Zheng, Hao Cai, E Garza, Ziniu Yu and Shengde Zhou 2010 From organic waste to biodiesel: Black soldier fly, Hermetia illucens, makes it feasible. Fuel, 1545–1548, doi:10.1016/j.fuel.2010.11.016.

[5] Ivana B Bankovic-Ilic, Ivan J Stojkovic, Olivera S Stamenkovic, Vlada B Veljkovic and Yung-Tse Hung 2014 Waste animal fats as feedstocks for biodiesel production. Renewable and Sustainable Energy Reviews, 238–254, http://dx.doi.org/10. 1016/j.rser. 2014.01.038.

[6] Divakar Shetty A S, Ravi Kumar R, Kumarappa S and Antony A J 2016 Study on Conversion of Municipal Plastic Wastes into Liquid Fuel Compounds, Analysis of CRDi Engine Performance and Emission Characteristics. IOP Conf. Series: Materials Science and Engineering, 1-10, doi:10.1088/1757-899X/149/1/012223.

[7] Hossein Esmaeili and Rauf Foroutan 2018 Optimization of Biodiesel Production from Goat Tallow Using Alkaline Catalysts and Combining Them with Diesel. Chemistry & Chemical Technology, 12(1), 120-126, https://doi.org/10.23939 /chcht 12.01.120.

[8] M Keihani, H Esmaeili and P Rouhi 2018 Biodiesel Production from Chicken Fat Using Nano-calcium Oxide Catalyst and Improving the Fuel Properties via Blending with Diesel. Physical Chemistry Research, 6(3), 521-529, DOI: 10.22036/pcr.2018.114565.1453.

[9] G Sujaykumar, Ravikumar R, Sushilendra Mutalikdesai, Basavaraj Pujar, Vinayak Baddi, Dhareppa Siddapur, Santosha P V and Basavaraj Telagadi 2017 Comparative Study of Waste Cooking Oil and Cashew Nut Shell Oil BioFuel Blends with Diesel. Energy and Power, 7(3), 65-69, DOI: 10.5923/j.ep.20170703.02.

[10] B Freedman, E H Pryde and T L Mounts 1984 Variables Affecting the Yields of Fatty Esters from Transesterified Vegetable Oils. Presented at the American Oil Chemists' Society annual meeting, Chicago. 1638- 1643.

[11] A Gnanaprakasam, V M Sivakumar, A Surendhar, M Thirumarimurugan and T Kannadasan 2013 Recent Strategy of Biodiesel Production from Waste Cooking Oil and Process Influencing Parameters: A Review. Journal of Energy, 1-10, http://dx.doi.org/10.1155/2013/ 926392.

[12] Ravikumar R, Mosharib Ahmad Hashmi, Shankar Ganesh L, Sarvesh V Bikkannavar and Vivek D R 2019 Biofuel Production and Characterization from Waste Chicken Skin and Pig Fat. International Journal of Recent Technology and Engineering, 8(3), 3598-3603, DOI:10.35940/ijrte.C5312. 098319.

[13] J M Encinar, J F Gonzalez and A Rodríguez-Reinares 2007 Ethanolysis of used frying oil. Biodiesel preparation and characterization. Fuel Processing Technology, 513–522, doi:10.1016/j.fuproc.2007.01.002.
[14] Ravikumar R, Harish Kumar M, Kiran K and Gurumoorty S Hebbar 2019 Extraction and Characterization of Biofuel from Industrial Waste organic Pupae-Silkworm. International Journal of Recent Technology and Engineering, 8(3), 1603-1607, DOI: 10.35940/ ijrte.C4422.098319.
[15] Kaisan Muhammad Usman, Gyang Pam and Dangana M Kulla 2013 Physico-Chemical Properties of Bio-diesel from Wild Grape Seeds Oil and Petro-Diesel Blends. American Journal of Engineering Research, 2(10), 11-21.
[16] Ravikumar R, G Sujaykumar, Divakar Shetty A S and Basavaraj Kambale 2017 Investigation of Effect of Chicken Biodiesel Blended Diesel on Engine Performance. International Journal of Advances in Scientific Research and Engineering, 3(1), 192-196.
[17] R Ravikumar, Kiran K and S Hebbar Gurumoorthy 2020 Experimental Analysis of Biofuel Produced from Fat Derivatives of Bird and Animal as an Additive Fuel in CI Engine. International Journal of Renewable Energy Research, 10(3), 1226-1233.
[18] Michele Espinosa da Cunha, Laiza Canielas Krause, Maria Silvana Aranda Moraes, Candice Schmitt Faccini, Rosângela Assis Jacques, Suelen Rodrigues Almeida, Maria Regina Alves Rodrigues and Elina Bastos Caramao 2009 Beef tallow biodiesel produced in a pilot scale. Fuel Processing Technology, 570–575, doi:10.1016/j.fuproc.2009.01.001.
[19] Rafael A Soldi, Angelo R S Oliveira, Luiz P Ramos and Maria Aparecida F Cee sar-Oliveira 2009 Soybean oil and beef tallow alcoholysis by acid heterogeneous catalysis. Applied Catalysis A: General, 42-48, doi:10.1016/j.apcata.2009.03. 030.
[20] Richard G Nelsona and Mark D Schrock 2006 Energetic and economic feasibility associated with the production, processing, and conversion of beef tallow to a substitute diesel fuel. Biomass and Bioenergy, 584–591, doi:10.1016/j. biombioe.2005.09.005.
[21] Ravikumar R, G Sujaykumar, Divakar Shetty A S, Basavaraj Kambale and Aravinda Pattar 2017 Investigation of Effect of Tire Oil Blended Diesel on Engine Performance. Energy and Power, 7(3), 61-64, DOI: 10.5923/j.ep.20170703.01.
[22] R Chakraborty and H Sahu 2013 Intensification of biodiesel production from waste goat tallow using infrared radiation: Process evaluation through response surface methodology and artificial neural network. Applied Energy, 1-10, http://dx.doi.org/10.1016/j.apenergy.2013. 04.025.
[23] R Yahyaee, B Ghobadiann and G Najafi 2012 Waste fish oil biodiesel as a source of renewable fuel in Iran. Renewable and Sustainable Energy Reviews, 312–319, http://dx.doi.org/10. 1016/j.rser.2012.09.025.
[24] Abdelrahman B Fadhil and Latif H Ali 2013 Alkaline-catalyzed transesterification of Silurus triostegus Heckel fish oil: Optimization of transesterification parameters. Renewable Energy, 481-488, http://dx.doi.org/10.1016/j. renene.2013. 04.018.
[25] Teresa M Mata, Nelson Cardoso, Mariana Ornelas, Soraia Neves and Nídia S Caetano 2011 Evaluation of Two Purification Methods of Biodiesel from Beef Tallow, Pork Lard, and Chicken Fat. Energy Fuels, 4756–4762 dx.doi.org/10.1021/ef2010207.
[26] Vinayak Baddi, P V Santosha, G Sujaykumar and R Ravikumar 2017 Study of Physical Properties of Cashew Nut Shell Oil Bio-Diesel Blends. Journal of Research in Science, Technology, Engineering and Management, NCETERM-2017, 74-78.
[27] Ramegouda R and Joseph A A 2021 Effect of Compression Ratio on Performance and Emission Characteristics of Dual Spark Plug Ignition Engine Fueled With n-Butanol as Additive Fuel. Int. Journal of Renewable Energy Development, 10(1), 37-45. https://doi.org/10.14710/ijred. 2021.32364.

Experimental Investigation of Air Circulation Using Duct System in a Non-AC Bus Coach

Niranjana SJ[1], Shivalingappa S Kubsad[2], Nagaraj Y[3], Manjunath S[4]

Shivakumar S[5]

[1, 4, 5] Department of Mechanical Engineering, School of Engineering & Technology, CHRIST (Deemed to be University), Bengaluru- 560076, Karnataka, India

[2]Department of Mechanical Engineering, Atria Institute of Technology, Bengaluru- 560024, Karnataka, India

[3]Department of Mechanical Engineering, SJM Institute Technology, Chitradurga-577502, Karnataka, India

Email: sjniranjan86@gmail.com

Abstract. Public transport is the life line in many of the developing and under developed countries for the safe conveyance, i.e. also consider as economical. The major limitation in public transport (non-AC busses) Air Condition, is the lack of proper air circulation leading to suffocation and vomiting. The present research work emphasis on design and analysis of air flow duct system (non AC Busses) to increase the level of comfortance of the passengers, tools like solidworks software 2016 is used for 3D drawing, Hypermesh software 13.0 is for the discretization and ANSYS Fluent software 16.0 for the Computational Fluid Dynamic (CFD) analysis, from the experimental the airflow is found to be 10 m/s, and from the numerical analysis the airflow is found to be 9.8 m/s, by comparing the experimental and numerical results a negligible deviation of 2% is observed and it is within the limit.

Keywords: Air Circulation, Non-AC Bus, Passenger Comfort, Pollution.

1. Introduction

In most developing and developed countries, passengers travelling in public transport are common, especially in peak travel hours, difficult to provide seats for all passengers, so a high percentage of non-AC bus users across the country are standing in aisle. The analysis shows the air circulation for people standing in the aisle of non-AC buses is causing discomfort and suffocation [1], while majority of the airflow is concentrated on people sitting near the windows.

Suffocation is a general concept that includes many types of asphyxia, either due to the absence of air in the breathing area or due to obstruction of the external airways (smothering) or internal airways (Choking), or due to pressure on the chest or abdomen or the location of the body preventing respiratory movement (traumatic and positional asphyxia, respectively). Especially in sunny days have complain of headache, vomiting [2] etc.

Currently there is no system in non-ac buses that regulates or purifies air in order to help ease comfort of people standing in the coach, the only current alternative is to employ the use of AC buses but it is impossible to convert all non-AC buses to AC without increasing the cost of Journey substantially [3]. A simple and affordable solution to increase the uniform air flow that cools the entire bus is to impart duct system in non-AC buses. This results in improving the passenger the and hence improving passenger comfort and greatly reducing fatigue of travelling long distances [4-6]. When a bus is moving, there is air resistance pushing against wind and the air naturally wants to gush into the bus to equalize pressure [7]. This air pressure is utilized by using a duct that is placed in the optimal location using CFD analysis [8]. Therefore increasing the air supply enhances the breathing ability of the passenger thereby increasing the comfort level [9].

The blower as an auxiliary unit facilitates in proper air circulation for the passengers when the bus is stationary. The duct size and heating load in a Non AC bus coach is calculated based on the standards [10-15]. The present work focuses on the development is that it's aimed at meeting all the standards set by ISHRAE (Indian society of Heating, Refrigerating and Air conditioning manufactures association) and the use of natural resources, does not pollute likeAC does.

Development of duct system for proper air circulation in a non-AC bus coach according to ISHRAE standards will improve the airflow in a non-AC bus by providing a proper duct system.

2. Materials & Methodology

In the present work the experimental investigation for a duct system is carried out. The model was fabricated using the Galvanized Iron (GI) sheets, pop rivets, neoprene gasket and a blower used to generate the air. The generalized expressions used to obtain the amount of air required to increase the passengers comfort and to reduce the cabin temperature has been discussed below, Table 1. Illustrates the material considered for the duct fabrication along with its descriptions.

Table 1. Material with description

Material	Description/Specification
Pop Rivet	POP rivets are used to connect two pieces of material in a quick efficient manner with a hand riveter or pneumatic rivet gun.
Neoprene Gasket	Neoprene is a material made from synthetic rubber designed to be flexible, durable, resilient, and very resistant to compression and liquids such as water, oil, and other solvents.
Blower	A 14kw, High performance blower is used that is capable of producing 353CFM (Cubic feet per minute) and a maximum of 2500 RPM (Revolutions per minute).
Volume Control Damper (VCD)	Automatically maintains a desired volume of airflow independent of pressure changes, used in Heating, Ventilation, coolers etc.

The fabrication of duct was carried out using GI material. The mechanical properties of GI material considered in this present work have been tabulated in Table 2.

Table 2. Material properties of GI

Grade	Modulus of Elasticity(E)	Bulk Modulus(G)	Poisson's Ratio (μ)	Shear Modulus (τ)	Density (ρ)
IS 1079	200 GPa	340 GPa	0.29	80 GPa	7.8 g/cc

2.1 Process parameters.

The scaled model was considered and tested experimentally using a high speed blower and an anemometer to record flow rate at the outlet of ducts. The results were compared with the values obtained from the CFD analysis, Table 3. Illustrates the different inlet flow velocities of duct.

Table 3. Experimental Input velocity

Sl No.	Velocity (m/s)
1	30
2	20
3	18
4	15

2.2 Generalized expression to obtain the quantity of air required for optimal passenger comfort.
Volume of air was measured in CFM (Cubic feet per minute) as shown in equation 1[15]. It is a measurement of the velocity at which air flows in/out of a space. The amount of air required to increase the passenger comfort is very much essential when dealing with HVAC (Heating Ventilation and Air Condition) or designing of ducts. Since, the obtained volume of air facilitates in designing the ducts and airflow control valves for better circulation of air.

$$\begin{aligned}&\text{The heat disipated in the bus} = \\ &\sum(\text{Total heat dissipated by the passanger+} \\ &\text{Heat dissipated from the roof+Glass heat+} \\ &\text{Heat from the engine and infiltrationload})\end{aligned} \quad (1)$$

The Passenger Metabolic load is the amount of heat generated from the passenger is calculated by using the equation 2[15].

$$Q_{met} = \text{Sum of } (M \times BSA) \quad (2)$$

Where M is the passenger metabolic heat production rate and BSA (Body Surface Area) that was obtained using the equation 3[15]. Based on activity levels and occupation for a driver and a sitting passenger, the values can be considered as 85 W/m^2 and 55 W/m^2 [15]

$$BSA = 0.202 \times W^{0.425} \times H^{0.725} \quad (3)$$

Where,
W = Weight of passenger in kg
H = Height of passenger in m
W=60 Kg and H=1.5m considered for driver and passenger (Indian adult average height and weight is considered in this present work)
Infiltration is an unintended or accidental introduction of outside air into a non-AC bus, usually through windows in the bus envelope and through the use of doors for passage. It is obtained by using the equation 4 [15].
Air Infiltration Load = Amount of air leaked =V/2 m^3/hr,
Where,
V = vehicle speed in km/h Assuming 72 kmph,
Air volume leaked = 36 m3/hr.
= 36 x Air density x (Enthalpy of Infiltered Air -Enthalpy of Cabin Air) (4)
Sensible heat given for roof sheet load is the amount of heat generated due to roof top of the non-AC bus and is obtained using the equation 5[15].

$$Q = A \times U \times \Delta t \quad (5)$$

Where,
A= Area m2,
U = Heat transfer coefficient W/m^2K from equation 6 [12],
Δt = Change in temperature K,

$$U = \frac{1}{\frac{1}{h_1} + \frac{\Delta x}{k} + \frac{1}{h_2}} \quad (6)$$

Where,
h_1 and h_2 = convection heat transfer coefficient of outside and inside surface in W/m2K from equation 7[15],
K= Thermal Conductivity in W/mK,
Δx = Thickness in m,

$$h = 0.6 + 6.64\sqrt{v} \quad (7)$$

IOP Conf. Series: Materials Science and Engineering 1065 (2021) 012006 doi:10.1088/1757-899X/1065/1/012006

Where,
V is the velocity of the vehicle in m/s.
Sensible heat given for portion of glass load is defined as the amount of heat generated when the surface of the glass is exposed to direct sunlight. In the same manner the window glass of the non-AC busses generates huge amount of heat inside the bus cabin leading to passenger's discomfort. The expression considered for obtaining the sensible heat is depicted in equation 8 [15][10].

$$Q = A \times U \times \Delta t + \left[\text{Diffused radiation} + \text{Direct radiation}) \times \cos(i) \right] \times A \times \tau \quad (8)$$

Where,
i = Sun's incidence angle on glass,
τ = transmissivity of glass.
(U =36.98 W/m2K from the expression and K= 0.756 W/mK for glass, $\Delta x = 0.04$ m)
Heat generated by blower motor is the heat from the engine during traveling is expressed in equation 9

= Blower Wattage x Load factor / Efficiency (9)

Area of Duct required for a flow rate is given by

$$\text{Area} = \frac{CFM}{V} \quad (10)$$

3. Results and Discussion

3.1 Airflow calculation

The analytical results are obtained based on the following assumption, air flow is consider only from the duct system and the air entering from the windows are not consider in the present work. However to obtain the amount of heat generated in a bus, the following assumptions were considered, direct radiation from the glass, roof top, engine and heat from the passenger.
Minimum amount of air required to travel in a non-AC bus with seating capacity (MiTR: Ashok Leyland make 28 seat school bus) of 28 passengers is obtained from the equation 1
Passenger metabolic load value is obtained using equation 2 i.e.,

$$Q_{met} = 2502.5W$$

Air Infiltration value is obtained using equation 4 i.e.,

Air infiltration = 902.8 W

Sensible heat given for Roof sheet load value is obtained using equation 5 i.e.,

Q = 320.76 W

Sensible heat given for portion of glass load value is obtained using equation 8 i.e.,

Q = 3234.37 W

Heat generated by blower motor from equation 9

Q = 202 W

Total Heat load on a bus is = 2502.5 + 902.8 + 320.76 + 3234.37 + 202

Q = 7162.43 W

For a safer side considering at the peak hour and peak load condition 20% of additional value is considered for the calculation.

Q = 2.44 Tr (Tonnage)

Total CFM required is = 2.44 x 315 = 768.6 ≈ 800 CFM

Area = 1.33 feet^2, or Area = 0.123 m^2 from equation 10

Cross sectional area of duct required for the comfortable traveling in a non-AC bus is 1.33 ft^2 or 0.123 m^2, but in the present work a square duct of cross-sectional area 0.0056 m^2 is considered.

3.2 Designing of Scaled Duct

For the conduction of the experiment a scaled model of a duct with a suitable scaling factor of 1:20 was considered for designing and fabrication. The tools such as solid works for drafting, Hyper-mesh for discretization and ANSYS Fluent for CFD analysis were considered in this present work. The main duct with sub branch duct is shown in the Figure1 and Figure 2 respectively (all dimensions are in millimeters) having the cross section 75mm x 75mm. the galvanized Iron material was considered to prevent the corrosion. The assembly of the duct was carried out with the help of rivets, gasket and VCD. The analytical work is conducted

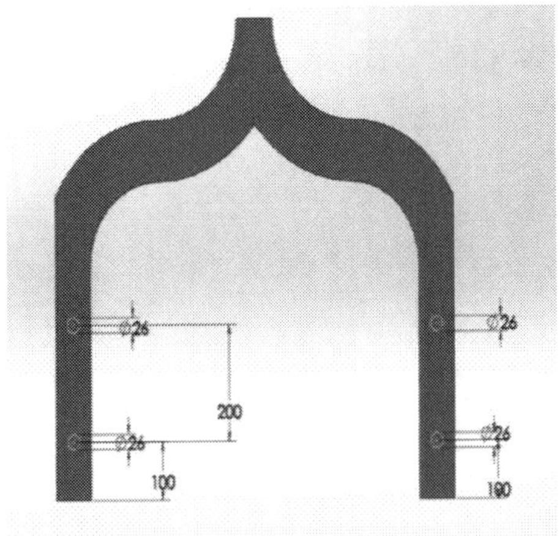

Figure 1. 3D Model of scaled

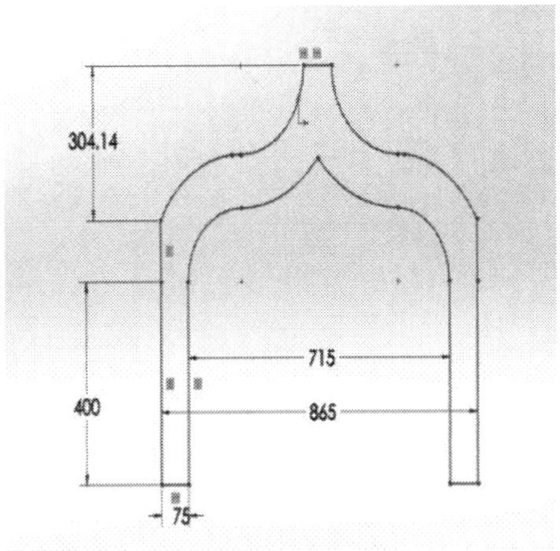

Figure 2. Drawing of scaled duct with specification.

3.3 Discretization

The discretization was carried out for the scaled duct model, the tetrahedral elements is used for the CFD analysis, tetra mesh is preferred in CFD because these elements can fit better complex geometry and it shown in Figure 3 and Figure 4.

Figure 3. Mesh Model of scaled Duct

Figure 4. Discretization of scaled duct model

3.4 CFD analysis

Analysis was carried out using ANSYS Fluent software, duct model is subjected to varying inlet velocity, and the results are tabulated in Table 4. The boundary condition was applied to permit the flow of air to enter and exit the duct system. Incompressible flow velocity at inflow and outflow was considered in this present work. Air from the inlet of main duct passes through the symmetric sub braches as shown in Figure 5. The entire duct system is located below the bus roof with regulating unit for the uniform air circulation. The filters are provided for the air purification. These filters are made up of spun fiberglass material or cloth enclosed in a cardboard frame.

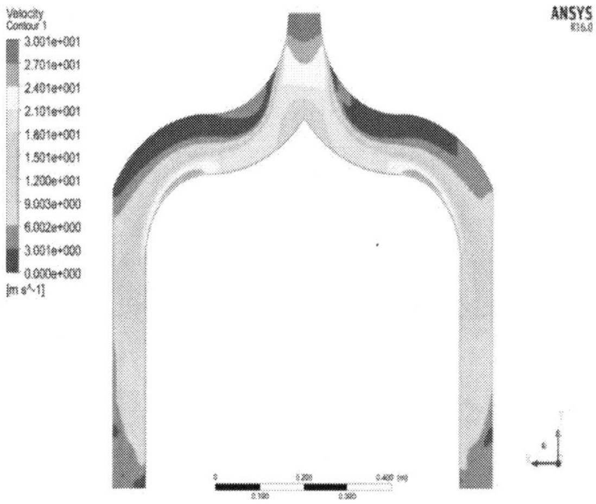

Figure 5. Analysis of scaled duct model

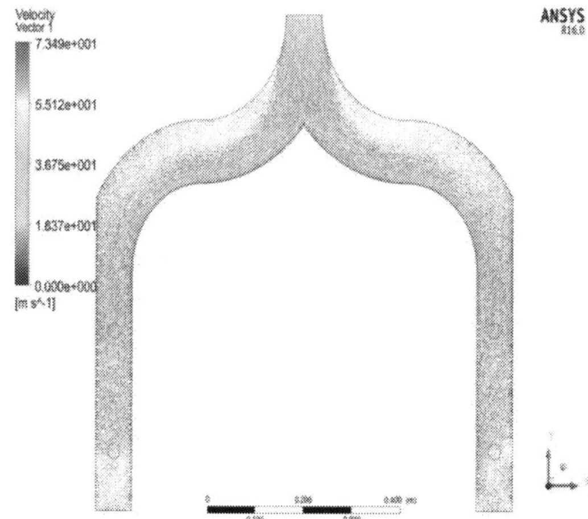

Figure 6. CFD results of scaled duct

3.5: Results of CFD analysis summary

Table 4 shows the result of CFD analysis, the duct model is subjected to varying velocity, in this study analysis is carried out for velocity of 15, 18, 20 and 30 m/s. and the velocity at the outlet is noted at four different positions, i.e., branch 1 and 2 of outlet A and B.

Table 4. Result of velocity outlet when subjected to varying input velocity.

Blower Speed Inlet m/s	Branch 1 Outlet A m/s	Branch 1 Outlet B m/s	Branch 2 Outlet A m/s	Branch 2 Outlet B m/s
30	9	10	12	10
20	8	7	10	7
18	8	6	8	7
15	6	5	7	6

3.6 Fabrication of scaled duct model

The scaled model of duct is fabricated and was assembled in the laboratory using pop rivet and gasket, to ensure leak-proof joint in the duct system and it is made up of GI (Galvanized Iron) sheet to avoid corrosion. The assembled duct is shown in Figure 7 and Figure 8. The material properties of GI sheet considered for the fabrication is depicted in Table 2. Figure 9 shows the duct system with uniform air circulation from the blower. The assembled duct model is fitted with volume control damper in the main duct, it ensures equivalent amount of airflow is happening on either side of duct i.e. sub branches. At the outlet of duct VCD is placed to control the amount of air required for a passenger comfort.

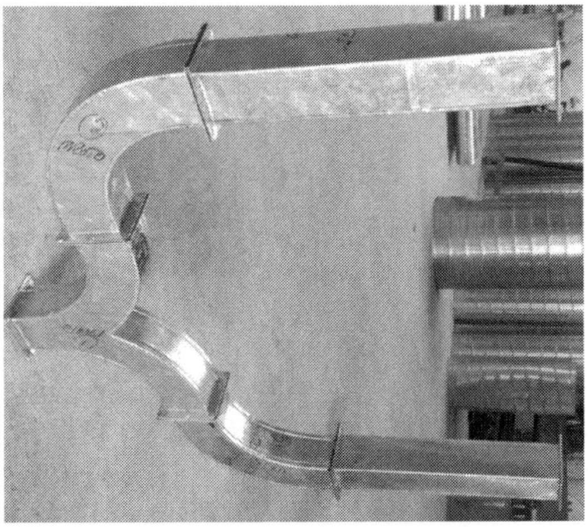

Figure 7. Fabrication of Scaled duct

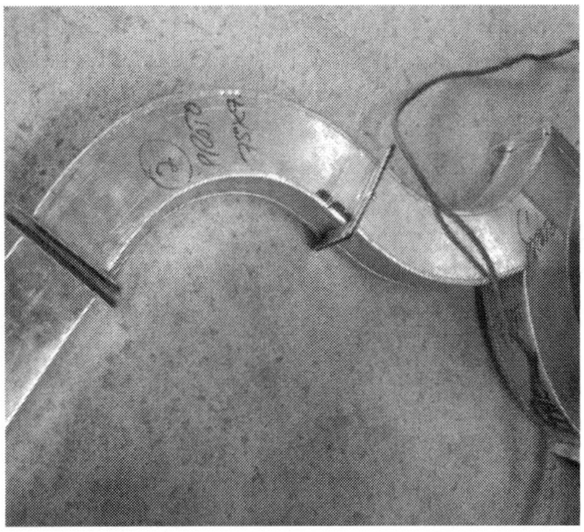

Figure 8. GI sheet used for fabrication of duct

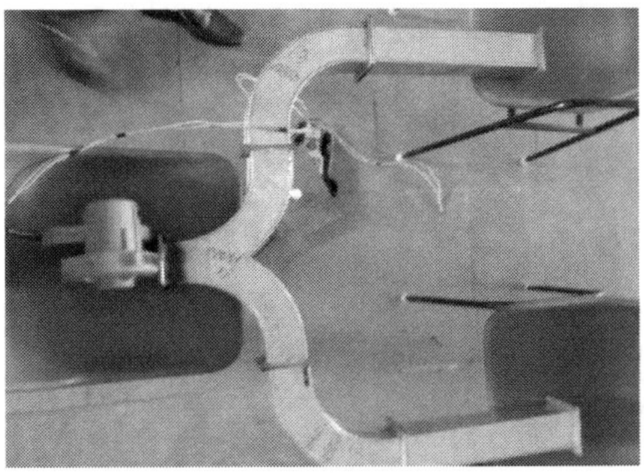

Figure 9. Experimental test setup of scaled duct with blower and other facilities.

3.7 Experimental results of the fabricated scaled model

Table 4 shows the experimental results conducted with the varying velocity. The duct was subjected to 15, 18, 20 and 30 m/s of velocity at the inlet, and the outlet velocity measured at the branch 1, 2 of section A and B using anemometer.

Table 5. Experimental Results.

Blower Speed Inlet m/s	Branch 1 Outlet A m/s	Branch 1 Outlet B m/s	Branch 2 Outlet A m/s	Branch 2 Outlet B m/s
30	10	9	12	10.2
20	7.8	8.5	10.4	9
18	7.9	8	7.5	7.7
15	5.7	5.5	6.9	6.7

4. Conclusion:

From the present experimental and numerical investigation the following conclusions were drawn.

- The minimum amount of air required for the comfortable traveling in a non-AC bus using duct system was found to be 8594.91 W or 2.44 Tr (Tonnage).
- Design and Analysis was carried out to obtain the outlet velocities at various branches of the duct system.
- The numerical results (velocities) were validated for the fabricated scaled model and values are found to be well agreed with each other.

References:

[1] Niranjana S J, Dr. Shivalingapa S Kubsad, Ravichandran G, Dr. Santhosh N, "A Comprehensive understanding of Airflow in Non-Air-conditioned Bus Coaching System", International Journal of Engineering and Advanced Technology (IJEAT), ISSN: 249 – 8958, Volume-9 Issue-3, February, 2020. doi.org/10.35940/ijeat.c5317.029320.

[2] Prithiv John, B Sriram, Senthil Kumar R, S Vinoth Kumar, and Prakash Ramasamy, (2013) "Ventilation Improvement in a Non-AC Bus", SAE International, 2013-01-2457. doi.org/10.4271/2013-01-2457.

[3] Vignesh S, Vijay Ram C and Sachin P, "Evaluation of Bus Ventilation Methods Using CFD", SAE International, 2013-26-0043 (2013). doi.org/10.4271/2013-26-0043.

[4] SR kale, SV veeravalli, HD punekar and MM yelmule, (2007) "Air flow through a non-airconditioned bus with open windows", Sadhana Vol. 32, Part 4, August 2007, pp. 347–363. doi.org/10.1007/s12046-007-0029-3.

[5] Marcus Thomas, Rajnish N Sharma, Michael Kilduff, (2006) "Aerodynamic Effects of Different Ventilation Methods on Buses", Forth International symposium on computational wind Engineering (CWE2006), Yokohama.

[6] J. Petzall , P. A. Torlund , T. Falkmer , P. Albertsson & U. Bjornstig, (2008) "Aerodynamic design of high-sided coaches to reduce cross-wind sensitivity, based on wind tunnel tests", International Journal of Crashworthiness, Tylor and Francis, ISSN: 1358-8265 (Print) 1754-2111 (Online) Journal homepage: http://www.tandfonline.com/loi/tcrs20.

[7] L.D.Jathar, K.C.Nikam, Dr. S. L. Borse, (2014) "Determination of Coefficient of Lift And Lift Force on Sedan Car Body by Changing Different Parameters Using Open Foam", International Journal on Recent Technologies in Mechanical and Electrical Engineering (IJRMEE), ISSN: 2349-7947, Volume: 1 Issue: 5.

[8] Siddhesh Kanekar, Prashant Thakre and E Rajkumar "Aerodynamic study of state transport bus using computational fluid dynamics", IOP Conf. Series: Materials Science and Engineering 263 (2017) 062052 doi:10.1088/1757-899X/263/6/062052.

[9] ASHRAE Standard, "Thermal Environmental Conditions for Human Occupancy," ASHRAE Standard 55, Rev. 2012.

[10] Ramesh Paranjpey, "Class room series-Air conditioning and Refrigeration journal", Indian society of Heating, Refrigerating and Air Conditioning Engineering, ISHRAE, (ISH002HVAC)

[11] ISO Standard, "Ergonomics of the Thermal Environment- Determination of Metabolic Heat Production", ISO Standard 8996, Rev. 2004.

[12] Mohammad Ali Fayazbakhsh and Majid Bahrami, "Comprehensive Modeling of Vehicle Air Conditioning Loads Using Heat Balance Method",SAE International, Volume-01,Issue 1507, 2013, doi.org/10.4271/2013-01-1507.

[13] Jignesh K. Vaghela, Dr. Ragesh G. Kapadia, "The Load Calculation of Automobile Air Conditioning System", International Journal of Engineering Development and Research, Volume 2, Issue 1, ISSN: 2321-9939, 2014.

[14] Niranjana Shravanabelagola Jinachandra, Shivalingappa Sadashivappa Kubsad, Manjunath Sarpabhushana, Shivakumar Siddaramaiah, Thejaraju Rajashekaraiah, "Modeling and computational fluid dynamic analysis on a non-AC bus coach system", Heat Transfer, Wiley Publications, 2020, DOI: 10.1002/htj.21857.

[15] OumSaad Abdulsalam, Budi Santoso, Dwi Aries, "Cooling Load Calculation and Thermal Modeling for Vehicle by MATLAB", International Journal of Innovative Research in Science, Engineering and Technology, Vol. 4, Issue 5, May 2015, DOI: 10.15680/IJIRSET.2015.0405076.

IOP Conf. Series: Materials Science and Engineering 1065 (2021) 012007 doi:10.1088/1757-899X/1065/1/012007

Numerical Investigation of wavy channel using several Nanofluids

Bharat Naik[1,*], Anand K. Hosmani[1], Shirish M. Kerur[1], Jagadeesh S. Pattanashetti[1], Vinayak Ratan[1], Chandra Bhushan[2]

[1]Department of Mechanical Engineering, Jain College of Engineering, Belagavi, Karnataka, India

[2]Department of Mechanical Engineering, Babu Banarasi Das University Lucknow, India

Email: bharatnaik.bgm@gmail.com

Abstract. This study investigates the effect of nano-fluids in enhancing heat transfer. These nano-fluids contain nanoparticles (Ag, CuO, Fe, Al_2O_3 and Tin) in volume fraction ranging between 2%-5%. Two-dimensional numerical simulation has been conducted on wavy channel using ANSYS Fluent. Two dimensional governing equations such as mass, momentum and energy equations are solved using SIMPLE algorithm and discretized using Finite Volume Method. The results show that there will be enhancement of heat transfer rate as the nanoparticle volume concentration increases and from the present study it is clear that silver-water nanofluid gives better results as compared to other nanoparticles.

1. Introduction

Heat transfer is an important phenomenon which plays a vital role in various fields of engineering applications such as IC engines, power plants, refrigeration system, heating and ventilation air conditioning systems, aerospace sector etc. Increasing the rate of heat dissipation is an important task in design and development of these systems, which can be achieved by the usage of extended surfaces on to the periphery of the hot surface. Conducted a numerical simulation of wavy micro channel using two phase models based on Lagrangian approach for base fluid and Eulerian approach for the nanoparticle and identified that the deviation of two approaches is due to the non-homogeneous dispersion of nanoparticle in the main domain. Also shown that, Nusselt number increases as the volume fraction increases and decrease in the diameter of the particle [1].Carried an experimental study on forced convection flow of different nanofluids with varying flow rates through a corrugated wavy channel for the measurement of heat transfer coefficient and the pressure drop across the channel and the results so obtained indicates the augmentation of heat transfer with the large drop in pressure[2].Investigated the heat transfer performance of the micro channel using the nanofluid with nonporous graphene in an aqueous solution as base hot fluid and with the varying flow rates and waviness the enhancement of the heat transfer coefficient is achieved at the expense of pressure drop[3]. Presented the work on numerical analysis of fluid flow and heat transfer between air and the wavy channel, studied the effect of pressure drop on the flow velocity and computed the local Nusselt number and the heat transfer coefficient. Further, compared the results so obtained with that of the 2D rectangular channel such that in an optimum range of Reynolds number, the wavy channel gave better enhanced results than the other [4]. performed a numerical investigation in a channel with lower and upper corrugated plate with a constant heat flux conditions using nanofluids, used FVM approach for the CFD simulation with k-ε Turbulent model for the Reynolds number in the range 5000-20000.For varying volume fractions and diameters of the SiO_2 nanoparticle, the simulation results yield the enhancement of Nusselt number and heat transfer rate [5]. Studied the influence of TiO_2 nanoparticle in water on the performance of a horizontal wavy duct, considering laminar flow of fluid by CFD analysis. Simulation results identifies that, concentration of nanoparticles and the Reynolds number plays a vital role in the enhancement of Nusselt number and heat transfer. However, the drop in pressure also intensifies leading to high pumping power [6].In the present study a 2D sinusoidal wavy channel is considered and simulations are carried out by using ANSYS Fluent. Different Nanofluids like Ag, CuO, Al_2O_3, Fe and Tin are being used with the volume concentration 2%-5% in water as a base fluid to predict the influence of these nanofluids on the heat transfer characteristics of the water flowing over the channel. By the application of necessary boundary conditions for laminar flow field, governing equations are solved to determine the heat transfer rate and its implications on the drop in pressure in the flow field.

2. Model description

The model of a wavy channel used in the present analysis is given in Fig.1 (a).Wavy channel consists of two sinusoidal wavy walls and is bounded by straight sections at inlet and outlet.

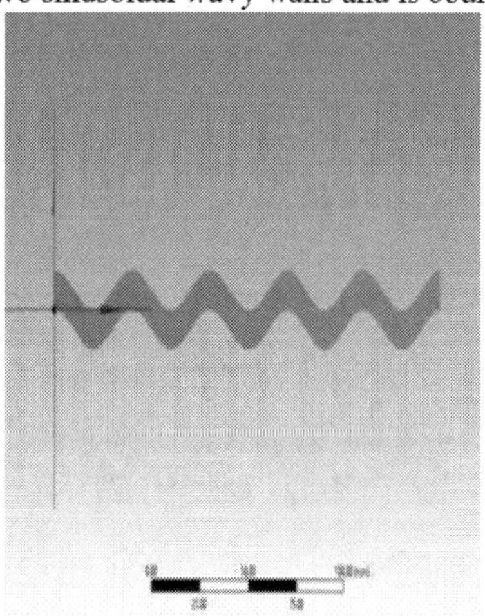

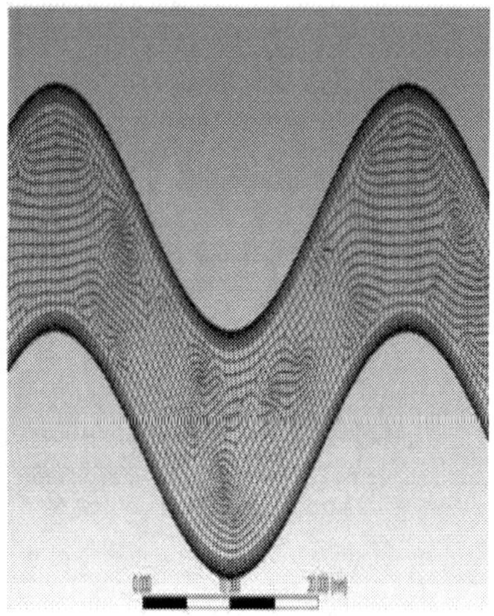

Figure 1a. Geometric model of a wavy channel **Figure1b.** Detailed view of meshed region showing the inflation layers near the walls of wavy passage with $\Phi = 0°$ phase shift

Table 1. Geometry parameters of the wavy channel.

Parameter	Value
Length of Channel (L)	0.2 (m)
Width of Channel (H)	0.01 (m)
Wavelength (λ)	0.04 (m)
Hydraulic diameter (Dh)	0.02 (m)
Phase Shift (Φ)	0°

2.1 Grid Independency Test

Five different meshes were tried, to evaluate the number of elements essential. To test the grid independence, the grid sizes 8932, 10500, 12467 and 15672.It is found that after 15672 elements, further increase in elements provides less than 2% variation in pressure drop which is taken as a criteria for grid independence.

2.2 Governing equations

The flow is laminar, steady state, incompressible and two dimensional. The mass, momentum and energy equations can be shown, respectively, as:

$$\frac{\partial u}{\partial x}+\frac{\partial v}{\partial y}=0 \tag{1}$$

$$\rho_{nf}\left(u\frac{\partial u}{\partial x}+v\frac{\partial u}{\partial y}\right)=-\frac{\partial p}{\partial x}+\mu_{nf}\left(\frac{\partial^2 u}{\partial x^2}+\frac{\partial^2 u}{\partial y^2}\right) \tag{2}$$

$$\rho_{nf}\left(u\frac{\partial v}{\partial x}+v\frac{\partial v}{\partial y}\right)=-\frac{\partial p}{\partial y}+\mu_{nf}\left(\frac{\partial^2 v}{\partial x^2}+\frac{\partial^2 v}{\partial y^2}\right) \tag{3}$$

$$u\frac{\partial T}{\partial x}+v\frac{\partial T}{\partial y}=\frac{k_{nf}}{\rho_{nf}c_{nf}}\left(\frac{\partial^2 T}{\partial x^2}+\frac{\partial^2 T}{\partial y^2}\right) \tag{4}$$

The viscosity of nanofluid is determined from the Drew and Passman correlation [7] using the following equation.

$$\mu_{n_f}=\mu_{bf}(1+2.5\phi) \tag{5}$$

The density is determined from the Choi correlation [8] using the following equation

$$\rho_{nf}=(1-\phi)\rho_{bf}+\phi\rho_p \tag{6}$$

The het capacity is determined from the Xuan and Roetzel correlation [9] using the following equation

$$\left(C_p\right)_{nf}=\frac{(1-\phi)(\rho C_p)_{bf}+\phi(\rho C_p)_p}{\rho_{nf}} \tag{7}$$

Thermal Conductivity is determined from the Maxwell-Garnett's correlation [10] using the following equation.

$$\frac{k_{nf}}{k_{bf}}=\frac{k_p+2k_{bf}+2\left(k_p-k_{bf}\right)\phi}{k_p+2k_{bf}-\left(k_p-k_{bf}\right)\phi} \tag{8}$$

2.3 Boundary conditions

Fully developed and uniform temperature of the inflow were applied as boundary condition at the channel entry. At lower and upper wavy wall slip conditions and constant wall temperature were applied. At the outlet boundary, the exit pressure was fixed as atmospheric.

2.4 Validation

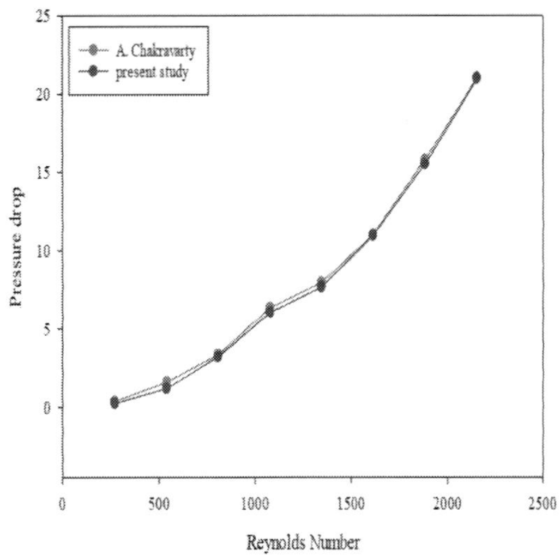

Figure 2. Comparison between present study and A. Chakravarty [4]

Fig.2 shows the comparison between our study and the study presented by A. Chakravarthy. The graph is plotted between Reynolds number and the pressure drop. It can be seen that there is a very close corelation between our study and the earlier study. It can be observed that the deviation at a few places exists, but it is less than 2% which is acceptable. Hence the work carried out by us and A. Chakravarthy are matching. This validation is done for air that is used as the working medium. Since our work very closely matches with the earlier work, it concludes that our methodology followed is correct. Hence we proceeded ahead to carry out similar simulation using nano-fluids as the working medium.

3. Results and Discussions

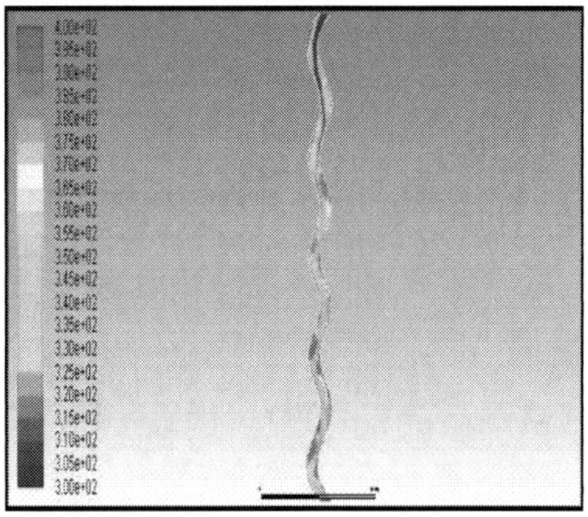

Figure 3a. Static temperature contour

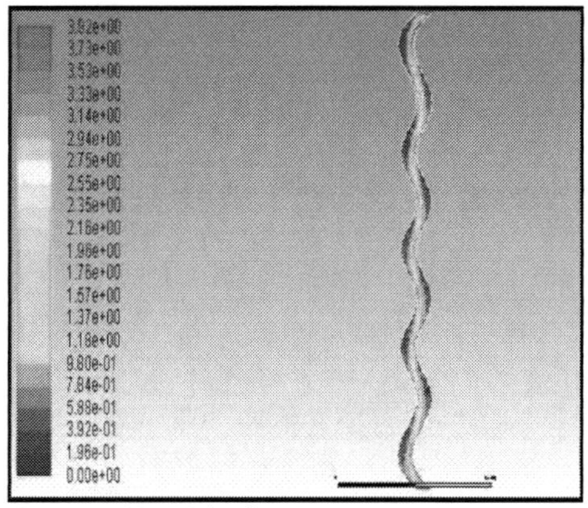

Figure 3b. Velocity Contour

In order to do the simulations for determination of fluid flow regime, Reynolds number based on hydraulic diameter of the channel was calculated. The hydraulic diameter Dh for a sinusoidal wavy channel is calculated as: Dh = 2 * h, where h is the height of channel. The default width of 1 m for 2D geometry was taken for the calculation of hydraulic diameter and Reynolds number. Further, the simulations for determination of pressure drop were performed for the laminar flow regime. The inlet velocities were varied from 0.2 to 1.6 m/s in a step of 0.2 and the effect of corresponding Reynolds number on the pressure drop and heat flux in the wavy channel is as shown in below Figures.The figure 4 (a), (b), (c), (d) and (e) represent the influence of Reynolds number on drop in pressure and the wall heat flux for various nanofluids with water as a base fluid. For varying the volume concentration of nanofluids from 2% to 5%. The results show that, drop in pressure intensifies up to 33.33% for Ag nanoparticle,18.18% for CuO nanoparticle, 17.39% for Iron as a nanoparticle, up to 20.83% for Al_2O_3 and 4.10% for Tin as a nanoparticle respectively in the laminar regime (Reynolds number up to 2000).Similarly, the enhancement in the wall heat flux up to 10.14% for Silver, 8.57% for CuO, 9.35% for Iron, 8.42% for Al_2O_3 and 11.85% for Tin as a nanoparticle is observed by the variation of their volume concentration from 2% to 5% respectively.

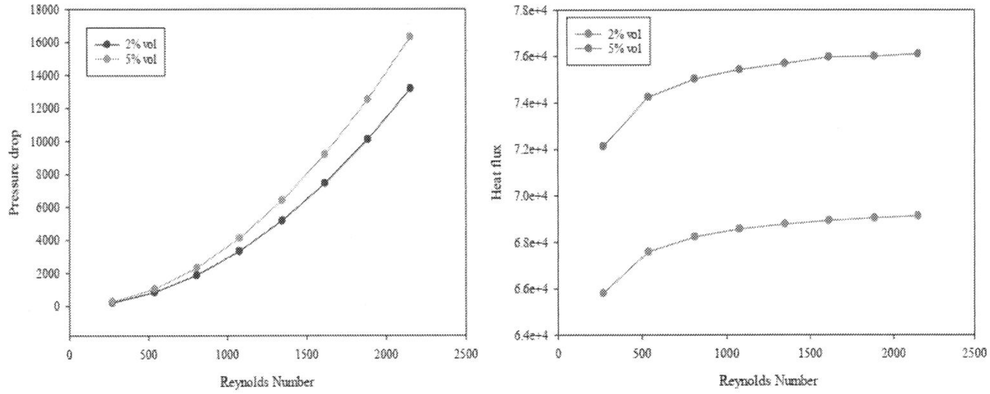

Figure 4a. Effect of Ag-water nanofluid on drop in pressure and wall heat flux in the laminar flow regime (Reynolds no. between 0 to 2000)

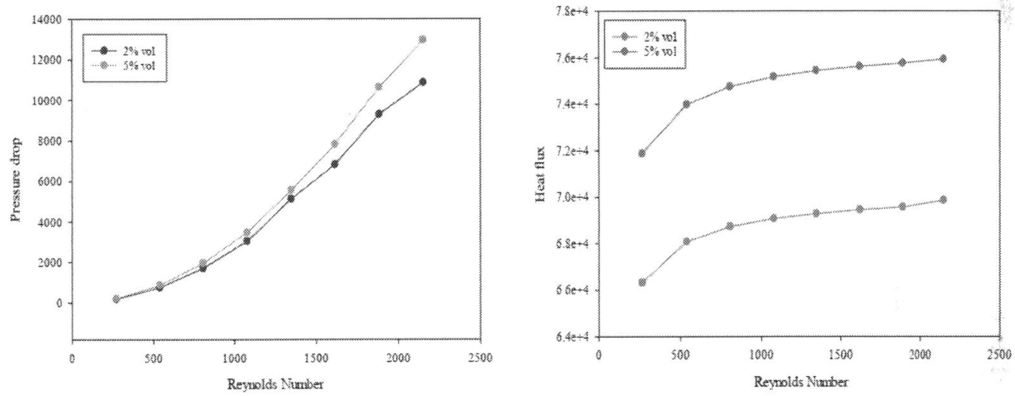

Figure 4b. Effect of CuO-water nanofluid on drop in pressure and wall heat flux in the laminar flow regime
(Reynolds no. between 0 to 2000)

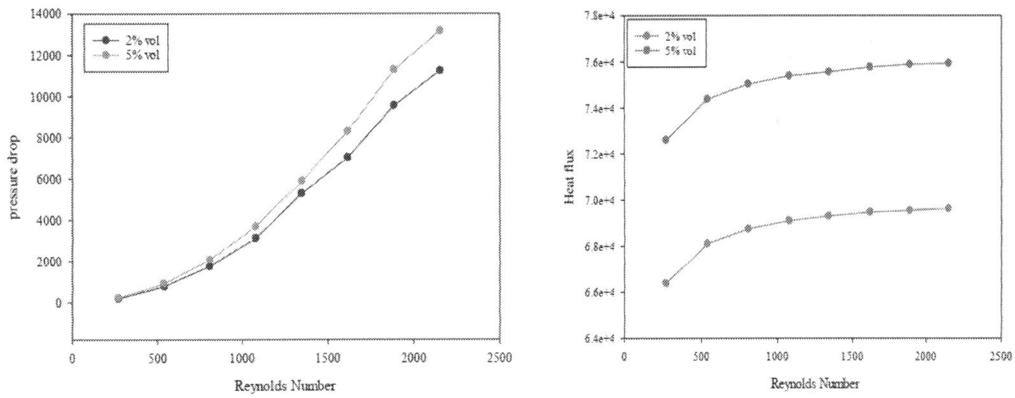

Figure 4c. Effect of Fe-water nanofluid on drop in pressure and wall heat flux in the laminar flow regime (Reynolds no. between 0 to 2000)

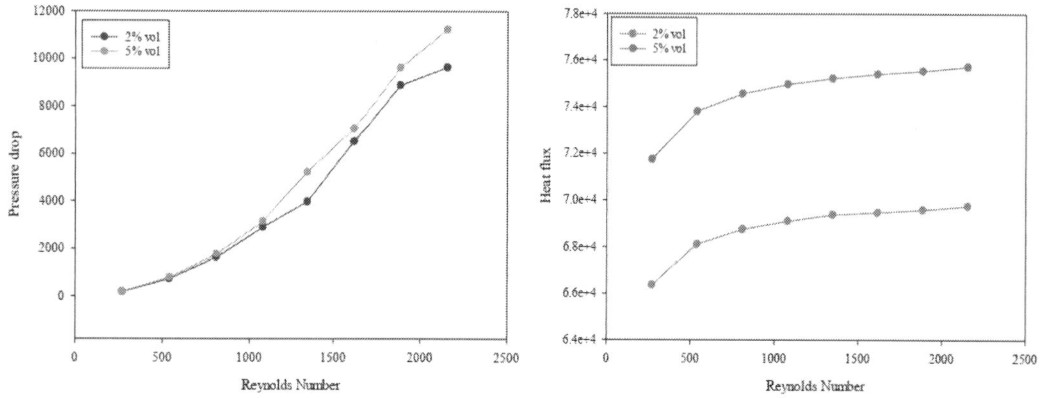

Figure 4d. Effect of Al_2O_3-water nanofluid on drop in pressure and wall heat flux in the laminar flow regime
(Reynolds no. between 0 to 2000)

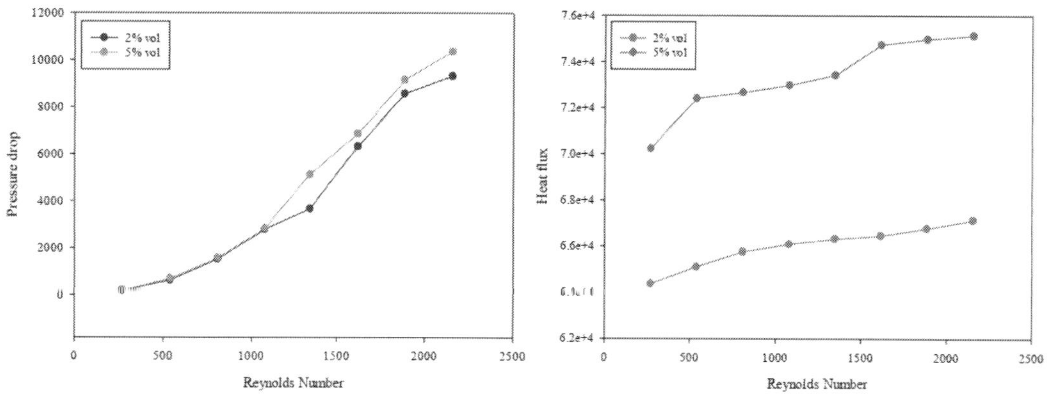

Figure 4e. Effect of Tin-water nanofluid on drop in pressure and wall heat flux in the laminar flow regime (Reynolds number between 0 to 2000)

4. Conclusion

Numerical simulation on a 2D sinusoidal wavy channel was performed using ANSYS Fluent software to understand and predict the impact of various nanofluids on the heat transfer characteristics. Based on the results so obtained, it is important to note that, increasing the volume concentration from 2% to 5% enhances the heat transfer rate at the wall of a wavy channel. Meanwhile, the drop in pressure also increases as result of which the pumping power required would be more.In particular, silver nanoparticle in water gives better heat transfer characteristics then the remaining. As an overall outcome of this study insist, one to make use of nanoparticle to enhance the thermal conductivity of the base fluid such that the heat transfer rate could be accordingly amplified with a compromise on the drop in pressure.

5. Index

λ	wavelength of channel (m)	Subscripts		
D_h	hydraulic diameter of channel (m)		b_f	base fluid
Φ	phase shift (deg)		nf	nanofluid
C_p	specific heat capacity		p	particle
ϕ	volume fraction			

6. References

[1] Rostami J, Abbassi A and Harting J 2018 Heat transfer by nanofluids in wavy channels,Advanced Powder Technology 29 (18),P.925-933
[2] Khoshvaght-Aliabadi M ,Zamzamian S A and Hormozi F 2014 Wavy channel and nanofluids effect on performance of a plate fin heat Exchanger, International Journal of Thermophysics and Heat transfer
[3] Anvari A, Javahardeh K and Meibodi M E 2018 Performance evaluation of a new nanofluid through micro channel heat exchangers, Chemical engineers' transactions, P. 973-978
[4] Chakravarti A 2017 Numerical Analysis of fluid flow and heat transfer in 2D sinusoidal wavy channel, students conference, czech technical university in Prague
[5] Majdi H and Abed A M 2014 Effect of Nanofluid on performance of a corrugated channel Within out of phase arrangement, International Journal of Scientific and Technology Research 3(1).
[6] Gadhaban S A, Aun S H A and Jehhef K A 2020 Numerical Modelling of fluid flow and heat transfer using TiO2- water nanofluid in wavy channel, IOP Conference Series-Material science and Engineering 881 012162
[7] Drew D A and Passman S L 1999 Theory of Multicomponent Fluids, Springer
[8] Das S K,Choi S U S, Patel E H 2006 Heat Transfer in Nanofluids-A Review Heat Transfer Engineering 27(10),P. 3-19
[9] Xuan Y and Wilfried R 2000 Conceptions for Heat Transfer Correlation of Nanofluids, International Journal of Heat and Mass Transfer 43 (19),P.3701-3707
[10] Maxwell J C 1881 Treatise on Electricity and Magnetism, second ed., Clarendon Press, Oxford University, UK.
[11] Naik B, Soudagar M E M, Thomas M, Dasurkar K, Alloli O, Hossain F and Alloli M 2020 Numerical Investigation of double pipe heat exchanger with different Nano-fluids, IOP Conference series: Earth and Environmental Science 573 012030

Recent Advancement in Electrospun nanofibrous mats with emphasis on their applications

Amith.V [1], Sridhar.R [2], Gangadhar Angadi [2], Narashima Murthy H.N [2]

[1,2] Department of Mechanical Engineering, R.V. College of Engineering, Bengaluru

Email id- amithrvce4@gmail.com

Abstract. PVDF, which has four phases, is one of the most commonly used dielectric polymers. Among them the piezoelectric property that is accessible via an electrospinning synthesis method is only seen in its β phase. In this research solution was prepared at a concentration by taking the Polyvinylidene fluoride about 2.5 gm was dissolved in Dimethylformamide (DMF) and Acetone with a ratio of 6:4.The prepared solution has been loaded to the syringe pump and it has been electrospinned by using the electrospnning set up by having the parameters at a flow rate of 0.5ml/hr with a power supply of 20kv,at a distance of 100mm.The SEM Images of PVDF electrospuned nanofibrous mats clearly indicates that the beads were formed which decreased in number when concentration was increased whereas the diameter of beads got increased in the range of 647.39 to 1208.35nm. Average diameters of the Nano fiber obtained were in the range of 34.21 to 107.35nm in diameter, with a minimum diameter of 25.50nm to a maximum of 135.85nm. PVDF based Electrospuned nanofibers mats, have been tested by applying the Impact Energy, Due to the impact force, the sound frequency of 86-89 dB was induced, the generated acoustic waves were tracked and sensed by electro-sponsored PVDF Mats based sensors. During the trial, the voltage reached a limit of 100 mV and disappeared quickly due to the low power generated by the PVDF Mats based sensors. The created Lamb wave can be easily detected by this PVDF sensor, which allows this system to characterize the structural element's properties and to detect cracks and structural defects as well.

Keywords: Electrospnning,Nanofibers, Polyvinylidenefluoride, SEM

1.Introduction

With improvement in the human life, the requisite of the new materials with special properties for many different applications ranging from energy harvesting, water filtering and other consumer goods can be strongly sensed. Polymeric materials offer this opportunity to scientists and engineers for designing these new materials. In this regard, precise understanding of structure and properties are very crucial. Electrospinning is the process of fibre generation with the use of electrostatic force to draw charged polymer solutions to fibre diameters in order of some nanometres [1-7].Over the years electrospinning has gained popularity and many research has been happening. The amount of research papers published has considerably increased from 2000-2010 and all most all fields of engineering are being tapped [8-10]. Currently there are a few methods for the development of micro and nanofibers for polymers and ceramics. These include melt-blowing, gel spinning, drawing, template phase separation and spun-bound fibres production which all rely on mechanical forces to extrude the predecessor solutions through spinnerets. Typical fibre diameters formed by these methods are in the order of 10-150 microns. Although fibres formed by these techniques finds engineering applications, but their orders of magnitude are less effective for thin film applications, which depends high specific

IOP Conf. Series: Materials Science and Engineering 1065 (2021) 012008 doi:10.1088/1757-899X/1065/1/012008

surface area for solar cells, antifungal, bacteria filtration for nano-crystalline NaCl on electrospun nanofibers compared to spun-bound fibers~10μm in diameter accentuating the necessity of nanometre dimensions [11-13]. Electrospinning sol-gel process used to be substitute method to fabricate continuous fibres of high surface area to volume proportion. The basic sub-systems required are a high voltage power supply, a spinneret with a storage container and a method-controlled delivery (pneumatic or mechanical). With these basic components and understanding of the processes involved, it is possible to manufacture nanofibers with controlled and uniform diameters ranging from 10nm to 100nm [14-15]. Electrospinning process involves the application of an electrostatic potential between the spinneret and a collector. A fluid is progressively pumped through the spinneret. The fluid is normally a liquid solution where the solvent can evaporate within the spinning. The droplet is held by its own surface tension at the spinneret tip, until it gets electrostatically charged. The polymer fluid accomplishes a cone like shape called a Taylor cone. At the point when the electrostatic force overcomes the surface tension of the droplet, the droplet gets unstable, and a liquid jet is ejected towards the collector plate, and the solvent evaporates before deposition of the polymer on the collector plate. As the jet dries in flight, the mode of current flow changes from ohmic too convective as the charge moves outwards to the surface of the fibre. The jet is the extended by whipping instability brought on by electrostatic repulsion started at small bends in the fibre, until it is finally deposited on the collector plate. The elongation and reduction in diameter of the fibre resulting from this bending instability leads to the formation of uniform fibres with nanometre-scale diameters PVDF is a chemical stable thermoplastic. The PVDF is a semi-crystalline piezoelectric polymer which a complex structure which has a distinct five crystalline phases of α, β, γ, δ and ε. The β-phase crystallization has the highest dipole moment per unit cell (8 X 10-30 C.m) related to the other phases of crystallization, from the β-phase of crystallization in PVDF material gives the piezoelectric properties in that polymer. The β-phase of crystallization which is having a technological interest in proving the ability of providing the maximum pyro and piezoelectric properties. [16-22]

2. Fabrication of electrospinning machine

A horizontal electrospinning system is fabricated to produce nanofibers by the researchers. The fabrication and assembly of the components was a major challenge. These challenges included, design of the electrospinning component, materials, fabrication of individual components, assembly of the system and its optimization, reliability testing, ultimately testing of the technique using the final system. The specification of the system was voltage ranging from 0-30kV, flowrate capacity 0.5ml to 60ml/h, maximum tip to collector distance of 150mm. An aluminium drum collector of high grade is fabricated to collect the film on it. The assembled system is shown in Figure below to spin the liquid solution horizontally on the drum collector. A syringe (5cc) with a highly electrically charged metal needle (0.5mm orifice diameter), was used to contain and inject the polymer and was mounted on the described syringe pump.

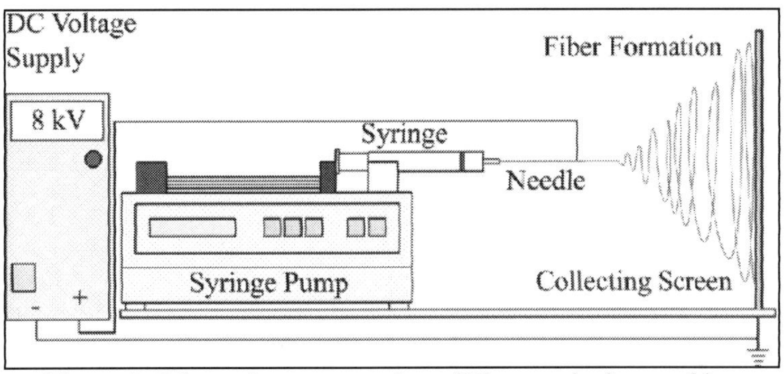

Figure 1: Schematic representation of Electrospinning machine

IOP Conf. Series: Materials Science and Engineering 1065 (2021) 012008 doi:10.1088/1757-899X/1065/1/012008

3. Preparation of Solutions

The Poly (vinylidene fluoride) (Aldrich, 530,000 g mol-1) about 2.5 gm was dissolved in Dimethylformamide (DMF) and Acetone with a ratio of 6:4. PVDF polymer which is having a better solubility in DMF was stirred for 1 hour at 50°C in a hot plate magnetic stirrer as shown in figure to form a homogeneous solution in figure

Figure 2: Magnetic Stirring of PVDF Solution

The Solution taken was then used for spinning in drum collector. We obtained a thick coating in PVDF on the drum collector. The PVDF solution is loaded into Syringe Pump and it has been electrospineed at a flow rate of 0.5ml/hr with a power supply of 20kv, at a distance of 100mm.

4. Rectangular Collector Plate

Figure 3: Rectanglar Plate Collector

A Rectangular collector plate is made of copper material. Its dimension is 110*110. Initially all the trail is done on the rectangular plate only to fix the distance between the syringe system and the collector plate stand. The dimensions of the plate is length – 110mm, Breadth-110mm, Thickness - 3.00mm

5. Rotating Drum Type Collector Plate

A Rotating drum type collector plate is made of aluminum material. The whole system consists of drum, stand, shaft, pulleys and electric motor.

IOP Conf. Series: Materials Science and Engineering 1065 (2021) 012008 doi:10.1088/1757-899X/1065/1/012008

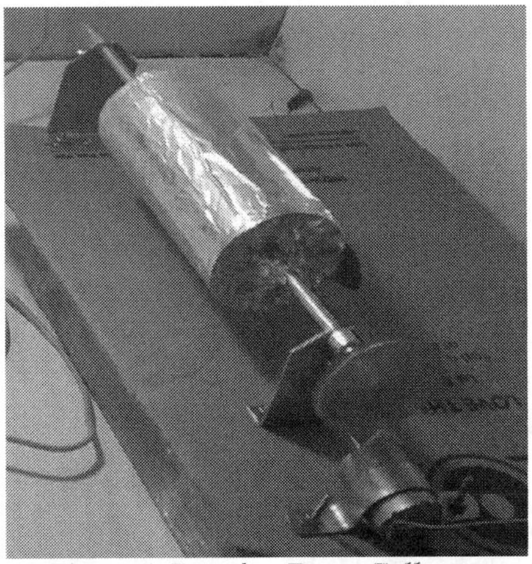

Figure 4: Rotating Drum Collector

6. SEM Images of PVDF Nanofiberous Mats

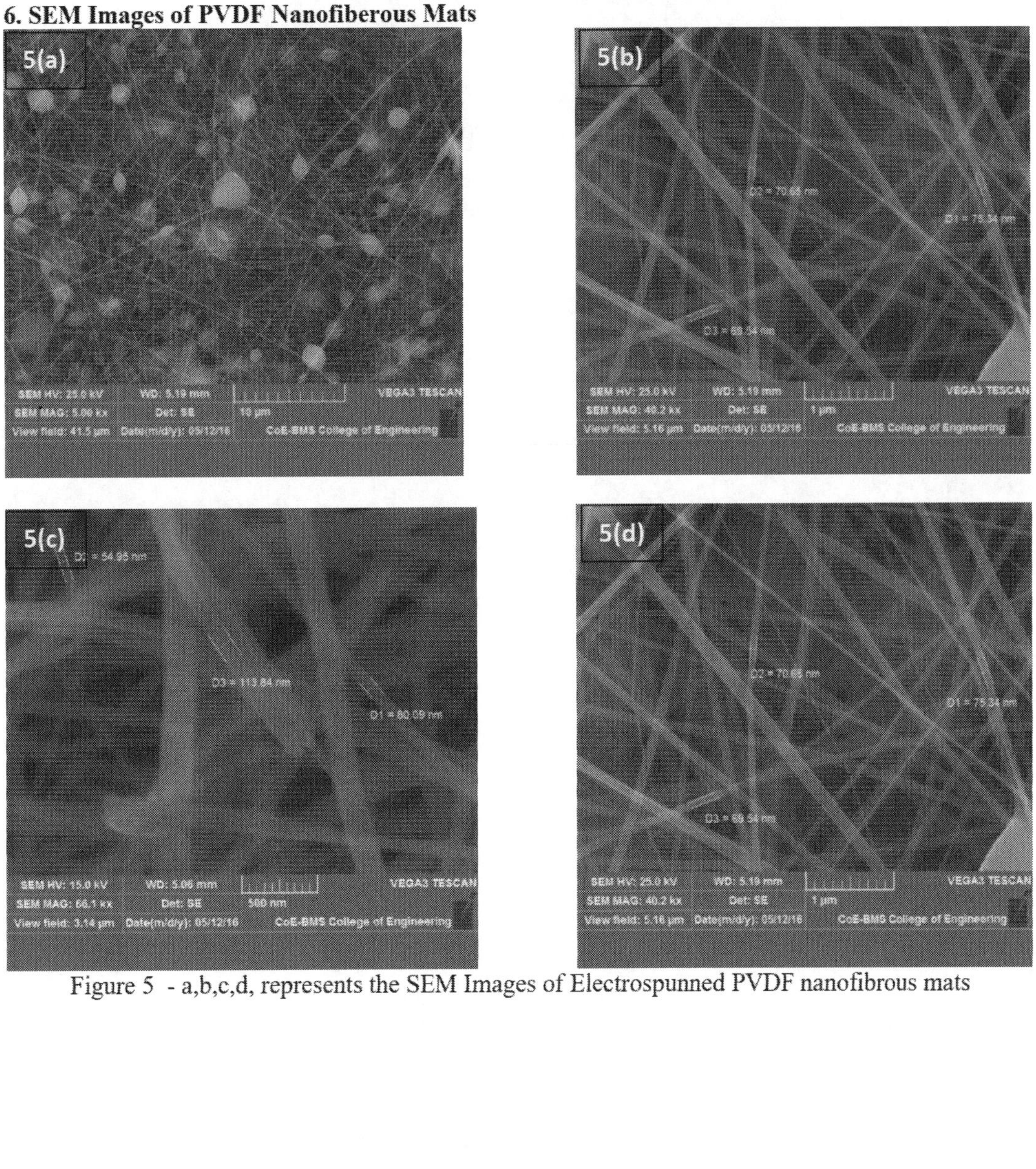

Figure 5 - a,b,c,d, represents the SEM Images of Electrospunned PVDF nanofibrous mats

The electrospun fibers were taken to the SEM in order to get a detailed look at the resultant fiber alignment and to measure fiber diameters. The accelerating voltage for the SEM was set to different setups for different samples of the fiber generated by using electrospinning process, with SEM HV ranging from 15-25kV with a working distance of 5.05-5.19 mm, view field of 2.94-69.2μm High magnification Images of 5-70.7Kx were taken to measure the diameter of the electrospun material. By using the SEM images, an average of fiber diameter was calculated by measuring the diameter of corresponding fiber from each individual. Beads were formed which decreased in number when concentration was increased whereas the diameter of beads got increased in the range of 647.39 to 1208.35nm. Average diameters of the Nano fiber obtained were in the range of 34.21 to 107.35nm in diameter, with a minimum diameter of 25.50nm to a maximum of 135.85nm.

7. Application

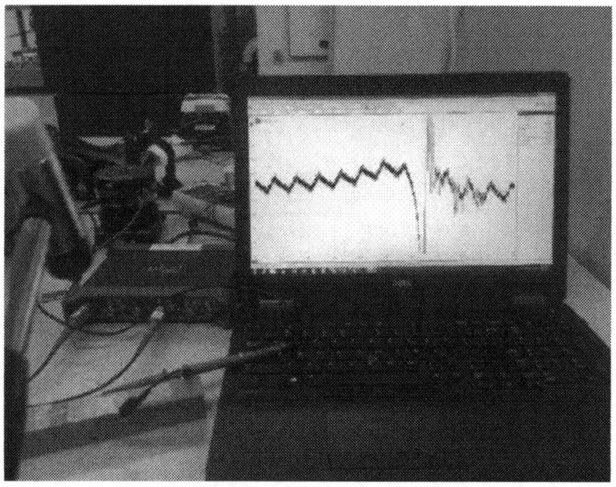

Figure 6 – Acoustic Emission Test on a cement paste beam

The sensor, which is PVDF based Electro-sponsored nanofiber mats, was tested by adding Impact Energy. Due to the impact force, the sound frequency of 86-89 dB was induced, the generated acoustic waves were tracked and sensed by electro-sponsored PVDF Mats based sensors. During the trial, the voltage reached a limit of 100 mV and disappeared quickly due to the low power generated by the PVDF Mats based sensors. The created Lamb wave can be easily detected by this PVDF sensor, which allows this system to characterize the structural element's properties and to detect cracks and structural defects as well.

8. Conclusion

The PVDF Electropsuned Nanofiber layers with nano and micro sized fiber diameters were prepared by a controlled electrospinning process. Average diameters of the Nano fiber obtained were in the range of 34.21 to 107.35nm in diameter, with a minimum diameter of 25.50nm to a maximum of 135.85nm. The prepared Electrospunned PVDF based nano fibrous structure is a very promising and prospective material for the pressure active sensors.

9. References

[1] Somayeh Mohamadi,''Preparation and Characterization of PVDF/PMMA/Graphene Polymer Blend Nanocomposites by Using ATR-FTIR Technique'' Polymer group, School of Chemistry ,2012, pp 213-230.

[2] Ladan Akbarpour Nikghalb, Gurinder Singh, "t", Journal of Applied Pharmaceutical Science, Vol. 2(10), oct-2012, pp.170-175 "Advances on modelling in tissue engineering", Springer Dordrecht Heidelberg London New York, vol. 20, 2011

[3] Katia Rodrıguez, Scott Renneckar, Paul Gatenholm, "Electrospinning cellulosic nanofibers for biomedical applications: structure and in vitro biocompatibility",Journal of Cellulosevol.19, 2012, pp.1583–1598

[4] Iman Esmaeilzadeh, Vahid Mottaghitalab, "A feasibility study on semi industrial nozzle-less electrospinning of cellulose nanofiber",Int J Ind Chem, vol.6, 2015, pp. 193–211
[5] Xiaojie Cui Luming Li, "Controlled assembly of poly (vinyl pyrrolidone) fibers throughan electric-field-assisted electrospinning method",Appl Phys A, vol.103, 2011, pp.167–172
[6] In Gi Kim, Jung-Ho Lee, Afeesh Rajan Unnithan, "A comprehensive electric field analysis of cylinder-type multi-nozzle electrospinning system for mass production of nanofibers",Journal of Industrial and Engineering Chemistry.Vol 31, 2015, pp.251–256.
[7] Hui W U, Wei PAN, "Electrospinning of ceramic nanofibers: Fabrication, assembly and applications",Journal of Advanced Ceramics, vol.1(1), 2012, pp.2-23.
[8] Z.H. Liua, C.T.Pana, C.Y.Su, L.W.Lin, Y.J.Chen, J.S.Tsai, Sens. 2014, vol 211, pp.78–88
[9] Ramesh Neppalli "Study On Structure And Morphology Of Polymer Nanocomposites" 2012, page 42-50.
[10] Peter Goldstein, "Electrospinning of Ceramic and Nanocomposite Nanofibers", University of Florida,2004.
[11] "Advances on modelling in tissue engineering", Springer Dordrecht Heidelberg London New York, vol. 20, 2011.
[12] Katia Rodrıguez, Scott Renneckar, Paul Gatenholm, "Electrospinning cellulosic nanofibers for biomedcal applications: structure and in vitro biocompatibility",Journal of Cellulosevol.19, 2012, pp.1583–1598.
[13] Khajavi R, Abbasipour M, "Electrospinning as a versatile method for fabricating core shell, hollow and porous nanofibers", Scientia Iranica journal, 2012, 19(6), pp.2029–2034
[14] Gregorio R. Determination of the alpha, beta, and gamma crystalline phases of poly (vinylidene fluoride) films prepared at different conditions. Journal of Applied Polymer Science. Vol 100 2006, pp.3272–3279.
[15] Harsha Malshe, "Design, fabrication, and testing of and electrospinning apparatus for the deposition of pmma polymer for future biomedical applications", Vol. 12, 2011.
[16] Panda P, "Ceramic nanofibers b electrospinning technique- review", The Indian Ceramic Society, vol. 66, 2007, pp.65-76
[17] Khajavi R, Abbasipour M, "Electrospinning as a versatile method for fabricating coreshell, hollow and porous nanofibers", Scientia Iranica journal, 2012, 19(6), pp.2029– 2034
[18] Yun-A Seo, Hem Raj Pant, "Fabrication of highly porous poly (e-caprolactone) microfibers
[19] Kim KW, Lee K H, Khil M S, Ho Y S and Kim H Y 2004 "The effect of molecular weight and the linear velocity of drum surface on the properties of electrospun poly (ethylene terephthalate) nonwovens Fiber" Polym. 5 122
[20] Fukada E. History and recent progress in piezoelectric polymers. IEEE Transactions on Ultrasonics, Ferroelectrics and Frequency Control, 2000, vol 47, pp 1277–1290
[21] Haitao Niu, Tong Lin, Xungai Wang," Needleless Electrospinning. I. A Comparison of Cylinder and Disk Nozzles", DOI 10.1002/app.30891
[22] Matthews J A, Wnek G E, Simpson D G and Bowlin G L 2002 "Electrospinning of collagen nanofibers Biomacromolecules" 3 232

Immobilization of copper sulfide flotation tailings through the use of geopolymers, Tacna – Perú.

C Morales-Aranibar[1*], N Linares-Gutiérrez[1], L Morales-Aranibar[2]

[1]Metallurgical Engineer from the Jorge Basadre Grohmann National University, Professor at the Professional School of Metallurgical and Materials Engineering of the Jorge Basadre Grohmann National University - Tacna.

[2] Director of Research Institutes of the National Intercultural University of Quillabamba - Cusco

E-mail: cagemoar123@gmail.com

Abstract. Currently, mining-metallurgical industrial activities generate significant amounts of mining tailings, and innovative environmentally friendly methods are required for their final disposal or sustainable immobilization. The objective of the study was to immobilize copper sulphide flotation tailings through the geopolymerization process. The samples were obtained from the Toquepala mining unit, - Tacna (Peru), the ash from the ENGIE-Moquegua hydroelectric plant. The results in the first phase of the characterization of the ash (FA) and the mining tailings (MT) through the chemical analysis EDX, mineralogical DRX and morphological SEM, show aluminosilicate content, in the second phase the mixtures of 55% MT, 37, 5% FA and 7.5% MK activated at 10 M at 35 days of curing, demonstrated resistance of 24.6 Mpa according to mechanical resistance tests, the mixture of 100% MT activated at 10 M at 35 days of curing, it is effective by itself for the formation of the geopolymer, obtaining resistances of 22.68 Mpa, the SEM analysis demonstrated formation of tetragonal structures typical of a geopolymer, the TCLP analysis demonstrated a significant reduction of the contaminating elements Ba, Cd, Cr, Pb, opening a new field for tailings treatment.

1. Introduction

Mining and metallurgical activities produce approximately two-thirds of the tailings from the volume of ore mined using the latest techniques [1]. The storage of mining tailings generates operational and logistical costs for their storage and subsequent treatment for final disposal, seeking to obtain a physically and chemically stable product [2], this being the major environmental problem associated with mining development. The final product obtained can be destined to different uses such as the use of these materials for road paving applications [3].

The mining tailings from the flotation process have significant contents of heavy metals such as copper, lead, zinc, iron, cadmium, arsenic [4] and [41], the tailings have a great potential to acidify the water when they are entered into adjacent water bodies, this acidification prevents the effluents are suitable for the leaching process, so it is required to immobilize these pollutants to stabilize or solidify in confinement avoiding their release into the environment [5].

The geopolymer is constituted of a wide range of materials forming chains or networks of inorganic molecules [6]. These materials form a matrix constituted of an aluminosilicate source, usually in the form of metakaolin, fly ash or slag, activated through an alkaline sodium/potassium silicate solution at room temperature or slightly higher [7].

The speed of the activation reaction depends on the particle size distribution, since a mechanical activation (such as a finer grinding) will be reflected in a better chemical activation because the alkaline medium will increase the contact with the aluminosilicate materials [8] and [39]. Geo-polymerization involves the chemical reaction under highly alkaline conditions rich in Al and Si starting materials, producing Si-O-Al-O type polymer bonds [9].

One variable that has a very strong impact is the activating solution, which can be of alkaline hydroxides and silicate hydroxides dissolved in the basic activator. Depending on the type of alkaline activator, a certain composition of the final product will be obtained and therefore different mechanical properties [10] and [40].

It has been identified that, in order to obtain geopolymer products with high resistance, it is necessary to consider the SiO_2/AI_2O_3; $M_2O/SiO2$; H_2O/M_2O and M_20/AI_2O_3 ratios, where M represents the alkaline metals present (K, Na, Ca, Ba and Li). From the composition point of view, [11] reported that the most convenient ratio between oxides for geopolymer processing should be around 3.3 to 4.5 (SiO_2/AI_2O_3), 0.20-0.48 (Na_2O/SiO_2) and 0.80 - 1.60 (Na_2O/AI_2O_3).

Copper sulfide is a contaminant that can generate acidic waters and the tailing ponds that store these compounds are potential threats to the environment since they can generate environmental liabilities, in addition to generating cost overruns for maintenance and immobilization. Sulfide contaminants can be encapsulated minimizing the risk of environmental impact and reducing costs associated with storage/containment. The general objective of the study was to analyze the immobilization of copper sulfides through geo-polymerization as an innovative process of tailings from mining activity. The specific objectives were: i) This study seeks to immobilize copper sulfide flotation tailings, using the geo-polymerization process to encapsulate contaminants that can generate acidic waters, being an innovative method for tailings pond closure and in the treatment of environmental liabilities left by mining, so that they are economically viable and environmentally safe to analyze FA, ii) MT, iii) chemical, iv) physical, etc.

2. Experimental

2.1. Sample selection

The Toquepala mine is operated by Southern Peru Copper Corporation of Tacna, together with the Cuajone mine of Moquegua and the Ilo refinery. Mine production is mainly copper. Smaller amounts of molybdenum and other minerals are also produced. The geology of Toquepala and Cuajone presents a complete porphyritic system with high presence of SIO_2 in its composition [12], giving indications that geopolymers can be formed.

Mine tailings (MT) were collected at the Southern Peru Copper Corporation mine, in the Ilabaya district, Jorge Basadre province, Tacna region. The tailings channel feeds the Quebrada Honda tailings pond located between the coordinates -17.311460 South Latitude and -70.719489 West Longitude, at an altitude between 3100 and 3600 meters above sea level. The sample of fly ashes (FA) was provided by the company ENGIE responsible for the thermoelectric plant of Ilo city, the metakaolin (MK) is commercial of the brand HAMEX. The samples were worked in the laboratory of mineralurgy of the National University Jorge Basadre Grohmann of Tacna.

The geopolymerization reaction can be divided into three stages: dissolution/hydrolysis, restructuring, and polycondensation/gelatinization [6]. These three mechanisms can occur simultaneously and are reversible to some extent [13].

2.2. Analysis of MT and FA

Elemental analysis of MT and FA, Tailings and ashes were analyzed to see chemical, mineralogical and morphological composition.

Chemical analysis, was determined by instrumental analysis EDX - Energy Dispersive X-Ray Spectroscopy in order to know the chemical composition of the MT and FA, with these data the molar ratios of aluminum silicates present in the MT and FA samples were determined [14].

Mineralogical analysis was determined by X-ray diffraction analysis which shows us the mineralogical phases present in the MT and FA [15].

Morphological analysis was carried out by means of analysis in the SEM, by means of scanning electronic microscopy to determine its morphology. It was also determined the density by means of the pycnometer method, humidity by means of [16], the particle size study was carried out by means of granulometric analysis of MT and FA samples. [17].

2.3. Preparation, analytical characterization, mechanical and chemical tests of the geopolymer

The tailings were air dried to 11.19 % moisture, then the MT was mixed with the FA for a 10-minute period in the mixer. Immediately, NaOH activating solution was added, in concentrations of 1M, 5M, 10M and 15 M, with a liquid solid ratio of 5 to 1, based on the non-randomized mix design specified by the user [18], after the mixture was compacted in a 50 mm side cubic tube, this is dried at room temperature for 28 days according to the standards referenced by [19] Cements, in our case 7 more days were added for curing.

Resistance tests were carried out under the [20] standard, to determine the mechanical resistance of the geopolymers formed to different types of mixtures and activating solution [21] and [22].

Microscopy analysis through SEM (Scanning Electron Microscope) and XRD, was performed to observe the geopolymer formations in its tetragonal mesh and the mineralogical phases present in the best mixture [23] and [24].

TCLP test (Toxicity characteristic leaching procedure) for the determination of total metals, was performed by the ICP Method 200.7, where the mixture of MT, FA and pure MK and the best sample of formed geopolymer are compared [25] and [26].

3. Results and discussion

3.1. Elemental analysis of MT and FA

The chemical compositional analysis presents a (20.44 % of Al_2O_3 and 53.39 % of SiO_2) for the MT, and (22.11 % of Al_2O_3 and 51.76 % of SiO_2) for the FA, high percentage of elements of importance for the formation of the geopolymer, which when interpreting them their molar relations present an optimal association of particles for the formation of the geopolymer (Table 1).

Table 1. Determination of molar relations

Number of order #	RV %	CV %	MK %	SIO_2 %	AL2O3 %	Na2O %	SIO_2/AL_2O_3	Na_2O/SIO_2	Na_2O/AL_2O_3
1	70	30	0	53.501	21.829	1.486	3.943	0.074	0.181
2	40	60	0	53.732	21.808	2.512	3.964	0.074	0.182
3	60	30	10	56.274	19.644	1.440	4.608	0.082	0.235
4	30	60	10	56.505	19.623	2.466	4.632	0.082	0.236
5	60	37.5	2.5	54.252	21.278	1.731	4.102	0.076	0.193
6	45	52.5	2.5	54.368	21.267	2.244	4.113	0.076	0.193
7	55	37.5	7.5	55.639	20.185	1.708	4.434	0.080	0.220
8	40	52.5	7.5	55.754	20.175	2.221	4.446	0.080	0.220
9	55	45	0	53.617	21.819	1.999	3.953	0.074	0.181
10	65	30	5	54.888	20.737	1.463	4.258	0.078	0.205
11	35	60	5	55.119	20.716	2.489	4.280	0.078	0.207
12	45	45	10	56.390	19.634	1.953	4.620	0.082	0.235
13	50	45	5	55.003	20.726	1.976	4.269	0.078	0.206
14	30	70	0	53.809	21.801	2.854	3.971	0.074	0.182
MT	100	0	0	53.270	21.850	0.460	3.922	0.074	0.180
FA	0	100	0	54.040	21.780	3.880	3.991	0.074	0.183

In the XRD analysis for MT and FA they present a high mineralogical phase in aluminosilicates presenting phases as silicon oxide, muscovite for MT and silicon oxide (Figure 1 and Figure 3)., silicon oxide aluminum for FA (Figure 3).

Morphological analysis for MT and FA denotes particulate metal material for MT and FA, which is typical of metallurgical mining activities and thermoelectric plants (Figure 2 and Figure 4). The particle size analysis performed on MT and FA samples show an average particle size below 106 μm for MT and 60 μm for FA, which greatly facilitates the mixing of the components for the formation of the geopolymer, the relative density for MT is 2.86 and for the FA is 2.21, which shows low density materials. They present 11.19 % humidity for the MT and 16.20 % humidity for the FA.

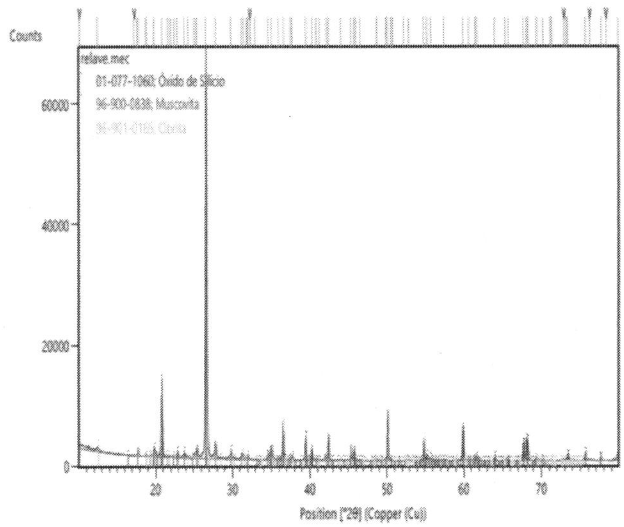

Figure 1. Results of mineralogical analysis of MT

Figure 2. Results of morphological analysis of MT

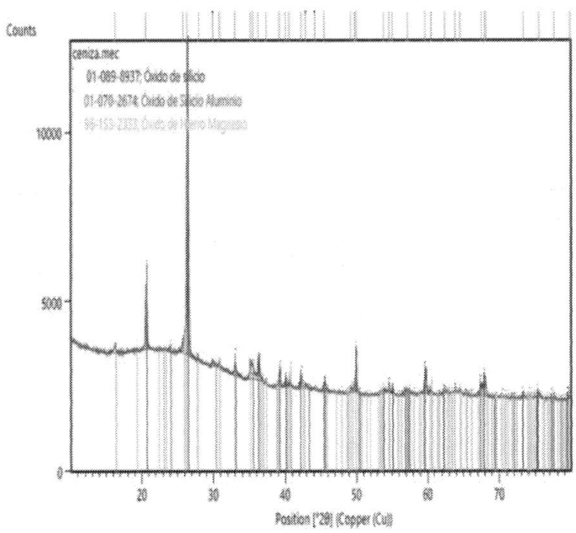

Figure 3. Results of mineralogical analysis of FA

Figure 4. Results of morphological analysis of FA

3.2. Mechanical and morphological characterization of the geopolymer.

The mechanical resistance tests were closely linked to the molarity of the activating solutions and the curing time of the geopolymer, obtaining the best results in 1 M, 5 M, 10 M and 15 M activating solutions, denoting a better treatment of the geopolymer formed by (55 % MT, 37.5 % FA and 7.5%

IOP Conf. Series: Materials Science and Engineering 1065 (2021) 012009 doi:10.1088/1757-899X/1065/1/012009

MK) obtaining a mechanical resistance of 24.6 Mpa at 35 days of curing in a 10M activating solution (Table 2).

Table 2. Comparison of the mechanical resistance of geopolymers to different molarities of activating solution

N°	MT (%)	FA (%)	MK (%)	1 M MPa	5 M MPa	10 M MPa	15 M Mpa	10 M to 35 days Mpa
1	0,700	0,300	0,000	1,45	2,90	10,58	7,33	12,80
2	0,400	0,600	0,000	1,16	1,98	06,19	5,67	10,34
3	0,600	0,300	0,100	1,07	4,42	13,42	12,99	23,62
4	0,300	0,600	0,000	1,24	1,90	07,74	6,85	11,52
5	0,600	0,375	0,025	2,73	5,08	12,30	13,52	21,43
6	0,450	0,525	0,025	2,29	3,83	11,57	11,70	21,02
7	0,550	0,375	0,075	2,82	5,57	14,08	12,87	24,60
8	0,400	0,525	0,075	2,21	3,72	11,74	11,88	19,20
9	0,550	0,450	0,000	1,61	3,76	12,23	12,37	14,26
10	0,650	0,300	0,005	3,18	4,68	13,56	12,63	23,06
11	0,350	0,600	0,005	1,47	3,48	09,68	9,79	18,40
12	0,450	0,450	0,010	1,32	3,72	11,54	11,67	23,42
13	0,500	0,450	0,005	2,64	3,43	11,71	11,25	19,44
14	0,300	0,700	0,000	1,11	1,87	04,19	3,73	07,67

A high strength of the geopolymer formed by 100% in tailings at 10 M to 35 days of curing was evidenced, yielding a strength of 22.68 MPa, and at 15 M to 35 days solution yielded a strength of 19.26 MPa, demonstrating an effectiveness of the use of the tailings by itself for geopolymer formation.

A white salt was evidenced on the surface of the geopolymers which was abundantly found in the geopolymers in 1 M and 5M activating solution, these same salts were found in the geopolymers formed with 10 M and 15 M solutions respectively, but in a minimal proportion, for which these salts were analyzed in the XRD showing mineralogical structures like the thenardite that is found in greater proportion.

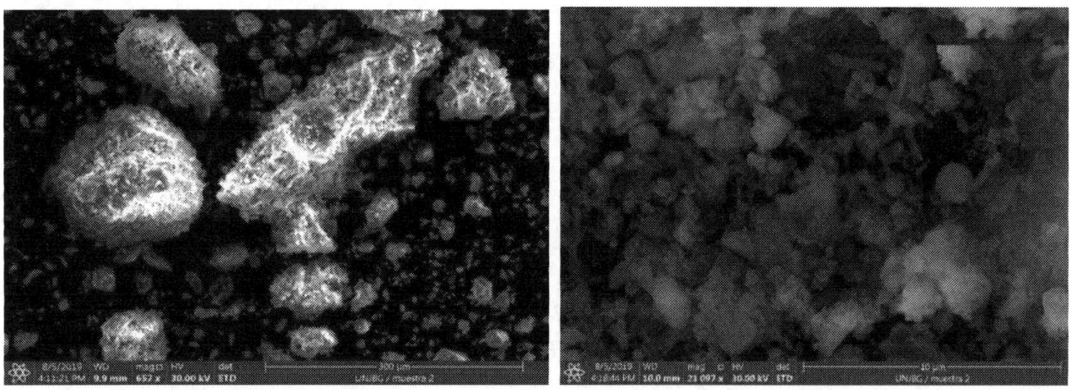

Figure 5. Micrographic results of geopolymer formation.

The results of the elemental analysis by SEM at 300 μm and 10 μm show formations of aluminosilicates in a tetragonal mesh which demonstrates a correct formation of the geopolymer (figure 5).

3.3 Results of the TCLP (Toxicity characteristic leaching procedure)

There is a large reduction in the levels of contaminating elements of the geopolymer formed, which can be seen in the (Table 3), which compared to the Environmental Quality Standards (EQS) for Soil Supreme Decree No. 011-2017-MINAM with the exception of Hg. The inorganic elements such as As, Ba, Cd, Cr, Pb are kept much lower even for agricultural soils (Table 3), besides them the analysis showed the reduction of heavy metals (Table 4).

Table 3. Comparison of geopolymer with the Environmental Quality Standard for soil in Peru

Inorganics	Agricultural soil ppm	Residential land ppm	Extractive industrial commercial soil ppm	Geopolymer ppm
As	50	50	140	0.439
Ba	750	500	2000	0.188
Cd	1.4	10	22	0.001
Cr	**	400	1000	0.004
Pb	70	140	800	0.01

Table 4. Analysis of the TCLP test

	Sample 1 ppm	Sample 2 ppm		Sample 1 ppm	Sample 2 ppm
Ag	<0.002	<0.002	Cr	<0.004	<0.004
Al	4.17	0.91	Cu	0.469	0.005
As	0.478	0.439	Fe	1.12	0.33
B	3.122	2.243	Mg	16.21	9.79
Ba	0.307	0.188	Mn	1.277	0.012
Be	0.0066	<0.0003	Pb	0.2	<0.01
Bi	<0.02	<0.02	Mr	4.3157	2.3054
Ca	493.42	493.42	Zn	0.279	0.016
Cd	<0.001	<0.001	Cr	<0.004	<0.004
Ce	<0.02	<0.02	Cu	0.469	0.005
Co	<0.002	<0.002	Fe	1.12	0.33

3.4. Discussion

Chemical compositional analysis for MT and FA shows a high proportion of aluminosilicates, taking into account the industrial origin of the samples. The amount of aluminosilicates varies between mining operations, due to the geology of the tailings source, as evidence [28] and [29] Molar ratios present in mixtures such as pure MT and FA show a high possibility of geopolymer formation. There is evidence that mining tailings contain significant amounts of aluminosilicates for the immobilization of these wastes as a geopolymer.

The molarity of the activating solution is very critical for the formation of geopolymers, being observed in the low mechanical resistance for the mixtures of 1 M and 5 M showing little reactivity between the mixtures forming salts on the surface as the thenardite, the mixtures of 10 M show an optimal reactivity for the formation of geopolymers compared to the activated mixtures to 15 M which present a medium resistance as evidence [13].

[29], has found that water content, reaction time, curing temperature, composition and concentration of the activating solution, the $Si0_2/Al_20_3$ ratio and the particle size of the raw material affect the dilution and precipitation rates of amorphous aluminosilicates and the mechanical performance of the geopolymer. The use of water for the formation of this geopolymer was low mainly because of its high muscovite content, This material is hydrophilic and plastic which gives binding properties of water absorption from the environment taking into account that the presence of moisture in the environment of Tacna according to the [30] of 90% which allows the geopolymer formed to be

a revelation for the use of this material as engineering material taking into account that Tacna is located in a desert itself the desert of Atacama which would allow a low water consumption.

The climate is an important factor for the formation of the geopolymer [31], taking into account that Tacna is a city with a high presence of moisture in the environment, The geopolymer was cured according to [19] in the open air for 28 days, but it was observed that it still presented moisture to the touch, so we waited a week more to dry more effectively and improve their mechanical properties for that sample type is evident that it needs more time to be cured and improve their mechanical properties. However, in recent studies the temperature can vary between 313.15 K (40 °C), [32], 333.15 K (60 °C), [33] and 358.15 K (85 °C) [34] and [35]. In addition to temperature, curing time has been an important factor in the development of good mechanical properties in geopolymer systems. Recent research has found that the time varies from 2 to 24 hours [36], [37] and [38].

It was experimentally demonstrated that the geopolymer made at 100% in MT in 10 M activating solution obtained a good mechanical resistance, so it can be used by itself for the formation of geopolymers without the need of other aluminosilicate contributors such as FA and MK, It can be applied as a compacting material in mine closures or mining roads, taking into account the mechanical resistance obtained and the immobilization of the contaminating components in its tetragonal geopolymer mesh, so it is recommended more studies of the application of geopolymer in engineering activities as a construction material and environmental restoration.

The surface formation of salts such as thenardite in some mixtures of geopolymer to solution of 1 M and 5 M shows the low reactivity of these with the elements present, which explains the formation of these products such as these salts, was observed in much smaller proportion in the samples of geopolymer formed with activating solution at 10 M and 15 M.

By micrographic analysis, the formation of the geopolymer is evidenced, observing the characteristic tetragonal mesh of geopolymers, which shows the fixation of the elements in their tetragonal mesh, which is demonstrated with the TCLP analysis with the reduction of the polluting elements, in which the reduction was observed for the contaminating elements contemplated in the [27] with the exception of As and Hg such as Ba, Cd, Cr, Pb which are present in the geopolymer in good quantities even for agricultural soils.

The analysis shows that meta-kaolin does not suggest any important incidence for the formation of the geopolymer, so its active use for the formation of the geopolymer is discarded. The FA show a secondary behavior since the analysis shows that the MT can form geopolymer, but with less mechanical resistance.

4. Conclusion

The immobilization of copper sulphide flotation tailings activated in NaOH solution was demonstrated, where the best mechanical behaviour for MT immobilization is formed by the mixture (55% MT, 37.5% FA and 7.5MK) at 35 days of curing in the open air in 10 M NaOH activating solution demonstrating a mechanical resistance of 24.60 Mpa, the mixture of MT 100% activated to 10 M NaOH cured for 35 days in the open air, demonstrates an effectiveness of the use of tailings by itself for the formation of the geopolymer, showing a mechanical resistance of 22.68 MPa, The geopolymer formed shows an important reduction of metallic contaminating elements such as As, Ba, Cd, Cr, Pb among others, being the geopolymer effective to immobilize physically and chemically the contaminating elements, besides possessing a considerable high mechanical resistance, for its use in future construction applications such as roads, rural roads, foundations for landfills and mine closures.

Aknowledgement

The authors were thankful to Dr. Héctor Rodríguez Paupico, Vice-Rector of Research at the Universidad Nacional Jorge Basadre, for his support in the different Academic-Administrative procedures for the development of this research work. To Dr. Jesús Plácido Medina Salas and Dr. Francisco Gamarra Gómez for their support in the SEM and XRD analyses and Gianna Melissa Luis Jalanoca, who has a degree in foreign language, translator and interpreter for her support in translating this paper.

References

[1] Ministerio de energía y minas 1993 Guía Ambiental Para el Manejo de Relaves Mineros.

[2] La Rotta Á and Torres M 2017 Mining and its health and environmental impacts. The case of Potosí in Bogotá

[3] Sangiorgi C, Tataranni P, Lantieri C and Castro J 2016 Reuse of mining waste into innovative alkali-activated-based materials for road pavement applications.

[4] Giannopoulou I. and Panias D 2006 Development of geopolymeric materials from industrial solid wastes article presented at the second international conference "Advances in Mineral Resources Management and Environmental Geotechnology", Hania, Grecia.

[5] Wilson D, Balkau F and Thurgood M 2005 Manual de Formación en gestión de residuos peligrosos para países en vías de desarrollo.

[6] J 1999. Chemistry of Geopolymeric Systems, Terminology In: Proceedings of 99 International Conference, eds. Joseph Davidovits, R. Davidovits & C. James, France.

[7] Davidovits J 2011 Geopolymer Chemistry & Applications. 3rd Edition. Institut Géopolymère. Saint-Quentin, Francia.

[8] Montes V 2014 Cementos Alcalinos: Materiales de Construcción Ecológicos, Revista CINTEX, Vol. 19, pp.109-125.

[9] Komnitsas K and Zaharaki D 2007 Geopolymerisation: A review and prospects for the minerals industry. Minerals Engineering, pp.1261–1277.

[10] Palomo A and Blanco M 1999 Granizo M.L, Puertas F., Vázquez T. and Grutzeck M.W.; Chemical stability of cementitious materials base on metakaolin. Cem. Concr. Res. Vol. 29.

[11] Provis L and Van Deventer J 2009 Structure, processing, properties and industrial applications. Woodhead Publising Limited. 1 ed. ISBN: 978-1-85573-607-8.

[12] Instituto Geológico Minero y Metalúrgico INGEMMET 2017 Geología y controles Tectonomagmáticos de los Sistemas Porfiríticos en el Arco Magmático Occidental Sur de Perú.

[13] J and Deventer V 2001 Effect of silicate activator pH on the leaching and material characteristics of waste-based inorganic polymers"

[14] Barrie E, Cappuyns V, Vassilieva E, Adriaens R, Hollanders S, Garcés D, Paredes C, Pontikes Y, Elsen, J and Machiels L 2015 Potential of inorganic polymers (geopolymers) made of halloysite and volcanic glass for the immobilisation of tailings from gold extraction in Ecuador Applied Clay Science 109-110:95-106.

[15] Komnitsas K, Zaharaki D, Vlachou A, Bartzas G and Galetakis M 2015 Effect of synthesis parameters on the quality of construction and demolition wastes (CDW) geopolymers. Advanced Powder Technology 26. pp. 368-376.

[16] ASTM D-2216 – 98 Método de prueba estándar para la determinación en laboratorio del contenido de agua (humedad) de suelos y rocas por masa.

[17] Mackenzie K, Waijarean N, Asavapisit S, Guy, N and Jamesonv L (2017). "Synthesis and properties of geopolymers based on water treatment residue and their immobilization of some heavy metals" Artículo publicado en Journal of Materials Science 52(12).

[18] Montgomery D 2005. Diseño y análisis de experimentos segunda edición Limusa Wiley, ISBN 9681861566, 9789681861568

[19] Norma Técnica Peruana Cementos NTP 334.009 2005. Cementos Portland. Requisitos

[20] ISO 1920-4:2005 Testing of concrete — Part 4: Strength of hardened concrete

[21] Waijarean N, Asavapisit S and Sombatsompop K 2014 Strength and microstructure of water treatment residue-based geopolymers containing heavy metals, Construction and Building Materials 50:486-491.

[22] Davidovits J 2017 Geopolymers: Ceramic-like inorganic polymers, J. Ceram. Sci. Technol., vol. 8, no. 3, pp. 335–350, 2017.

[23] Duxson P, Fernández A, Provis J, Palomo A and Van Deventer J 2007 Geopolymer Technology: The Current State of the Art.

[24] Robayo-Salazar R, Mejia J, and Mejia de Gutierrez R 2017 Eco-efficient alkali-activated cement based on red clay brick wastes suitable for the manufacturing of building materials. J. Clean. Prod. 2017, 166, 242–252.

[25] Xu H, Van Deventer J, and Lukey G 2001 Effect of alkali Metals on the preferential Geopolymerization of Stilbite/kaolinite Mixtures Industrial and Engineering Chemistry Research

[26] Kinnunen P, Ismailov A, Solismaa S, Sreenivasan H, Räisänen M, Levänen E and Illikainen, M 2018 Recycling mine tailings in chemically bonded ceramicsv—A review. J. Clean. Prod. 2018, 174, 634–649.

[27] Ministerio del ambiente. 2017. Estándares de Calidad Ambiental (ECA) para Suelo Decreto Supremo N° 011-2017-MINAM

[28] Hardjito D, Wallah S, Sumajouw D and Rangan B 2004 Properties of Geopolymer Concrete with Fly Ash as Source Material: Effect of Mixture Composition. Paper to the Seventh CANMET/ACI International Conference on Recent Advances in Concrete Technology. Las Vegas, USA

[29] Komnitsas K, Zaharaki D and Perdikatsis V 2009 Effect of synthesis parameters on the compressive strength of low-calcium ferronickel slag inorganic polymers, Journal Hazardous Mataríais, vol. 161, pp.760-768.

[30] Servicio Nacional de Meteorología e Hidrología del Perú SENAMHI

[31] Rahie H, Simons W, Van Mele B and Biesemans M 1997 Low-Temperature Synthesised Aluminosilicate Glasses Part 111: lnfluence of the Composition of the Silicate Solution on Production, Structure and Properties". J. Mater. Sci. Vol. 32, No. 22.

[32] Yip C, Lukey G, and van Deventer J 2003. Processing, microstructure and properties of geopolymeric materials. 1 05th Annual Meeting and Exposition of the American Ceramic Society, Nashville TN

[33] Barbosa V, MacKenzie K and Thaumaturgo, C 2000 Synthesis and Characterisation of Materials Based on Inorganic Polymers of Alumina and Silica: Sodium Polysialate Polymers. International Journal of Inorganic Materials 2(4): 309-317.

[34] Palomo A and Alonso S 2001. Calorimetric study of alkaline activation of calcium hydroxidemetakaolin solid mixtures. Cement and Concrete Composites, vol. 31, no. 1, pp. 25-30.

[35] Kastiukas G, Zhou X and Castro J 2017 Preparation conditions for the synthesis of alkali-activated binders using tungsten mining waste. J. Mater. Civ. Eng. 2017, 29, 1–9.

[36] Lee W, and Van Deventer J 2004. The interface between natural siliceous aggregates and geopolymers. Cem. Concr. Res. Vol. 34, No. 2.

[37] Cheng T, and Chiu J, 2003. "Fire-resistant geopolymer produced by granulated blast furnace slag." Miner. Engg. Vol. 16, No. 3.

[38] Samarakoon M, Ranjith P, Rathnaweera, T and Perera, M 2019 Recent advances in alkaline cement binders: A review. J. Clean. Prod. 2019, 227, 70–87.

[39] Abdollahnejad Z, Mastali M, Woof B and Illikainen M 2020 High strength fiber reinforced one-part alkali activated slag/fly ash binders with ceramic aggregates: microscopic analysis, mechanical properties, drying shrinkage, and freeze-thaw resistance, Constr. Build. Mater. 241.
[40] Xiao R, Ma Y, Jiang X, Zhang M, Zhang Y, Wang Y, et al., 2020. Strength, microstructure, efflorescence behavior and environmental impacts of waste glass geopolymers cured at ambient temperature. J. Clean. Prod. 252, 119610.
[41] Wang A, Liu H, Hao X, Wang Y, Liu X and Li Z 2019. Geopolymer synthesis using garnet tailings from molybdenum mines. Minerals 9 (1), 48.

IOP Conf. Series: Materials Science and Engineering 1065 (2021) 012010 doi:10.1088/1757-899X/1065/1/012010

Effect of Silicon Content on wear and Hardness of Al-Si Alloys

Gururaj Hatti[1], Vishwanath V H[2] and K R Dinesh[3]

[1] Mechanical Engineering department KLS VDIT, Haliyal-581329, VTU, Belagavi, India.

[2] Mechanical Engineering department, B.L.D.E.A's V. P. Dr. P. G. Halakatti College of Engineering & Technology, Vijayapur-586103, Karnataka, India, VTU, Belagavi, India.

[3] Mechanical Engineering department, GEC, Hassan, Karnataka, VTU, Belagavi, India.

E-mail: hatti.gururaj@gmail.com

Abstract. In the current work the Al-Si composite is prepared using stir casting method with variation in the percentage(4%, 6% and 8%) of the Si particles in it. stir casting method is one of the cheapest method for production of Al alloys. The prepared composites of different percentages of Si are investigated for it wears behaviour using Pin-on-Disk machine and other properties like hardness. Characterization of Worn surface of the composite is done using SEM. Among the metallic matrices, the aluminum-based matrix remains the most studied metal matrix material for the production of MMCs. This is largely due to the large variety of properties provided by aluminum-based matrix composites at low manufacturing costs. While many Metal Matrix Composites (MMCs) are desirable for use in various industrial applications, Aluminum Matrix Composites (AMCs) are the most commonly used in specialized applications because they combine reasonable strength , low density, durability, machinability, availability , performance and price. In aerospace applications, the excellent tribological characteristics of Al-Si and Al-Sn alloys have led to their widespread use, especially in plain bearings, cylinder liners and internal combustion pistons. It is seen that Si is uniformly distributed in the Al matrix and hardness increases to 71BHN with 8%Si in Al matrix. Further analysis revealed that the microstructure homogeneity of the samples is strongly affected its hardness and Wear rate decreases with increase in Si.

1. Introduction

As of late aluminium composites are generally utilized in car businesses. This is especially because of the genuine need to weight putting something aside for more decrease in fuel utilization. The common alloying components are Cu, Mg, Mn, Si and Zi. The surfaces of aluminium combinations have an outstanding dry polish due to the arrangement of a protective coating of aluminium oxide [1-5]. Aluminum composites of 4xxx, 5xxx and 6xxx, containing significant basic added substances of Mg and Si, are currently being used to substitute steel boards in numerous automotive firms. Because of such reasons, these combinations were subject of a few logical investigations in a previous couple of years. Aluminum alloys are characterised by their principal elements in alloy[6-10]. As its main casting alloy part, silicon is found in the 4xxx group. Silicon in metallic alloys is perfect. It increases the fluidity of the melt, decreases the temperature of the melt, decreases shrinkage during solidification, and is very economical as a raw material. Density of the silicon is low (2.34 g cm^{-3}), which will help to minimise the cast component's overall weight. Silicone has a low aluminium solubility, precipitates like almost pure silicone, and is strong and thus increases the resistance to abrasion. Aluminum-silicon alloys form a eutectic with a silicon content of 12.6 wt. % and a eutectic temperature of 577 ° C. That refers to a standard casting alloy composition since its melting temperature is as low as possible [11-14].

Aluminum-Silicon alloys are a higher priority for the manufacturing sector. Since they have a high ratio of strength to weight, high wear resistance, low hardness, low coefficient of thermal expansion, etc. Due to low shrinkage and high fluidity, which results in increased ability for casting and welding. In compliance with the Aluminum Association Wrought Alloy, Al-Si alloys are numbered 6xxx alloys.

The 6xxx series has two main uses, welding of filler alloys and forging. The excellent flow properties given by the presence of a relatively high silicon material are attributed to these two applications. The effect of silicon in al-Si alloys is as follows: i. Presence of silicon reduces the heat transfer ii. Magnetic properties are going to reduce slightly iii. Silicon diminishes the lattice parameter dramatically iv. Due to the hardness of the silicon, the surface finish is low, but there are several literatures and it is obvious based on the previous section that there is an increased potential for further study of al-Si alloys, particularly their mechanical properties [15-17].

Al-Si is an essential alloy for many transportation components (pistons, cylinder liners, etc.) due to its special characteristics. Al-Si casting alloys are by far the most durable of all casting alloys in the manufacture of automotive components specially engine pistons. Since they are the hardest Al alloys to be cast and machined due to higher Si content, industrial applications for hypereutectic alloys are comparatively limited. Once Al is alloyed with a high Si material, it adds a large amount of heat that needs to be drained from the alloy to solidify it during the casting operation. There are significant variations in the size of the primary Si particles within the various areas of the casting system, resulting in major changes in the sample's mechanical properties. Si's primary crystals can be refined to increase tolerance to wear and hardness properties. Hypereutectic alloys are not very cost-effective to produce for these purposes because they have a large range of solidification that contributes to poor casting efficiency and needs additional processing methods to monitor the microstructure [18-20].

2. Experimental

2.1 Preparation of Composites

In the present, the samples examined were MMCs based on 6061 aluminium alloy (having 0.4 per cent Mg and 0.75 per cent Si) and containing graphite particles. In preparing the composite for the first set a known quantity of Al6061 as matrix, Silicon in varying proportions as reinforcement i.e.4%,6%,8% by weight and specimens are prepared under stir casting method figure 1. In order to produce the composite Al6061 was heated in the induction furnace and the temperature raised to 700° C and then the mixture was stirred with a suitable speed using stirrer coated with zirconium. After certain time specific quantity of silicon was added to the vortex figure 2 as a set of composite mean while stirring was continued and mixed isothermally. Finally, stirrer was removed from the slurry and the molten mixture was poured into metallic mould figure 3 which were preheated earlier. After solidification set of composites were removed from the metallic mould.

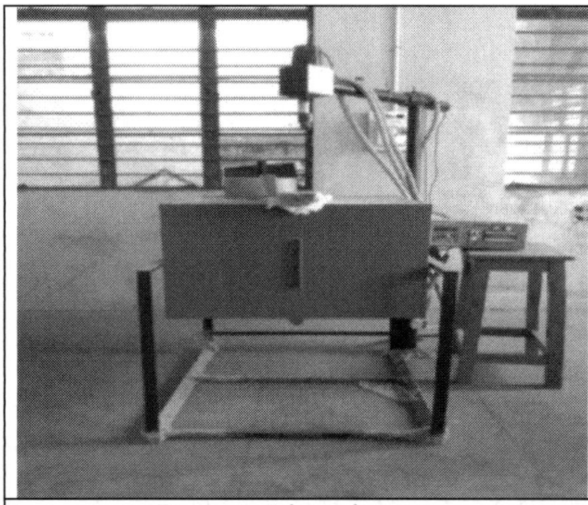

Figure 1. Stir casting setup

Figure 2. Vortex

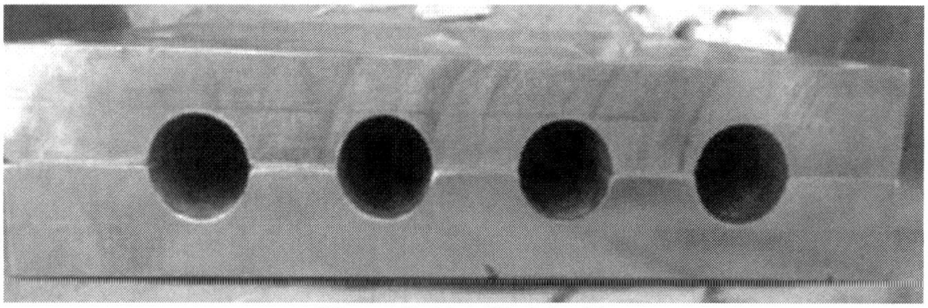

Figure 3. Metallic Mould

Machining of composites to ASTM specifications for conducting dry sliding wear test on pin-on - disc machine by applying parameters such as load, speed, distance and microstructures studied under SEM.

2.2 Preparation of Specimen for testing and parameters

The specimens that undergo dry sliding wear test is conducted by using pin on disc machine (modal; wear and friction monitor.TR-20). The pin on disc machine, the disc is made up of EN32 steel disc the pin was held against the counter face of a rotating disc track diameter 120mm. Normally the load consider was 1kg,5kg and the sliding distance was 754m for (Al6061/ Si). Initially the specimen was polished with the 600grit emery paper to obtained flat surface. Initial weight w1 will be taken by the electronic balance machine. Fix the pin on disc with 60mm track diameter. Put the load 1kg and start the machine and speed and velocity will be calculated and set the machine. The frictional force readings take by every minute. After completing of the 1000 distance final Wight w_2 was taken and volume loss will be calculated by, w_1-w_2 =delta w.

Volume loss =delta (w)/density, Density=mass/volume. The test will be repeated for remaining specimen.

In the present work wear test is carried out by Subsequent variables

1. Load
2. Sliding speed
3. Sliding distance

In this present experiment the parameters are as in Table 1.

Table 1. Parameters Considered for wear test

Load (N)	Speed V (m/s)	Distance (M)
9.81	2.5	754
49.05	2.5	754

Al-Si sampl
es of various weight configurations were mechanically polished using standardized metallographic procedures prior to examination.

3. Results and Discussion

Tests like Hardness test & wear test are carried on Al-Si alloys. Microstructure analyses are also carried out on worn surface specimens. The outcomes of these experiments are recorded, analysed and debated.

3.1 Wear test

Wear tests of Al-Si alloys were conducted with differing applied loads, sliding speeds and sliding distances.

3.1.1 Calculation.

Wear rate =volume loss/sliding distance
Force friction (ff) =Enter (n)
Coefficient friction (CoF) =Frictional force/load
At 2.5m/s
V= ΠDN/60
N=400rpm
Time=5min

Table 2. 96% Aluminium and 4% silicon composite wear rate,force friction(ff) and coefficient friction

Load (N)	Distance (M)	Speed (rpm)	W_1 (gm.)	W_2 (gm.)	Delta W	Density	Vol. loss	Wear	ff	CoF
9.81	754	2.5	3.758	3.306	0.452	0.003	133.09	0.176	5.1	0.519
49.05	754	2.5	3.306	2.197	1.108	0.003	326.21	0.432	27.1	0.552

Calculation of the Volume loss , wear rate and coefficient of friction for 96% Aluminium and 4% silicon composite. Similarly it is calculated to other composites.

1) vol loss = delta / density.
= 0.452/0.003
= 133.09

2) wear rate = vol loss / distance.
= 133.09/754
= 0.176

3) coefficient of friction = friction force / load
= 5.1/9.81
= 0.519

Table 3. 94%Aluminium and 6% silicon composite wear rate,force friction(ff) and coefficient friction

Load (N)	Distance (M)	Speed (rpm)	W_1 (gm)	W_2 (gm)	Delta W	Density	Vol. loss	Wear	ff	CoF
9.81	754	2.5	4.4132	4.2314	0.1818	0.00251	72.372	0.095	15.4	1.569
49.05	754	2.5	4.231	3.63	0.6014	0.002	239.41	0.317	26.7	0.544

Table 4. 92%Aluminium and 8% silicon composite wear rate,force friction(ff) and coefficient friction

Load (N)	Distance (M)	Speed (rpm)	W_1 (gm)	W_2 (gm)	Delta W	Density	Vol. loss	Wear	ff	cof
9.81	754	2.5	4.058	3.960	0.098	0.0025	38.329	0.050	8.3	0.846
49.05	754	2.5	3.960	3.429	0.530	0.0025	207.18	0.274	30.6	0.623

The details of the wear tests on all samples with different loads (9.8, 49.05 N) are shown in Table 2, 3 and 4 and the activity of all samples is shown in Fig. 4. It is noted that as the percentage of silicon increases, the drop in height due to wear decreases. For Al-4 % of Si load alloy 49.05N, the volume loss is 326.21mm^3, while for Al-6 % of Si load alloy 49.05N, the volume loss is 239.41mm3 and for Al-8 % of Si load alloy 49.05N, the volume loss is 207.18mm^3. Linked volume failure trends for all other loads are found. The addition of hard silicon particles applied to the alloy may be attributed to reduced height loss of a growing percentage of silicon. [21] have studied the effect of grain refiner and or modifier on the wear behavior of hypoeutectic (Al–0.2, 2, 3, 4, 5 and 7Si) and eutectic (Al–12Si) alloys using a Pin-On-Disc apparatus under dry sliding conditions. It is important that Al-Si alloys solidify with fine equiaxed α-Al in hypoeutectic/fine primary Si particles in hypereutectic and fine eutectic Si.

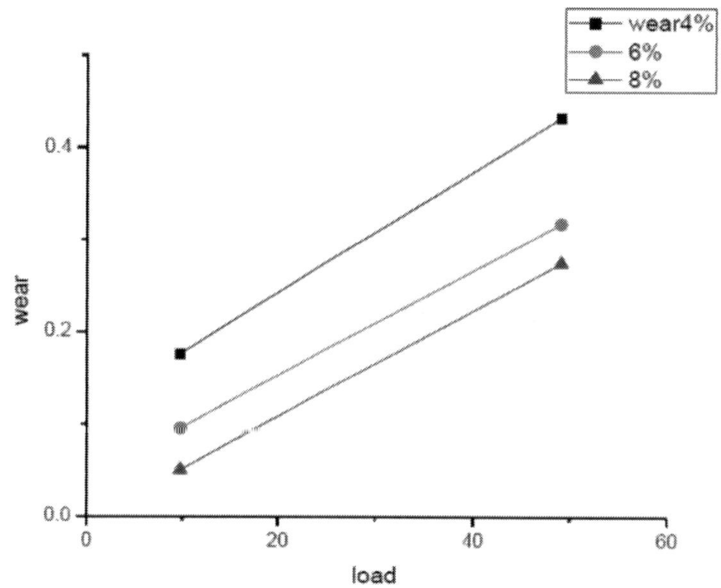

Figure 4. Wear at various loads

3.2 Brinell Hardness test

Hardness, which is defined as a amount of the resistance of a material to external indentation, can be thought of as a function of the stress required to create certain specific forms of surface deformation. The hardness test for all samples is performed using the Brinell hardness test.

Table 5. Brinell hardness number of different Al-Si alloys

Composition	Avg BHN
Al-4%Si	53
Al-6%Si	63
Al-8%Si	71

Figure 5 indicates the Brinell hardness number of Al-4 % Si, Al-6 % Si and Al-8 % Si with a silicon percentage.

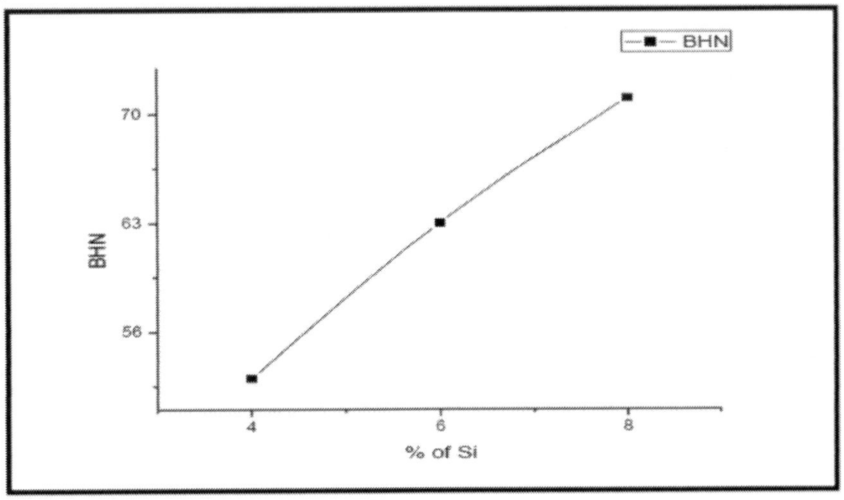

Figure 5. Brinell hardness number of Al-4% Si, Al-6% Si and Al-8% Si

The Brinell hardness values are found to be 53, 63 and 71, respectively, for Al-4 % Si, Al-6 % Si and Al-8 % Si. This indicates that Al-Si hardness improves with an improvement in the silicone percentage. This may be because of a hard increase in the volume of silicon.

3.3 Worn morphology
Samples of alloys from the maximum wear examination. 49.05 N load, 25 rpm sliding velocity and 754 m sliding distance are obtained and their worn surfaces are investigated by an electron scanning microscope. Figure 6 shows micrographs with high magnification, respectively.

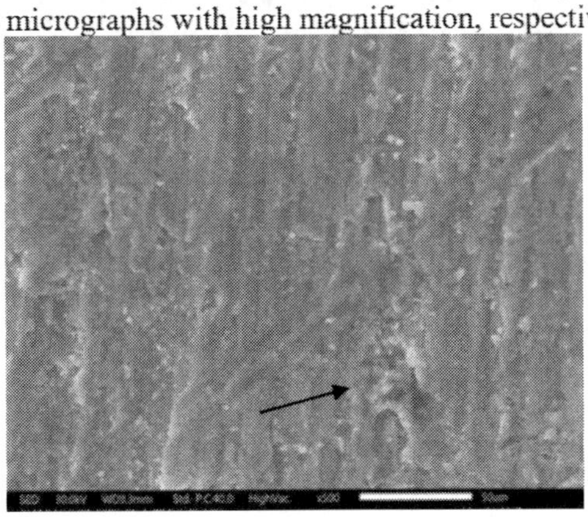

a

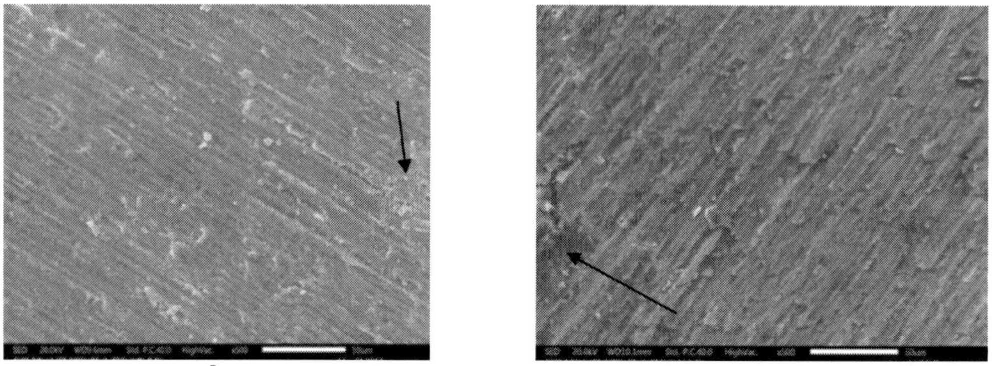

b c

Figure 6. (a) Al-4%Si alloy (b) Al-6%Si alloy (c) Al-8%Si alloy

The worn out surfaces of the samples were observed under scanning electron microscope to examine the mechanism of wear. Micrograph magnification with low SEM , Fig. 6, show the points for the right grade. The score depth in the case of the Al-4 % Si alloy is greater compared to Al-6 % Si and the lowest in Al-8 % Si. This indicates that the wear rate is the highest in Al-4% Si and the lowest in Al-8% Si. The appearance of the mark can be due to the trapped abrasion of the particles, the rugged asperity of the hardened steel counter face, or the hardened deposits of the counter face. During wear, hard-distributed particles or damaged materials are more likely to be dislodged. The pinholes thus created serve to nucleate and grow as possible sites for cracks. As cracks expand and get intertwined, a surface is lost. These are indicated using arrow marks in the figure 6 .

Conclusion

The conclusions obtained from the examinations conducted out are as follows:

Aluminum-silicon alloys prepared have a homogeneous silicon distribution in the casting. The Al-Si composite hardness increases with an increase in the quantity of silicon present. For an increase in the silicon percentage, volume loss due to wear decreases. Wear is seen to increase at higher sliding speed and at higher applied load. Wear behavior is dependent on applied load, sliding speed mainly.

References

[1] Sozhamannan G G, Balasivanandha P and Venkatagalapathy V S K 2012 *J. of Surface Engineered Materials and Advanced Technology* **2** 11-15

[2] Hashim J, Looney L and Hashmi M.S.J 1999 *J. of Materials Processing Technology* **92-93** 1-7

[3] Himanshu Kalaa, Merb K.K.S and Sandeep Kumarc 2014 *Procedia Materials Science* **6** 1951–60

[4] Rajeshkumar Gangaram BhandareÀ and Parshuram M. Sonawane 2014 *International J. of Current Engineering and Technology* **3** *148-155*

[5] Suman Kant and Ajay Singh Verma 2017 *International J. of Mechanics and Solids* **9** 61-9

[6] Sachin P, Shivasharanayya S and Karthik S 2016 *J. of Materials Science & Surface Engineering* **4** 463-66

[7] Muhammad Hayat Jokhio, Muhammad Ibrahim Panhwar, And Mukhtiar Ali Unar 2011 Mehran University Research Journal Of Engg. & Technology **30** 53-64

[8] V Rama koteswara Rao, J Rangaraya Chowdary, Balaji, Sai Krishna, P R Bhavabhuthi, Sreevatsava, Abhiram 2016 *International J. of Scientific & Engineering Research* **7** 742-748

[9] Bharat Kumar, Anil Parmar, Dhaval Ghoghalia, Mandhata Yadav and Samarth Bhaduwala 2015 *International J. of Innovative Research in Science, Engineering and Technology* **4** 7253-9

[10] Khalid Almadhoni and Sabah Khan 2015 *IOSR J. of Mechanical and Civil Engineering* **12** 22-40

[11] R Kumar, J Jegan and L Initha 2014 *IPASJ Int. J. of Mechanical Engineering* **2** 17-25

[12] Dinesh Pargunde, Dhanraj Tambuskar and Swapnil S. Kulkarni 2013 *Int. J. Adv. Engg. Res. Studies* **2** 49-51
[13] S Palani., R Irudhayaraj, A Alagu Sundara Pandian , M Selvam. and S Shanmugan 2016 *Int. J. of Control and Automation* **9** 3797-803
[14] S P Nikanorova, M P Volkova, V N Gurina, Yu.A Burenkova, L I Derkachenkoa, B K Kardasheva, L L Regelb and W R Wilcoxb 2005 *Materials Science and Engineering* **3** 63–9
[15] Raj Kumar Singh, Amit Telang and S Das 2016 *American Journal of Engineering Research* **5** 133-7
[16] Mousa Javidani and Daniel Larouche 2014 *J. Int. Materials Reviews* **59** 132-158
[17] V Vijeesh and K. Narayan Prabhu 2014 *Trans Indian Inst. Met.* **67** 1–18
[18] WU Shu-sen , T U Xia and Y. Fukuda M EB 2003 *Trans. Nonferrous Met. Soc* **6** 1285
[19] Hemant kumar Neopaney and Santanu Chakraborty 2014 *IOSR J.l of Mechanical and Civil Engineering* **11** 54-63
[20] Chikezie W. Onyia1, Boniface A Okorie, Simeon I. Neife and Camillus S. Obayi 2013 *World Journal of Engineering and Technology* **1** 9-16
[21] Kori, S.A., Chandrashekharaia T M 2007 *Wear* 263 745-755

Machining and forecasting of square profile areas using artificial neural modelling at different slant angles by WEDM.

I.V.Manoj[1], Narendranath S[1]

Department of Mechanical Engineering, National Institute of Technology Karnataka, Surathkal 575025, India.

E-mail: vishalmanojvs@gmail.com

Abstract. Wire electric discharge machining non-conventional process that removes materials by thermal erosion. Tapering in WEDM has many applications in machining accurate geometric profiles. In the present investigation, a slant type taper machining was performed to produce taper profiles with the help of the slant fixture. The machining parameters such as wire guide distance, corner dwell time, wire offset and cutting speed override was employed to find the variation in the taper profile area. A simple square profile of 1mm, 3mm and 5mm was machined at different slant angles namely 0°, 15° and 30°. It was observed that each parameter yielded different profile areas. The artificial neural network was used for the forecasting the areas of 1mm, 3mm and 5mm square for different parameters. The optimum artificial neural network model was experimentally validated and the errors were ranging from 0-10%.

1. Introduction

Wire electric discharge machining (WEDM) is an electro-thermal process where the sparks are generated between the workpiece and electrode due to high potential difference. It is a non-contact machining process where all the conductive and hard materials can be precisely machined [1-3]. This forms a versatile way for machining for different materials like nickel, titanium alloys, composites etc. which can be used in many for complex machining [4-6]. Nickel-based superalloy which possesses a challenge for machining due to its high-temperature strength and low thermal diffusivity [7, 8]. Using Conventional machining for these alloys may cause surface defects and damages recorded both on the tool and as well as in workpiece [9, 10]. So wire EDM is the most promising machining techniques as it depends only on the conductivity of the material. The tapered cutting is unique machining used in dies, forming cutters, splicing moulds, vanes and blades etc. [11, 12]. During traditional taper cutting, there are many disadvantages as a flexible wire electrode is bend to an angle. Many researchers like Kinoshita et al. [13], Plaza et al. [14], Sanchez et al.[15] has identified the disadvantages like wire guide wear, friction, wire breakage, angular inaccuracies, poor surface due to insufficient flushing etc. have been observed during taper machining of materials.

Artificial neural network (ANN) is a versatile method that is used for both optimisation [16, 17] and prediction/forecasting. Some of the researchers like Manoj et al. [18] have attempted a different approach to obtain taper and have predicted profile areas for a triangular profile at different taper angles by an (ANN). Singh and Misra [19] have concluded ANN with a back-propagating neural network gives accurate results in the prediction of material removal rate and surface roughness than response surface method. Saha et al. [20] have used ANN for prediction of cutting speed and surface roughness while WEDM of tungsten carbide-cobalt (WC-Co) composite material. Sarkar et al. [21] have predicted trim cutting speed, average surface roughness and dimensional sift using ANN model like Bayesian regularization and early stopping method and an effective cutting speed which gives maximum productivity and surface finish was selected.

The tapering operation in WEDM was used in machining of many complex parts for various applications. In the present investigation, Hastelloy-X is used as investigating material. Hastelloy-X is used in different applications like turbine parts, flame holders, turbocharger impellers etc. which requires slant or taper form of machining. For obtaining taper structure, a unique slant fixture was employed for machining of tapered profiles. This slant fixture avoids the bending of wire, inaccuracies, guide wear etc. as in traditional taper machining. A simple square profile was used slant in the machining of Hastelloy-X using WEDM at 0°, 15° and 30° slant angles. The input parameters such as wire guide distance, corner dwell time, wire offset and cutting speed override were used. ANN was employed for the prediction of taper areas at all the slant angles for different parameters. The model was experimentally validated for intermediate parameters.

2. Experimental Details

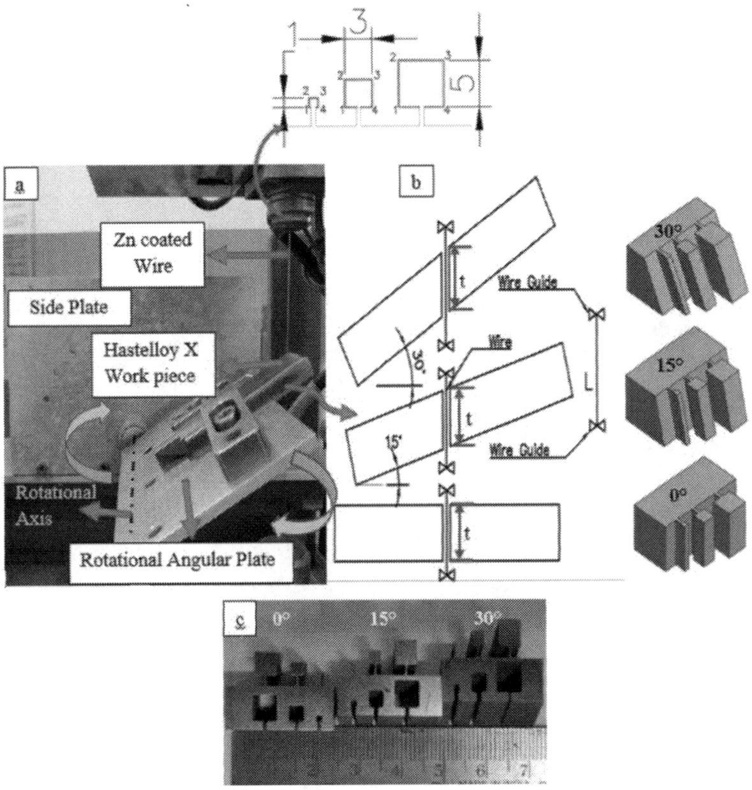

Figure 1. Experimental methodology.

Fig.1 shows the method followed to achieve the tapering with help of in-house developed fixture. The fixture made of aluminium H9 series was made to rest on the WEDM table. The Electronica 'ELPULS 15' made WEDM was employed. The zinc-coated copper wire as the electrode and deionized water as the dielectric medium was made used to machine Hastelloy-X. The workpiece placed in the rotating angular plate which is locked to the required angle. The angular plate is fixed to the fixture resting on WEDM table as shown in fig.1. The programmed profile is coded and uploaded to the CNC of WEDM so the wire guide movement follows the shape. The different angles were achieved as shown in fig. 1

(b) and the orthogonal view of the tapered components can also be seen. The fig.1 (c) shows the tapered components of 10mm cut at different angles.

3. Parameters

The input parameters like wire guide distance, corner dwell time, wire offset and cutting speed override were made use as machining parameters. Based on the initial experiments and machining capabilities theses parameters were defined as in table 1 [18, 22, 23]. As the interactions were insignificant or having a minimal effect on the area, therefore L16 orthogonal array was used. For each angle, 16 components were machined and areas of each 1mm, 3mm and 5mm squares were measured by Scanning electron microscope (SEM), Image J software and co-ordinate measuring machine (CMM).

Table 1 WEDM Parameters

Constant Parameters				
EDM Parameters	Settings			
Pulse off time (μs)	44			
Servo Feed (mm/min)	20			
Wire Feed (m/min)	6			
Pulse on time (μs)	115			
Servo voltage (V)	40			
Machining Parameters				
	40	50	60	70
Wire guide distance (mm)	75	85	95	105
	100	110	120	130
Cutting speed override (%)	31	54	77	100
Wire offset (μm)	0	40	80	120
Corner dwell time (s)	0	33	66	99

4. Results and Discussion

Table 3 shows the recorded areas and predicted areas at different machining parameters. The optimal neural network in fig.2 was decided by test runs consisting of numerous combinations of hidden and output layers. The neural network (type: feed forward backprop) was used to train, test and validate the data for 1mm, 3mm and 5mm square areas. From the 48 iterations, the ANN toolbox in Matlab takes 70% for training, 15% for validation and 15% for testing the model. Back-propagation neural network was employed for training of ANN optimal network [24]. The logarithmic-sigmoid function was employed for the hidden layer and pure-line function was used for the output layer. The mean square error (MSE) performance function with a combination of 5-10-1-1 architecture was employed as shown in fig 2. The regression plots in Fig.3 and R values from table 2 shows the training, testing and validation of the optimal model.

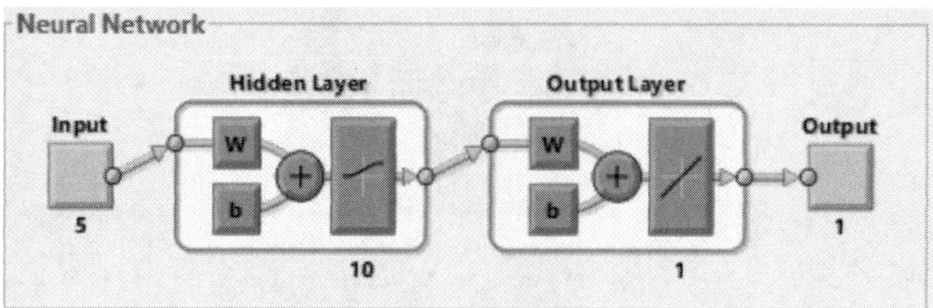

Figure 2. Optimum neural network for used for forecasting.

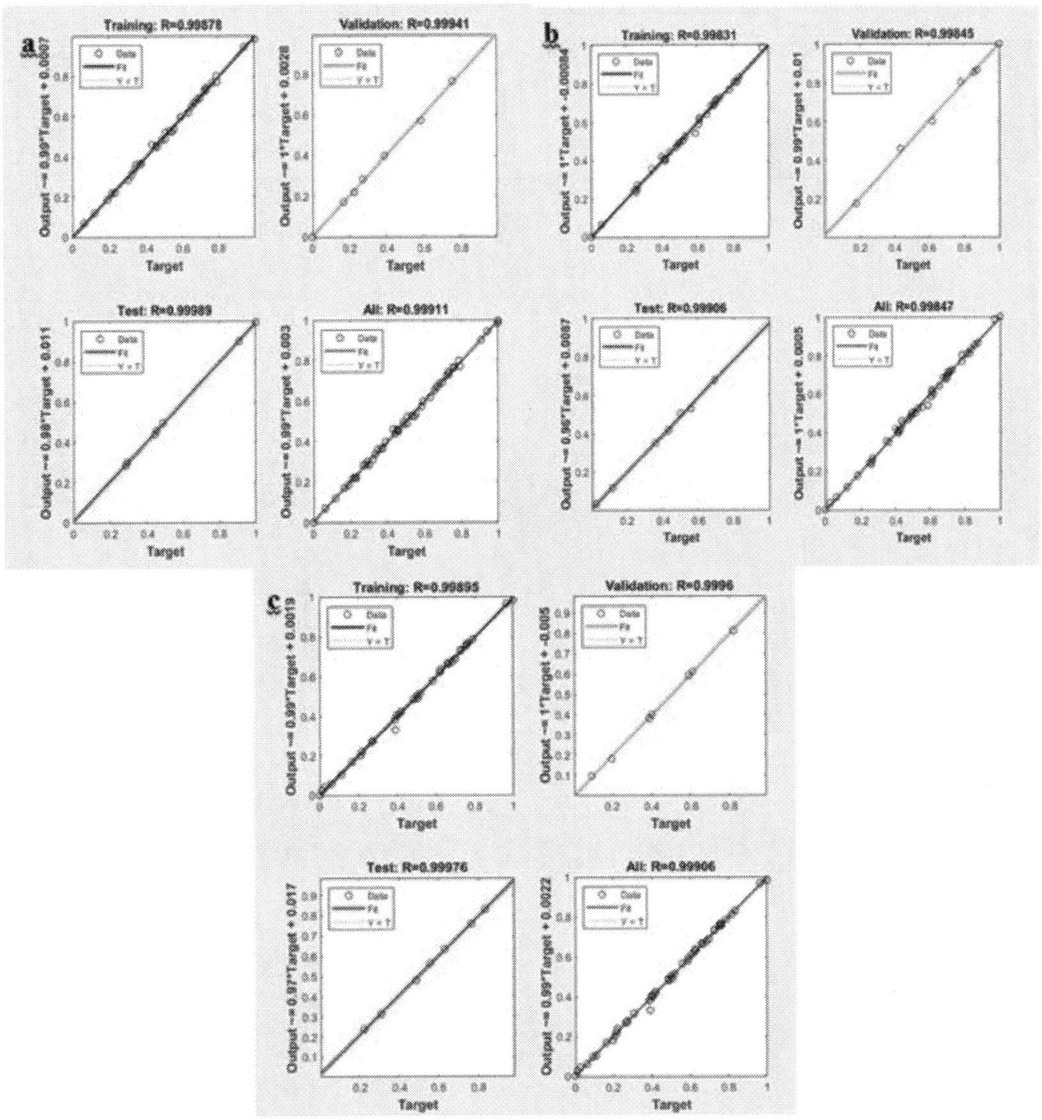

Figure 3. Regression Plots for (a) 1mm (b) 3mm and (c) 5mm square profiles.

Table 2 R values from Regression plot

Sl. No.	Square Areas	Training	Validation	Testing	All (R)
1	1mm	0.99878	0.99941	0.99989	0.99911
2	3mm	0.99831	0.99845	0.99906	0.99847
3	5mm	0.99895	0.9996	0.99976	0.99906

IOP Conf. Series: Materials Science and Engineering 1065 (2021) 012011 doi:10.1088/1757-899X/1065/1/012011

Table 3 Machining and estimation of profile areas

Sl. No.	WDG (mm)	CDT (s)	WO (μm)	CSO (%)	Slant Angle	Predicted Area of the square in mm²			Area of the square in mm²		
						1mm	3mm	5mm	1mm	3mm	5mm
1	40	0	0	31		0.783	7.594	23.108	0.773	7.66	23.086
2	40	33	40	54		0.767	7.619	22.865	0.786	7.622	22.969
3	40	66	80	77		0.982	8.364	23.207	0.986	8.265	23.112
4	40	99	120	100		1.041	8.640	24.097	1.059	8.797	24.131
5	50	0	40	77		0.695	7.394	22.347	0.690	7.392	22.323
6	50	33	0	100		0.660	7.193	22.025	0.664	7.194	21.914
7	50	66	120	31		1.029	8.603	24.279	1.033	8.689	24.284
8	50	99	80	54		0.768	8.221	23.437	0.765	8.249	23.432
9	60	0	80	100	0°	0.696	7.647	22.989	0.701	7.672	23.032
10	60	33	120	77		0.826	8.224	23.671	0.822	8.138	23.633
11	60	66	0	54		0.589	7.014	21.922	0.588	6.989	21.826
12	60	99	40	31		0.796	7.710	22.398	0.806	7.676	22.425
13	70	0	120	54		0.935	8.152	23.394	0.938	8.206	23.392
14	70	33	80	31		0.857	8.011	22.786	0.865	7.956	22.766
15	70	66	40	100		0.536	6.913	22.130	0.529	6.877	22.13
16	70	99	0	77		0.460	6.797	21.775	0.460	6.787	21.754
17	75	0	0	31		0.951	8.981	25.497	0.953	9.019	25.486
18	75	33	40	54		0.947	8.878	25.383	0.957	8.872	25.369
19	75	66	80	77		1.163	9.262	25.633	1.166	9.261	25.612
20	75	99	120	100		1.241	9.532	26.444	1.239	9.454	26.431
21	85	0	40	77		0.854	8.512	24.698	0.855	8.555	24.723
22	85	33	0	100		0.849	8.306	24.203	0.833	8.293	24.213
23	85	66	120	31	15°	1.200	9.557	26.565	1.198	9.603	26.584
24	85	99	80	54		0.957	9.224	25.807	0.926	9.218	25.832
25	95	0	80	100		0.891	8.848	25.294	0.881	8.905	25.332
26	95	33	120	77		0.987	9.263	25.971	1.006	9.262	26.033
27	95	66	0	54		0.766	8.187	24.093	0.751	8.214	24.126
28	95	99	40	31		0.995	8.518	24.827	0.986	8.484	24.825
29	105	0	120	54		1.106	9.126	25.834	1.100	9.094	25.792
30	105	33	80	31		1.026	8.914	25.212	1.049	8.872	25.166
31	105	66	40	100		0.699	8.393	24.352	0.709	8.411	24.330
32	105	99	0	77		0.644	7.994	23.781	0.640	8.012	24.154
33	100	0	0	31		1.251	9.106	25.893	1.253	9.160	25.946
34	100	33	40	54		1.260	9.232	26.221	1.247	9.192	26.205
35	100	66	80	77		1.477	9.745	26.723	1.476	9.765	26.760
36	100	99	120	100		1.534	10.21	27.772	1.539	10.20	27.869
37	110	0	40	77		1.205	8.799	25.644	1.215	8.862	25.582
38	110	33	0	100		1.127	8.581	24.754	1.143	8.594	24.845
39	110	66	120	31	30°	1.521	10.16	27.672	1.538	10.09	27.656
40	110	99	80	54		1.291	9.544	26.411	1.316	9.549	26.384
41	120	0	80	100		1.181	9.164	26.239	1.181	9.172	26.209
42	120	33	120	77		1.285	9.634	26.838	1.276	9.638	26.863
43	120	66	0	54		1.078	8.479	24.739	1.091	8.489	24.728
44	120	99	40	31		1.286	9.135	25.570	1.286	9.146	25.563
45	130	0	120	54		1.432	9.703	26.400	1.440	9.706	26.433
46	130	33	80	31		1.323	9.414	26.359	1.309	9.456	26.381
47	130	66	40	100		1.023	8.443	24.901	1.009	8.447	24.915
48	130	99	0	77		0.955	8.187	24.222	0.950	8.187	24.195

Table 4 shows the validation parameters that were chosen randomly and machined. Table 5 shows the error percentage that was calculated for ANN predicted values comparing the experimental values. The machined components were measure and the areas were recorded. From the optimal model, the predicted areas were generated by loading the machining parameters to the model. It can be observed that the errors varied from 0-9.16% in case of prediction from ANN.

Table 4 Validation parameters for the optimum ANN model

Sl. No.	WGD (mm)	CDT (s)	WO (μm)	CSO (%)	Slant Angle (Degree)
1	55	40	100	45	0°
2	80	45	180	80	0°
3	90	50	75	65	15°
4	120	60	150	50	15°
5	115	80	50	90	30°
6	145	20	130	70	30°

Table 5 Variation of ANN predicted and experimental areas

Sl. No	Area of the square in mm²								
	1mm			3mm			5mm		
	ANN Predicted areas	Measured areas	Error %	ANN Predicted areas	Measured areas	Error %	ANN Predicted areas	Measured areas	Error %
1	0.977	0.903	7.574	8.7	8.34	4.138	25.698	23.344	**9.160**
2	0.915	0.927	1.311	7.869	8.621	9.556	24.899	23.807	4.386
3	0.957	0.948	0.940	9.846	9.016	8.430	27.89	25.562	8.347
4	0.998	0.95	4.810	9.449	9.051	4.212	26.564	25.306	4.736
5	1.296	1.216	6.173	9.757	8.942	8.353	24.289	25.466	4.846
6	1.22	1.224	**0.328**	9.079	9.815	8.107	25.586	26.245	2.576

5. Conclusions

The taper was be achieved from a slant type taper fixture a unique way different from the traditional method. The areas were machined on Hastelloy-X and the following conclusions can be drawn.

1. The ANN with BPNN algorithm gave the most effective of the method for forecasting the areas as the errors vary from 0-5% irrespective of dimensions.
2. The ANN optimal models have been validated experimentally having validation errors varying from 0-9.16%.

References

[1] Nayak B B and Mahapatra S S 2016 Optimization of WEDM process parameters using deep cryo-treated Inconel 718 as work material *Eng. Sci. Technol. Int J.* **19** 161-170.

[2] Soni H, Narendranath S, and Ramesh M R 2020 An experimental study of influence of wire electro discharge machining parameters on surface integrity of TiNiCo shape memory alloy. *J. Mater. Res* **32(16)** 3100-8.

[3] Chaudhari R, Vora J, Parikh D M, Wankhede V and Khanna S 2020 Multi-response Optimization of WEDM Parameters Using an Integrated Approach of RSM–GRA Analysis for Pure Titanium. *J. Inst. Eng. India Ser. D* **101** 117–126.

[4] Ramesh M R, Hargovind Soni, Narendranath S, 2019 Advanced machining of TiNiCo shape memory alloy for biomedical application *Emerg. Mater. Res.* **8(1)** 14-21.

[5] Mandal A, Dixit A R, Das A K and Mandal N 2016 Modeling and Optimization of Machining Nimonic C-263 Superalloy using Multicut Strategy in WEDM. *Mater. Manuf. Process.* **31** 860-868.

[6] Manjaiah M, Narendranath S and Basavarajappa S Gaitonde V N 2018 Investigation on material removal rate,surface and subsurface characteristicsin wire electro discharge machining of Ti50Ni50-xCux shape memory alloy. *Proc IMechE Part L: J Materials: Design and Applications* **232** 164–177.
[7] Choudhury I A and El-Baradie M A 1998 Machinability of nickel-base super alloys: a general review. *J. Mater. Process. Technol.* **77** pp. 278–284.
[8] Ezugwu E O, Wang Z M and Machado A R 1999 The machinability of nickel-based alloys: a review *J. Mater. Process. Tech.* **86** pp.1-16.
[9] Thakur A, Gangopadhyay S and Mohanty A 2016 Investigation on Some Machinability Aspects of Inconel 825 During Dry Turning *Mat. Manu. Process.* **30(8)** pp. 1026-1034.
[10] Dahu Zhu, Xiaoming Zhang and Han Ding 2013 Tool wear characteristics in machining of nickel-based superalloys *Int. J. Mach. Tool. Manu.* **64** pp.60-77.
[11] Hsue A W and Su H 2004 Removal analysis of WEDM's tapering process and its application to generation of precise conjugate surface. *J. Mater. Process Technol.* **149** 117–123.
[12] Selvakumar G, Jiju K B, Sarkar S and Mitra S 2016 Enhancing die corner accuracy through trim cut in WEDM. *Int. J. Adv. Manuf. Technol.* **83** 791–803.
[13] Kinoshita N, Fukui M and Fujii T 1987 Study on Wire-EDM: Accuracy in Taper-Cut. *CIRP Annals* **36** 119-122.
[14] Plaza S, Ortega N, Sanchez J A, Pombo I and Mendikute A 2009 Original models for the prediction of angular error in wire-EDM taper-cutting. *Int. J. Adv. Manuf. Technol.* **44** 529–538.
[15] Sanchez J A, Plaza A, Ortega N, Marcos M and Albizuri J 2008 Experimental and numerical study of angular error in wire-EDM taper-cutting. *Int. J. Mach. Tool. Manu.* **48** 1420–1428.
[16] Suresh R, Joshi A G and Manjaiah M 2020 Experimental Investigation on Tool Wear in AISI H13 Die Steel Turning Using RSM and ANN Methods. *Arab. J. Sci. Eng.* https://doi.org/10.1007/s13369-020-05038-9.
[17] Joshi A G, Manjaiah M, Suresh R and Mahesh B Davangeri 2020 ANN Modeling of Surface Roughness and Thrust Force During Drilling of SiC Filler-Incorporated Glass/Epoxy Composites. *Nature-Inspired Optimization in Advanced Manufacturing Processes and Systems.* https://www.taylorfrancis.com/books/e/9781003081166/chapters/10.1201/9781003081166-10.
[18] Manoj I V and Narendranath S 2020 Variation and artificial neural network prediction of profile areas during slant type taper profiling of triangle at different machining parameters on Hastelloy X by wire electric discharge machining. *Proc. Inst. Mech. Eng. E.* **234(6)** 673–683.
[19] Singh B and Misra J P 2019 Surface finish analysis of wire electric discharge machined specimens by RSM and ANN modelling. *Measurement* **137** 225-237
[20] Saha P, Singha A, Pal S K and Saha P 2008 Soft computing models based prediction of cutting speed and surface roughness in wire electro-discharge machining of tungsten carbide cobalt composite. *Int. J Adv. Manuf. Technol.* **39** 74–84.
[21] Sakar S, Ghosh K, Mitra S and Bhattacharyya B 2010 An Integrated approach to Optimization of WEDM Combining Single-Pass and Multipass Cutting Operation. *Mater. Manuf. Process* **25** 799-807.
[22] Manoj I V, Joy R and Narendranath S 2019 Investigation on the Effect of Variation in Cutting Speeds and Angle of Cut during Slant Type Taper Cutting in WEDM of Hastelloy X. *Arab. J. Sci. Eng.* **45** 641–651.
[23] Joy R, Manoj I V and Narendranath S 2020 Investigation of cutting speed, recast layer and micro-hardness in angular machining using slant type taper fixture by WEDM of Hastelloy X. *Mat. Today: Proceedings* **27** 1943-1946
[24] Soni H, Narendranath S, and Ramesh M R 2018 ANN and RSM modeling methods for predicting material removal rate and surface roughness during WEDM of Ti50Ni40Co10 shape memory alloy

IOP Conf. Series: Materials Science and Engineering 1065 (2021) 012012 doi:10.1088/1757-899X/1065/1/012012

Experimental analysis of Fracture Toughness of EN9 and EN31 Steel under double layer of electroplating materials with different varied thickness

D Kari Basavaraja [1,2], C Sowmya [3], and Ravindra Naik [4]

[1] Assistant Professor, Dept. of Industrial and Production Engineering

[2] Sri Jayachamarajendra college of Engineering, Mysuru

[3]Assistant Professor, Dept. of Industrial and Production Engineering

[4] Assistant Professor, Dept. of Mechanical Engineering.

karibasavaraja@sjce.ac.in

1. Abstract

With the advent of technology, there is a need to invent newer plating to improve the strength of the material. In this paper, two materials namely Steel Grades EN9 and EN 31 are considered. These materials are having quite a good industrial applications and are being used in many machine parts manufacture. Some of the widely used applications are in the manufacture of Crank shafts, Cylinders, Gears, Sprockets, aerospace parts and body of the ships and automobile parts mainly its surfaces which require high impact resistance and also wear resistance for increased life of the components. The non-electro plated materials are also having the property of higher abrasion resistance. In this paper, the preparation of the specimen, electro plating techniques adopted for getting varied thickness and their critical characteristics in improving the property of the material are discussed further, a statistical model is proposed to obtain the bench mark for its effective implementation of these materials for its industrial use.

2. Objective

The objective is to prepare 2 specimens namely EN9 and EN31 material and to carry out electro plating operations to vary the plating thickness. Further, the IZOD and Charpy tests have to be conducted with varying notch shapes and are best suited for the industrial applications.

3. Methodology

The specimens are prepared using heat treatment process and the electro plating are done under varying cutting conditions. These specimens are cut as per the configuration/specification of IZOD and Charpy tests. Various strength parameters are checked using the impact-testing device. Mathematical models are proposed to bench mark the process parameters.

4. Set up of the specimens

The following properties are considered for the preparation of the specimen.

4.1 Mechanical properties of EN 9 & EN 31 steels

- Materials having high degree of hardness with compressive strength and abrasion resistance.
- Through proper grain flow orientation maximum impact strength and fatigue resistance. Possible in a metal.

- There are no internal gas pockets or voids that may cause unexpected failure under conditions of high stress or impacting Steels Designated on the Basis of Mechanical Properties.

The mechanical property of EN 9 & EN31 steels as shown in table 1

Table 1. Mechanical property of EN9 Steel

Max Stress	600-700 N/mm^2
Yield Stress	310-355 N/mm^2 Min
Hardness	170-255 Brinell

4.2. Chemical composition of EN 9 & EN31 steels:

The Chemical composition of EN 9 & EN31 steels as shown in table 2

Table 2. Chemical composition of EN 9 & EN31 steels

EN STEELS	C	Mn	Si	S	P	Cr	Ni	Mo
EN-9	.50 - .60	.50 - .80	.05 - .35	.040 MAX	.040 MAX	-	-	-
EN-31	.90 - 1.20	.30 - .75	.10 - .35	0.04	0.04	1.00 - 1.60	-	-

4.3 Size and specification of testing specimens

4.3.1 Charpy Test Specimens

Table 3 shows the specification of Charpy test specimen.

Table 3 Size and specification of Charpy testing specimens

Size	55×10×10mm
Types of notch	'V' notch and 'U' notch
Depth of 'V' notch	2mm with 0.25mm radius.
Angle of V' notch	45⁰
Depth of 'U' notch	4.5mm with 1mm radius.

4.3.2 Izod test specimen

The specification of Izod test specimen is selected as per ASTM standards as shown in table4.

Table 4 Size and specification of Izod testing specimens

Size	75×10×10mm
Type of notch	'V' notch
Depth of 'V' notch	2mm with 0.25mm radius.
Angle of 'V' notch	45⁰

4.3.3 Making of Charpy test specimen

Fig: 1 (a)

Fig: 1 (b)

Fig: 1 (a). V-notch Charpy test specimens, Fig: 1 (b). U-notch Charpy test specimens

IOP Conf. Series: Materials Science and Engineering 1065 (2021) 012012 doi:10.1088/1757-899X/1065/1/012012

5. RESULTS AND DISCUSSION

Impact test has been carried on EN9 and EN31 Steel under different electroplating material with thickness under Normal, Nickel electroplating, Silver electroplating, Chromium electroplating and Zinc electroplating condition and the results are tabulated as follows.

5.1. SILVER ELECTROPLATING

5.1.1. Solution composition

The silver electroplating process is carried out by selection of solution. Chemical composition is as shown table 5.

Table 5 Solution composition of silver electroplating

Silvermix Bright Silver Salt	200.0 g/l
Argolume Carrier Additive	10.0 ml/l
Argolume Brightener	30.0-40.0 ml/l

5.1.2 Operating conditions

The operating conditions of silver electroplating is shown in Table 6.

Table 6 Operating condition during silver electroplating

Cathode current density	1.0-2.5 A/dm^2
Anode current density	1.0-2.5 A/dm^2
Agitation	Cathode rod movement
Anodes	Pure silver
Temperatures	20-32 °C
pH (Electrometric)	12.0-12.5

5.1.3 Solution Preparation

The silver solution was prepared in a storage tank, and then thoroughly cleaned with water. The lining of the tank, filters are leached with 10% Caustic Potash Solution and immersed in this condition for 12 hours. The tank was cleaned again with fresh water and then cleaned in deionized water. The following steps are followed for the preparation of solution,

- The auxiliary tank was filled with two-thirds full of deionised water.
- Silvermix Bright Silver Salt is added by stirring till all the salt are completely dissolved.
- Activated Carbon was added and stirred for few hours and then allowed to settle for a period of 12 hours
- This solution was filtered and, transferred to the plating tank and then deionized water is added to make up the level
- The filter unit is cleaned and filtered until the solution becomes clear.
- A calculated quantities of Argolume Carrier additive and Argolume brightener were added and stirred. The prepared solution for plating operation as shown in fig13.

Fig 2: materials before electroplating in to solution bath.

5.1.4 Plating procedure

The pre-treatment procedure before plating in Argolume process is the same as for standard silver plating. A silver strike is required to assure good bonding of silver deposits to copper, brass and copper based alloys. A Strike silver bath composition is shown in below table 7. A silver electroplating as per the double layer 2μ and 4μ is shown in the figure 14 a & 14 b.

Table 7: Strike silver bath composition.

Silver content	1.5 - 2.0 g/l
Free Potassium Cyanide	100.0 g/l

Fig.14 (a) Fig.14.(b)

Fig.3. (a): 2 micron silver electroplating of Izod and Charpy specimens.

Fig.3. (b): 4 micron silver electroplating of Izod and Charpy specimens.

5.1.5 Deposit Thickness in Time (Based on 100% Cathode Efficiency)

The deposition of electroplating thickness of a material w.r.t. time (in min) is shown in table 8.

Table 8 Deposit Thickness Time (in min)

	Current density Amp/dm² with time in minutes			
Thickness (Mic)	**0.5 Amp/dm²**	**1.0 Amp/dm²**	**2.0 Amp/dm²**	**3.0 Amp/dm²**
2	5.76	2.88	1.44	0.96
4	11.52	5.76	2.88	1.92

5.2 CHROMIUM ELETROPLATING

5.2.1 Bath solution composition

The Chromium electroplating process is carried out by suitable selection of solution. Chemical composition is as shown table 9

Table 9 Bath solution composition of Chromium Eletroplating

	OPTIMUM	**RANGE**
Grobrite Chrome Salt	250 g/l	220 - 300 g/l
Chrome Conducting Salt	5 g/l	4 - 10 g/l

5.2.2 Operating conditions

For the Chromium electroplating process, the following operating condition is considered and is indicated in Table 10.

Table 10 Operating condition of Chromium Electroplating

Density	19 - 23°Be
Temperature	45 - 50°C
Cathode Current Density	10 - 30 A
Anode Current Density	5 - 15 A
Voltage	4 - 12 volts

5.2.3 Solution Preparation

The Grobrite chrome solution was prepared with some precautions. The salt has to be dissolved completely.

The procedure to be followed in it preparation is as follows.

- The auxiliary tank was filled with two-thirds full of deionised water.
- Grobrite chrome salts is added by stirring till all the salt are completely dissolved.
- Activated Carbon was added and stirred for few hours and then allowed to settle for a period of 12 hours.
- This solution was filtered and, transferred to the plating tank and then deionized water is added to make up the level
- The filter unit is cleaned and filtered until the solution becomes clear.
- A calculated quantities of Argolume Carrier additive and Argolume brightener were added and stirred. The prepared solution for plating operation as shown in fig13.

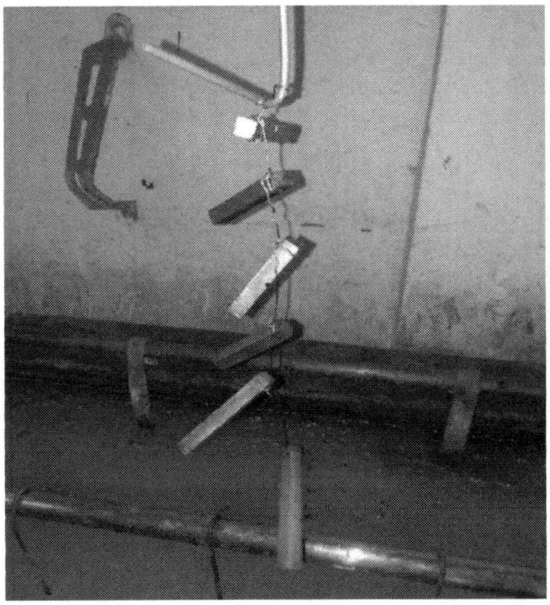

Fig 4: materials before electroplating in to solution bath.

5.2.4 Plating procedure

The pre-treatment procedure before plating in Chromium plating is the same as hexavalent chromium plating. A Chromium strike is required to assure good bonding of Chromium deposits to copper, brass and copper based alloys. A Strike Chromium bath composition is shown in below table 11. A Chromium electroplating as per the double layer 2μ and 4μ is shown in the figure 16 a & 16 b.

Fig.16. (a)

Fig.16. (b)

Fig.4. (a): 2 micron chromium electroplating of Izod and Charpy specimens.
Fig.4. (b): 4 micron chromium electroplating of Izod and Charpy specimens.

5.2.5 Deposit Thickness inTime (in min)

The deposition of electroplating thickness of a material w.r.t. time (in min) is shown in table 11.

Table 11 Deposit Thickness in Time (in min)

	Current density Amp/dm² with time in minutes			
Thickness (Mic)	**8.0 Amp/dm²**	**16.0 Amp/dm²**	**25.0 Amp/dm²**	**31.0 Amp/dm²**
2	28.0	12.0	8.0	6.0
4	56.0	24.0	16.0	12.0

5.3 Nickel Electroplating

5.3.1 Solution composition

The Nickel electroplating process is carried out by selection of solution. Chemical composition is as shown table 12.

Table 12 Bath solution composition of nickel electroplating

	Optimum	**Range**
Nickel Salt	350 g/l	300 - 400 g/l
Additive22M	6. 0 ml/l	4 - 10 ml/l
Spectra 999	0.3 ml/l	0.2 - 0.4 ml/l
Superglo A	1.0 ml/l	0.8 - 1.6 ml/l

5.3.2 Operating condition

In Nickel electroplating process the following operating condition is maintained as per standard of electroplating is as shown in table13

Table 13 operating condition of Nickel Electroplating

	Optimum	**Range**
Temperature	55°C	50-65°C
pH (Electrometric)	4.2	3.8-4.5
Cathode Current Density Amp	4.0	2.5-8.0
Anode Current Density Amp	2.0	1.0-3.0
Voltage	5.5	4.0-12.0
Agitation	Low pressure air /cathode rod agitation	
Filtration	Continuous	changes per hour
Anodes	Electrolytic nickel rolled anodes, nickel rounds in titanium anode baskets	

5.3.3 Solution preparation

Fresh Nickel solution is prepared as under

- The rubber lined tank was filled with two-thirds full of deionised water.
- Bright nickel salt is added by stirring till all the salt are completely dissolved.
- Activated Carbon was added and stirred for stir the solution vigorously for 2 hours and then allowed to settle for a period of 12 hours.
- This solution was filtered and, transferred to the plating tank and then deionized water is added to make up the level
- The filter unit is cleaned and filtered until the solution becomes clear.
- A calculated quantities of Argolume Carrier additive and Argolume brightener were added and stirred. The prepared solution for plating operation as shown in fig5.
-

Fig 5: materials before electroplating in to solution bath.

5.3.4 Nickel plating tank

The plating tank should be of mild steel and should be lined with an approved grade of semi hard or hard rubber. The quality of rubber is very important for good performance of solution and hence it is advisable to consult the Growel Laboratory / Local Representatives before lining the tank. Thermal lagging is recommended though not essential, as it is already insulated inside by rubber lining.

5.3.5 Plating procedure

The pre-treatment procedure before Nickel plating in Nickel plating is A Nickel strike is required to assure good bonding of Nickel deposits to copper, brass and copper based alloys. A Strike Nickel bath composition is shown in below table 11. A Chromium electroplating as per the double layer 2μ and 4μ is shown in the figure 5 a & 5 b.

Fig.18. (a) Fig18. (b)

Fig.5 (a) : 2 micron nickel electroplating of Izod and Charpy specimens.
Fig.5.(b) : 4 micron nickel electroplating of Izod and Charpy specimens.

5.3.6 Deposit Thickness in Time (in min)

The deposition of electroplating thickness of a material w.r.t. time (in min) is shown in table 14

Table 14 Deposit Thickness in Time (in min)

	Current density Amp/dm² with time in minutes			
Thickness (Mic)	**1.0 Amp/dm²**	**1.5 Amp/dm²**	**2.0 Amp/dm²**	**2.5 Amp/dm²**
2	10.4	6.8	5.2	4.4
4	20.8	13.6	10.4	8.8

5.4 ZINC ELECTROPLATING PROCESS

5.4.1 Solution make-up

The Zinc electroplating process is carried out by selection of solution. Chemical composition is as shown table 15

Table 15 Bath solution of Zinc Electroplating

	Optimum	Range
Zincalite NCZ 106 salt	120 g/l	110 - 130 g/l
Uniglo 2012 BrightenerA	10.0 ml/l	8 - 12 A ml/l
Uniglo 2012 Brightener B	2.0 ml/l	1.5 - 2.5 ml/l
Uniglo 2012 Brightener C	2.0 ml/l	1.5 - 2.5 ml/l

5.4.2 Operating conditions

In Zinc Electroplating process the fallowing operating conditions is maintained as per standards of Electroplating is as shown in table 16.

Table 16 Operating condition of Zinc Electroplating

	OPTIMUM	RANGE
Zinc Metal	10 g/l	7.0 - 15.0 g/l
Sodium Hydroxide	120 g/l	100 - 150 g/l
Temperature	25 °C	20 – 40 °C
Current Density	2 Amp/dm²	1 - 3 Amp/dm²
Voltage	5 Volt	3 - 6 Volt
Current Loading	0.7 Max Amp/l	
Anode to Cathode Area	1 : 1 to 2 : 1	

5.4.3 Solution preparation

To make-up 100 litres of solution add 12 kgs. Zincalite NCZ 106 salts.

1. To a 100 litre storage tank add about 25 litres of cold water.
2. Add the salt to the cold water. Heat is generated by the dissolution of salt and it is recommended that salt is added in small increments with continuous stirring. Wear protective clothing and shield the face when handling the salt.
3. Thoroughly stir to dissolve the solid and add slowly partial amount of water.
4. Allow this solution to settle for 2 to 3 hours and then filter the solution into the plating tank.
5. All the above procedures can be avoided while using Uniglo Make-Up solution. In this case, it is recommended to take 25 lit. of Uniglo Make-Up solution and then dilute it to 100 lit.
6. Add water to make-up for the final volume. Stir thoroughly and analyse for zinc metal and Caustic soda.
7. Add Uniglo 2012 brightener components as follows
   Uniglo 2012 Brightener A - 10.0 ml/l
   Uniglo 2012 Brightener B - 2.0 ml/l
   Uniglo 2012 Brightener C - 2.0 ml/l
8. Electrolyze the bath at 0.05 amp/l for 4 to 8 hours.
9. Use the solution for normal plating, with regular brightener replenishment.

Fig 6: materials before electroplating in to solution bath

5.4.4 Plating procedure

The quality of deposit obtained depends upon the proposed pre-treatment sequence. Use of non-ionic urfactants is highly recommended throughout the preparation sequence. Frequent checks of cleaner, acid pickle concentrations and temperature are recommended. An alkaline dip containing 25 to 45 g/l of sodium hydroxide is highly recommended immediately prior to the plating tank to neutralize any acid film on the surface of the metal.

Process Sequence:

For steel, the following sequence is recommended:

Immersion clean- Groclean MS / Steelex K-20 / Metaprep Act

Pickle- Hydrochloric acid (50%)

Electro clean- Ginbond 808 Activation- Hydrochloric Acid (5%) (only when necessary).

The figure 20 shows the after completion of zinc electroplating as per the double layer 2μ and 4μ.

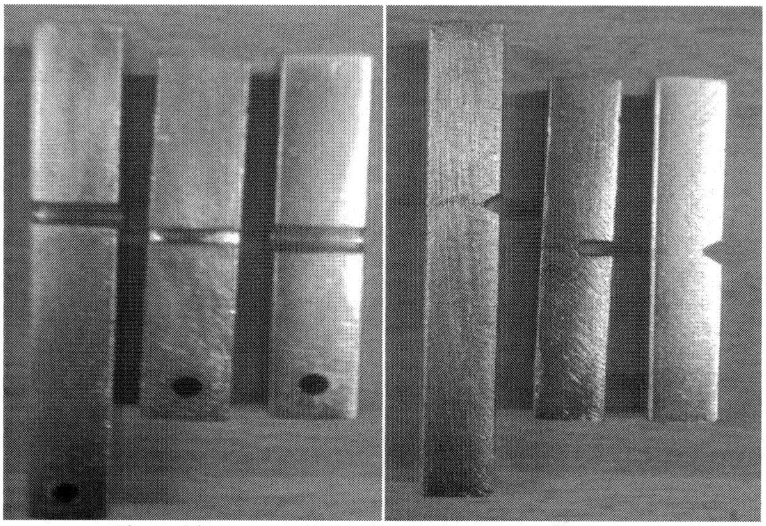

Fig..6 (a) Fig.6 (b)
Fig.6 (a): 2 micron zinc electroplating of Izod and Charpy specimens.
Fig.6. (b): 4 micron zinc electroplating of Izod and Charpy specimens.

5.4.5 Deposit Thickness in Time

The deposition of electroplating thickness of a material w.r.t. time (in min) is shown in table 17.

Table 17: Deposit Thickness Time (in min)

	Current density Amp/dm² with time in minutes			
Thickness (Mic)	**1.0 Amp/dm^2**	**1.5 Amp/dm^2**	**2.0 Amp/dm^2**	**2.5 Amp/dm^2**
2	7.4	5.8	4.2	3.4
4	14.8	11.6	8.4	6.8

5.5 Collection of Data and Analysis

The Impact test has been carried out by selection of different Electroplating materials with varied thickness of 2μ & 4μ,these specimens were prepared as per ASTME23 Standards with both 'U' &'V' notche specimens is as shown in table 18.

Table 18: Information table of experimental work

Sl.no	Type of test specimen	Types of notch	Thickness of plating	Electroplating materials
1	Charpy (55 x10x10 mm)	U-notch V-notch	Double layer coating with thickness of "2μ" & "4μ"	1. Nickel (ni) 2. Silver (ag) 3. Chromium (cr) 4. Zinc (zi)
2	Izod (75x10x10 mm)	V-notch	Double layer coating with thickness of "2μ" & "4μ"	1. Nickel (ni) 2. Silver (ag) 3. Chromium (cr) 4. Zinc (zi)

The Impact test has been carried out by Charpy Impact test method by different materials condition with varied Electroplating material thickness.

IOP Conf. Series: Materials Science and Engineering 1065 (2021) 012012 doi:10.1088/1757-899X/1065/1/012012

There are 36 Charpy Impact test specimen is carried out to determine its Fracture Toughness strength is as shown in table 19

Table 19 shows the number specimens used for the Charpy impact testing by 2µ 4µ

Type of Notch	Material Condition								
	Normal	Nickel		Silver		Chromium		Zinc	
		2µ	4µ	2µ	4µ	2µ	4µ	2µ	4µ
'U'-notch	2	2	2	2	2	2	2	2	2
'V'-notch	2	2	2	2	2	2	2	2	2
Total	4	4	4	4	4	4	4	4	4
Grand total = 36 specimens									

There are 18 Izod Impact test specimen is carried out to determine its Fracture Toughness strength is as shown in table 20.

Table 20: shows the number specimens used for the Izod impact testing by 2µ & 4µ

Type of Notch		Material Condition							
	Normal	Nickel		Silver		Chromium		Zinc	
		2µ	4µ	2µ	4µ	2µ	4µ	2µ	4µ
'V'-notch	2	2	2	2	2	2	2	2	2
specimens		2	2	2	2	2	2	2	2
Total	2	2	2	2	2	2	2	2	2
Grand total = 18 Specimens									

As per ASTME23 Standards, the Normal specimens was maintained by 'U'-notch 2 specimen & 'V'-notch 4 specimens, in the same Electroplating materials of Nickel 'U'-notch for 2µ is 2 testing specimens & for 4µ is 2 testing specimens, similarly for Nickel 'V'-notch for 2µ is 4 testing specimens & for 4µ is 4 testing specimens. In the same Electroplating materials of Silver 'U'-notch for 2µ is 2 testing specimens & for 4µ is 2 testing specimens, similarly for Silver 'V'-notch for 2µ is 4 testing specimens & for 4µ is 4 testing specimens. In the same Electroplating materials of Chromium 'U'-notch for 2µ is 2 testing specimens & for 4µ is 2 testing specimens, similarly for Chromium 'V'-notch for 2µ is 4 testing specimens & for 4µ is 4 testing specimens. In the same Electroplating materials of Zinc 'U'-notch for 2µ is 2 testing specimens & for 4µ is 2 testing specimens, similarly for Zinc 'V'-notch for 2µ is 4 testing specimens & for 4µ is 4 testing specimens.

Total specimens used for Charpy Impact testing =36

Total specimens used for Izod Impact testing =18

There are total impact testing specimens = 54, is as shown in table 25.

EXPERIMENTAL TABLE 21: Fracture toughness of different material condition

Sl no	Material condition	No of specimens	Trial. no	Impact energy (in J/mm²) Charpy test V notch	Charpy test U notch	Izod test V notch
			1	1.631	1.756	1.298
1	normal	6	2	2.031	1.767	1.201
			1	2.083	1.546	1.125
2	nickel (2 µ)	6	2	1.798	1.672	1.143
			1	2.619	2.080	1.557
3	nickel (4 µ)	6	2	2.590	2.121	1.483

			1	1.916	1.266	1.202
4	silver (2 μ)	6	2	1.872	1.217	1.244
			1	1.918	1.818	1.462
5	silver (4 μ)	6	2	1.929	1.343	1.564
			1	1.713	1.443	1.371
6	Chromium(2 μ)	6	2	1.705	1.347	1.198
			1	1.909	1.961	1.61
7	chromium (4 μ)	6	2	1.741	2.210	1.466
			1	1.907	1.320	1.106
8	zinc (2 μ)	6	2	1.813	1.515	1.04
			1	2.111	1.689	1.500
9	zinc (4 μ)	6	2	2.376	1.7189	1.421
10	total no of specimens	54				

As per ASTME23 standards the normal specimen was maintained by u notch 2 specimens' v notch 4 specimens in the same electroplating material of nickel v notch for 2 microns 4 testing specimens and u notch 2 testing specimens and for 4 microns 4 testing specimens and u notch 2 testing specimens

Electroplating material of silver v notch for 2 microns 4 testing specimens and u notch 2 testing specimens and for 4 microns 4 testing specimens and u notch 2 testing specimens electroplating material of Chromium v notch for 2 microns 4 testing specimens and u notch 2 testing specimens and for 4 microns 4 testing specimens and u notch 2 testing specimens electroplating material of zinc notch for 2 microns 4 testing specimens and u notch 2 testing specimens and for 4 microns 4 testing specimens and u notch 2 testing specimens. As shown in table 25 and the total number of specimens are tested is 54.

5.6 CHARPY IMPACT TEST RESULTS OF 'U'-NOTCH NORMAL, NICKEL, SILVER CHROMIUM AND ZINC 2μ SPECIMEN.

The Charpy impact test is made for 'U' notch double layer 2μ as per ASTME23 Standard and different electroplating materials is as shown in table 22.

Table 22: Impact strength of different electroplating materials.

Electroplating material	Impact strength
Normal	1.5
Nickel	1.609
Silver	1.241
Chromium	1.395
Zinc	1.417

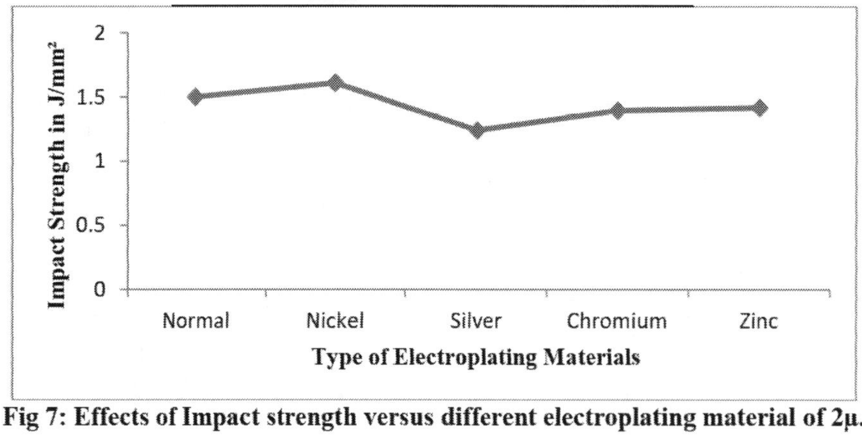

Fig 7: Effects of Impact strength versus different electroplating material of 2μ.

Above figure represents variation of Impact strength versus different Electroplating materials. The figure reveals that the thickness of electroplating by 2μ means double layer. If thickness of Electroplating material increases the Impact strength also increases. In this figure Nickel Electroplating have more Impact strength 1.609 J/mm² to other Electroplating materials, the lowest minimum Impact strength is Silver Electroplating material is 1.241 J/mm².

5.7 CHARPY IMPACT TEST RESULTS OF 'U'-NOTCH NICKEL, SILVER, CHROMIUM AND ZINC 4μ SPECIMEN.

The Charpy impact test is made for 'U' notch double layer 4μ as per ASTME23 Standard and different electroplating materials is as shown in table 23.

Table 23: Impact strength of different electroplating materials.

Electroplating material	Impact strength
Normal	1.5
Nickel	2.1
Silver	1.58
Chromium	2.086
Zinc	1.702

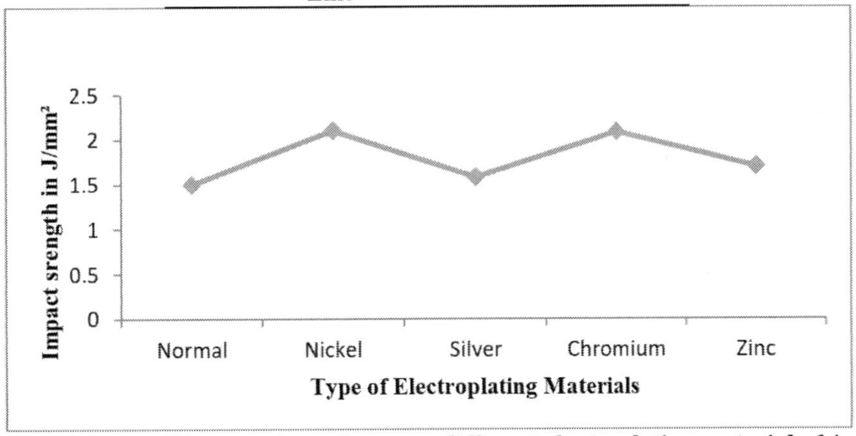

Fig 8: Effects of Impact strength versus different electroplating material of 4μ.

Above figure represents variation of Impact strength versus different Electroplating materials. The figure reveals that, the thickness of electroplating by 4μ means double layer. If thickness of Electroplating material increases the Impact strength also increases. In this figure Nickel Electroplating have more Impact strength 2.1 J/mm² to other Electroplating materials , the lowest Impact strength is Normal material is 1.5 J/mm².

5.8 CHARPY IMPACT TEST RESULTS OF 'V'-NOTCH NICKEL, SILVER, CHROMIUM AND ZINC 2μ SPECIMEN.

The Charpy impact test is made for 'V' notch double layer 2μ as per ASTME23 Standard and different electroplating materials is as shown in table 24.

Table 24: Impact strength of different electroplating materials

Electroplating material	Impact strength
Normal	1.83
Nickel	1.941
Silver	1.894
Chromium	1.709
Zinc	1.86

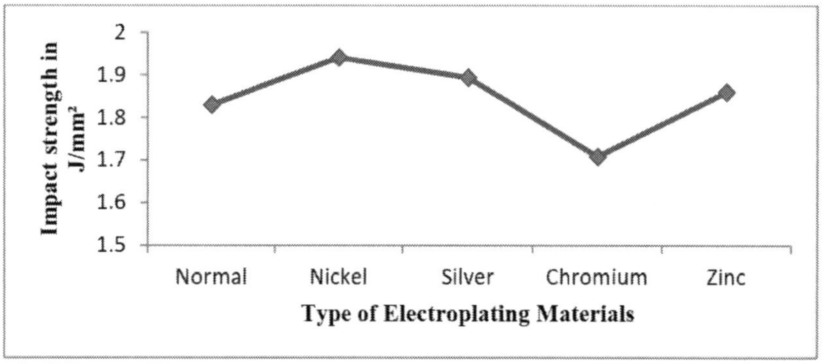

Fig 9: Effects of Impact strength versus different electroplating material of 2μ.

Above figure represents variation of Impact strength versus different Electroplating materials. The figure reveals that the thickness of electroplating by 2μ means double layer. If thickness of Electroplating material increases the Impact strength also increases. In this figure Nickel Electroplating have more Impact strength 1.941 J/mm^2 to other Electroplating materials, the lowest Impact strength is Chromium Electroplating material is 1.709 J/mm^2.

5.9 CHARPY IMPACT TEST RESULTS OF 'V'-NOTCH NICKEL, SILVER CHROMIUM AND ZINC 4μ SPECIMEN.

The Charpy impact test is made for 'V' notch double layer 4μ as per ASTME23 Standard and different electroplating materials is as shown in table 25.

Table 25: Impact strength of different electroplating materials

Electroplating material	Impact strength
Normal	1.83
Nickel	2.605
Silver	1.924
Chromium	1.825
Zinc	2.243

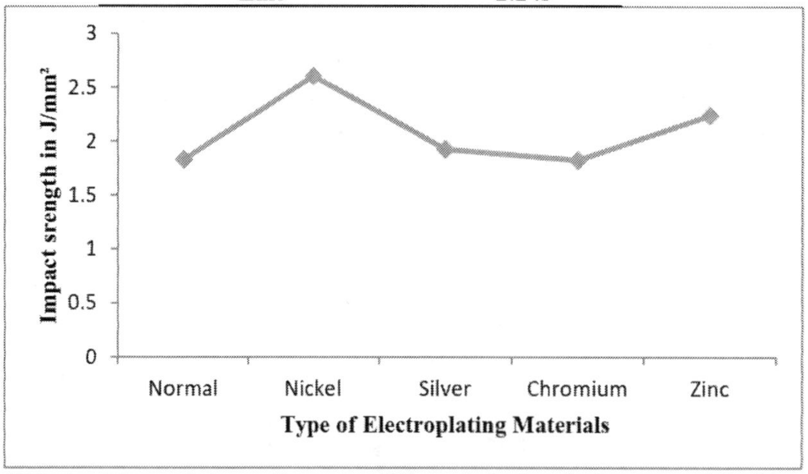

Fig 10: Effects of Impact strength versus different electroplating material of 4μ.

Above figure represents variation of Impact strength versus different Electroplating materials. The figure reveals that the thickness of electroplating by 4μ means double layer. If thickness of Electroplating material increases the Impact strength also increases. In this figure Nickel Electroplating have more Impact strength 2.605 J/mm² to other Electroplating materials, the lowest Impact strength is Chromium Electroplating material is 1.825 J/mm².

5.10 IZOD IMPACT TEST RESULTS OF 'V'-NOTCH NICKEL, SILVER CHROMIUM AND ZINC BY 2μ SPECIMEN.

The Izod impact test is made for 'V' notch double layer 2μ as per ASTME23 Standard and different electroplating materials is as shown in table 26.

Table 26: Impact strength of different electroplating materials

Electroplating material	Impact strength
Normal	1.249
Nickel	1.163
Silver	1.223
Chromium	1.284
Zinc	1.073

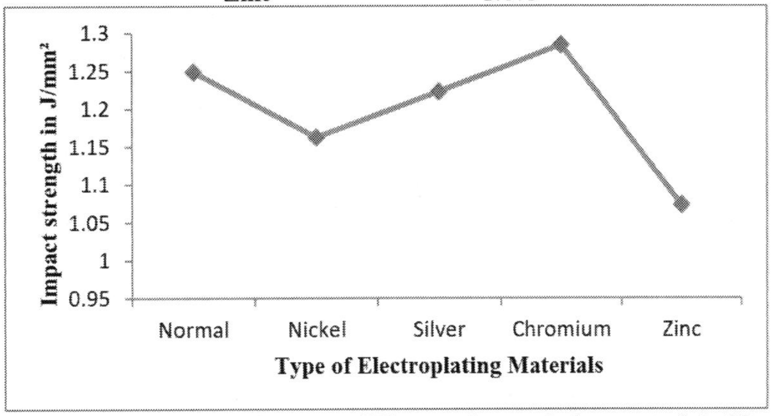

Fig 11: Effects of Impact strength versus different electroplating material of 2μ.

Above figure represents variation of Impact strength versus different Electroplating materials. The figure reveals that the thickness of electroplating by 2μ means double layer. If thickness of Electroplating material increases the Impact strength also increases. In this figure Chromium Electroplating have more Impact strength 1.284 J/mm² to other Electroplating materials, the lowest Impact strength is Zinc Electroplating material is 1.073 J/mm².

5.11 IZOD IMPACT TEST RESULTS OF 'V'-NOTCH NICKEL, SILVER CHROMIUM AND ZINC BY 4μ SPECIMEN.

The Izod impact test is made for 'V' notch double layer 4μ as per ASTME23 Standard and different electroplating materials is as shown in table 27.

Table 27: Impact strength of different electroplating materials

Electroplating material	Impact strength
Normal	1.249
Nickel	1.52
Silver	1.513
Chromium	1.538
Zinc	1.461

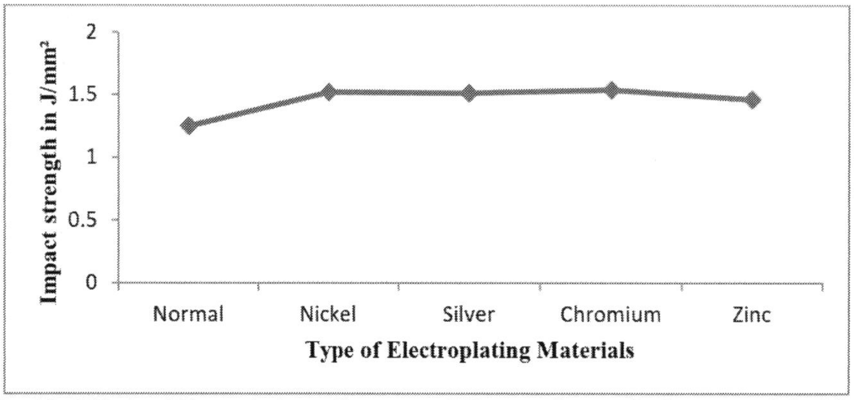

Fig 12: Effects of Impact strength versus different electroplating material of 4µ.

Above figure represents variation of Impact strength versus different Electroplating materials. The figure reveals that the thickness of electroplating by 4µ means double layer. If thickness of Electroplating material increases the Impact strength also increases. In this figure Chromium Electroplating have more Impact strength 1.538 J/mm² to other Electroplating materials, the lowest Impact strength is Normal material is 1.249 J/mm².

5.12 CHARPY IMPACT TEST RESULTS OF 'U'-NOTCH NICKEL, SILVER CHROMIUM AND ZINC BY 2µ & 4µ SPECIMEN.

The Charpy impact test is made for 'U' notch double layer 2µ and 4µ as per ASTME23 Standard and different electroplating materials and a combined test result is obtained as shown in table 28.

Table 28: Impact strength of different electroplating materials

	Impact strength	
Electroplating material	2µ	4µ
Normal	1.5	1.5
Nickel	1.609	2.1
Silver	1.241	1.58
Chromium	1.395	2.086
Zinc	1.417	1.702

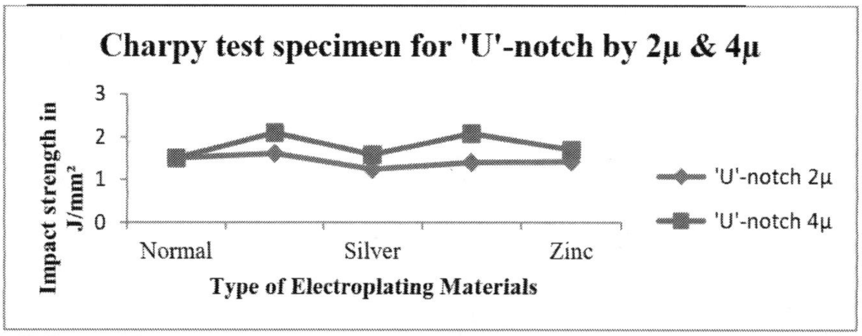

Fig 13: Effects of Impact strength versus different electroplating material of 2µ &4µ .

Above figure represents variation of Impact strength versus different Electroplating material by 22µ & 4µ conducting Charpy test for 'U' notch. Above figure reveals that increase the thickness of electroplating material, impact strength is also increases. In this figure 4µ Electroplating material having more Impact strength compared to 2µ electroplating material. Here the Nickel electroplating material having more impact strength compared to other materials, in the same way 4µ electroplating material nickel having more impact strength of 2.1 J/mm² to other Electroplating materials, the lowest Impact strength is Silver Electroplating of 2μ is 1.241 J/mm².

5.13 CHARPY IMPACT TEST RESULTS OF 'V'-NOTCH NICKEL, SILVER CHROMIUM AND ZINC BY 2μ & 4μ SPECIMEN.

The Charpy impact test is made for 'V' notch double layer 2μ and 4μ as per ASTME23 Standard and different electroplating materials and a combined test result is obtained as shown in table 29.

Table 29: Impact strength of different electroplating materials

	Impact strength	
Electroplating material	2μ	4μ
Normal	1.83	1.83
Nickel	1.941	2.605
Silver	1.894	1.924
Chromium	1.709	1.825
Zinc	1.86	2.243

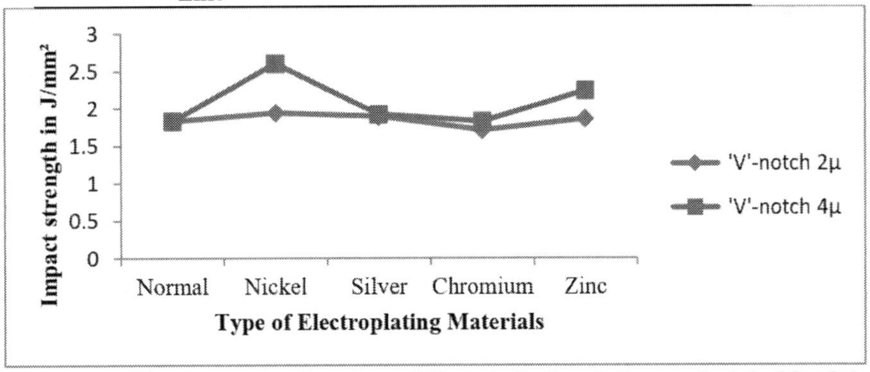

Fig 14: Effects of Impact strength versus different electroplating material of 2μ &4μ.

Above figure represents variation of Impact strength versus different Electroplating material by 2μ & 4μ conducting Charpy test for 'V' notch. Above figure reveals that increase the thickness of electroplating material, impact strength is also increases. In this figure 4μ Electroplating material having more Impact strength compared to 2μ electroplating material. Here the Nickel electroplating material having more impact strength compared to other materials, In the same way 4μ electroplating material nickel having more impact strength of 2.605 J/mm² to other Electroplating materials, the lowest Impact strength is Chromium Electroplating of 2μ is 1.709 J/mm².

5.14 IZOD IMPACT TEST RESULTS OF 'V'-NOTCH NICKEL, SILVER CHROMIUM AND ZINC BY 2μ & 4μ SPECIMEN.

The Izod impact test is made for 'V' notch double layer 2μ and 4μ as per ASTME23 Standard and different electroplating materials and a combined test result is obtained as shown in table 30.

Table 30: Impact strength of different electroplating materials

	Impact strength	
Electroplating material	2μ	4μ
Normal	1.249	1.249
Nickel	1.163	1.52
Silver	1.223	1.513
Chromium	1.284	1.538
Zinc	1.073	1.461

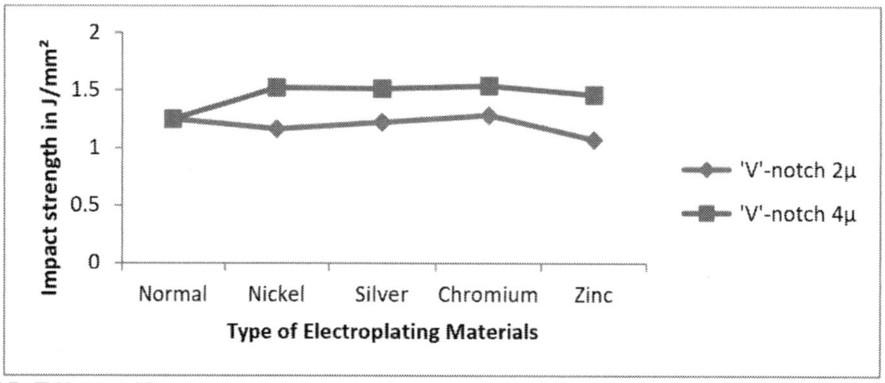

Fig 15: Effects of Impact strength versus different electroplating material of 2μ &4μ

Above figure represents variation of Impact strength versus different Electroplating material by 2μ & 4μ conducting Izod test for 'V' notch. Above figure reveals that increase the thickness of electroplating material, impact strength is also increases. In this figure 4μ Electroplating material having more Impact strength compared to 2μ electroplating material. Here the Chromium electroplating material having more impact strength compared to other materials, in the same way 4μ electroplating material Chromium having more impact strength of 1.538 J/mm² to other Electroplating materials, the lowest Impact strength is Zinc Electroplating of 2μ is 1.073 J/mm².

5.15 Summary representation of different electroplating material with different thickness

Electroplating material	Impact strength					
	CHARPY				IZOD	
	'U'-notch 2μ	'U'-notch 4μ	'V'-notch 2μ	'V'-notch 4μ	'V'-notch 2μ	'V'-notch 4μ
Normal	1.5	1.5	1.83	1.83	1.249	1.249
Nickel	1.609	2.1	1.941	2.605	1.163	1.52
Silver	1.241	1.58	1.894	1.924	1.223	1.513
Chromium	1.395	2.086	1.709	1.825	1.284	1.538
Zinc	1.417	1.702	1.86	2.243	1.073	1.461

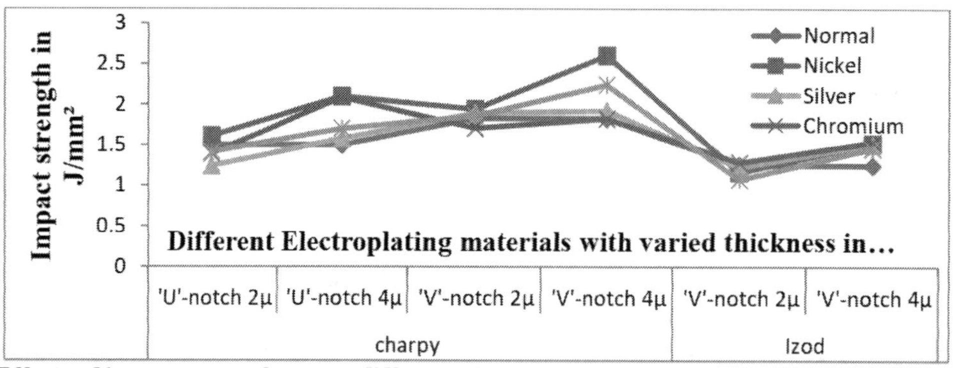

Fig 16: Effects of impact strength versus different electroplating material with varied thickness in microns

The above figure represents the variation of impact strength versus different electroplating materials and different coating conditions 2μ and 4μ for 'U' and 'V' notch specimen. Above figure reveals that Nickel having more strength by double layer compared to the other 2μ electroplating materials that means in 4μ double layered Nickel having more impact strength 2.605J/mm² (4μ), in Zinc electroplating material having least impact strength 2.243 J/mm² (4μ), compared to 2μ Nickel having impact strength 1.94 J/mm² and least impact strength of 2μ is Zinc it means 1.073 J/mm² The overall summary of figure represents Nickel having more impact strength than other electroplating material but least impact strength is Zinc electroplating material.

CONCLUSIONS

- Since fracture toughness behaviour of mode Ī is similar to the behaviour of impact energy. We can say Charpy impact testing is the best method for determination of fracture toughness of EN 9 & EN 31 grade steels both the Martials having higher corrosion resistance.
- As increases the electroplating thickness from 2 microns to 4 microns (double layers) impact strength is increases.
- As electroplating of thickness decreases its impact strength is also decreases.
- Since thickness of 4μ electroplating material having more fracture toughness value compared to thickness 2μ electroplating material.
- Up to critical value of thickness EN 9 & EN 31 graded steel can be best used as structural material for aerospace, marine, automobiles and other engineering applications

REFERENCES

[1] Nicholas Gates, Ali Fatemi (2016) *"Fatigue crack growth behavior in the presence of notches and multiaxial nominal stress states"*, science direct, Engineering Fracture Mechanics vol. **165,** pages 24-38.

[2] Jianhui Lin, Chong Wang, Shouxu Wang, Yuanming Chen, Wei He and Dingjun Xiao. (2016) *"Initiation electroless nickel plating by atomic hydrogen for PCB final finishing"*, Science Direct, **vol.306,** pages 117-123.

[3] Sanjib Kundu, Suman Kalyan Das and Prasanta Sahooa.(2014) *"Properties of electroless nickel at elevated temperature-a review"*, science direct.vol.97.

[4] Tanu Sharma, Ralf Bruning, Tang Cam Lai Nguyen, Tobias Bernhard and Frank Bruning.(2015) *"Properties of electroless Cu films optimized for horizontal plating as a function of deposit thickness"*, science direct **vol.140**, pages 38-46.

[5] Popoola, A. P. I., Loto, C. A., Osifuye, C. O., Aigbodion, V. S. and Popoola, O. M. (2016) *"Corrosion and wear properties of Ni-Sn-P ternary deposits on mild steel via electroless method"*, Alexandria Engineering Journal **vol.55,** pages 2901-2908.

[6] Mohan Kumar, S., Pramod, R., Shashi Kumar, M. E., Govindaraju, H. K. (2014) *"Evaluation of Fracture Toughness and Mechanical Properties of Aluminum Alloy 7075, T6 with Nickel Coating"*, Science Direct: Procedia Engineering **vol.97,** pages 178 – 185.

[7] Ming-Dong Wei, Feng Dai, Nu-Wen Xu, Tao Zhao (2016) **"***Stress intensity factors and fracture process zones of ISRM-suggested chevron notched specimens for mode I fracture toughness testing of rocks"* science direct, vol **168,** pages 174-189

[8] Xian-Kui Zhu (2016) **"***Review of fracture toughness test methods for ductile materials in low-constraint conditions"* science direct, vol **139-140,** pager 173-183

[9] Kazumasa Shimamoto, Yu Sekiguchi, Chiaki Sato (2016) ***"Mixed*** *mode fracture toughness of adhesively bonded joints with residual stress"*, science direct **vol102-103**, pages 120-126

[10] J. Elambasseril, R.N. Ibrahim (2012) *"Validity requirements of circumferentially notched tensile specimens for the determination of the interfacial fracture toughness of coatings"* science direct, vol **43,** pages 2415–2422

IOP Conf. Series: Materials Science and Engineering 1065 (2021) 012013 doi:10.1088/1757-899X/1065/1/012013

Productivity enhancement of Evacuated Tubes solar still of different water depth: Thermal modelling and an Experimental analysis.

Sunil More[1], Joseph Daniel[2]

Research Scholar[1], Associate Professor[2], Department of Mechanical Engineering, School of Mechanical Engineering (SMEC), VIT University-Chennai, Tamilnadu, India

Email-sunil.more33@gmail.com

Abstract. In this research article, the author has conducted the comparative thermal experimental analysis and investigation on Conventional still and Modified Evacuated Tubes solar still with different water volume of 30 and 50 Litre in Basin, in respect of Energy Balance Equation. The various variable elements have been taken into account and duly calculated for both the stills i.e. active still and passive still, based on thermal modelling. The variables calculated comprise inner glass surface temperature, distillate yield, basin water temperature etc. The entire study has been carried out at Akison's Solar Equipments Pvt. Ltd., Shirwal, Maharashtra, India (latitude18.14° N: longitude 73.97° E), with adequate instrumental infrastructure setup appropriate for the purpose. The investigation concluded that there is substantial escalation of. 242% and 246% at 0.03 and 0.05m of water depth respectively in the yields obtained from Modified Evacuated Tubes solar still in comparison of Conventional still. It has been observed that theoretical outcomes from thermal modelling and the outcomes attained from experiments done for simple and modified stills are in respectable consensus.

1. Introduction.

Water establishes the primary component for the endurance of life on planet earth and the reduction of drinkable water source, attributed to various reasons such as rapid increase in population, industrial revolution, climate changes resulting in frequent droughts, underground pollution, is quite alarming. Earth surface comprises 71% of water and out of that, most of it, say around 98.8% of it, is not fit for human consumption. That leaves only 2.5% of total the water available on earth fit for drinking. [1]

There are different methods are available to get distilled water but these various methods of water distillation require fossil fuels and electricity. Solar distillation is one of the simplest and eco-friendly technique which uses solar energy as renewable form of energy and it is also non-polluting. So, it is effectively used in costal and remote areas where electricity is unavailable to reduce water scarcity problem.

Solar Still is an environment friendly green energy device, categorized into Active and Passive Solar Still, in which solar energy is primarily employed for purification of water via distillation practice. The difference between Active and 'Passive Solar Still being that in former, in comparison to the latter, additional thermal energy is also utilized to gain potable water. The outcome of the Passive Solar Still is quite less and to increase the desirable outcomes, a lot of research work has been poured into the issue to develop various techniques.

Dimri et al [2] carried out experimental work on the Conventional solar still with copper, glass and plastic as 'condensing cover material' and came to conclusion that the highest outcome was obtained with copper due to its higher thermal conductivity. Rai and Tiwari [3] performed experiment on 'Solar Still' accompanied with 'Flat Plate Collector' and observed that the productivity got accelerated by 24%. In the experiment carried out by Velmurugan et al.[4], a mini solar pond combined with a Conventional Still was constructed and the observations came out that the outcome rate of water in a Solar Still with sponge and combined with a Mini Solar Pond and Ordinary Solar Still were 3.14 and 4.65 kg/m^2/day respectively, escalating productivity thereby by 48%. Badran et al and Tiris et al [5, 6] utilized a Flat Plate Collector with a Single Basin Solar Still and observed 52% escalation in the production of drinkable water. Dashtban M et al [7] experimented using the Latent Heat Thermal Energy Storage System in a specially designed Weir Type Cascade Solar Still, to increase the outcome. Prabahar et al [8] studied Modified Double Slope Solar Still. An escalation by 14% in outcome was noticed by utilizing Flat Plate Collector', an escalation of 10% with sponge and 17% by sponge and Flat Plate Collector.

H. Kargar Sharif Abad et.al.[9] introduced a Solar Desalination System by using a Pulsating Heat Pipe and noticed 875 ml/(m^2.h) increase in outcome. Z. M. Omara et al, [10] experimented on developed Solar Still with Solar Concentrator. The productivity was increased by 244% and 347% respectively without and with preheating of the brackish water. An inclined solar system was modelled and simulated by Hikmet S.Aybar [11] .The result showed that the system generated distilled water of 3.5-5.4 Litre/m^2 and also maximum temperature of hot water reached to 60^{0}C. Hitesh N. Panchal [12] worked on double basin solar still combined with Vacuum Tubes and Black Granite Gravel placed in basin water to enhance the productivity. Anil Kr. Tiwari et.al [13] performed experiment on Conventional Solar Still for different water volumes with cover inclination angle of 30^0 and also thermal modelling of the system was carried out. The experimental yield was found in good agreement with theoretical values obtained from thermal modelling. Hitesh N. Panchal [14] worked on the Evacuated Tube Solar Still and noticed a considerable increase in the average distillate output. Rahul Dev et.al [15] conducted experiment on an Inverted Absorber Solar Still (IASS) and Conventional Solar Still (SS) at diverse levels of water depth and they found that higher basin water temperature and yield was obtained with IASS as compared with simple still.

Under Natural Circulation Mode. K. Sampathkumar et al [16] experimentally and theoretically studied Evacuated Tube Solar Still. The results concluded that the performance of Evacuated Tube Solar Still is much higher than the Conventional Solar Still throughout the year. It was observed by Selcuk Selimli et al [17] that the productivity was considerably enhanced by about 62.5%, when Solar Vacuum Tube was integrated with Simple Solar Still. An experiment was carried out by Ali.F. Muftaha et al [18] on Stepped Solar Still integrated with internal and external reflectors, external condensers along with absorber materials, which yielded 29% escalation in the daily outcome of modified solar still. K. Sampathkumar [19] worked on Vacuum tube solar still and noticed that daily outcome was increased by 49.7% and it increased further to 59.48% when coupled with black stones and vacuum tubes. Lilian Malaeb et al [20] studied analytically the modified still with rotating drum to optimize the performance and concluded a noticeable escalation in the productivity was observed in comparison of a Conventional Solar Still.

Hence to enhance the yield from Solar Still, various methods are tried by researchers like use of Flat Plate Collector, concentrating collector, heat pipes, reflectors and internal and external condensers, air

IOP Conf. Series: Materials Science and Engineering 1065 (2021) 012013 doi:10.1088/1757-899X/1065/1/012013

blower with solar still. Also provide vacuum condition in still, incorporate the phase change materials, sensible heat storage and Nano materials, wick materials and dyes in basin water. Evacuated tube solar still is more efficient as there is no loss of energy through vacuum. The solar radiations are positively tracked continuously through- out the day due to its cylindrical shape. The incident angle of sunlight on the cylindrical tubes is at 90^0 thought the day; hence the peak absorption is always for it so it's thermal performance is higher than other collectors.

The core objectives of the research are:

- Thermal modelling of Conventional Solar Still and Evacuated Tubes Solar Still with natural circulation for different water depth in Basin.
- Validation of the values like hourly inner glass surface temperature, basin water temperature and distillate yield with Experimental values.
- To evaluate the potential of active solar still by comparing its performance parameters with conventional still.

2. Experiment.

The experiment set was designed and installed at Akison's Solar Systems Pvt. Ltd. at Shirwal (Latitude18.14° N: Longitude 73.97° E), Maharashtra (India) and as shown in Fig.1.For the experiment Solar Still Basin used with inner side painted in black up to 10mm high so that it could absorb maximum solar radiation. It is made up of galvanized iron plates, 1.5mm thick and 1m × 1m area, with bottom and all sides of it are comprised of Rockwool insulation (50mm thick) with thermal conductivity of 0.045W/mK to achieve minimum heat loss from it. A toughened glass (4mm thick) was fixed, at upper side of Solar Still at 19^0 angles, in accordance with horizontal axis (Latitude of Shirwal), which acts as condensing surface. The glass was tightened with solar still by a silicon rubber sealant to prevent the vapour leakage from the system. The condensate water from inclined glass cover was collected in U shape mild steel channel which is located at lower side of solar still. A rubber pipe was utilized to accumulate distillate in measuring jar from collection tray. A hole had been drilled in Solar Still body to attach 'K Type Constantan Thermocouples' to measure basin, water and inner glass cover temperature respectively. The same material and Dimensions was used for active solar still. Total 13 holes of diameter 0.063 m were made at the bottom end of front wall of the Solar Still to attach the one end of same number of Evacuated tubes with specifications of length 1.8m, outer diameter 0.058 m, inner diameter 0.047m and glass thickness of 0.0016m, with the help of rubber gaskets, at an angle of 30^0 with horizontal surface, and whose other ends were fixed at a metal frame connected to the Solar Still stand again at 30^0 angle.

3. Instrumentation.

- For the accurate measurement of radiation from sun incident on the 'Solar Still' surface, a calibrated solarimeter was used with a range of 0-1200 W/m^2 & accuracy of $\pm 1W / m^2$.
- For accurate measurement of water temperature in the basin, inner glass cover and ambient temperature, 'K type Thermocouples 'were used with accuracy of $\pm 0.1^0 C$.
- For measurement of wind velocity, an 'Anemometer' with range of 0-15m/sec & accuracy of $\pm 0.2m/\sec$ was used.
- A plastic jar of 1 litre capacity with an accuracy of $\pm 10ml$ was used to collect the distillate yield.

Extensive experiments were carried out on both Simple solar Still and Active Solar Still with water depths precisely 0.03m and 0.05m during May 2019 from 9 am to 6 pm on clear sunny days.

Dimensions of experimental set up are shown in Table No.1

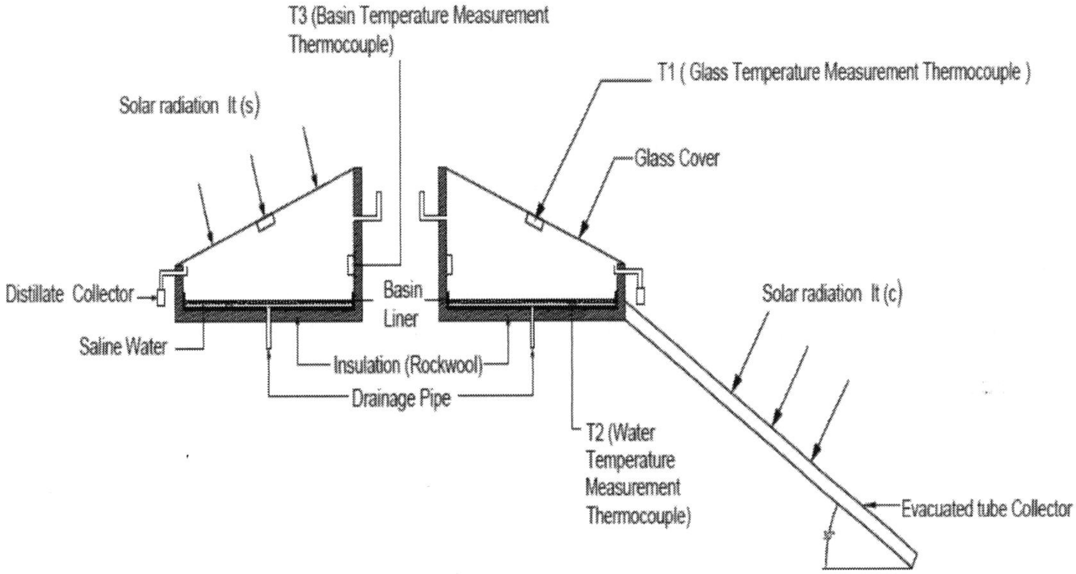

Figure 1. Schematic diagram of the experimental setup.

Table 1. Dimensions of experimental set up

Particulars	value
Simple Solar Still and Active Solar Still Area. (As and Aa)	$1m^2$
Evacuated Tubes Area integrated with Solar Still (A_I)	$3.455\ m^2$
Absorber area of evacuated tubes. (A_{ET})	$1.099m^2$
Solar still outer area.	1 m× 1m
Glass cover inclination to horizontal.	19^0
Length of each evacuated tubes	1.8 m.
Inner diameter of evacuated tubes	0.047 m
Outer diameter of evacuated tubes	0.058 m
Glass thickness in each evacuated tube.	0.0016m

Table 2. List of various 'Constants and Designs' parameters employed in theoretical modelling.

Parameters	Value	Parameters	Value
α_b	0.36	C_w	4190 J /Kg ^{0}K
α_g	0.05	h_w	250 W/m^2K
α_w	0.34	t	3600 s
F_R	0.831	σ	5.67×10^{-8} W/ m^2K
M_w	30 and 50 kg	$(\alpha\tau)_e$	0.8
ε_{eff}	0.82	U_{LC}	2.44 W/ m^2K
L_i	0.05m	h_{tg}	5.7+3.8 V_a, V_a=4 m/sec
K_i	0.045 W/ mK	L	2.25×10^5 J/Kg

4. Thermal modelling:

The Thermal Model of Conventional Solar Still and Evacuated Tubes Solar Still is developed by utilizing First Law of Thermodynamics.

For analysis the following assumptions are made:

- No vapour leakage in Solar Still.
- Glass cover, basin liner and insulation are having negligible heat capacity.
- Steady state condition is reached at every hour.
- Across the glass cover of the Solar Still and Basin water of the 'Solar Still, no temperature gradient exists.
- No change in the water level inside the Solar Still Basin throughout the experiment. It is maintained constant.
- Inside the glass cover, it has been assumed that only film type condensation of water occurs.

4.1 Energy Balance of Glass Cover

$$\alpha'_g I_{effs} + [q_{cw} + q_{ew} + q_{rw}] = q_{cg} + q_{rg}$$

$$\alpha'_g I_{effs} + h_{tw}(T_w - T_g) = h_{tg}(T_g - T_a) \quad (1)$$

The rate at which glass absorbs solar energy and sum of energies transferred from water surface via radiation, convection & evaporation to glass are equivalent to energy lost by glass to atmosphere via convection & radiation.

$$h_{tw} = h_{cw} + h_{ew} + h_{rw}$$

The 'Convective Heat Transfer Coefficient' between water and glass [21] has been given by

$$h_{cw} = 0.884 \times [(T_w - T_g) + \frac{(P_w - P_g) \times (T_w + 273)}{268.9 \times 10^3 - P_w}]^{1/3} \quad (2)$$

The 'Evaporative Heat Transfer Coefficient' between water and glass [21] has been given by

$$h_{ew} = \frac{16.273 \times 10^{-3} \times h_{cw}(P_w - P_g)}{T_w - T_g} \quad (3)$$

The 'Radiative Heat Transfer Coefficient' between water and glass [21] has been given by

$$h_{rw} = \varepsilon_{eff} \times \sigma \times (T_w + 273)^2 + (Tg + 273)^2 + (T_w + T_g + 546) \quad (4)$$

Where ε_{eff} = The 'Effective Emissivity' between the water surface and the glass cover has been taken [22] as 0.82.

The total 'Heat Transfer Coefficient' between glass and atmosphere [23] is given by

$$h_{tg} = 5.7 + 3.8(v) \quad (5)$$

V has been taken as 4 m/sec for numerical calculations.

The 'glass cover temperature' T_g is derived from energy balance equation as

$$T_g = \frac{\alpha'_g I_{effs} + h_{tw} T_w + h_{tg} T_g}{h_{tw} + h_{tg}} \quad (6)$$

4.2 Energy Balance for Basin Liner

The energy transferred to water by convection and conduction energy lost from bottom and sides of the basin are equivalent to the rate of solar energy absorbed by basin

$$\alpha'_b(1-\alpha'_g)(1-\alpha'_w)I_{effs} = q_w + q_b$$

$$\alpha'_b(1-\alpha'_g)(1-\alpha'_w)I_{effs} = h_w(T_b - T_w) + h_b(T_b - T_a) \quad (7)$$

The 'Convective Heat Transfer' between basin and water is given by [4]

$$q_w = h_w(T_b - T_w) \quad (8)$$

The 'Convective Heat Transfer Coefficient' between basin and water is taken as 135 W/m^2 K. [4].
The 'Conductive Heat Transfer' between basin and atmosphere is given by (22)

$$q_b = h_b(T_b - T_a) \quad (9)$$

The 'Basin Liner Temperature' T_b is derived from 'Energy Balance Equation' of the basin liner

$$T_b = \frac{\alpha'-bI_{effs} + h_w T_w + h_b T_b}{h_w + h_b} \quad (10)$$

Where $\alpha'-b = \alpha'_b(1-\alpha'_g)(1-\alpha'_w)$

4.3 Energy balance of water mass for 'Simple Solar Still'

The heat deposited due to solar energy in water and the heat transferred to glass cover is equivalent to the radiation energy absorbed by water mass in basin and convective heat gained from basin liner.

$$\alpha'_w(1-\alpha'_g)I_{effs} + q_w = (MC)_W \frac{dT_W}{dt} + [q_{cw} + q_{ew} + q_{rw}] \quad (11)$$

Substituting Equations 6 and 10 in Equation 11, the following Differential Equation is obtained.

$$\frac{dTw}{dt} + aTw = f(t) \quad (12)$$

Where $a = \frac{UA_{effs}}{(MC)_w}$ & $f(t) = \frac{IA_{effs} + UA_{effs}Ta}{(MC)_w}$

Expressions for UA_{effs} & IA_{effs} are given in the Appendix

To obtain the approximate analytical solution, the following assumptions have been made. The time interval $\Delta t(0 < t < \Delta t)$ is small

a is constant during the time interval Δt

1) The function $f(t)$ is constant i.e. $\overline{f(t)} = f(t)$ for the time interval between 0 and t

In equation 12, by using initial condition at $t = 0$, $Tw(t = 0) = Two$, the expression for water temperature is derived as

$$Tw = \frac{\overline{f(t)}}{a}[1 - e^{-at}] + Twoe^{-at} \quad (13)$$

The following expression is used for deriving the hourly yield:

$$m_{ew} = \frac{h_{e,w-g}(T_w - T_g)}{L} \times 3600 \times A_s \quad (14)$$

4.4 Energy Balance of water mass for Solar Still with evacuated tubes

The Radiation Energy absorbed by water mass in basin and Convective Heat gained from basin liner and heat energy supplied to basin water from Evacuated tubes is equivalent to the heat stored in water and heat transfer to glass cover.

$$\alpha'_w(1-\alpha'_g)I_{effs} + q_w + Q_u = (MC)_W \frac{dT_W}{dt} + [q_{cw} + q_{ew} + q_{rw}] \quad (15)$$

The Thermal Energy supplied to the Solar Still via Evacuated Tubes has been given by [22]

$$Q_u = F_R[\,(\alpha T)_e\ I_{effe} - U_{LE}\ (\frac{A_L}{A_{ET}})\,(T_w - T_a)\,] \quad (16)$$

Substituting equations 6 and 10 in Equation 15, the following differential equation has been obtained.

$$\frac{dTw}{dt} + aTw = f(t)$$

Where $a = \frac{UA_{effe}}{(MC)_w}$ and $f(t) = \frac{IA_{effe} + UA_{effe}Ta}{(MC)_w}$

Expression for UA_{effe} and IA_{effe} are given in the Appendix.

To derive the approximate analytical solution, the same assumptions were made as in the case of Simple Solar Still. By using initial condition at $t = 0$, $Tw(t=0) = Two$

The expression for Water Temperature has been obtained as

$$Tw = \frac{\overline{f(t)}}{a}[1 - e^{-at}] + Twoe^{-at}$$

MATLAB 14' which is primarily used for computing various Heat Transfer Coefficient has been utilized to develop the Thermal Model. In order to compute various Heat Transfer Coefficients, the values of various designs and climatic parameters are prerequisites and which are duly provided in Table 2. The values of Heat Transfer Coefficients are further utilized for computing the theoretical values of inner glass surface temperature, water temperature and hourly yield, by giving initial values of glass and water temperature. The values of solar radiation and ambient temperature have taken with respect to daily observation. Figure 2 represents a flow chart for computation of theoretical values.

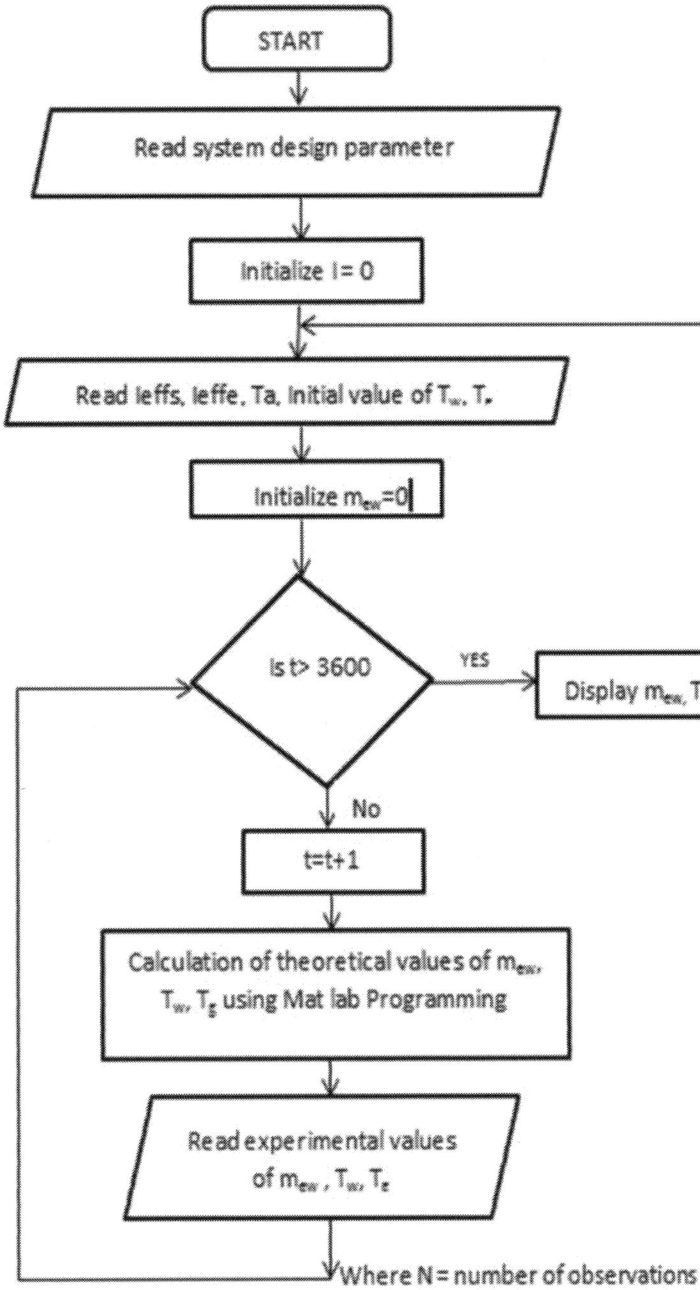

Figure 2. Flow Chart representing 'Computer Model of Solar Still'

Table 3. Ambient (T_a), Water (T_w) and Inner Glass cover Temperature (T_g) for 0.03-meter water depth solar still.

Time of the Day	Ambient Temperature T_a(°C)		Water Temperature T_w (°C)		Inner Glass cover Temperature T_g(°C)	
	Passive Still (31-05-19)	Active Still (29-05-19)	Passive Still (31-05-19)	Active Still (29-05-19)	Passive Still (31-05-19)	Active Still (29-05-19)
9	33	36	40	44	42	45
10	37	39	47	55	49	54
11	40	40	58	69	58	62
12	41	41	65	76	63	70
13	43	42	72	84	67	75
14	42	42	74	88	68	79
15	41	43	72	89	64	81
16	40	42	69	86	59	77
17	35	39	58	82	48	72
18	33	35	53	75	42	63

Table 4. Ambient (T_a), Water (T_w) and Inner Glass cover Temperature (T_g) for 0.05-meter water depth solar still

Time of the Day	Ambient Temperature T_a (°C)		Water Temperature T_w (°C)		Inner Glass cover Temperature T_g(°C)	
	Passive Still (01-06-19)	Active Still (30-05-19)	Passive Still (01-06-19)	Active Still (30-05-19)	Passive Still (01-06-19)	Active Still (30-05-19)
9	32	35	31	41	35	40
10	33	36	37	49	42	47
11	35	38	45	60	46	55
12	36	39	53	70	55	63
13	37	40	59	78	58	71
14	40	41	64	83	60	75
15	41	43	**65**	85	59	78
16	39	41	64	**86**	56	77
17	38	40	61	82	53	73
18	36	35	57	75	49	65

5. Result and Discussion.

In this experiment, various values derived from Theoretical Modelling were tested, verified and validated with experimental results for various water depths for both Active and Passive Solar Still. The hourly variations of water and glass temperatures for Evacuated Tubes Solar Still and Conventional Solar Still with 0.03 m water depth have been demonstrated in Figure 3 and 4, wherein it has been revealed that the maximum temperature of water (89 °C) and glass (81 °C) are obtained in Evacuated Tubes Solar Still at 15 hours, which are considerably higher than the Simple Solar Still's water (74 °C) and glass (68 °C) temperatures at 14 hours for the same water depth.

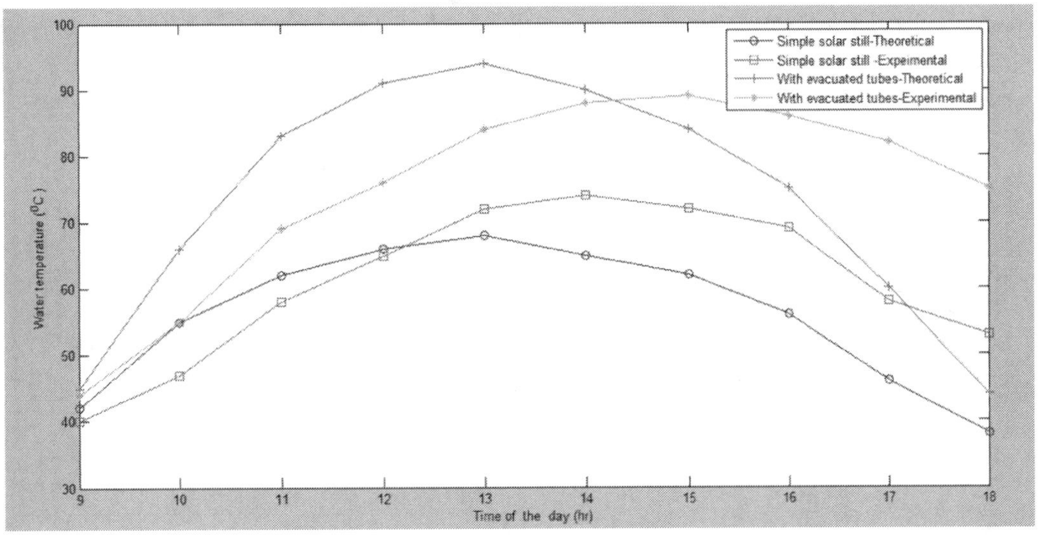

Figure 3. Hourly variation of theoretical and experimental water temperature for 0.03 m water depth solar still.

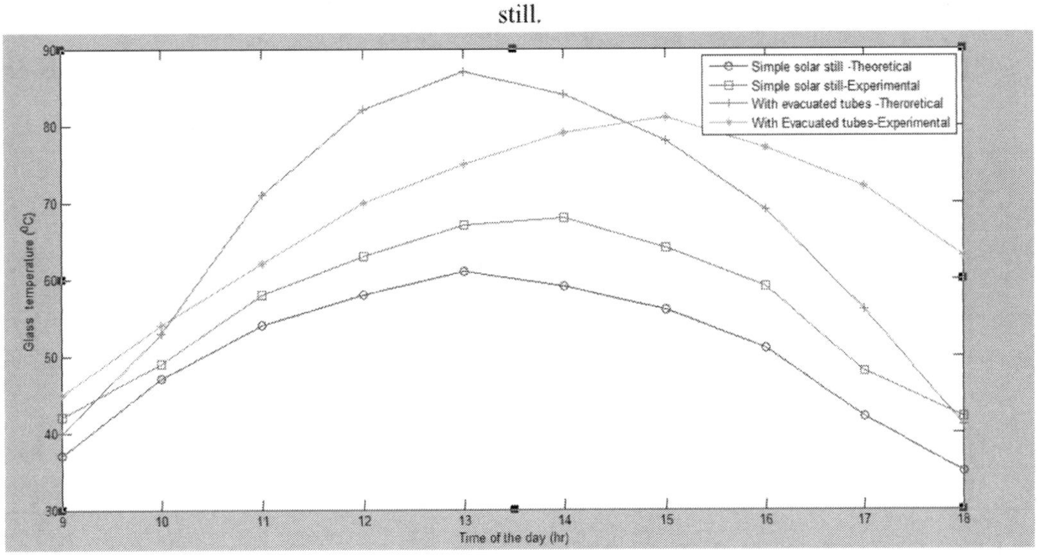

Figure 4. Hourly variation of theoretical and experimental glass temperature for 0.03 m water depth solar still.

From figure 5 and 6 ,we can perceive that the maximum temperatures of water (86° C) and glass (77° C) have been obtained in Evacuated Tubes Solar Still at 16 hours, which are considerably greater than the Simple Solar Still water(65°C) and glass (59 °C) temperatures at 15 hours for 0.05 m water depth. Water and glass temperature values are higher in Evacuated Tubes Solar Still for both the water depths. It is because the Evacuated Tubes provide extra thermal energy to basin water. Also it could be observed from figure 1 to 4 that the theoretical values of water and glass temperatures support the corresponding values with experimental results and are in good agreement.

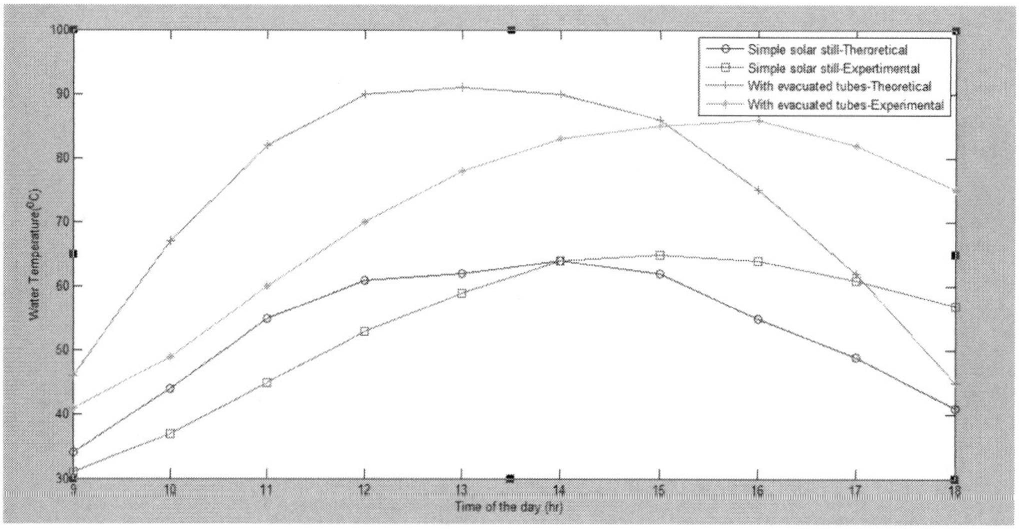

Figure 5. Hourly variation of theoretical and experimental water temperature for 0.05 m water depth solar still.

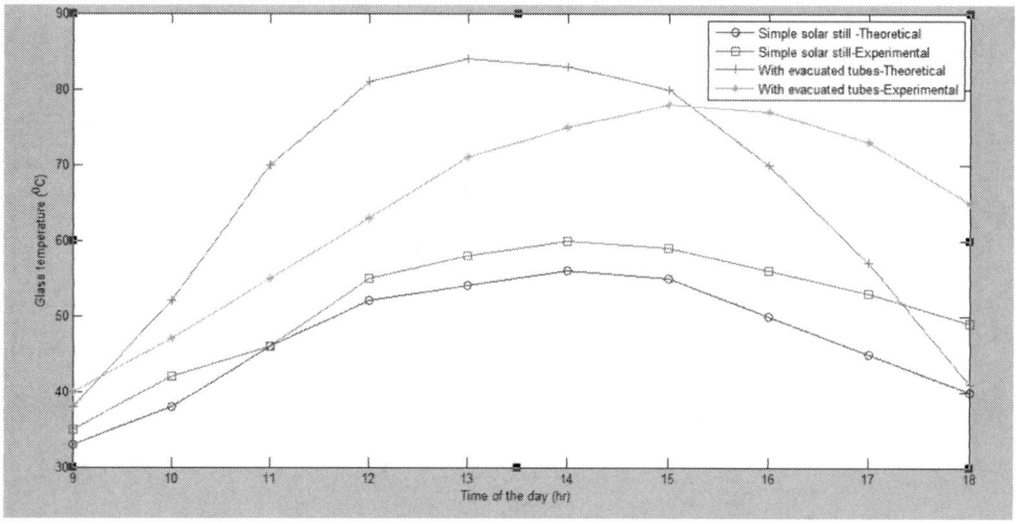

Figure 6. Hourly variation of theoretical and experimental glass temperature for 0.05 m water depth solar still.

It also has been discovered that in early hours, the hourly theoretical values of glass and water temperatures are higher, as in theoretical calculations heat losses are not considered. But in the evening hours, the experimental values of water and glass cover temperature are higher as they hold the heat as stored energy. It also has been observed that as the water depth increased from 0.03 to 0.05 m for both Active and Passive Solar Still, the maximum water and glass temperature attainment time was shifted to 1 hour later as more time is required to achieve maximum temperature for more water mass.

Figure 7 and 8 reveals the hourly variation of convective, evaporative, radiative heat transfer coefficients, respectively, for both Simple Solar Still and Evacuated Tubes Solar Still of 30 Litre water in Basin. The results show that the Evaporative Heat Transfer Coefficient escalates with time and in the afternoon hours a decreasing trend is observed in its value attributed to lower temperatures in the evening.

It can be duly observed from figures 7 and 8 that, the values of Evaporative (137W/m^{2}K), Convective (3.07 W/m^{2}K) and Radiative (8.93W/m^{2}K) heat transfer coefficient are high in Solar Stills integrated with Evacuated Tubes than the Simple Solar Still having values of Evaporative (41W/m^{2}K), Convective (2.30 W/m^{2}K) and Radiative (7.16W/m^{2}K) heat transfer coefficient respectively. Due to higher temperature difference between water and glass in the Evacuated Tubes Solar Still attributed to additional thermal energy provided by evacuated tubes, the values of different heat transfer coefficients are higher in Evacuated Tubes Solar Still.

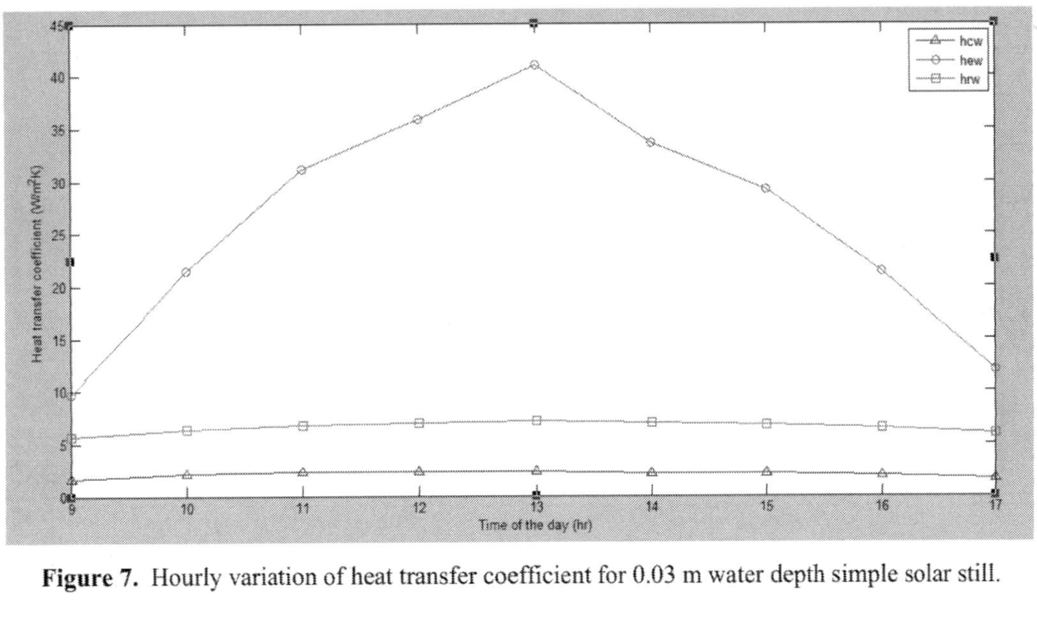

Figure 7. Hourly variation of heat transfer coefficient for 0.03 m water depth simple solar still.

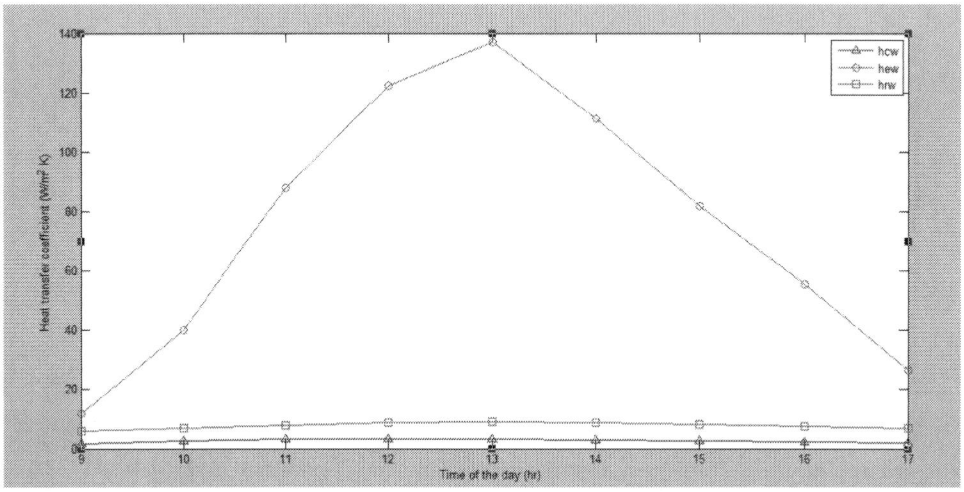

Figure 8. Hourly variation of heat transfer coefficient for 0.03 m water depth solar still with evacuated tubes.

Figure 9 and 10 represents the hourly variation in respect of yield obtained from Simple Solar Still and Active Solar Still for 0.03 m and 0.05 m water depths respectively. It has been observed that as water depth was increased, the total yield from the Solar Still was decreased for both stills. The higher water depth increases water mass in basin, which results in more time consumption for evaporation. For this reason, the basin water temperature and evaporation rate are found less in higher water depth Solar Still, which leads to decrease in its output. The maximum values of hourly output for Evacuated Tubes Solar Still were 0.88 Litre/m^2/ hr. and 0.770 Litre/m^2/ hr for 0.03 and 0.05 water depths respectively.

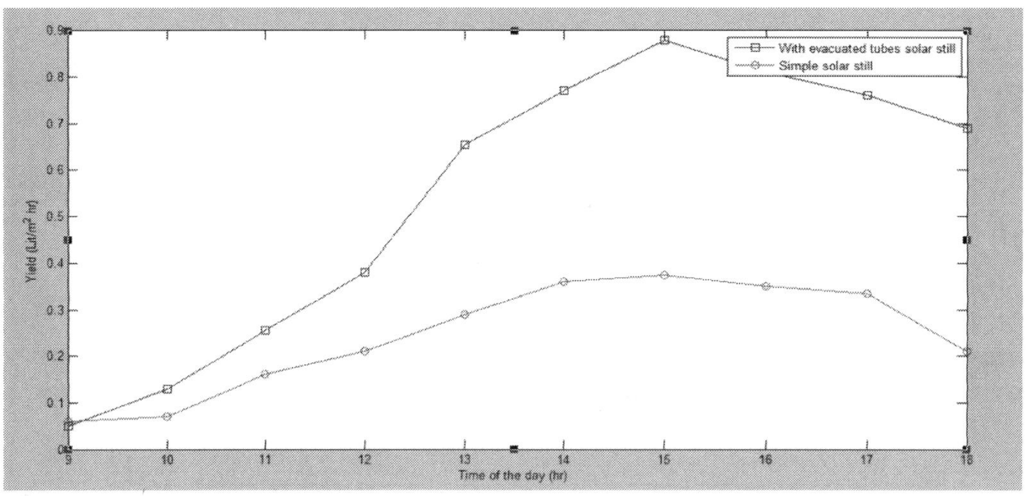

Figure 9. Hourly variation of experimental yield for 0.03 m water depth solar still.

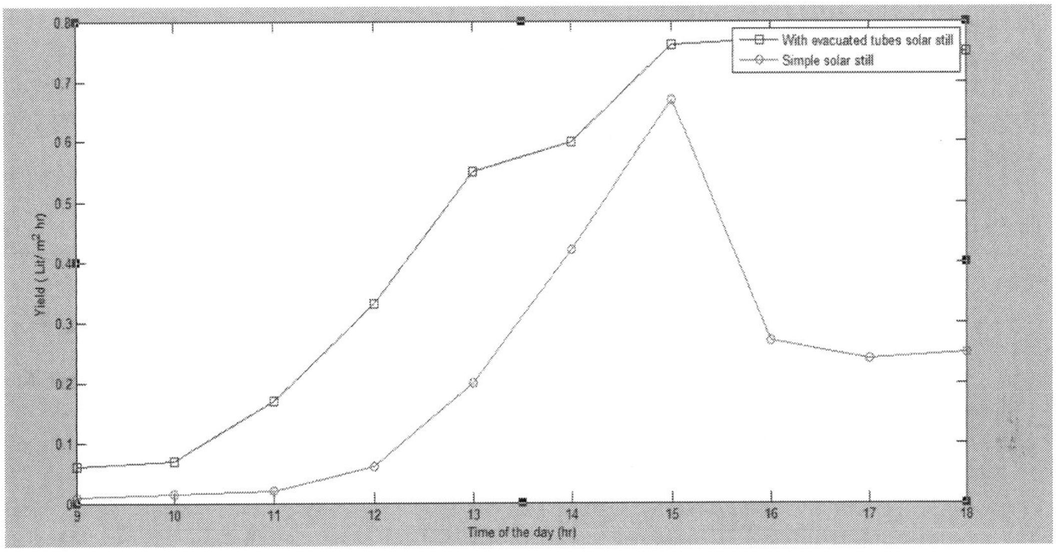

Figure 10. Hourly variation of experimental yield for 0.05 m water depth solar still.

6. Conclusion.

The Thermal Modelling of Conventional Still and Evacuated Tubes Solar Still has been experimentally confirmed and duly validated for 30 and 50 Litre of water in the research. The following points are highlighted on the basis of experimental and theoretical results:

1. The average water temperature of an Evacuated Tubes Solar Still was found to be increased by 15° C and 21° C for 0.03m and 0.05 m water depths respectively.
2. Due to additional heat energy supplied by evacuated tubes, the basin water temperature was found to be increased in Active Solar Still which further leads to increase in its productivity.
3. The yield of Evacuated Tubes Solar Still has been found as 7.61 & 7.535 Litre/m^2/day for 0.03m and 0.05 m water depths respectively, while for Simple Solar Still, it was 3.15 and 3.055 Litre/m^2/day for 0.03m and 0.05 m water depth respectively.
4. The total and evaporative heat transfer coefficient for an Evacuated Tubes Solar Still has been found very high, attributed to the higher water temperature difference as compared to conventional Still for both the respective water depths.
5. A fairly good agreement has been found between theoretical results obtained from Thermal Modelling and the Experimental results for both the Solar Stills.

7. References

[1] Gleick, P.H. Water in Crisis: A Guide to the World's Freshwater Resources; Oxford University Press: New York ,NY, USA, 1993

[2] Dimri V. ,Sarkar B., Singh U.,and Tiwari G.N. 2008. Effect of condensing cover material on yield of an active solar still: an experimental validation. Desalination 227: 178-189.

[3] Rai.S.N., and G.N. Tiwari,1983 Single basin solar still coupled with flat plate collector .Energy Conversion and Management 23 (4) :145-149.

[4] Velmurugan, V, Srithar, K., Solar Stills Integrated with a Mini Solar Pond Analytical Simulation and Experimental Validation, Desalination 216 (2007), 1-3, pp. 232-241

[5] Ali A. Badran, Ahmad A. Al-Hallaq, Imad A. Eyal Salman, Mohammad Z. Odat. A Solar Still Augmented with a Flat Plate Collector Desalination 172 (2005) 3 pp.227-234.

[6] Tiris C. Tiris M. Y Erdalli M. Sohmen Experiental Studies on a Solar Still Coupled with a Flat-Plate Collector and a Single Basin Still Energy Conversion and Management 39 (1998) 8, pp. 853-856.

[7] Dashtban M, Tabrizi FF,Thermal analysis of a weir-type cascade solar still integrated with PCM storage. Desalination 2011 279 415–22.

[8] J.Prabahar, Dr.T.Balusamy,Varghese M John, Enhancement of Productivity on Double Slope Single Basin Solar Still by Flat Plate Collector and Sponge Applied Mechanics and Materials Vols. 813-814 (2015) pp 690-694.

[9] H. Kargar Sharif Abad , M. Ghiasi , S. Jahangiri Mamouri , M.B. Shafii, A novel integrated solar desalination system with a pulsating heat pipe Desalination 311 (2013) 206– 210

[10] Z.M. Omara a, Mohamed A. Eltawil. Hybrid of solar dish concentrator, new boiler and simple solar collector for brackish water desalination. Desalination 326 (2013) 62–68.

[11] Hikmet S.. Aybar , Mathematical modeling of an inclined solar water distillation system Desalination 190 (2006) 63–70

[12] Hitesh N.Panchal ,Enhancement of distillate output of double basin solar still with vacuum tubes Journal of King Saud University-Engineering Sciences (2015) 27,170-175

[13] Anil Kr. Tiwari, G.N. Tiwari , Thermal modeling based on solar fraction and experimental study of the annual and seasonal performance of a single slope passive solar still: The effect of water depths Desalination 207 (2007) 184–204

[14] Hitesh Panchal.,Anuradha Awasthi, Theoretical modelling and experimental analysis of solar still integrated with evacuated tubes Heat and Mass Transfer (2016)

[15] Rahul Dev , Sabah A. Abdul-Wahab , G.N. Tiwari, Performance study of the inverted absorber solar still with water depth and total dissolved solid. Applied Energy 88 (2011) 252–264

[16] K. Sampathkumar, T. V. Arjunan, M. Eswaramoorthy, and P. Senthilkumar, Thermal Modelling of a Solar Still Coupled with Evacuated Tube Collector under Natural Circulation Mode. Energy Sources, Part A 35:1441–1455, 2013.

[17] Selcuk Selimli, Ziyaddin Recebli, Semra Ulker, Solar vacuum tube integrated seawater distillation - an experimental study facta universities , Vol. 14 No 1 2016 pp. 113 – 120.

[18] Ali.F. Muftah, K. Sopiana, M.A. Alghoulc, Performance of basin type stepped solar still enhanced with superior design concepts. Desalination (2017).

[19] Sampathkumar Karuppusamy, An Experimental Study on Single Basin Solar Still Augmented with Evacuated tubes. Thermal Science: Year 2012 Vol. 16 No.2 pp. 573-581

[20] Lilian Malaeb, Kamel Aboughali , George M. Ayoub , Modelling of a modified solar still system with enhanced productivity. Solar Energy 125 (2016) 360–372.

[21] D. W. Medugu , L. G. Ndatuwong, Theoretical analysis of water distillation using solar still International Journal of Physical Sciences Vol. 4 (11) pp. 705-712, November, 2009.
[22] Tiwari G.N.,Dimri V.,Singh U.,Chel A., and Sarkar B., 2007 Comparative thermal performance evaluation of an active solar distillation system International Journal of Energy Research 31:1465-1482
[23] Duffie J.A. and W.A. Beckman 1991. Solar engineering of thermal processes. Wiley Publication.

8. Appendices.

For the purpose of numerical analysis of Simple Solar Still and Evacuated Tubes Solar Still, the below mentioned Heat Transfer Equations have been employed.

The design parameters of Evacuated Tubes Solar Still has been calculated in the following expressions.

$$I_{Aeffe} = A_{ET}F_R(\alpha\tau)_e + (\alpha\tau)_{effe}I(t)_{effe}$$

$$Where(\alpha\tau)_{effe} = [\alpha'_b \frac{h_w}{(h_w + h_b)} + \alpha'_w + \alpha'_g \frac{h_{tw}}{h_{tw} + h_{tg}}]$$

The expression of U_{Aeffs} for Simple Solar Still

$$U_{Aeffs} = U_b + U_t$$

$$U_b = \frac{h_w h_b}{h_w + h_b}$$

$$U_t = \frac{h_{tw} h_{tg}}{h_{tw} + h_{tg}}$$

Where

The expression of U_{Aeffe} for Solar Still integrated with Evacuated Tubes

$$U_{Aeffe} = U_b + U_t + A_L F_R U_{LC}$$

9. Nomenclature.

English Letters

A_L	Area of solar still, m^2
A_{ET}	Diameter of outer glass ×total length of the tubes, m^2
C_w	Specific heat capacity of Basin water, J/kg K
F_R	Heat removal factor
h_b	Basin liner overall heat transfer coefficient, W/m^2K
h_w	Convective heat transfer coefficient from basin liner to water, W/m^2K
h_{cw}	Convective heat transfer coefficient from water to glass cover, W/m^2K
h_{ew}	Evaporative heat transfer coefficient from water to glass cover, W/m^2K
h_{rw}	Radiative heat transfer coefficient from water to glass cover, W/m^2K

IOP Conf. Series: Materials Science and Engineering 1065 (2021) 012013 doi:10.1088/1757-899X/1065/1/012013

h_{tg} Total heat transfer coefficient from glass cover to ambient, W/m^2K
h_{tw} Total heat transfer coefficient from water to glass cover, W/m^2K
I Intensity of solar radiation, W/m^2
Ki Thermal conductivity of insulation material, W/m K
L Latent heat of vaporization, J/kg
L_i Thickness of insulation material, m
L_g Thickness of glass covers, m
M_w Mass of water in the basin, kg
Mew Total distillate output from passive solar still at end of each day (kg)
P_g Partial vapor pressure at inner surface glass temperature, N/m^2
P_w Partial vapor pressure at water temperature, N/m^2
Q_u Useful thermal energy gain by vacuum tubes, W/m^2
t Time, s
T_a Ambient air temperature, K
T_b Basin temperature, K
T_g Glass inner surface cover temperature, K
T_w Water temperature, K
U_b Overall bottom heat loss coefficient, W/m^2K
U_t Overall top heat loss coefficient, W/m^2K
v Wind velocity, (m/s)

Greek Letters
α Fraction of energy absorbed
τ Absorptance-transmittance product
σ Stefan Boltzman constant,
ε Emissivity

Subscripts
b Basin liner
e Evacuated tubes
eff Effective
g Glass cover
s Solar still
w Water
0 Initial

DESIGN AND FABRICATION OF EXTRUSION MACHINE FOR RECYCLING PLASTICS

Sagar Kumar[1], Sooraj R[2], Vinod Kumar M V[3]

[1,2]Students, Department of Mechanical Engineering, Canara Engineering College Benjanapadavu, Mangalore

[3]Assistant Professor, Department of Mechanical Engineering, Canara Engineering College Benjanapadav Mangalore

[1]skmech999@gmail.com

ABSTRACT. The extruder is designed for producing filament that can be used for 3-D printers. The 3-D printer requires a filament having dimension ranging from 2-2.5mm. The quality expected is to obtain continues filament having the specified dimensional range which can be fed in to the 3D printer. Screw extrusion process is adopted to produce filament to fixes cross section. PET pellets were heated between 230 to 250 and pushed through a nozzle. The material experiences compressive and shear force. A continues filament of diameter ranging from 2 to 2.5mm is obtained without any air bubbles and roughness.

I. INTRODUCTION.

Extrusion process produces thin filaments which is suitable for 3-D printing process. In the extrusion of PET plastic, raw PET material in the form of flakes is fed into the hopper mounted on the barrel of the extrusion machine. The PET fakes get transported by the action of gravity as it passes through the hopper. The flakes enter through the feed throat and comes in contact with the screw. The screw rod is rotated using a motor at low rpm and the barrel chamber is heated by ceramic heater for a temperature range of 230 to 250C. The PET flakes as it passes through the screw rod gets heated and melted which then finally gets extruded from the nozzle designed for the size of the filament. Initially in most of the process, a heating profile is set for the barrel in which there are three heater zones which gradually increases the temperature from low to high. This allows the plastics pallets to melt gradually as they are pushed through the barrel and it will reduce the risk of overheating which may cause degradation in polymers.

The basic mechanism involved in this process is of a screw which transports the raw plastic pellets from a hopper through a heating metal pipe where the plastics are melted. The raw plastic pellets is fed into the screw from the hopper with the help of gravity. Then the molten plastic is forcefully passed into the pipe with the help of the nozzle at the end of the pipe to form the filament. Finally, the extruded plastics which is drawn from the nozzle to determine the final diameter of the filament which is drawn from the nozzle [2].

IOP Conf. Series: Materials Science and Engineering 1065 (2021) 012014 doi:10.1088/1757-899X/1065/1/012014

II. Implementation in 3D printer

3D printer utilizes the extruder to perform the type of additive manufacturing referred to as Fused Deposition Modeling. Fused Deposition Modeling also known as FDM begins with a software process that includes the processing of an STL (stereolithography file format). The STL file mathematically slices and orients the particular model for the physical build process. The part is then produced by extruding small beads of thermoplastic material to form layers as the material immediately hardens after extrusion from the J-Head hot end. There is typically a worm-drive or drive wheels (powered by the stepper motor) that feed the filament into the nozzle at a controlled rate. Inside the nozzle, the thermoplastic is then heated past its glass transition temperature by the heated resistor and thermistor, deposited by the extrusion head onto the build platform or base (FDM Technology: Stratasys). If required, support structures may also be generated. In this instance, the machine may dispense multiple materials to serve different purposes. It may dispense one material to build the model up and another material to serve as the support structure or base upon which the model is built. The movement of the nozzle can be in both horizontal and vertical directions which follows the tool path with the help of modelling software. In relevance to the 3D printer design, the extruder along with the motor is placed directly onto the z-axis and the stepper motor is plugged into an Arduino Mega 2560 microcontroller. The Arduino is then instructed by the user through the CAM program the appropriate dimensions and parameters upon which to build the model. The model is then built from the bottom up approach, which basically made one layer at a time.

III. LITERATURE REVIEW

1. Historical Development of Extruder

The extrude process means that the machine to push out the material through an opening which is the product as the extradite. The meaning of extruder is all about mixing. The inventor of this extruder machine was Charles Hancock and his co-workers in the year 1870s and developed in 1890s by applying the fundamental principle of extrusion in USA. In Europe the extrusion machine is developed by Francis Shaw and Paul Troester in 1900.

Extrusion technology was initially invented in the year 1870, this is done by setting up extruder system by the accurate temperatures in extrusion process. Due to this the plastic deformation also evolved and also side by side thermoplastic polymers evolution has also been developed. The main reason for the evolution of the extrusion machine is due to plastic which is having wide range of the polymers. From the year 1960 the evolution of the extrusion machine made successful in manufacturing part. The basic design of the extrusion machine does not change only the working and mechanical procedure during the productions. Intially, due to this machine the demand for the plastic product raised by the development of the extruder machine, the next type of extruder machine started after the generation of die making. Due to this most changes had to be done to the machine in order to obtain the required outputs .The compressions as well as the pressures of the machine were studied together. Nowadays the extruder machine had been changed in many ways, it is changed by the modern engineers in classical form to the modern form. The timing of the extruder screw was increased by co rotating to counter rotating principle. Every design were noted in order to make implementation. The produce pressure were analysed more in order to obtain better extrusion process.

Recently, the extrusion machines are fully automated and there are huge development in extrusion system. This extrusion machine have high demand in pipes to the objects that printed in 3D printer by using the extrusion process by using the latest new technology. The generation of this machine is done in two branches, the first branch has high efficiency and second has the finished product which is to be modified. But more than this wide range of extrusion technologies which is in hold and in development which is still running to produce eco-friendly polymers such as PLA [1].

2. Extrusion process

The process of the extrusion is mainly classified into five main parts:

1. Feeding
2. Melt

3. Melt conveying
4. Mixing
5. Filament forming

The extruding assembly generally consists of screw, barrel, heat control unit and the die during the extrusion process. The working of the extrusion machine is due to the gravitational force the plastic is fed into the extruder in the form drops into the rotational screw of extruder. Then the screw starts to rotate in order to that the friction is generated in order to generate heat and the plastic materials start to melt. In order to melt extra materials additional heating barrels are attached in the extruder. The electric heat control acts as shielding effect which does not allow heat to go out where the molten plastic is injected to the die. It also acts as thermal energy to the process.

Then the melted plastics is passed through the small hole in order to remove the impurities and then passes through the die to the final product. Extrusion process is mainly for large scale industries. In this manufacturing process it is very important to keep right operating control inputs and temperature measurements for required polymer and resin. If we followed the instructions then only we are able to get final products. During this extrusion process, there might be many errors in the product, for example if the temperature is not maintained properly the final product which will get is not as good as it should be.

2.1. Hopper
This is the part which is present at the starting stage were the plastic is fed into the system which works by gravity where granules reaches the hopper. The hopper is generally conical in shape so that the raw materials can move easily. Most hoppers are joined with help of nut and bolt so that it can be removed for different shape and size of material.

2.2. Screw
The screw assumes an imperative function in expelling. The screw is expected to push the material which is feed to the framework from the container. The pivoting screw pushes the granules in to the barrel. The screw is the basic part to improve the ill-advised plan if the screw may result hazards and inappropriate item result. The speed of the screw can be resolved from the control unit.

2.3. Barrel
The granules are warmed through the five channels of warming utilizing PTC or fired radiators in barrel. The granules are from strong state to fluid state combining. This is otherwise called warming and blending zone. The contact power among screw and granules makes it simple to liquefy on the grounds that the barrel temperatures get sufficiently high to soften the plastic. The liquid material through the barrel gets into the openings of kick the bucket where the state of item is given.

2.4. Nozzle
The extruder contains mainly a breaker plate which is used to brake the materials which is directly connected to the nozzle. Usually, the gear pumps are generally placed between the nozzle in order to maintain a uniform pressure which results in producing uniform dimensions of cross section. So it acts as an seal between the nozzle and the extruder and it contains a hole where the plastics is forcefully sent through it after rotating movement process. It also reduces the impurities while passing through the nozzle. We can change the nozzle according to the required size of the filament. Therefore, the nozzle is a key unit in this process.

3. Control unit
The control unit is the gadget wherein every gadget is associated with work. The simple or computerized control unit can control the creation esteem. The speed and the warmth in the barrel can be effectively kept up by the assistance of this control unit. Basically it is the arrangement of electronic associated together to make expelling simple. The creation esteem and the force taken by the engine are additionally commonly found in presentation board of control unit [1].

4. VISCOSITY IN EXTRUSION
Thickness is the amount that portrays a liquid's protection from stream. Liquids oppose the general movement of drenched items through them just as to the movement of layers with varying speeds inside

them. Officially, thickness is the proportion of the shearing pressure (f/A) to the speed angle ($\Delta vx/\Delta z$ or dvx/dz) in a liquid.

Polymer consistency is significant in expulsion to comprehend the handling window, the job temperature plays in thickness, and the significance of shear rate during preparing. The thickness versus shear rate bend shows enormous contrasts with temperature changes and just little contrasts with shear rate changes. To bring down thickness during expulsion, it is more successful to diminish the liquefy temperature. Going to a higher shear screw in one or the other single or twin-screw expulsion doesn't significantly change the gum consistency. Higher shear rate actuates shear warming, which brings down the polymer thickness and can prompt gum corruption. Some gum frameworks show both solid temperature and solid shear reliance. In these frameworks, while both temperature and shear effectsly affect thickness, changes in shear rate influence consistency more than changes in temperature [1].

5. **OBJECTIVES**

The objective of this task is to plan and construct an expulsion machine that makes 3D printing fiber from PET bottle pellets. The application is for any business utilizing 3D imprinting in a non-industrial nation with the plan to make the 3D printing practical and efficient. The deliverable is an economical and tough machine that admissions polyethylene terephthalate (PET) plastic water bottle shreds, dissolves and blends them, and afterward expels them as homogeneous fiber. Although, PET plastic is hard to reuse, it was picked in light of the fact that it is the most ordinarily accessible waste plastic [3].

IV. METHODOLOGY

1. PROCEDURE OF EXPERIMENT

The development and activity of the plastic extruder is to a great extent dependent on existing plans utilized in both mechanical and diversion applications. The essential component is contained a screw that transports crude plastic pellets from a container through a warming zone in a metal line where the plastic is liquefied. The crude plastic pellets are gravity-taken care of from the container into the screw. Inside the line the liquid plastic is constrained through a spout toward the finish of the line to frame a fiber. The expelled plastic can be attracted from the spout to decide the last breadth of the fiber. The spout is molded to frame the expelled plastic into the ideal cross-segment. Single screw extruder is an extruder which have just one screw in the framework. It is regularly utilized for the straightforward and general materials. As of late, individuals have done heaps of exploration for more steady and stable expulsion. In extruder, material is blended inside the barrel till the length of the expulsion machine [2].

2. **EXTRUSION**

2.1 **Install Extruder Nozzle**

Install a bolt nozzle of diameter 2.5mm so that the filament is drawn easily.

2.2 **Set Material Temperatures**

Intially the temperature of the machine is very low. To melt the very high temperatures are required. So suitable temperature is needed to be maintained.

2.3 **Feed Hopper**

After the equipment in the machine gets heated up fill the hopper with raw materials. Always clean the hopper with the brush so that the controls and inside wall of the hopper will be clean. Do not fill the hopper fully as the screw starts to rotate there may be chance to overflow of the melted plastics.

2.4 **Guide Filament**

Tenderly guide the fiber into a characteristic twisting, either heading turns out great. To get a predictable fiber, try not to pull on the fiber. Contacting the fiber will cause irregularities in the fiber measurement. Contacting the fiber while its expelling presents curves and changes in the width.

Figure.2.2: Digital temperature indicator

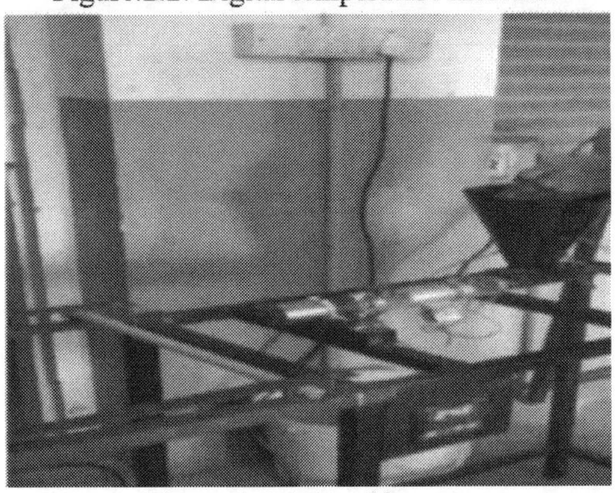

Figure.2.4.1: Feed hopper

The principle point of this cycle is to spool it without any problem. Intially toward the starting the beginning of twisting is hardest, yet whenever it is spiraled it turns out to be simple.

Figure.2.4.2: Guide Filament sample

2.5 **Measure diameter**

After the filament gets hardened and made spool the measurement is done in the filament coil itself. The temperature also varies for different sizes of filament. If the size of filament is too large high temperature is required, if the size of filament is small low temperature is required.

3. EXTRUDER MACHINE

The extruder machine these days is generally cutting-edge. The extruder is characterized into different various sorts depending the quantity of screws. There are fundamentally 3 sort's extruders

- Single screw extruder
- Twin screw extruder
- Multi-screw extruder.

3.1 Single screw extruder

Single screw extruder is the extruder machine which consists of only one screw. This is used only for one type of materials which is to be melted. Generally, many research have been done on this extruder in order to obtain higher accuracy by mixing up of different materials.

3.2 Twin screw extruder

This type of extruder is mainly used in powder processing plant. The extruder is automatic since it can clean itself and mixes mainly based on the ratio. The twin screw extruder is mainly used in industry for the pelletization process. The performance obtained from this extruder is more qualitative.

3.3 Multi-screw extruder

This is the extruder is used in order to get more output. In this extruder it has more numbers of screw. In this type of extruder there is no problem in mixing process. The output obtained from this process is very good and strong.

3.4 Fiber and 3D printer

3d printer is machine which takes fiber with the assistance of combined together constrating to get the items. It prints the fibers in layer by layer structure. The size of printing layer thickness shifts from 0.01 mm to 0.04 mm. The thickness of the fiber goes from 1.75 mm to 3 mm. The 3D printer causes the fiber to dissolve at high temperature and the prints it to get required shape and size of the article. All together get more strength of the item it ought to have a layer thickness of 0.01 mm. There are different sorts of the printers however it utilizes fluid resin based and powder gum based rather than fiber. Some sort of help material can be effectively eliminated by dissolving it in dissolvable to acquire the eventual outcome.

3.5 Stereo-Lithography

3D printing is a laser type of the sintering. This is the stage where the melted pellets plastics is cured with the help of the laser beam. The photosensitive polymers are made set during this cycle. The obtained complex polymers are obtained of high accuracy.

3.6 Electrical Heaters

Electrical heaters are essentially utilized in this machine by supplanting other liquid warmers or steam warming frameworks. These are principally utilized since it proficient and cost diminishing. It isn't costly to keep up correlation between warming framework. Electric radiators typically covers an enormous territory then other warming framework. It has become a specialist well-disposed loathing framework in expelling measure. The specific measure of current goes through the conductor which has certain obstruction, these opposition functions as hindrance in the stream and creates the warmth. These warmers are basically needed for liquefying of materials to took care of into next stage. In the extruder, on the off chance that the extruder machine is little and has less zones, at that point it have not many warmers and in the event that the machine is enormous and numerous zones, at that point it has more electric radiators.

3.7 PID Controller

PID is the gadget which gives the Information to SSR when to turn on and when to off. ID peruses the temperature of any framework with the assistance of indoor regulator. It capacities as the info board to the framework. Broadly useful of temperature regulators are utilized to control most common cycles in industry. Commonly, they arrive in a scope of DIN sizes, have different yields, and programmable yield capacities. These regulators can likewise perform PID control for incredible general control circumstances. They are generally positioned in the front board with the showcase for simple

IOP Conf. Series: Materials Science and Engineering 1065 (2021) 012014 doi:10.1088/1757-899X/1065/1/012014

administrator openness. These regulators have a pre-tune capacity to at first ascertain the PID temperature for a cycle, and a consistent tune capacity to continually refine the PID temperature. This takes into account quick arrangement, sparing time and diminishing waste."

3.8 **Thermocouple**

The main function of the thermocouple is used to measure the temperature. It acts like an sensors which are made from different materials and consist of two wires. When the current is passed to the wires it generates some voltage which determines the temperature of the system. These are mainly used since they are cheap and measures high temperature.

3.9 **Motor**

The most important motor used in this machine is DC motors. This motor can rotate both in clockwise as well as anticlockwise with the help of commands given to the motors. They can generate a torque up to three to five times in emergency situations. In this motor the energy is direct fed into the resistors so that it can work in various fields when required. The motor can work continuously without any brake and can come to state of rest without the help of switch. It mainly reduces the mechanical brake sizes.

V. DESIGN

1. BARREL

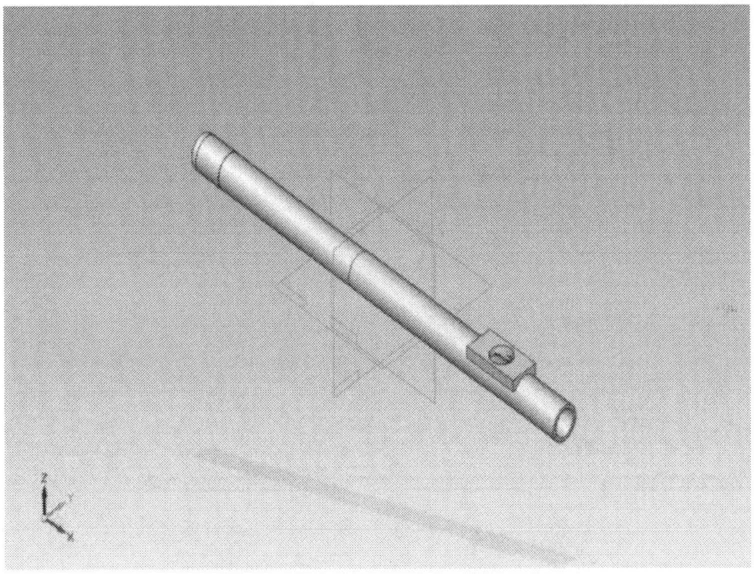

Figure.1.1: 3D model of Barrel

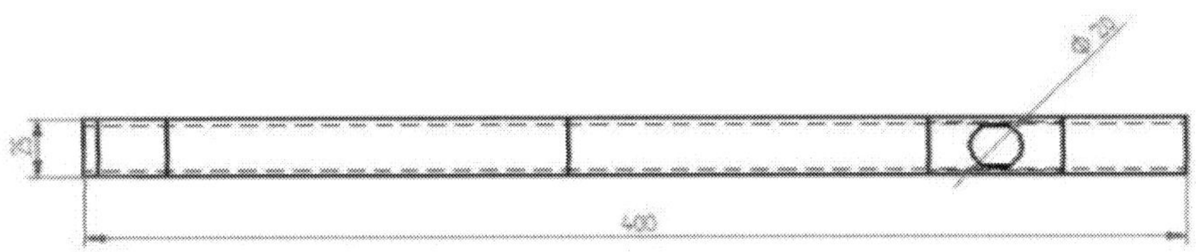

Figure.1.2: Dimensions of Barrel

2. HOPPER

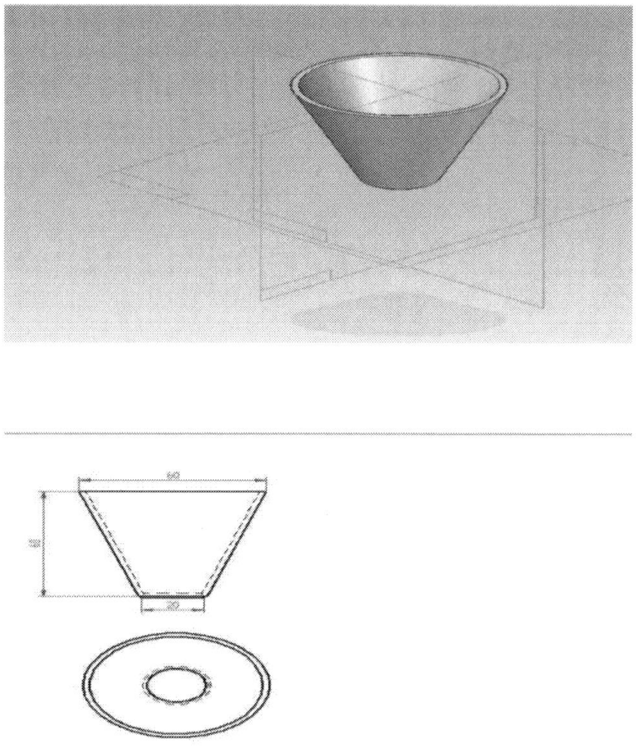

Figure.2: 3 D model and Dimensions of Hopper

Calculation Volume $= \pi h/3\ (R^2+Rr+r^2)$
$= \pi*60/3\ (30^2+30*10+10^2)$
$= 81681.40$ m

3. NOZZLE

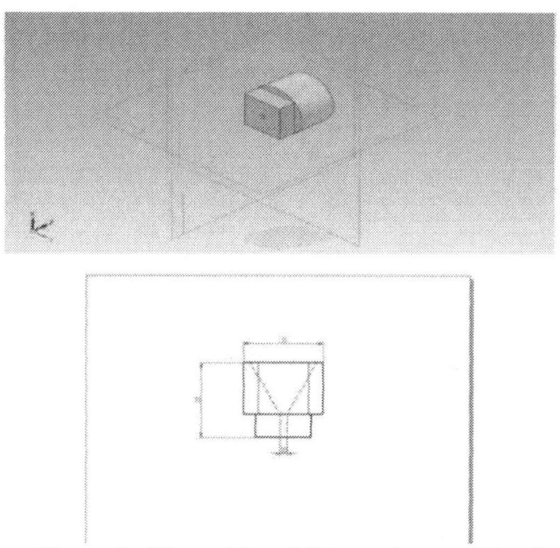

Figure.3: 3D model and Dimensions of Nozzle

4. SCREW ROD

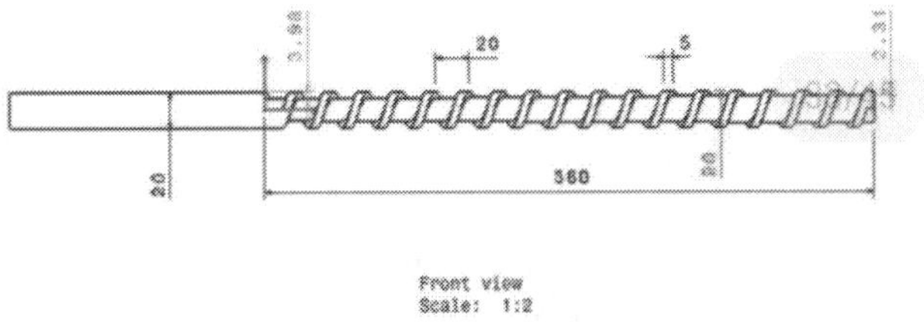

Figure.4: Dimensions of Screw Rod

5. SHREDDER BLADE

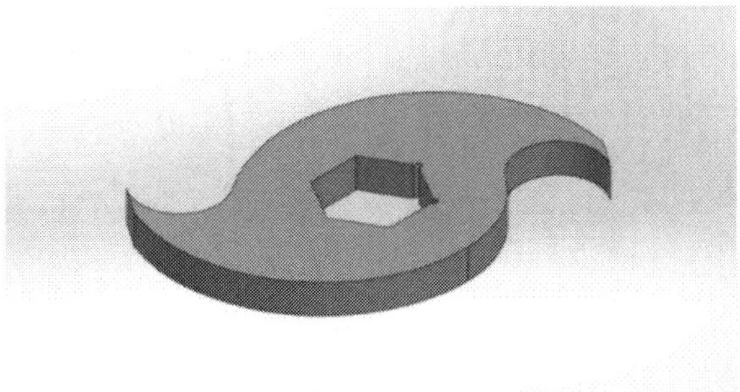

Figure.5: 3D model of Shredder Blade

6. SHREDDER CASING

Figure.6: 3D model of Shredder Casing

7.EXTUSION ASSEMBLY

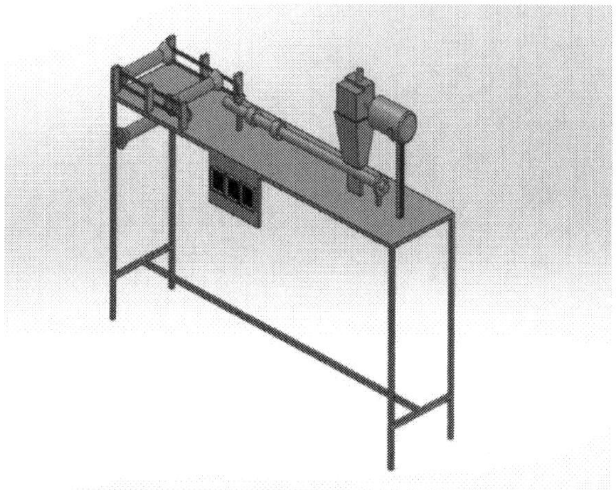

Figure.7: 3D model of Extrusion Assembly

VI. CONCLUSION

The optimum output is obtained when the temperature ranges from 230-250°C. By decreasing the heat conduction we can increase the efficiency of the extruder machine. In order to remove the excess materials which is present in the pipe must be cleaned before the screw rod gets cool down else it may get hardened. The excess plastic which is left inside can also be removed by rotating the screw rod in a clockwise manner.

Extrusion process happens at high temperature to dissolve the plastics where the extrusion rate is set roughly. In the event that it doesn't arrive at the extrusion rate inexact worth the crude plastic won't completely liquefy and little strong pieces will cause unpleasantness and air pockets in the subsequent fiber. This issue can be overwhelmed by diminishing the engine speed to inexact temperature. This will empowers to warm the plastic totally and results in homogenous and smooth plastic fiber. The outcome acquired from the experimentation that the great nature of fiber is gotten with engine DC power supply set to 6V and the temperature abundance to 230ºC, the liquefying purpose of the PLA pellets being used for try. The last measurement of the fiber drawn is dictated by both the size of the kick the bucket and the hot fiber which is drawn away from the source. For effective and powerful proceeded with activity of the extruder the container must be segregated from the warming zone so the crude pellets utilized at starting don't start to dissolve in the container and obstructs the delta port. With more prominent separation between the container and warming zone so a bigger volume of the plastics can be added so that the extruder can expel greater fiber without stress of the stopping up the delta of the warming line. Likewise, diminishing warmth conduction must between the warming line and the edge would improve the security of the extruder and furthermore builds proficiency with no force misfortune. To eliminate the abundance plastic which is available in the line must be cleaned before the warming line cools and again warmed the extruder to utilize once more [2].

REFERENCES

[1] Natural Fiber Composites For 3D Printing,2015, Pandey Kapil , www.theseus.fi .

[2] Filament Extruder, www.docplayer.net .

[3] Recycled Pet Pellets from Bottles , 2017 , L Costa Smith, www.scholarcommons.scu.edu .

[4] www.astraasal.jimdo.com .

[5] www.reportshop.co.kr .

[5] Ingeo Biopolymer 3001D, Nature Works LLC, retrieved January 8, 2012 from http://www.natureworksllc.com/Technical-Resources/3000-Series.aspx.
[5] Lee, Soo-Il, 1977, Plastic Extrusion and Apparatus, US Patent 411816.
[6] Chris Rauwendaal. (2013). Polymer Extrusion (5 ed.). Cincinnati, USA: Hanser Publications.

IOP Conf. Series: Materials Science and Engineering 1065 (2021) 012015 doi:10.1088/1757-899X/1065/1/012015

Performance Enhancement of CPC Using Graphical Ray Tracing

G H Bargale[1] and Dr N K Chougule[2]

[1]Research Scholar, [2]Professor in Mechanical Engineering, Department of Mechanical Engineering,

College of Engineering Pune, Maharashtra, India,

E-mail: khotgargi1008@gmail.com

Abstract. This paper presents the approach of finding an optimized location of a tubular receiver for a compound parabolic collector (CPC) with 23.06° acceptance angle. At this angle reflected rays concentrate at and directly above the main focus of a parabola. Ray tracing plays vital role in analysing rays falling on collector surface and reflected to the receiver. Using this analysis, solar collectors are often designed or improved to enhance the thermal performance. As compound parabolic collectors (CPC) are non-imaging kind and positioned in E-W direction, they are capable of assembling and reflecting an enormous quantity of radiation to the receiver with least tracking. Ray tracing technique was used to evaluate the performance of CPC. A graphical ray tracing (GRT) with SolidWorks was employed to observe the pattern of solar ray concentration. The main aim was to design CPC for low to medium temperature thermal process heat applications, specifically for Textile industries. It is observed that over one tubular absorbers will work better for CPC to absorb most solar rays and two tubular receivers lined up horizontally enhanced the CPC daily average optical efficiency by up to 15-20% compared to the single tubular receiver with the same diameter. Ray-Tracing analysis would possibly provide necessary contributions to beat design problems, complementing the experimental results obtained through thermal testing and allowing the accomplishment of extra thorough testing outputs with lower experimental requirements.

1. INTRODUCTION

Solar collectors are classified as a concentrating and non-concentrating type, further they are classified as imaging and non-imaging types of solar collector. Compound Parabolic Collector (CPC) is come under non imaging and advantages of these type of collectors are they don't need frequent tracking. Kalogirou, advocated CPC has been classified as a medium temperature application which would be suitable for applications such as textile, dairy etc.[1, 2]. Winston et al described the design and geometrical parameters for solar concentrator and demonstrated applications of cylindrical receivers at CPC and also demonstrated the edge-ray principle to investigate thermal performance of collector using the string method. [3]. Antonini et al modelled CPC for photovoltaic applications and illustrated that CPC absorbs good amount of solar radiation with minimum tracking of collector. [4], Mussard et al made several attempts by optical analysis to increase the concentration ratio and thermal performance [5]. Solar radiation falling of collector has been simulated as light rays falling of the surface. Behavior of rays can be analyzed using ray tracing method either manually or using any drafting or modelling tool in order to understand frequency of tracking to increase the thermal performance.

IOP Conf. Series: Materials Science and Engineering 1065 (2021) 012015 doi:10.1088/1757-899X/1065/1/012015

Non-imaging collector works on the fundamental principle of edge-ray. Rays when strike to the boundary gets reflected to the edges of a target area if optical surface is properly selected. Further it can be directed towards the point or line by designing proper CPC. Rabl A compared different type of solar concentrators with respect to different applications to the industry and also emphasized the study on optical and thermal properties of a CPC [6]. Duffie J A and Gaos et al stressed that the CPC is very close to be the ideal solar concentrator, as it reaches the highest concentration possible for any angle of acceptance [7, 8]. Wang et al derived formulae for calculating average reflections in a CPC. It was concluded that for an ideal CPC, only two parameters are required, acceptance angle and receiver diameter. With these values concentrator's width and height can be easily calculated [9].

Previous researchers studied CPC theoretically as well as experimentally and it was suggested that CPC can be used for medium temperature applications without frequent tracking. With this context this paper aims to study behavior of light rays on CPC for Textile applications.

2. CONSTRUCTION OF CPC & RAY TRACING:

CPC is broadly classified as symmetrical or asymmetrical. The acceptance angle of CPC plays vital role in designing the system for particular applications. In order to collect solar radiation, receiver is located at a focal point. There are different positions where receivers are placed, we can find out the proper position of absorber tube by ray tracing analysis. As compound parabolic collectors (CPC) are positioned in E-W direction, they're capable of assembling and reflecting an enormous quantity of radiation to the receiver with least tracking when the acceptance angle is 23.06^0 (When the acceptance angle is greater than or equal to solar swing angle, there is no need of frequent tracking, which is 17.5^0 for location Ichalkaranji, Maharashtra, India). For the acceptance angle of 23.06^0, the concentration ratio is 5 so for the 200mm absorber width, aperture width and height of the CPC are 1000mm and 2939.38mm respectively. Design CPC consists of calculation and selection of material for reflector and receiver. Generally material with high reflectivity and less absorptivity would be preferred for reflector and vice a versa for receiver. While constructing CPCs generally they are truncated to reduce area required and cost of the material without compromising the concentration ration. Design calculations arrive to truncate the CPC to 33% of original height.

CPC geometry constructed with highly reflective surface with minimum absorption in order to achieve maximum optical efficiency. The incident solar rays concentrated at the focus inside the collector due to concave profile of a reflector.

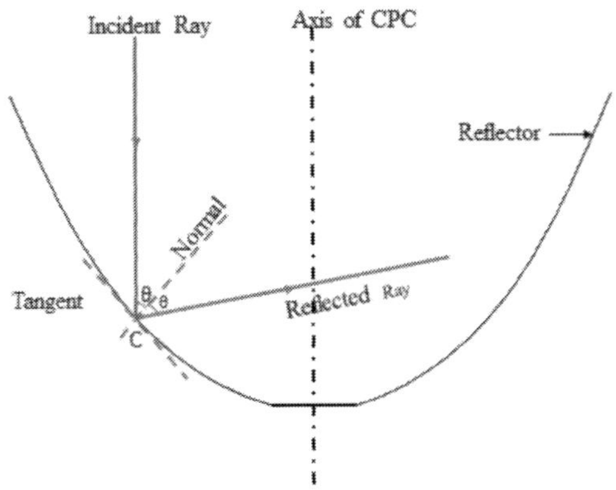

Figure 1. Reflection of solar ray on CPC geometry

Figure 1 depicts the reflection of solar rays on CPC geometry. Figure 1 demonstrates that incident ray falls at point C on reflector of CPC and reflected at an angle 2θ. Incident ray makes angle θ with the normal drawn at the point C, further it reflects with the same angle θ in the opposite direction normal to the tangent. This ray tracing analysis gives idea about location and diameter of a receiver.

3. RAY TRACING OF DIFFERENT CPC CONFIGURATION

As CPC is the combination of two parabolas, CPCs can be formed by different configurations. By combining two parabolas with shifting their focus linearly (Type A-with outer truncation & Type C-with inner truncation). By combining two parabolas such that their focus points coincide (Type B-with outer truncation & Type D-with inner truncation). Or asymmetric CPC (Type-E).

SolidWorks software was used to simulate the CPC as it is more flexible in designing and optimizing the CPC. Ray tracing is done in Solidworks by considering same no.of rays for all types and tracing the rays by their angle as shown in Fig.1.

SolidWorks was used to simulate CPC collectors with acceptance angle 23.06^0 and concentration ratio of 5. For all types, aperture width and height of CPC types are tabulated below in Table 1.

Table 1. Design calculations for different CPC configurations

	A	B	C	D	E
Absorber Width	200	200	200	200	200
Aperture Width	1183.74	815	2155.69	2377	1183.74
Truncated Height	979.20	979.2	985.61	1138.90	1052.22
Diameter of Tube receiver	58	58	58	58	58
All dimensions are in mm					

In order to analyse performance of CPC collector with different design sun hour angle of 0° considered. The surface area of reflector has been divided into equal 100 parts. In order to estimate maximum concentration of solar rays, GRT is done. Tracing of 100 rays is carried out for this work. Figure 2 shows how the ray tracing is done for Type A considering the angle θ at every incident point with single tube receiver of diameter 58mm.

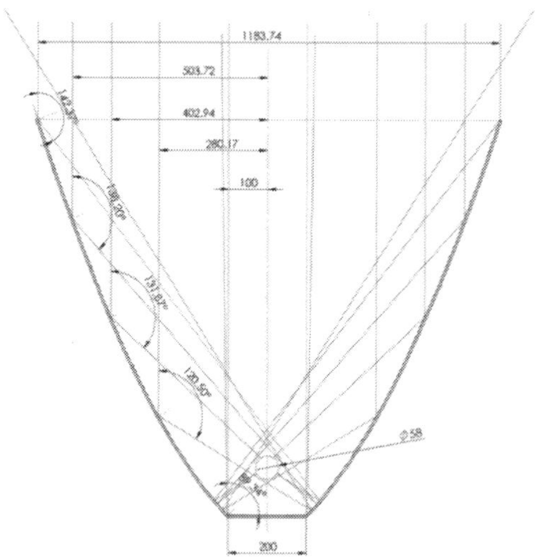

Figure 2. Ray tracing of Type A CPC with single tube absorber

Same procedure is followed for all above types and it was observed that most of the rays grabbed by receiver tubes directly and also after repetitive reflections of the rays. It is found from the ray tracing data shown in figure 3, Type C gives maximum coverage of rays but as it is formed by the inner truncation; its area is very large. On the other hand Type A gives good results for ray trace as well as it is formed by inner truncation so requires average area. Talking about Type B and Type D, they cover fewer rays as compared to remaining types. Whereas the Type E also covers good rays but as this is asymmetric CPC, this type is difficult to fabricate. So, Type A is selected as it covers good amount of rays with average area requirement and with ease of fabrication.

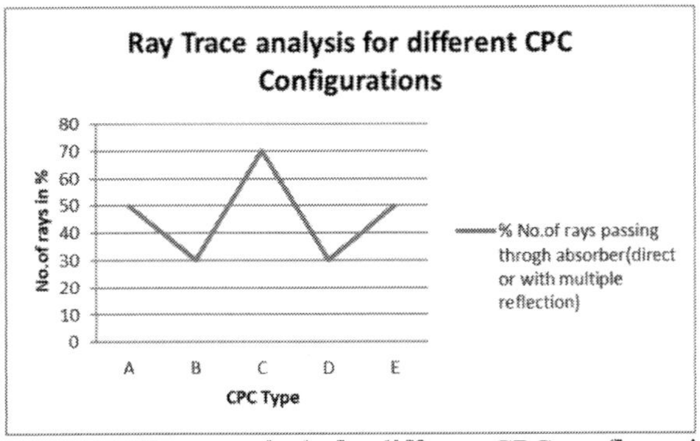

Figure 3. Ray Trace analysis for different CPC configurations

Further, Type A with single absorber and with two tubular absorbers is studied. figure 4 shows the ray tracing with two tubular receivers ((a) two receiver tubes placed horizontal each of 58mm diameter denoted by A-2TH & (b) two receiver tubes placed vertical each of 58mm diameter denoted by A-2TV).

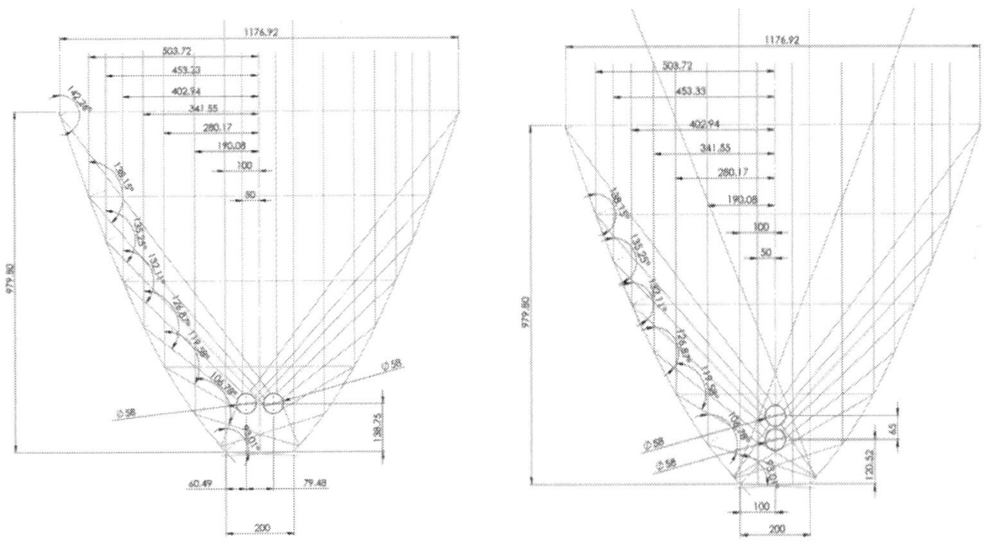

Figure 4. Ray tracing of Type A CPC with two tubular absorbers ((a) absorbers placed horizontally, (b) absorbers placed vertically)

It can be seen from ray trace analysis that the Type A CPC with two tubular receivers placed horizontally covers maximum solar radiation i.e. maximum rays falling on the reflector strikes the absorber tubes with direct or multiple reflections.

From Fig.5, it was observed that 50% of rays strike to CPC with the single tube receiver, 80% for two horizontal tube receivers of same diameter and 70% with two vertical tube receivers of same diameter. Optical efficiency of the CPC collector can be obtained by the ratio of radiation received by the receiver to the reflector of the collector.

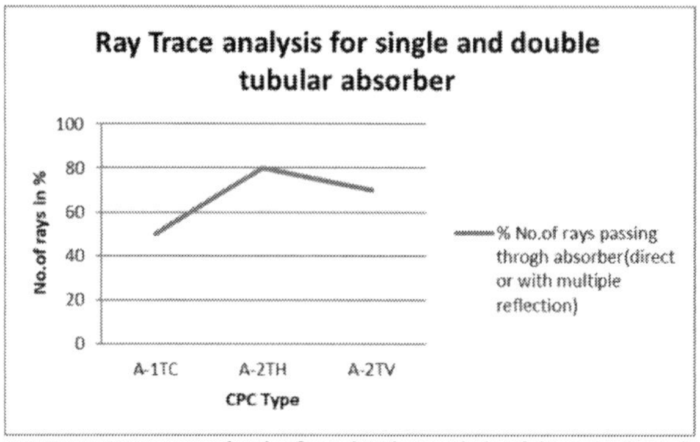

Figure 5. Ray Trace analysis for single and double tubular absorber

As all the ray tracings are carried out at 12 noon, inclination of solar rays with time is not considered here. All results shown here don't include time variation over a day. Also it is difficult to quantify thermal losses while calculating optical efficiency by ray tracing. So, optical efficiency is estimated to be 60% more in two horizontally placed tubular receivers CPC than single tubular receiver CPC from Ray trace only. Considering all losses, it will surely improve the optical efficiency up to 20% which would be further validated by CFD analysis and experimentation. It reveals favourable results and motivation for further research in CPC for different applications.

4. CONCLUSION

Ray tracing was done for 5 different constructions with acceptance angle of 23.06^0. Optimized location of a tube receiver for a compound parabolic collector (CPC) with 23.06° acceptance angle is found with the help of ray tracing. Graphical ray tracing is done with the help of SolidWorks following the property of reflection of solar ray. By considering different CPC geometries for concentration ratio of 5 and acceptance angle of 23.06^0, Type A gives the feasible results. About 50% of solar rays are absorbed by the receiver in Type A. Whereas, considering two tubular absorbers placed horizontally with same diameter for Type A CPC, would enhance the optical efficiency about 60%. As all the ray tracings are carried out at 12 noon, inclination of solar rays with time is not considered here. All results shown here don't include time variation over a day. Also it is difficult to quantify thermal losses while calculating optical efficiency by ray tracing. Considering all losses, it will surely improve the optical efficiency up to 15- 20% which would be further validated by CFD analysis and experimentation.

REFERENCES

[1] Kalogirou S A, 2003, Appl. Energy, *The potential of solar industrial process heat applications*, **76**, 337–361.
[2] Kalogirou S A, 1996, Solar Energy, *Design and construction of a one-axis sun-tracking system*, **57**, 465-469.
[3] Winston R, Miñano J C and Benitez, 2005, Academic Press: San Diego, CA, USA, *Nonimaging Optics,* ISBN 0-12-759751-4, 50-63.
[4] Antonini A, Stefancich M, Coventry J and Parretta A, 2013, Int. J. Opt. Appl., *Modelling of compound parabolic concentrators for photovoltaic applications*, **3 (4)**, 40-52.
[5] Mussard M and Nydal O J, 2014, Energy Proc. **57**, *Influence of solar tracking inaccuracy and sun rays modeling on the efficiency of a small-scale parabolic trough*, 1508-1515.
[6] Rabl A, 1976, Solar Energy, *Comparison of Solar Concentrators*, **18**, 93-111.
[7] Duffie J A and Beckman W A, 2013, John Wiley & Sons: Hoboken, NJ, USA, *Solar Engineering of Thermal Processes*, 4th ed, ISBN 978-0-470-87366-3, 337-342.
[8] Gaos Y S, Yulianto M, Juarsa M, Nurrohman, Marzuki E, Yuliaji D and Budiono K, 2017, AIP Conf. Proc, *The performance of solar collector CPC (compound parabolic concentrator) type with three pipes covered by glass tubes*, 1826, 020022.

[9] Wang Q, Wang J and Tang R, 2016, Energies, *Design and Optical Performance of Compound Parabolic Solar Concentrators with Evacuated Tube as Receivers*, **9**,795.
[10] Gudekar A S, Jadhav A S, Panse S V, Joshi J B and Pandit A B, 2013, Solar Energy, *Cost effective design of compound parabolic collector for steam generation*, **90**, 43-50.

IOP Conf. Series: Materials Science and Engineering 1065 (2021) 012016 doi:10.1088/1757-899X/1065/1/012016

Design and Development of Portable Electrical Resistance Furnace

Anand K. Hosamani[1, *], Bharat Naik[1], Shirish M. Kerur[1], Vinayak Ratan[1], Chandra Bhushan[2], Girisha L.[3], Yash Anvekar[1], Shivakumar s[4]

[1]Department of Mechanical Engineering, Jain College of Engineering, Belagavi, Karnataka, India

[2]Department of Mechanical Engineering, Babu Banarasi Das University Lucknow, India

[3]Department of Mechanical Engineering, PES Institute of Technology &Management, Shivamogga, Karnataka, India

[4]Department of Mechanical Engineering, KLS Gogte Institute of Technology, Belagavi, Karnataka, India

Abstract: Casting is a very old technique used for making various parts. There are various types of furnaces used for melting metals and alloys, working on different principles. But we Designed & Developed portable electrical resistance furnace to melt at low cost, environment friendly Metals. Further, the furnace is intended to be able to cling to the melt for necessary amount of time. Both electrical and mechanical aspects of furnace design were presented. Finally, we fabricated a furnace of around 9 kW evaluations with working temperatures closer to 1200°C. Constructions of furnace process are discussed and its heating performance was tested by melting aluminium, brass and its alloys. Its operation was discussed to suit the heating requirements for various metals and alloys.

1. INTRODUCTION

Stated that for a nation to relate, it should be able harness, convert its mineral resources and fabricate its equipment and machines locally. On these lines, this paper its aimed to design and fabricate a to low-cost electrical resistance furnace to fulfil the metal melting requirements in the academic laboratories and research [1]. Metals components which are cast, are widely use in human activities. Designed and constructed an electrical arc furnace using Soderberg electrode technique to melt the 5kg of steel or cast-iron scrap which is procured locally. Author performed on the furnace and confirmed that it takes 60minutes to heat up the furnace to 1200°C. To melt the 1st charge of 2kg steel/cast iron scrap at melting rate of 21.05g/min the furnace took 95min [2]. To fire ceramic products designed and constructed an electrical furnace with an automatic control [3]. Manufactured gas furnace using local source of material which had heating rate of 25° C/min and attained the temperature in the heating zone of 1000°C in burning pot and 700°C in the crucible [4]. Expressed that there are various heating methods are available. There might be numerous strategies for providing heat, delivered either by ignition of fuel or electric resistance. By considering cost, security, effortlessness and simplicity of fabrication authors have gone for an electrical resistance with indirect heating provisions. Further authors extended their work for fabricating stir casting furnace for melting aluminium [5]. Presented the design and fabrication of stir casting setup for preparing the aluminium, magnesium and copper-based metal matrix composites. In this setup, a nano particle preheating attachment at the top of the furnace was incorporated. It is a direct result of red hot condition with consistent temperature of nano particles infused by push bar into the liquid metal. The stirrer rod composed by variety of speed (0-2000rpm) for blending purpose [6]. Designed a novel fast quenched stir caster for handling Al–SiC composites in fluid and semisolid state [7]. Designed an electrical furnace. Some authors considered its appearance method of heating, extreme temperature, included cost, weight static and furnace that can be easily carried or moved. From the simulation, it is found that convection of heat flow collects the entire space in furnace and aluminium is completely melted. To melt complete 1kg of aluminium it took 45minutes, which is almost 62.55% quicker contrasted with the ordinary 2hours from utilising convectional technique. The cost of producing this revised plan of electric heater is considerably less expensive since the aggregate cost of material expected to manufacture this heater is just RM 5160. It gives the most temperature and moderate cost for an effective electric heater (pro-efficiency 78.53%) to be utilized as a part of the little scaled craft industries [8].

2. MATERIALS AND METHODOLOGY

The following are the basic logical steps involved in the design and fabrication of the proposed furnace.

1) Deciding the furnace specifications: The furnace specifications consider the following points:
    - Power requirement
    - Temperature to be achieved
    - Sizes of crucibles that the furnace can hold.
2) Preparation of the list of elements and components required for the furnace: A list of items which are going to be part of the furnace are prepared.
3) Making design calculations: Design calculations need to be carried out to satisfy the specifications of the items in the above list.
4) Geometry and layout of furnace: Since it is a portable furnace, the geometry and feasibility of transporting it has to be checked.
5) Fabrication of the furnace: Incorporating all the contents and meeting the design specifications, the furnace is fabricated.
6) Seasoning of the furnace: This is done in order to remove the existing moisture content and make the furnace suitable for further use.

2.1 MODEL & DRAWING

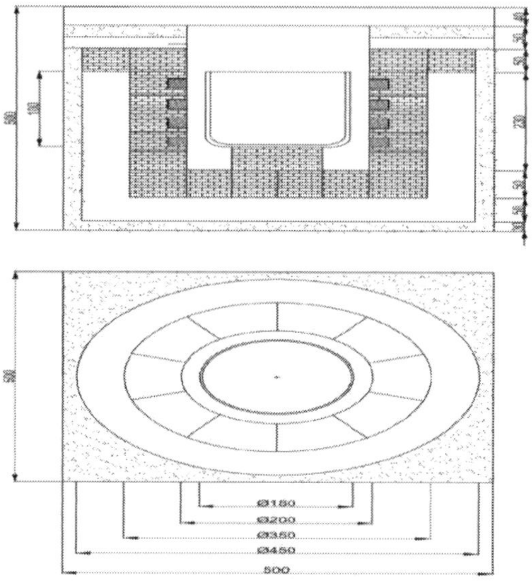

Figure 1. Design of the furnace

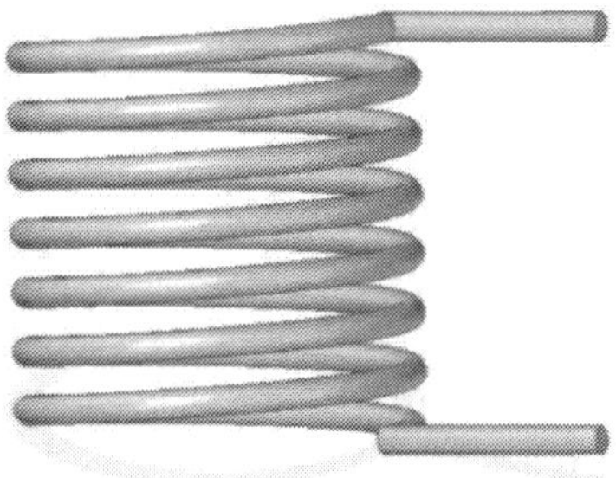

Figure 2. Heating Coil

Figure 3. Heating coil arranged in the refractory bricks

2.2 MAJOR COMPONENTS OF FURNACE

Silica Bricks: Silica bricks or fire brick is a block of refractory silica material used in lining of furnaces and kilns. Here we are using 40% silica bricks. Each brick weighs up to 4.2~4.5 Kg.

Properties of silica bricks:

- Good corrosion resistance of acid slag.
- Good stability under the high temperature.
- Low thermal conductivity.
- Good air tightness.
- True density of 2.35g/cc.

Commercially available silica bricks were purchased and arranged in circular way as shown in the figure below

Figure 4. Silica Brick

Figure 5. Ceramic wool

Ceramic wool: Ceramic wool or ceramic fibre blanket is made of ceramic needle like minute particles. It is light in weight, flexible, excellent in fire protection, contain very low thermal conductivity, and store low heat, thermal shock resistant and corrosion resistant. It is a commercially available in various densities, thicknesses, widths and lengths. The ceramic blanket used in the present furnace is 1 inch thick is shown in Fig.5

Properties of ceramic wool:

- Excellent handling strength.
- Excellent hot strength.
- Low thermal conductivity.
- Light weight.
- Thermal shock resistance.
- High heat reflectance.
- Excellent corrosion resistance.
- Excellent thermal stability.

2.3 SUPPLEMENTARY COMPONENTS:

Other components were used in order to fabricate the furnace. These components help in the improvement of strength and efficiency of the furnace. The various components used are:

1. Silica bricks
2. Ceramic wool
3. Thermocouple
4. Sheet metal cabin
5. Heat resistant wire
6. Refractory cement
7. Solid state relay+
8. Fully packed furnace
9. Auto transformer

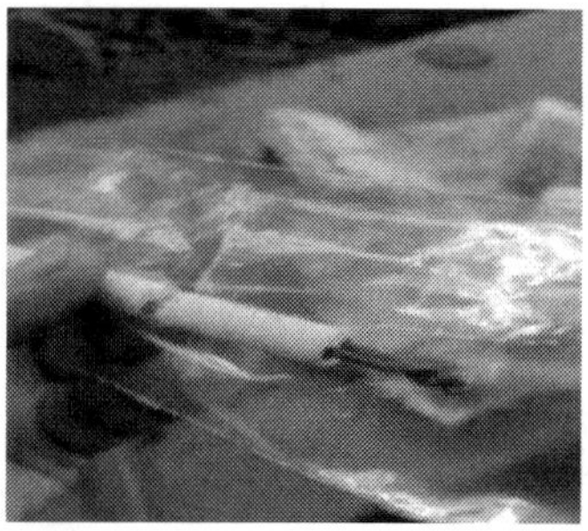

Figure 6. Thermocouple

Figure 7. Sheet metal cabin

Figure 8. Heat Resistant wire

Figure 9. Refractory cement

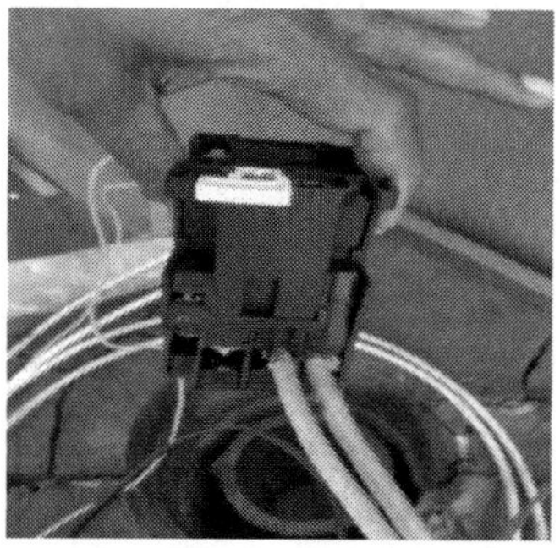

Figure 10. Solid state relay+

Figure 11. Fully packed furnace

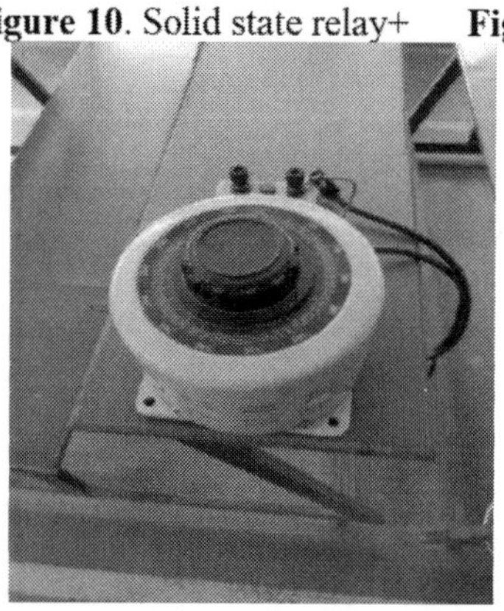

Figure 12. variable auto transformer

2.4 METALS THAT CAN BE MELTED

Table1. Metals and their melting temperature

Metals	Temperature in Degrees Centigrade
Brass	900-940
Aluminum	660
Brass (Rod)	990-1025
Cadmium	321
Copper	1084
Gray Cast Iron	1127-1200
Ductile iron	1149

2.5 WORK CARRIED OUT:

Initially we studied what type of heating element is suitable for our requirement by carrying out thorough literature survey. We then decided to use Kanthal A1 grade coil owing to its availability and economic feasibility. The heating element was then wound into to spring shape having diameter of 200mm, height of 180mm, pitch of 30mm and having 8 number of turns. Along with the heating element we procured two separate rods of 120mm length & 8mm diameter for making the necessary electrical connections. These two rods were soldered at the both ends of the heating element which is one at the top and one at the bottom. 20 silica bricks were purchased that had 40% silica in it. Grooves were cut in the bricks to accommodate the heating coil. Eight grooves were cut in the bricks by using the cutting machine. The bricks were then arranged properly in a circular manner and then polished on both side. The next step was to cover this assembly with the refractory cement to ensure stability of the furnace. Some bricks were cut to the smaller dimensions to create the base for the crucible. The remaining bricks were arranged as per the design. A sheet metal cabin of size 500mm×500mm×500 mm was fabricated to hold all the components in it. For lifting and transportation purpose, handles and wheels were attached to this sheet metal box. A lid is also provided to have easy access to the inner furnace.

Insulation was ensured by using ceramic wool at the suitable places. This does not allow the heat to escape from the cabin. A groove is cut on the top of any one brick to insert the thermocouple. Finally, all the electrical connections are made to supply the power to heating element. Control units such as solid-state switch, auto transformer etc. are placed outside the cabin to prevent them from getting damaged due to extreme temperatures. After providing the power, testing of the furnace was carried out.

Step-1

Step-2

Step-3

Step-4

Step-5

Step-6

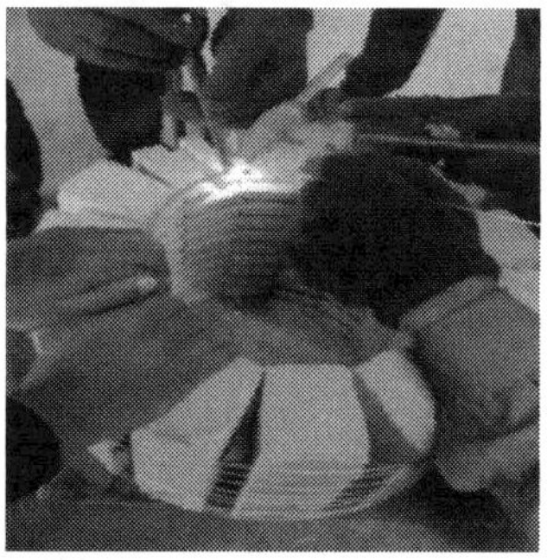

Step-7

Step-8

Step-9

Step-10

Step- 11

Step-12

Final assembled furnace:

Figure 13. Furnace setup

IOP Conf. Series: Materials Science and Engineering 1065 (2021) 012016 doi:10.1088/1757-899X/1065/1/012016

Table2. List of components used

Sl. No.	Components	Quantity
1.	Fire clay cement	1 Bag
2.	Silica bricks	20 Pc
3.	Box fabrication	1 No's
4.	Thermocouple	1 No's
5.	PID Controller	1 No's
6.	Heat Resistant Wire	6 Metres
7.	Crucible	1 No's
8.	Heating Element	8 Turns
9.	Glass Wool	4 mm^2
10.	Solid State Switch	1 No
11.	Dimmer	1 No
12.	Auto transformer	1 No

Table 3. Thermal conductivity of various materials

Sl. No	Material	Component	Thermal Conductivity
1	Cement	Fire clay cement	2.8 W/m °C
2	Ceramic Wool	Ceramic Wool	0.25 W/m °C
3	Crucible	Crucible	6 W/m °C
4	Putting Wool + Cement	Door	0.25 W/m °C
5	Silica Bricks	Silica Bricks	0.85 W/m °C

The find out the variation of thermal properties in the furnace following boundary conditions were applied as shown in Fig.14

- To heating coil temperature of 1473K was applied
- For better insulation (i.e. heat flow q =0), was applied to the ceramic wood
- Convection boundary conditions were applied on the outer surfaces of the furnace with heat transfer co-efficient of 50e6 W/m^2 K and ambient temperature of 300K.

3. RESULTS OF FINITE ELEMENT ANALYSIS

The furnace was meshed having 25459 nodes and 11243 elements with default settings of mesh. Fig.15 shows temperature distribution of the furnace with the maximum temperature of 1718.3K and minimum temperature of 144.81K at the outer surface of the furnace. This indicates that inside part of the furnace will be maintained at the maximum temperature. This temperature distribution is clearly visible in the cut section of the furnace as shown in Fig 16. Fig.17 Shows heat flux for the furnace with the maximum heat flux of 2.56e5 W/m^2.

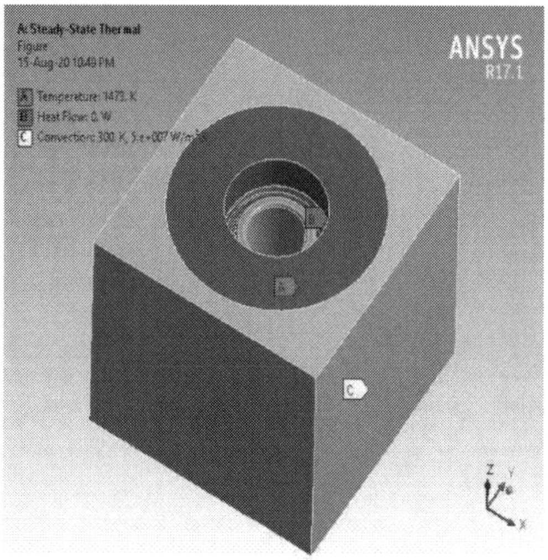

Figure 14.

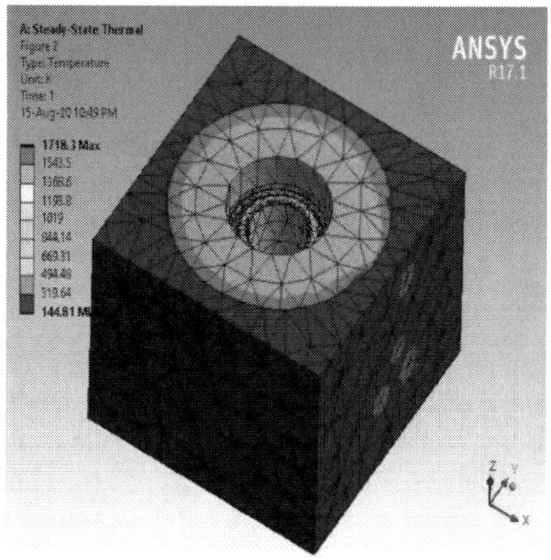

Figure 15.

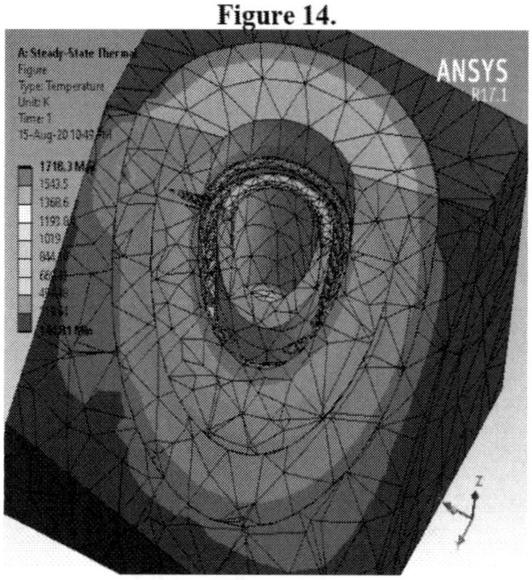

Figure 16.

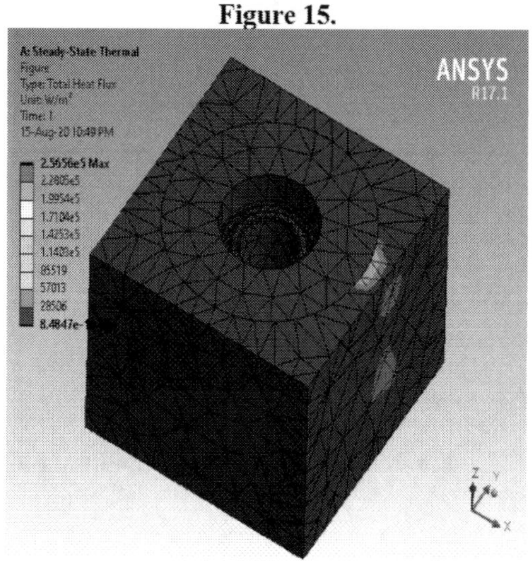

Figure 17.

4. CONCLUSION

- A low-cost metal melting furnace for melting cast iron and its alloys is designed and fabricated. Both mechanical and electrical parameters were considered in designing the furnace.
- This furnace is suitable for melting and holding small quantities of metals whose melting points are less than 1473K like cast-iron, Al, Zn, and their alloys for laboratory and research purposes. The heating rate, melting rate are comparable to existing standard furnaces attaining a temperature of well over 1273K. The work is completed up to 95% with all necessary equipment and precautions.
- The approximate error between the FEA results and the experimental results (actual temperature achieved was 1693K) was found to be 1.5% which is acceptable.

5. REFERENCES

[1] Bala k 2005 Design analysis of an electric induction furnace for melting aluminium scrap, *AU Journal of Technology*,**2**, P. 83-88.
[2] Oyawale F and Olawale D 2007 Design and prototype development of a mini-electric arc furnace, *Pacific Journal of Science and Technology*, **8 (1)**, P. 12-16.
[3] Bayindir R 2007 Design and construction of an electrical furnace to fire ceramic product, **66**, P. 135-140.
[4] Abed E.J 2013 Manufacture and performance of gas furnace. *International Journal Metallurgical & Materials Science and Engineering*, **3 (1)**, P. 109-118.
[5] Sahoo M and Rout S and Patra R 2015 Design and fabrication of a stir casting furnace set-up. *International Journal of Engineering Research and Applications*, **5 (7)**.
[6] Sekar K and Allesu K 2013 Design of stir casting machine, *American Int. Journal of research in science, Technology, Engineering & Mathematics*, P. 2328-3491.
[7] Sahoo N Introduction to refractory and insulating material.
[8] Ariff T and Zakaria.A 2015 Enhanced heating mechanism of electrical metal melting furnace in traditional foundry, **10**,9659.

Nanostructured Anti-Glare Coatings for Night Visibility

Madhusudhana R*[1], Sushma K L[1], Chandra Sekhar K[1] , L. Krishnamurthy[1], R Gopalakrishne Urs[2] and Sangamesha M A[3]

[1]Centre for Nanotechnology, Department of Mechanical Engineering, The National Institute of Engineering, Mysuru-08

[2]Department of Physics, The National Institute of Engineering, Mysuru-08

[3]Department of Chemistry, The National Institute of Engineering, Mysuru-08

Abstract. Anti-glare coatings (AGCs) have progressed into highly effective reflectance and glare minimizing components for several optical and opto-electrical equipment's. The extent in research and development with the contribution of nanotechnology in opto-electronics has improved the AGCs. The present study is related to the composition of a nanostructured antiglare coating. In general, antiglare films have been widely used to deal with glare but this is not proper for an outdoor usage in glasses. Hence, improvement in performance of night-driving glasses is necessary to reduce the negative effects of headlights glare. In this paper nanostructured anti-glare coating layers on glass substrate is done using Zirconium Oxide (ZrO_2). The obtained coatings are highly transparent and reflective in nature. Sol–gel method is adopted for the preparation of gel, dip-coating and spin coating technique was employed to develop coatings on glass substrates. The obtained coatings were further characterized by XRD to analyze the phase crystallinity; SEM was used to determine the surface morphology of the coatings, EDX to estimate the elemental composition of the coatings and UV-vis Spectroscope was used to determine the optical properties.

Keywords – Antiglare Coatings (AGc), Nanostructured, Sol-gel.

1. Introduction

Anti-reflective coatings have been used for many years to reduce reflection off the surface of transparent materials. At night, a lot of inconvenience is faced by the vehicle drivers due to the glare from the headlamps of approaching vehicles, this requires the treatment on the surfaces of vision glasses [1]. Surface treatment reduces reflection of glass surface by diffuse dispersing. In general, sparkling impacts happens due to various optical parameters. The main object of this study is to characterize the reduction of glare parameters which can be applied as antiglare treatment [2]. The use of antireflection coatings can significantly lessen the complete reflectance; there are different strategies for building surface-treatment structure like grinding, etching, holographic techniques, decorating process. Further, antiglare films are broadly set up by blending inorganic or natural light dissipating particles with glue pitches and covered onto a substrate [3]. In this study, we have employed a straight forward strategy for creation of AG coatings by sol-gel procedures using zirconium. Different physical and chemical techniques, including aqueous strategy, sol gel and co-precipitation have been utilized to integrate Zirconia. Among various procedures established, sol-gel technique is preferred because of its cost effectiveness and ease of process [4]. Zirconia is known as a material with high dielectric quality, high estimation of hardness, high softening point, high thickness and crack. Zirconium can be found in various crystal phases [5]. In this current study, Zirconium oxychloride is used as the precursor material for developing thin films by dip coating and spin coating techniques. The structural properties of the developed thin films were evaluated by using XRD (Bruker D2 Phaser). Optical Transmission and reflectance studies were carried in the wavelength range from 270 to 1100 nm by using (Shimadzu UV1800) UV–VIS spectrophotometer. The Surface properties of the films were characterized by using Scanning Electron Microscopy (SEM) and the elemental composition of the deposited films were carried out using EDAX.

2. Experimental Details

2.1 Materials:
Zirconium oxychloride ($ZrOCl_2.8H_2O$), Deionized water (H_2O),Isopropanol,Diethylamine ($(CH_3CH_2)_2NH$) and Glass substrate (Microscope Slides) were used.

2.2 Sample preparation:
Glass Substrates were ultrasonically cleaned using Isopropanol for about 8-9 minutes[6]After ultrasonication process, substrates were immersed in a solution of Isopropanol for about 15 min. In Parallel, a solution of 20 ml of Deionized water, 20 ml of Isopropanol is measured and mixed in a 200 ml beaker, 6.4gms of $ZrOCl_2.8H_2O$ is weighed and mixed into beaker and it is kept for continuous stirring by introducing a magnetic bit into the beaker for about 1 hr. 2 ml of Diethylamine is added drop wise and continuous stirring is continued for another 2 hrs at normal room temperature. Immersed glass substrates were then coated via Sol-gel Dip coating and Spin coating technique. In case of dip coating, substrates are dipped in the prepared solution and taken out from the solutions at uniform rate of dipping.In the case of spin coating, it's a production of thin films in a batch on flat substrates. After the coating is done, the substrate was dried for 2 days in a room temperature and the sample was annealed [7] for about 4 hrs at 350 °C.

3. Characterization

X-ray diffractometer analysis is carried out to know the phase crystallinity of the films [8]. Fig. 1. Shows the diffraction peaks of zirconia dioxide on the substrate using Spin coating technique. The strongest diffraction peak was seen at around 30°,50°,60° and we can see the clear crystallinity at (111) (220) (311).

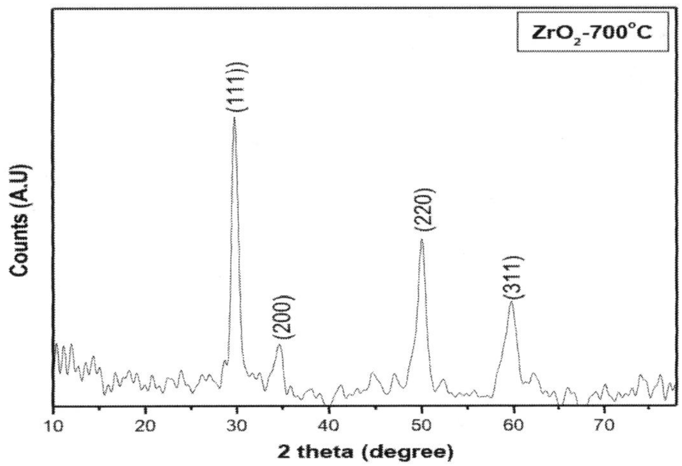

Figure 1. XRD diffraction peaks of zirconia dioxide on the Glass substrate using Spin coating.

Scanning Electron Microscope (SEM) was used to study the surface morphology of the sample by making use of focused beam of electrons [9]. Fig. 2 shows the SEM image of the ZrO_2 Sol-gel Dip coating thin film coating on glass substrate in which the growth of Nanodots and nanorods can be seen.

Formation of clump on the substrate is due to the dipping rate which effects the coating uniformity. In fig. 3 it is evident that the SEM image of the thin film coated on the substrate by using Sol-Gel spin coating technique at 10 µm range shows the growth of Nanorods and Nanodots on the glass substrates. A uniform coating of ZrO_2 substrate is taken into consideration and Annealed at 350 °C which shows good AR viability without any deterioration. The film shows almost indistinguishable anti reflectance bends which reveals that the thin films hold undefined nature up to 350 °C. Beyond this temperature, there is a slight addition and slow move of AR wave minima towards the lower frequency locale because of prompted crystallinity [5].

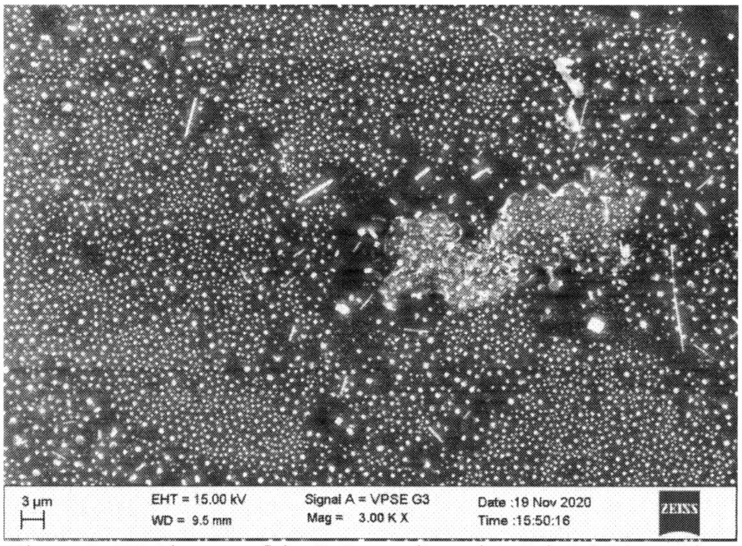

Figure 2. SEM images of the ZrO_2 coating using Sol-Gel Dip coating

Figure3. SEM images of the ZrO_2 coating using Sol-Gel Spin technique

EDAX is an analytical technique used for the elemental analysis or chemical characterization of a sample [10]. It relies on an interaction of some source of X-ray excitation and a sample. Fig. 4 and Fig. 5 shows elemental analysis of Sol-gel Dip coating and spin coating respectively through EDAX spectra. The elemental strong peak for Si and Zr and other elements like O, Ca. Cl, Na are seen since the slides are usually made of optical quality glass such as soda lime glass or borosilicate glass in which Si and Zr weight % are of about 23.4 and 14 respectively in dip coating and 23.7% of Si and Zr of about 17.5%

in spin coating.

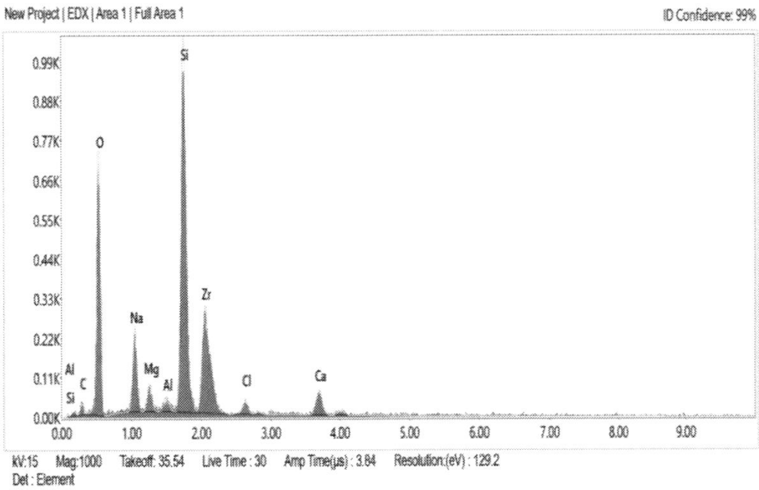

Figure 4. EDAX spectra of ZrO_2 coating using Sol-Gel Dip coating

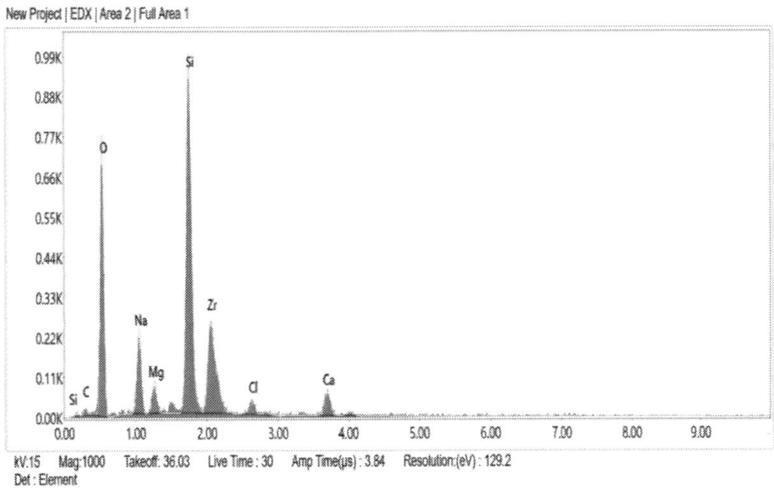

Figure 5. EDAX spectra of ZrO_2 coating using Sol-Gel Spin coating

The optical properties of ZrO_2 thin films on the glass substrate was analyzed by using UV-Vis Spectrometer [11,13].

Optical transmittance and reflectance curves of ZrO2 thin films deposited using spin and dip coating technique is shown in fig. 7. The films deposited using spin coating exhibited 85% transmittance and that of the films deposited using dip coating showed 75%T at higher wavelengths. Reflectance spectra of the films measured at normal light incidence are presented in Figure 8. From the fig. it is noticed that dip coated film has high reflectance when compared to spin coated film. This means that the significant difference in thicknesses and optical constants for these ZrO_2 films [14]. The absorption co-efficient for dip coated and spin coated films are at 368nm and 360nm respectively. Further, it can also be attributed to particle nature, homogeneous nature and optical absorption of the film.

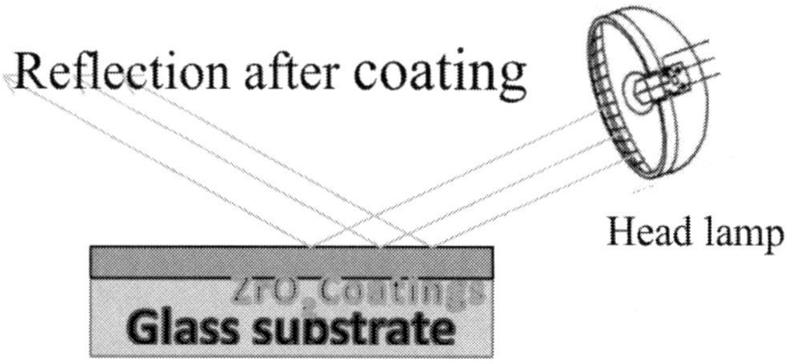

Figure 6. Schematic representation of Antiglare Thin Film

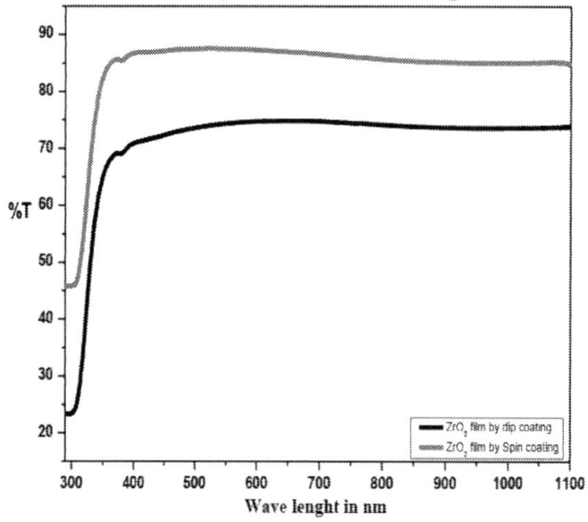

Figure 7. UV-Vis Transmittance spectra of zirconium dioxide

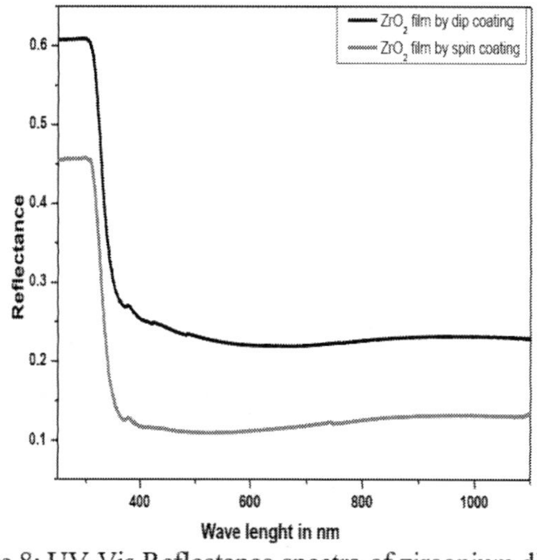

Figure 8: UV-Vis Reflectance spectra of zirconium dioxide

4. Conclusion

In the current study, Zirconium dioxide (ZrO_2) films were prepared by Sol-Gel Dip coating and Spin coating techniques on microscopic glass slides. Microscopic glass slides were used for the study since it is a cheap and reliable substrate to characterize displays with AG treatment regarding the sparkle effect

[15]. The present study also provides an efficient route for the synthesis of ZrO_2 nanostructures with controlled morphology. Surface studies confirm that the developed films are Nanorods and Nanodot like structures. Optical studies reveal that the films have good transparency and reflective properties. XRD confirms the formation of Zirconium dioxide and it can be concluded Zirconium oxide thin films coatings can be used for the reduction of glare.

References

[1]Valeo-TNS: Sofres 2015 *Study Conducted on 3000 French and German Drivers.*

[2] Thomas Fink, Udo Kruger, Dr. Ing. h.c. F. Porsche AG, Stuttgart, SID 2016 DIGEST **365** TechnoTeam Bildverarbeitung GmbH, Ilmenau "Pixel Crosstalk: A New Metric to Characterize DOI Loss Due to AG Treatments of Display Glasses".

[3] Jiaqi Zhu, Huihua Huang and Xiaohong Peng 2016 "Preparation and characterization of antiglare waterborne polyurethane" *RSC Adv.,* **6, 102368**.

[4] Hassan Yousaf, M. Azhar, Mahwish Bashir, Saira Riaz, Zohra N. Kayani, Shahzad Naseem, "Effect of capping agent on microwave assisted sol-gel synthesized zirconia coatings for optical applications " *International Journal for Light and Electron Optics* **222 (2020)** 165297.

[5] Sadaf Bashir Khan, Zhengjun Zhang, Shern Long Lee, "Annealing influence on optical performance of HfO2 thin films" *Journal of Alloys and Compounds,*

[6] A. Méndez-López, O. Zelaya-Ángel, M. Toledano-Ayala, I. Torres-Pacheco, J. F. Pérez-Robles and Y. J. Acosta-Silva, "The Influence of Annealing Temperature on the Structural and Optical Properties of ZrO_2 Thin Films and How Affects the Hydrophilicity" *Crystals 2020,* **10, 454**.

[7] Mun Teng Soo, Niki Prastomo, Atsunori Matsuda, Go Kawamura, Hiroyuki Muto, Ahmad Fauzi Mohd Noor, Zainovia Lockman, Kuan Yew Cheong, 2012 "Elaboration and characterization of sol–gel derived ZrO_2 thin films treated with hot water" *Applied Surface Science* **258** (2012) 5250– 5258.

[8] Latha Kumari and W. Z. Li, J. M. Xu and R. M. Leblanc ,D. Z. Wang, Yi Li, Haizhong Guo, and Jiandi Zhang, "Controlled Hydrothermal Synthesis of Zirconium Oxide Nanostructures and Their Optical Properties" *Crystal Growth and Design*, **Vol. 9**, 2009 3875.

[9] Mun Teng SOO, Go KAWAMURA, Hiroyuki MUTO,Kuan Yew CHEONG,Zainovia LOCKMAN, Ahmad Fauzi Mohd NOOR , Atsunori MATSUDA, 2011 "Design and synthesis of mesoporous ZrO2 thin films using surfactant Pluronic P123 via sol–gel technique " *Journal of the Ceramic Society of Japan* **119 [6]** 517-521.

[10] Suelen Aline de Lima Barros, Diana Gabriela Soares, Maria Luisa Leite, Fernanda Goncalves Basso, Carlos Alberto de Souza Costa, Gelson Luis Adabo, 2019 "Influence of Zirconia-Coated Bioactive Glass on Gingival Fibroblast Behavior " *Brazilian Dental Journal* **30(4)**: 333-341.

[11] Yi Du, Lunet E. Luna, Wui Siew Tan, Michael F. Rubner, and Robert E. Cohen, 2010"Hollow Silica Nanoparticles in UV_Visible Antireflection Coatings for Poly(methyl methacrylate) Substrates"*ACS NANO* **Vol. 4 . no. 7** . 4308–4316.

[12] J.M. Khoshman, A. Khan, M.E. Kordesch, "Amorphous hafnium oxide thin films for antireflection optical coatings" Surface & Coatings Technology 202 (2008) 2500–2502.

[13] Latha Kumari, Roger M Leblanc, Dezhi Wang, Haizhong Guo, 2009 "Controlled Hydrothermal Synthesis of Zirconium Oxide Nanostructures and Their Optical Properties" *American Chemical Society Crystal Growth and Design,* **Vol. 9, No. 9**,3875.

[14] Funda Sayılkan, Meltem Asilturk, Esin Burunkaya, 2009"Hydrothermal synthesis and characterization of nanocrystalline ZrO_2 and surface modification with 2-acetoacetoxyethyl methacrylate" *J Sol-Gel Sci Technol* 51:182–189.

[15] Qinyuan Zhang, Xiaoli Li, Jun Shen, Guangming Wu, Jue Wang, Lingyan Chen, 2000 "ZrO thin films and ZrO SiO optical reflection filters deposited 2 2 2 by sol–gel method " *Materials Letters 45 _2000.* **311–314**.

Synthesis and Characterization of Zirconium Oxide (ZrO_2) Films on AA5052 and Glass substrates

*Madhusudhana R[1], Navyashree K C[1], L. Krishnamurthy[1], R Gopalakrishne Urs[2]

[1]Centre for Nanotechnology, Department of Mechanical Engineering, The National Institute of Engineering, Mysuru-08

[2]Department of Physics, The National Institute of Engineering, Mysuru-08

Abstract: Aluminum alloy has gained lot of importance in Aerospace Industries and Engineering applications because of its strength, high resistance to heat, light weight and durability. In this paper we made an attempt to improve the AA5052 substrates properties by coating Zirconium Oxide (ZrO_2) by Sol-Gel Method. Glass substrates were also coated with Zirconium Oxide (ZrO_2) coating. The coated AA5052 and glass substrates will be annealed to check the thermal resistance for various time of thermal rig test. Micro crystalline properties and formation of ZrO_2 of the coated film is evaluated for surface morphological studies by using Scanning Electron Microscope (SEM) and the elemental composition studies done by EDAX (Energy Dispersive Absorption X-ray Spectrometer).
Keywords: Al alloy, Sol-Gel, Zirconium oxide (ZrO_2)

1. Introduction

Aluminum is one of the metals which are extensively used in day today lives from utensils to structural elements of airplanes. Pure aluminum is very soft and has no structural strength to manufacture the parts of an aircraft. It is also very ductile due to which it can be machined easily and strengthened further by alloying it with other metals such as copper, magnesium, manganese, zinc etc., for different application [1]. It is estimated that 75 to 80 percent of the materials used in modern-day aircraft is made up of aluminum alloy. Benefits of Aluminum alloy are it is extremely light weight material, high strength and has good corrosion resistance [2]. Different types of aluminum alloys are used for different applications. The commonly used Aluminum alloys in the aircrafts industry are AA1050, AA 2024, AA 5052, AA6061, AA7050, AA7075 and AA7475 [8]. These alloys are used in the wing's planes, rudder, doors and floors, exhaust pipes, engine turbines and the cockpit instrumentation [3,4].

Many products of today, ranging from refractory to medical products, pigments, electronics, coatings, and ceramics, have been based on zirconia due to its superior characteristics. Further, some of the other typical applications of zirconia include dies for hot metal extrusion, oxygen sensors, and membranes in fuel cells, deep well valve seats, and marine pump seals [5, 6].

ZrO_2 can be synthesized using several techniques i.e., Sol-Gel synthesis, Hydrothermal synthesis, Plasma sprayed deposition technique, and chemical vapor deposition etc. [7]. In our present work synthesis of the ZrO_2 is experimented using sol-gel technique; since it is economical, chemically pure, Nano-sized particles are formed with narrow sized distribution [8, 9].

IOP Conf. Series: Materials Science and Engineering 1065 (2021) 012018 doi:10.1088/1757-899X/1065/1/012018

2. Experimental Details

2.1 Synthesis procedure for ZrO_2.

In the present research work, Zirconium Oxide (ZrO_2) inorganic coatings are prepared by Sol-Gel method and coated on by dip and spin coating techniques respectively [10]. Here, Zirconium Oxide is used for the coating because it exhibits several excellent mechanical properties like strength, toughness, biocompatibility, high fatigue, high wear resistance, low thermal conductivity, high melting point and high resistance to corrosion. AA5052 substrate is used as it has high strength, high ductility and high corrosion resistance. The prepared gel is coated on the substrates by Dip coating and spin coating technique. The coated specimen is subjected to structural, surface and elemental analysis to evaluate the improved properties. ZrO_2 solution is prepared for the coating using Sol-gel technique. Firstly, 40 ml of ethanol is taken in a beaker and kept for stirring at 500 rpm with a magnetic stirrer for uniform dissolution. 10 ml of Nitric acid is added together and stirred for 15 minutes. Then, 9 ml of Zirconium butoxide is added into the mixture and stirred for 30 minutes. Later, 5.3 ml of Acetyl Acetone is also added and kept for stirring for 1 hour. A pale yellow ZrO_2 solution is obtained [1, 2].

In this experimentation, Aluminum alloy AA5052 and microscopic slides (glasses) are used as substrate materials. AA5052 substrate contains Al, Si, Fe, Cu, Mn, Mg, Cr, Zn, Ti in the ratio of 96.7, 0.25, 0.40, 0.10, 0.10, 2.2, 0.15, 0.10 respectively [10]. The surface of the aluminum alloy is made flat by using a scratch pad. Then both the substrates were cleansed in water and kept for ultra-sonication for 25 minutes. Later, the substrates were dried completely before the coatings were done [14].

Firstly, AA5052 substrates were dip coated manually using ZrO_2 gel. The substrates were coated for 1, 3, 5, 7 number of times respectively. The substrates were immersed in the solution for 15 seconds and air dried for 1 minute after each coating. The coating deposited on the substrate is then air dried for 3 days and kept for annealing at 350 ^{0}C for 2 hours.

Further, the solution is aged for 3 days and then deposited on glass substrates following the same method as mentioned for aluminum alloy substrates. Spin coating is also carried out on the glass substrates using the same ZrO_2 solution which was kept for aging as the same number of times mentioned above. The substrates were coated using a spin coater at 340 rpm for 10 seconds. 2 drops of the solution were placed on to the substrate each time the coating was carried out and the substrates were annealed at 350 ^{0}C for 2 hours [1].

3. Characterization Techniques

Scanning Electron Microscope (SEM) and Energy Dispersive X-ray Analysis (EDAX) were used to analyze the coatings deposited on the substrates for surface morphology and elemental composition.

Scanning Electron Microscope (SEM) is an electron microscope which scans the surface of any sample and produces the image by using electrons beam. It is used to analyze the surface morphology and the chemical composition of the sample.

Energy Dispersive X-ray Analysis (EDAX or EDX) is also called as Energy Dispersive X-ray spectroscopy (EDS) which is used to estimate the elemental composition of the sample using Scanning Electron Microscope (SEM). It also detects the elements having higher atomic number than boron even when it is in the concentration of 0.1%. Therefore, it also helps in evaluating and identifying the metal and in detection of the contaminants etc. [11, 13].

4. Results and Discussions

The scanning electron microscope (SEM) images of the coated AA5052 substrates (dip coated) and glass substrates (dip coated and spin coated) for different number of times are shown below. The thickness of the coating increases with increase in the number of dipping times [1].

We initially got the uniform coatings for each dip which were also homogeneous. But with respect to the above images obtained after annealing sintered structures, clusters, cracks were observed and the coatings were non uniform. Different structures are also observed. Bonding strength was poor between the coating and substrates which is evident from the spallation of coatings [15]. It may be caused due to excessive air drying and then heating [16].

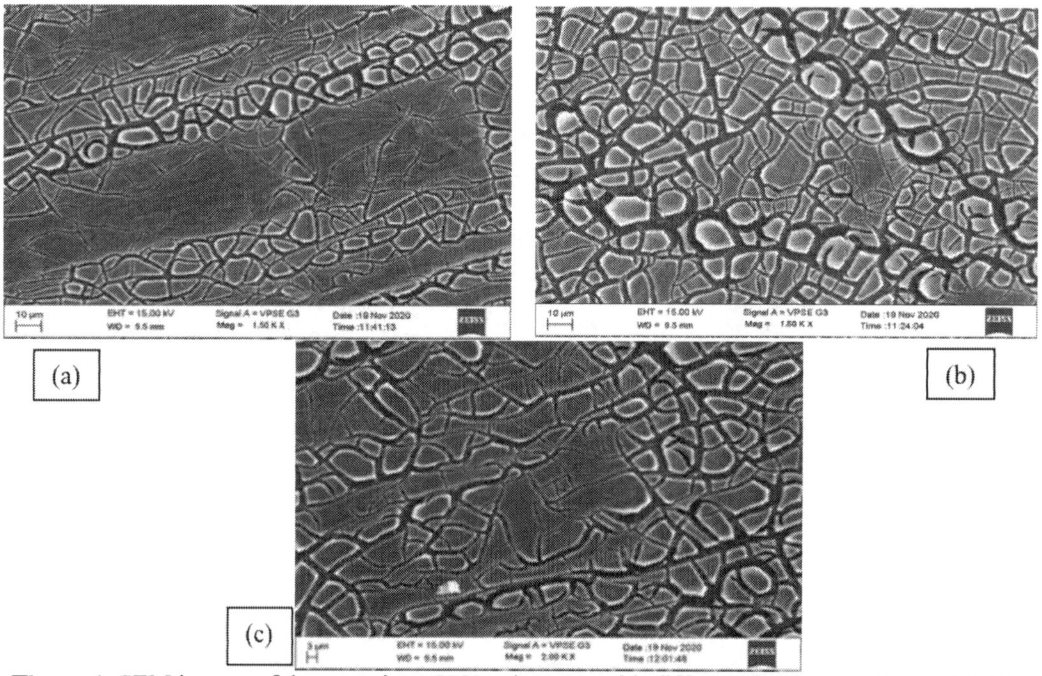

Figure 1. SEM images of the coated AA5052 substrates with different dipping numbers (a) 1 (b) 3 (c) 5

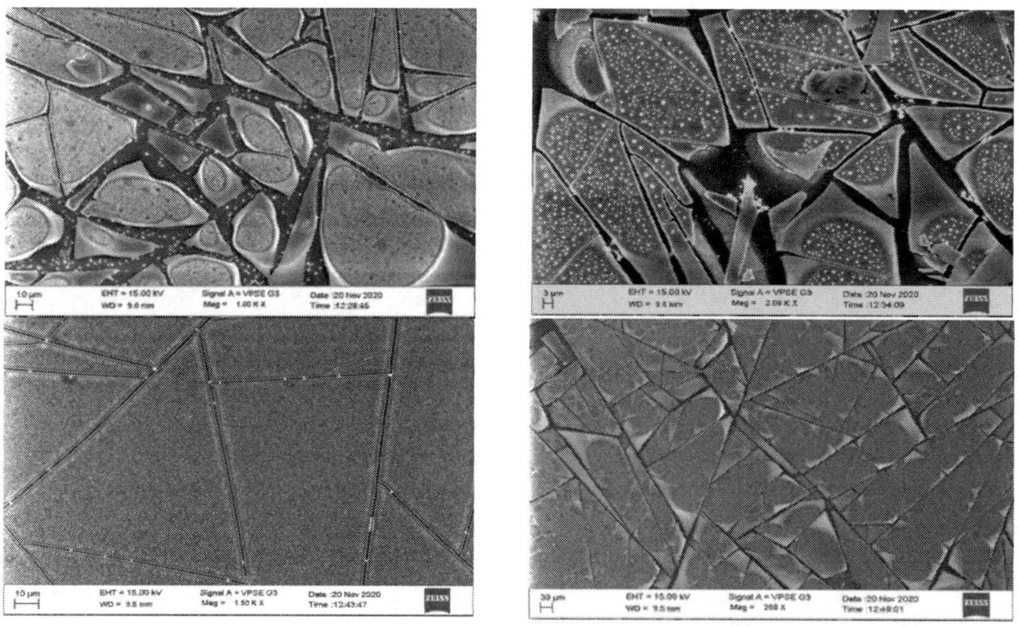

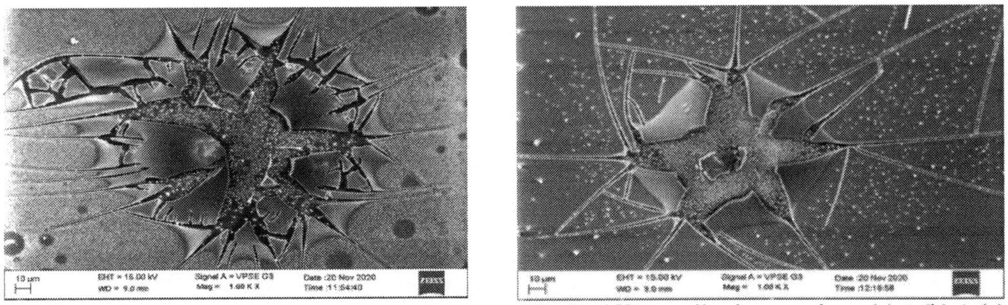

Figure 2. SEM images of the coated glass substrates with different dipping numbers (a) 1 (b) 1 (c) 3 (d) 5 (dip coated) and (e) 3 (f) 7 (spin coated)

The results obtained are dependent on a number of process parameters which includes roughness of the substrate, nature of the gel, number of dipping etc. Figure 1. (a) (b) (c) depicts the surface morphological images of 1,3,5 number of dipping respectively. It is evident that the coatings were uniform throughout before annealing the samples. After the samples were subjected to annealing time for 1 hour at a temperature of 350 ^{0}C, spallation of the coatings was observed. Crack initiation in the coatings of ZrO_2 was observed due to the penetration of heat until the substrate. In figure 2. (a) (b) (c) (d) (e) and (f) glass substrates which were dip coated as well as spin coated were subjected to thermal rig test for 2 hours respectively [12]. The coatings had a void at the center which indicated the epicenter of crack initiation as shown in figure 2. (e) and (f). Further in figure 2. (d) the structure indicated to be in the hexagonal shape.

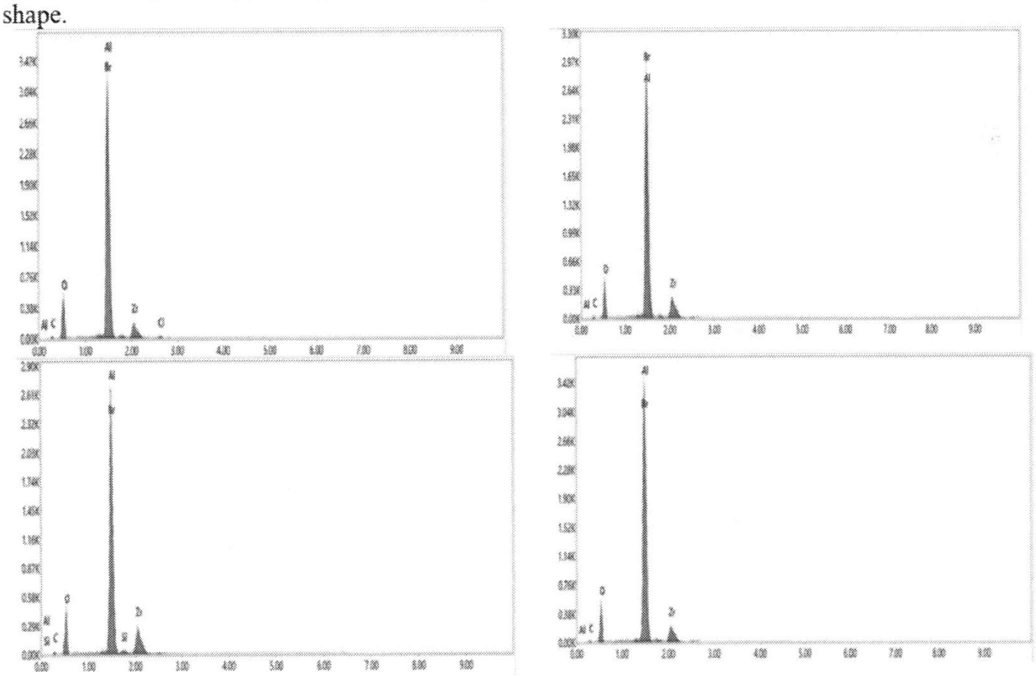

Figure 3. EDX analysis of the coated AA5052 Substrates with different number of times (a) 1 (b) 3 (c) 5 (d) 7

Figure 3. (a) (b) (c) and (d) represents the elemental composition of ZrO2 dipping on AA5052 which indicates the presence of Aluminum, Zirconium, Oxygen and Bromine. The weight percentage of Al was 35, 35.4, 32.9, 43.5, the weight percentage of O was 22.3, 22.5, 20.7 and 21.5, the weight percentage of Br was 25.9, 19, 26.1 and 17.4 and the weight percentage of Zr was 10.5, 16.5, 14.2, 12.3 for 1, 3, 5,

7 number of times respectively. It showed an uncertain variation in the weight percentage of the ZrO2 due to the breakage of coatings.

5. Conclusions

ZrO_2 coatings were successfully deposited on the AA5052 and glass substrates. Though the coatings were observed to be uniform and homogeneous initially, SEM images of the substrate depicted the cracks and spallation of the coatings occurred due to weak bond strength/adhesion between the film and the substrate. EDX analysis showed the elemental composition of the coated AA5052 substrate, which indicated an uncertain variation in the weight percentage of the Al, O, Br and Zr. Further, the morphological study revealed that the deposited films were initiated with cracks due to the exposure to thermal flux.

References

[1] Veanuga Ganapathy, Tedi Kurniawan, H Mas Ayu, Yuli Panca Asmara Rosdi Daud, Niki Prastomo and Asep Bayu Dani Nandiyanto 2018 *Journal of Engineering Science And Technology* **Vol. 13** No. 6 1713-21

[2] Y Joshua Dua, Matt Damrona, Grace Tang a, Haixing Zheng a, C J Chua and Joseph H. Osborne b 2001 *Progress in Organic Coatings* 41 226–32

[3] R Di Maggio, L Fedrizzi, S Rossi and P Scardi 1996 *Thin Solid Films* **286** 127-35

[4] Ivana Bacic, Helena Otmacic Curkovic, Lidija Curkovic, Vilko Mandic and Zrinka 2016 *Int. J. Electrochem. Sci.* **11** 9192 – 205

[5] V R Capelossi, M Poelman, I Recloux, R P B Hernandez, H G de Melo and M G Olivier 2014 *Electrochimica Acta* **Volume 124** 69-79

[6] Duhua Wang and Gordon P. Bierwagen 2009 *Progress in Organic Coatings* **Volume 64** 327-38

[7] A N Khramova, N N Voevodinb, V N Balbysheva and R A Mantzc 2005 *Thin Solid Films* **Volume 483** 191-96

[8] M L Zheludkevich, R Serraa, M F Montemor, I M Miranda Salvado and M G S Ferreira 2006 *Surface and Coatings Technology* **Volume 200** 3084-94

[9] Gilmar P Thim, Maria A S Oliveira, Evandro D A Oliveira and Francisco C L Melo 2000 *Journal of Non-Crystalline Solids*, **Volume 273**, 124-8

[10] Hehong Zhang, Xiaofeng Zhang ,Xuhui Zhao, Yuming Tang and Yu Zuo 2018 *Coatings* **8(11)** 397

[11] F Andreatt, L Paussa, A Lanzuttia, N C Rosero Navarro, M Aparicio, Y Castro A.Duran, D Ondratschek and L Fedrizzia 2011 *Progress in Organic Coatings* **Volume 72** 3-14

[12] Armaghan Naderi, Abolghasem Dolati, Abdollah Afshar and Genevieve Palardy 2020 *Materials Chemistry and Physics* **Volume** 122696

[13] Renata Batista Rivero Garcia, Fábio Santos da Silva and Elizabete Yoshie Kawachi 2013 *Colloids and Surfaces A: Physicochemical and Engineering Aspects* **Volume 436** 484-8

[14] F Andreatta, L Paussa, P Aldighieri, A Lanzutti, D Ondratschek and L Fedrizzi 2009 *Aluminium Surface Science & Technology* **Volume42** 293-8

[15] E A G Shillington and D R Clarke 1999 *Acta Materialia* **Volume 47** 1297-305

[16] M Shinozaki and T W Clyne 2013 *Acta Materialia* **Volume 61** 579-88

IOP Conf. Series: Materials Science and Engineering 1065 (2021) 012019 doi:10.1088/1757-899X/1065/1/012019

Synthesis and characterization of yttrium nitrate hexahydrate

Madhusudhana R*[1], Lovesome Benedict S[1], L. Krishnamurthy[1], R Gopalakrishne Urs[2] and L Vinaykumar[3]

[1]Centre for Nanotechnology, Department of Mechanical Engineering, The National Institute of Engineering, Mysuru-08

[2]Department of Physics, The National Institute of Engineering, Mysuru-08

[3]Department of Physics, JSS College of Arts, Commerce and Science, Mysuru-25

Abstract Nanostructured coatings are extensively employed in aerospace applications as they unveil unique and excellent properties because of their nano-sized structural behavior. This article focusses on the study of preparation of yttrium (III) nitrate hexahydrate-based nano-sized structural coating obtained by sol-gel method and the analysis of their properties. The properties that are suitable for aerospace applications such as thermal stability, oxidation resistance ability, heat resistant nature, phase stability were analyzed using Scanning Electron Microscope (SEM), Energy Dispersive X-ray Analysis (EDX) characterization techniques.

Keywords: Yttrium nitrate hexahydrate, nanostructured coatings, sol-gel.

1. Introduction

Nanostructured coatings are very essential in day to day applications as they offer unique properties than any other material composition due to their nano-regime. These coatings made of yttrium, zirconium, cerium which are implemented in aerospace applications such as turbine blade coatings, coatings on aircraft body, coatings on engine components as thermal barrier coatings etc. Here yttrium-based yttrium (III) nitrate hexahydrate is used as it offers high corrosion resistant property when compared to other material composition. So these coatings are generally developed by various methods such as plasma spraying techniques, vapor deposition techniques and also through various chemical procedures. In this article sol-gel wet chemical synthesis method has been employed to develop yttrium (III) nitrate hexahydrate based nanostructured coatings as this process produces highly homogenous, high quality [1] and layers associated with uniform phase [4]. This surface modification based sol-gel technique provides enhanced sliding wear resistant property [5,13] and high adherent coatings as compared to the nanostructured coatings developed by other conventional methods [6]. The main properties of these coatings to be considered are low thermal conduction, high thermal, chemical inertness, phase stability, high oxidation, heat resistant ability and high refractive index for high temperature applications [3]. The effect of various calcination temperatures involved in sol-gel results in varied thickness of the coatings [7,14]. The phase composition and changes are mainly dependent on percentage of yttrium present in the material [8]. Here the nano-coating of freshly prepared yttrium based coating was developed by dip coating and spin coating as it provides control over layer thickness and uniform deposition respectively [10,16].

Characterization techniques such as Scanning Electron Microscope (SEM) and Energy Dispersive X-ray Analysis (EDX) were employed to study and analyze various properties such as phase stability, thermal stability, heat resistant properties of yttrium (III) nitrate hexahydrate based nanostructured coatings.

2. Experimental Details

2.1 Synthesis of yttrium coating by sol-gel technique:

Y_2O_3 nano-structured coatings are developed by sol-gel chemical wet synthesis process in which 10g of $Y(NO_3)_3.6H_2O$ powder is dissolved in 200ml of distilled water. Further, it is mixed and heated at 40°C for 10 min with an rpm of 250 - 260 using a magnetic stirrer for the proper dissolution of the solution. After 5 min, 6ml of ethylene glycol solution is added dropwise to the solution. Later, the mixed solution is subjected to thermal heating and mixing at 100°C for 2 hours with rotational speed of 360 – 370 rpm using magnetic stirrer to achieve Y_2O_3 nanostructured gel formation. Then, the formed gel solution is kept separately for 30 min at room temperature [2]. Again 5ml of ethylene glycol solution is added, heated and mixed at 100°C for 1 hour with a rotating speed of 960 rpm using magnetic stirrer and allowed for aging of the solution for 18 hours.

Further to develop nano-coating on the glass substrate, dip coating and spin coating techniques were employed. Initially, the glass substrates (microscope slides) were cleaned by subjecting it to ultra-sonication for about 480 seconds by immersing it in isopropyl alcohol (IPA) solution. The nano-structured coatings are developed in four different combinations, first dip coating is performed during which dipping the four different glass substrates (named as x, xx, xxx, xxxx) into the solution for 1time, 2times, 3times and 4times respectively for the dipping duration of 10s and with the time lag of 30s between each dip for the proper analysis of the coating. During spin coating, same like dip coating 4 different glass substrates (named as 1,2,3 and 4) are considered, in which each spin action is performed on all 4 substrates for 10s of spin for 1 spin, 2 spins, 3 spins and 4 spins respectively by adding 1drop of prepared gel at each spin with the rotating speed of 350 rpm. The coated glass substrates were subjected to annealing at 350°C for 5 hours to obtain uniform coating on the substrate.

3. Characterization

Scanning Electron Microscope (SEM), a type of characterization method that uses a beam of electrons which comes in contact with the atoms of the sample and provides the details and information about the surface morphology and composition. Here, CARL ZEISS SEM images exhibited porous and worm- like structures that were considered to define the thermal conductivity and phase stability of the nano- coating. Energy Dispersive X-ray Analysis (EDX) is used to study the elemental and chemical composition of the sample. It works on the principle of XRD where the generated x-rays comes in contact with the atom of the sample and provides the information about chemical composition. Here the variation in the weight percentage composition of yttrium is analyzed by EDAX AMETEK.

4. Result and Discussion

The SEM images of both dip coated and spin coated coatings were analysed. The images of dip coated samples (figure 1. low magnified SEM image of dip coated sample and figure 2. high magnified SEM image of dip coated sample) were associated with porous structures that were required to define the phase and thermal stability of the coating.

Usually when we consider nano-sized structures or small –sized structures, their thermal conductivity property is mainly influenced by the nature of porosity [15]. As the porosity of the nano structured coating increases it results in decrease in thermal conductivity property that helps in applying this coating in high temperature applications [9]. Here figure 1 is 30µm resolution image, that shows highly porous nano-structured coating obtained by dip coating for 2 times. Figure 2 is

10µm resolution image, that also showed the development of prominent porous features of the 4 times dip coated sample.

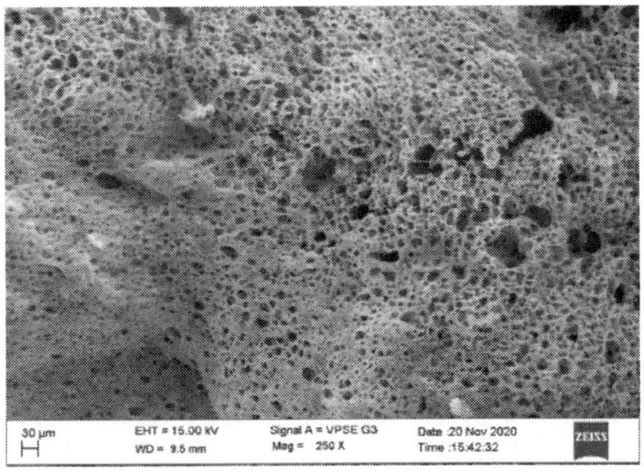

Figure 1. Low magnified (30µm) SEM image of dip coated sample (xx).

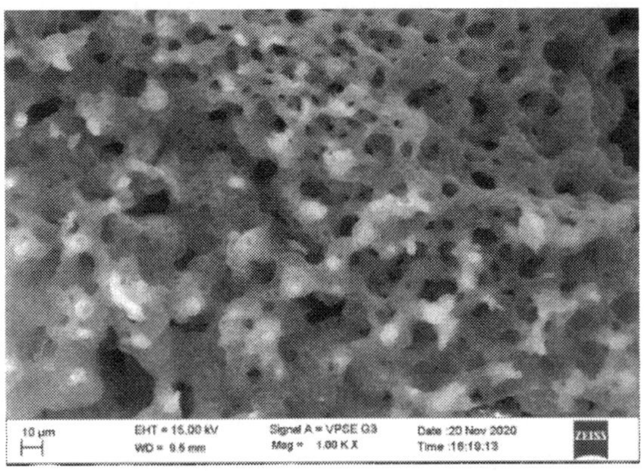

Figure 2. High magnified (10µm) SEM image of dip coated sample (xxxx).

So the above SEM image results indicated that, the coatings obtained by sol-gel method through dip coating exhibited porous structure for every different dipping combinations which was very essential in attaining low thermal conductivity property.

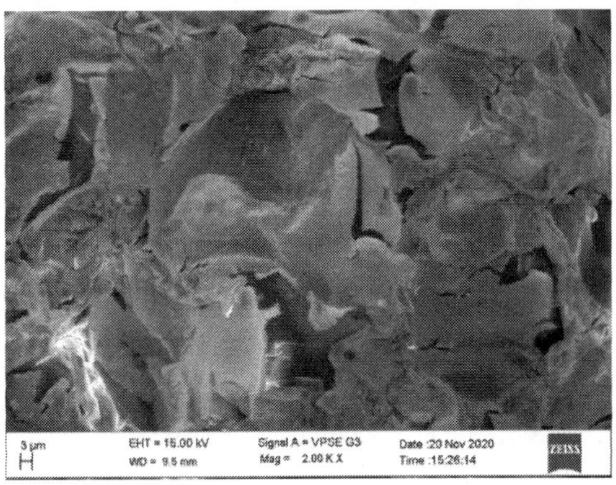

Figure 3. SEM image of 4 times spin coated sample

IOP Conf. Series: Materials Science and Engineering 1065 (2021) 012019 doi:10.1088/1757-899X/1065/1/012019

Figure 3 indicates the surface morphology of yttrium after spin coating. As can be inferred, the obtained coatings includes some agglomerates containing adhered sol-gel derived nanoparticles [11,12]. The yttrium coating surface consists of impacted coating splats and some small adhered particles which can originate from splashing of molten and partially upon impact on the substrate surface [17,18]. Moreover figure 3 indicates the spallation of the coatings upon introduction to thermal flux in an oven at a temperature of 350°C. Residual stresses in the splats which originated from thermal expansion co-efficient mismatch of yttrium coating and the substrate material led to the formation of horizontal intra-splat cracks during cooling [19]. On the other hand, the crack initiation was attributed to the lower adhesion strength between the splats and the substrates [20,21].

The EDX analysis showed the variation in the weight percentage of yttrium composition in the coating.

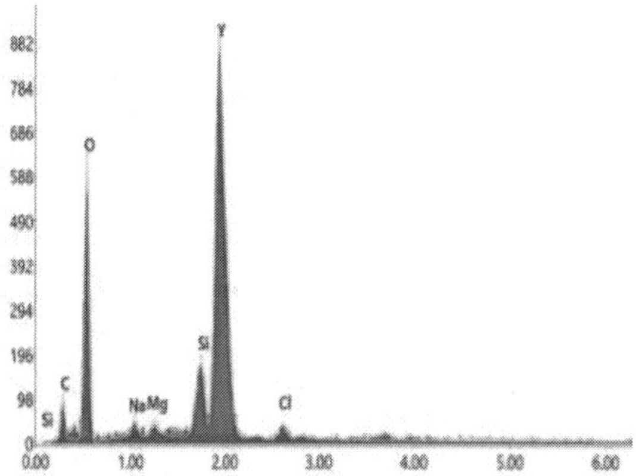

Figure 4. EDX image of 4times dip coated sample

Figure 4 shows the EDX analysis of 4times dip coated sample, that exhibited 42.1 weight% of yttrium composition which confirmed the presence of yttria coated by dip coating technique. The plot also revealed the presence of oxygen content with small traces of Si and Cl.

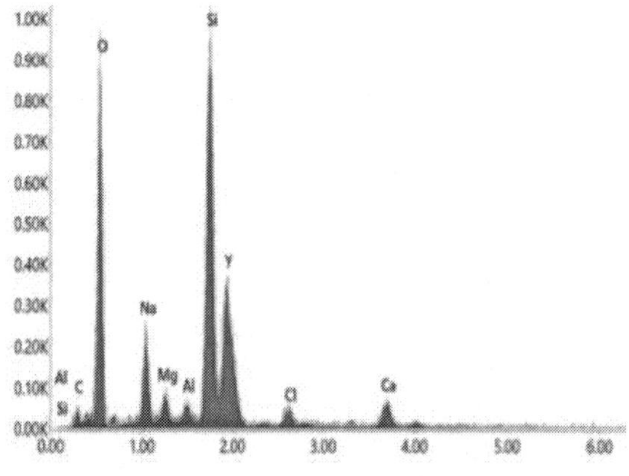

Figure 5. EDX image of 4times spin coated sample

Figure 5 depicts the EDX image of 4 times spin coated sample, image shows the presence of only 18.7 weight% of yttrium composition in the sample as compared to figure 4. The image also depicts the highest content of silicon and oxygen which is about 20.1 weight% and 40.8 weight% respectively. The presence of silicon indicates the background counts of the substrate.

5. Conclusion

Sol-gel assisted nanostructured yttrium nano particles were synthesized which was made in the form of gel so as to produce coatings. The coatings were analysed for SEM which exhibited porous structures and agglomerated adhered nanoparticles. Spallation of the coatings was observed as they were exposed to high temperature of about 350°C. The cracks developed due to cooling action which resulted in spallation of the coatings which is attributed to poor adhesion between splats and the substrates. The EDX results revealed the presence of yttrium content of about 42.1 weight% and 18.7 weight% in both dip and spin coated coatings respectively. Also the EDX results showed the presence of oxygen, silicon and chloride composition.

References

[1] R Subasri, K. R. C. Soma Raju 2020 Handbook of Advanced Ceramics and Composites: Defence, Security, Aerospace and Energy Applications 1413-1442
[2] Keyvani, M. Bahamirian, B. Esmaeili, 2020 ceramic international vol 46 21284-21291
[3] Satish Tailor, Manoj Singh, A. V. Doub 2016 Journal of Cluster Science 24(4) 1097-1107
[4] Kuo Jiang, Songbai Liu, Xin Wang 2017 Ceramics International 43(15) 12633-12640
[5] Dipak Kumar, KN Pandey 2020 Sadhana 40(1) 1-12
[6] Dipak Kumar, KN Pandey 2020 Metals and Materials Internaltional, 1-3
[7] Ali Majedi, Fatemeh Davar, Alireza Abbasi, Ali Ashraf 2016 Journal of Inorganic and Organometallic Polymers and Materials 26(5) 932-942
[8] Artem S Mokrushin, Elizaveta P Simonenko, Nikolay P Simonenko, Kirill A Bukunov, Vladimir G Sevastyanov, Nikolay T Kuznetsov 2019 Ionics 25(3) 1259-1270
[9] Hatim Machrafi, Georgy Lebon, 2015 Physics letters A 379 968-973
[10] Baomei Wen, Yizhong Huang, John J Boland 2008 The Journal of Physical Chemistry C 112(1)
[11] 106-111
[12] Qi Zhang, Chengjian Wang, Milin Fu, Jianli Wang, Shiping Zhu 2017 Polymer Chemistry 8(36)
[13] 5474-5480
[14] Karsten Vogtt, Gregory Beaucage, Micheal Weaver, Hanqiu Jiang 2017 Soft Matter 13(36) 6068- 6078
[15] NP Simonenko, EP Simonenko, VG Sevastyanov, NT Kuznetsov 2016 Russian Journal of Inorganic Chemistry 61(6) 667-673
[16] NP Simonenko, EP Simonenko, VG Sevastyanov, NT Kuznetsov 2016 Russian Journal of Inorganic Chemistry 61(6) 667-673
[17] Thierry Grosdidier, Gang Ji, Nathalie Bozzolo 2006 Intermetallics 14(7), 715-721
[18] Ravinder Kaur, AV Singh, Kiran Sehrawat, NC Mehra, RM Mehra 2006 Journal of non-crystalline solids 352(23-25) 2565-2568
[19] Jinlong Wang, Minghui Chen, Lanlan Yang, Li Liu, Shenglong Zhu, Fuhui Wang, Guozhe Meng 2016 Applied Surface Science 366 245-253
[20] F Rovere, PH Mayrhofer, A Reinholdt, J Mayer, JM Schneider 2008 Surface and Coatings Technology 202(24) 5870-5875
[21] Jie Lu, Ying Chen, Han Zhang, Chunshan Zhao, Xiaofeng Zhao, Fangwei Guo, Ping Xiao 2019 Corrosion Science 159 108145
[22] Harish C Barshilia, Archana Chaudhary, Praveen Kumar, Natarajan T Manikandanath 2012 Nanomaterials 2(1) 65-78
[23] F Rovere, D Music, JM Schneider, PH Mayrhofer 2010 Acta materialia 58(7) 2708-2715

IOP Conf. Series: Materials Science and Engineering 1065 (2021) 012020 doi:10.1088/1757-899X/1065/1/012020

Physical, chemical and surface morphological characterization of single areca sheath fiber

Suresh P S[1,2], Dilip Kumar K[3], Dhanalakshmi S[4], and Basavaraju B[5*]

[1]Department of Mechanical Engineering, Alva's Institute of Engineering and Technology, Mijar - 574225, Visvesvaraya Technological University, Belagavi, Karnataka, India.

[2]Research Scholar, Department of Mechanical Engineering, Shree Devi Institute of Technology, Mangalore - 574142, Visvesvaraya Technological University, Belagavi, Karnataka, India.

[3]Department of Mechanical Engineering, Shree Devi Institute of Technology, Mangalore - 574142, Visvesvaraya Technological University, Belagavi, Karnataka, India.

[4]Department of Chemistry, KLE Technological University, Hubballi - 580031, Karnataka, India.

[5]Department of Chemistry, Alva's Institute of Engineering and Technology, Mijar - 574225, Visvesvaraya Technological University, Belagavi, Karnataka, India.

*basavaraju_b@yahoo.co.in

Abstract: The technological advance in various streams necessitated the increased demand for lesser weight and minimal cost materials. The concerns to ensure environmental sustainability, those newer materials were expected to be decomposable. In the current study, chemical, physical and mechanical and morphological aspects of areca sheath/frond fiber are investigated. It is observed that, the cellulose, hemicelluloses and lignin wt. % are 65.02, 8.26 and 18.62 respectively. The XRD study confirms the percent crystallinity of 81.33% and the crystallinity index of 0.77. The fiber further exhibited the tensile strength of 59.62±2.20 MPa. The TGA study reveals the higher degradation rate at 322.2 °C and the residual mass at 500 °C is 33.23%. The morphological study is done through scanning electron microscopy (SEM). The other properties such as density (0.9 g/cm^3), the fiber length, diameter, and moisture uptake are also reported. From the very promising results, the fiber can be used as a reinforcement agent and the composites may find potential applications in automobile panels, interiors, structural light weight bearing sheets, etc.

1. Introduction:

The propensity to save the environment can be witnessed with great amount of eco-friendly approaches by the human beings. The vivid research activities in the field of natural fiber composites during the recent past few years can be attributed to it [1-3]. The inclination towards the natural fibers as a reinforcement in the polymer composites is because of its intrinsic qualities viz., reduced weight, readiness to decompose, non-polluting, toxic free, minimal cost, renewability, and good mechanical properties [4-7]. The researchers exploited its inherent qualities in devising the composites out of these fibers and found its suitability for various applications in automobile, aerospace, sport equipment, construction and many other sectors.

The fibers used in the manufacturing of the composites are from the various parts of a plant. The concentration of fibers may be different for different plants and the common sources are bark, stem, fruit, leaf stalk, and frond. The source from where the fibers are extracted from will have direct influence on its biochemical features and physical characteristics such as density, diameter of fiber, surface roughness, and thermo-stability [8, 9]. The natural fibers consist of three primary constituents, namely cellulose, hemicelluloses and lignin [10]. The cellulose has a unique geometry with respect to a fiber type. Generally, the cellulose content present in it will have a significant influence on the mechanical properties [11]. The mechanical properties of composites like tensile strength, Young's modulus, elongation at break are dependent on the cellulose content of the fiber [12]. The fraction of components such as cellulose and lignin will affect the thermo-stability of the fiber and hence its composites [13]. The hemicelluloses, lignin and pectin present in the fiber will act as an adhesive to bind the cellulose structure in the fiber [14]. The features of the fiber that governs its mechanical properties includes its crystal structure, the degree of crystallinity, included spiral angle of fibrils, degree of polymerization, void content, lumen size and the chemical constituents [15]. Other factors such as the form of the fiber (powder, short, or continuous), inclination of the fiber, and fiber fraction will also alter the mechanical properties of a composite, when the fibers are used as reinforcement. The other aspects like the density, aspect ratio, and fibril orientation, will significantly affect the mechanical behavior [16]. For example, the aspect ratio is one of the critical functions which govern the stress transmission. If the diameter of the fiber is less, then the fiber and matrix adhesion will be good in the composites, which in turn help in the effective stress transmission. Water uptake character is also discussed. If the fiber absorbs more moisture due to the hydroxyl groups present in it, the fibers present in the composites will swell and therefore the adhesion with the matrix will fail resulting in poor mechanical properties [17].

Various perennial fibers from giant reed (Arundo donax L.), Cissus quadrangularis, Grewia tiliifolia, banana, hemp, sisal, etc. and novel natural fibers like Saccharum Bengalense Grass (Sarkanda), Furcraea Foetida leaf fiber, Coccinia grandis.L, Sida cordifolia.L., etc. are being explored to find its potential in environment friendly applications [18-23]. With a quest to find the potential and its suitability as a reinforcement agent, the areca sheath lignocelluloses fibre is investigated. The other unique reasons concerned to the study on areca sheath fiber are: The fiber is abundantly available at the site of research situated in Mijar, Dakshina Kannada, Karnataka, India; To the best of authors' knowledge, no similar literature is available on the current study. The study is concerned with physical properties, chemical constituents, thermo-stability, tensile properties, crystalline properties and morphological aspects.

2. Research Methodology

2.1 Materials

The parameters such as the fiber property, the type of polymer used as a matrix material, the strength of adhesion between the fiber and the matrix govern the different properties of the composite materials. The climatic condition and the age of the plant from which the fibers are extracted influence the chemical constituents in it which contribute to the various properties of the fibers [24]. All the fibers have cellulose in common in them as a principal structural component. Most fibers also contain the structural polymers such as hemicellulose and lignin. With these three major constituents the fibers are collectively named as lignocellulosic fibers. The cellulose structure in the fiber comprises of more hydrophilic groups. The conflict in hydrophilic nature of the natural fibers and the hydrophobic nature of the matrix interrupt the interfacial adhesion between them. Therefore, before introducing a fiber in to the manufacturing of composites, the characteristics of the fibers are to be studied thoroughly and hence this study. The areca sheath fibers were procured from Go Green products, Chennai. The fiber characterization is done as per the ASTM guidelines.

2.2 Thermal studies

The composites stability associated with the temperatures above the surrounding temperatures are of practical interest [25]. STA 409 PL Luxx thermal analyzer is employed which is operated from room temperature into high temperature of 500 °C with a rate of heating maintained at 10 °C/min. Inert atmosphere of nitrogen is used.

2.3 X-ray diffraction investigation

High modulus fibers are preferred for the composite application. The crystallinity of the fibers influences the moduli of the fibers and hence to achieve the desired mechanical properties on the fiber composites, consideration of higher crystallinity is an essential factor. To study the crystallinity of the areca sheath fiber, the powdered areca sheath fiber specimen is tested under X-ray generator (Bruker D8 Advance), Ni filter and scintillation counter as 40 kV and 50 mA. The radiation source Cu-Kα line (1.54 Å) is used. The intensity of the diffraction is recorded in 2θ angle range (between 3° and 60° at 2θ scale at 1 second step size).

2.4 Morphological study

The areca sheath fibers sputtered with 10 nm gold layer are observed under JEOL JSM-T330A SEM (scanning electron microscope).

2.5 Quantification of chemical constituents:

Following methods were incorporated in the investigation of the chemical constituents present in the areca sheath fiber [26-28].

2.5.1 Quantification of extractives

The organic extractives of air dried areca sheath fibers (5 g) were removed in Soxhlet extraction heater for 5 h, using a mixture of ethanol and toluene. After taking the fibers out from the extractor, they were rinsed with ethanol and hot water. Then it is oven dried at a 60 °C. The extractives were quantified comparing the weight loss and the initial weight in terms of percentage.

2.5.2 Quantification of lignin

Certain properties like the hardness, bleachability, color etc. are connected with the amount of lignin present in the fiber. As lignin is hydrophobic in nature, it reduces the water absorption. The lignin quantification is done with the following procedure. 15.0 mL of 72% sulphuric acid incrementally added while stirring to the beaker containing 2.0 g of extracted areca sheath fiber sample. Sulphuric acid is used to destroy the cellulose and to enable the lignin isolation. To ensure the complete solution, the material is stirred frequently for 2 h at 25±1 °C. Once the specimen is dispersed, 300 mL of distilled water is added. The solution is then boiled for 2 h and allowed to get the insoluble lignin to get settled keeping the flask in an inclined position. After about 24 h, the supernatant solution is siphoned off and the lignin was transferred and washed with hot water to remove the acid residual. The crucible containing the lignin is then dried at 105±3 °C to constant weight. Then it is cooled in desiccator and weighed.

2.5.3 Quantification of holocellulose

Holocellulose is highly hydrophilic in nature and causes in more water absorption. A 3.0 g of extractive-free dry fiber samples are placed in a 250 ml Erlenmeyer flask. Further, 150 ml of distil water is added in to the flask along with the acetone acid 0.5 ml and 1.0 g sodium chlorite. The content in the flask is heated up to 70 °C in a water bath for 1 h, following an additional dosage of 0.5 ml of acetone and 1.0 g of sodium chlorite were added. Heating is continued for next 1 h. Repetition of the process is done 4-6 times until all the lignin content is completely removed. Finally, the content in the flask is allowed to cool for about 24 h and the residue is washed using acetone. It is then dried in the hot air oven at 105 °C for 24 h and solid residue left out is weighed.

2.5.4 Quantification of α-Cellulose

Approximately 3 g holocellulose sample is kept in a 250 cc beaker and 35 ml of 17.5% carbonate-free solution of sodium hydroxide (20 °C) is added to it. The solution is allowed to stand for about 5 minutes. The sample is macerated by gently stirring with a glass rod. The sodium hydroxide is intermittently (once in 5-6 min) added in 10 ml portions and a total of 40 ml of sodium hydroxide is added during this interval. Then a 35 ml of distil water is added and the beaker is covered with a watch glass for an hour. The residue is filtered and successively rinsed with 100 ml of 8.5 % sodium hydroxide, distil water, 10% acetic acid and finally with water. The α –cellulose is then oven dried up to constant weight and allowed get cooled in desiccator and then weighed.

2.5.5 Quantification of hemicellulose

The difference between the fractions of holocellulose and the α-cellulose is the fraction of hemicellulose.

3. Results and Discussions

3.1 Chemical constituents of areca sheath fiber

The experimental results of the major constituents of areca sheath fibers are in agreement with various other studies reported. The experimental values of cellulose, hemicelluloses and lignin are observed to be in wt.% of 65.02, 8.26 and 18.62 respectively. The result is very much compromising with other studies [29, 30]. The cellulose content is very much closer to the Jute (61-71 wt.%), Hemp (68 wt.%), Ramie (68-76 wt.%), Abaca (53-63 wt.%), and Sisal (65 wt.%) [31].

3.2 Fiber length

Fiber length in a composite is very significant parameter which imparts certain properties to it. Generally, the parameters pertaining to a fiber such as the higher crystallinity, higher cellulose content, higher aspect ratio and lower micro-febrile angle will cause in enhanced mechanical properties of the composites [32]. The fiber length is performed with a hundred areca sheath fibers. The fibers were extended and the length is measured between the ends, ensuring no stretch outs incurred. The lengths of the fibers are ranging between 800 mm and 1400 mm.

3.3 Determination of linear density and the diameter of the fiber

The measurement of fineness, linear density of the fiber is the amount of mass per unit length and determined in line with ASTM-D1577–01. A single fiber which is long enough is cut in to small pieces of known length and it is then weighed. By measuring the weight (0.2352 g) and length of specimen fibers (110 cm), linear density is calculated with the following Equation 1 and found to be 0.2138 g/m (213.8 Tex) [33].

$$\text{Linear Density (in Tex)} = \frac{\text{Weight of fiber (in gram)}}{\text{Length of fiber (m)}} \times 1000 \tag{1}$$

The optical microscope (LEICA DM750M) is used to determine the fiber diameter. The average of 10 known length single fibers and average of 100 fiber diameters are used to determine the volume density (ρ_f) with the Equation 2 [34]. With the lengths varying from 80 cm to 140 cm and diameter varying between 400 µm to 700 µm the fiber density of areca sheath fiber is estimated to 0.90 g/cm^3.

$$\text{Volumetric Density } (\rho_f) = \frac{\text{M}}{\text{L}\left(\frac{\pi d^2}{4}\right)} \text{ g/cm}^3 \tag{2}$$

Where, M is the weight of the fiber (g), L is the length of the fiber (cm) and d is the diameter of the fiber (cm).

The density of areca sheath fiber is very closer to Hemp (1.1-1.6 g/cm^3), Kenaf (0.6-1.5 g/cm^3), Bamboo (1.2-1.5 g/cm^3), and Coir (1.15-1.6 g/cm^3) [35].

3.4 Thermal analysis

The areca sheath lignocellulosic fiber is subjected to various techniques of thermal analysis and the physical property and its reaction products are determined as a function of temperature. Consequent change in mass versus the temperature (TGA curve) will provide facts on its thermal stability and/or composition of the specimen/ intermediate compound and of any residue left out. The DTG curve depicts the decomposition reaction rate for the weight losses over time. The analytical technique DTA which provides supplementary information such as transformations occurred during the heating program such as glass transition, crystallization, melting and sublimation is included. TGA and DTA curves pertaining to the thermal degradation of areca sheath fiber is shown in figure 1. The DTG curve of raw areca fibers is shown in figure 2.

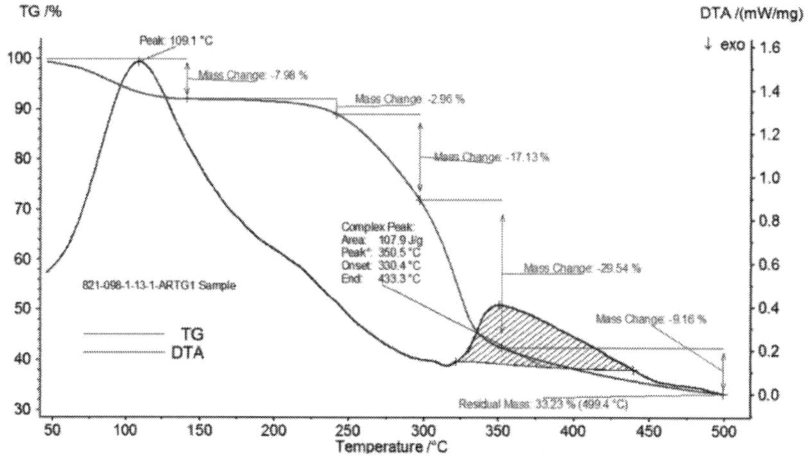

Figure 1. TGA and DTA curves.

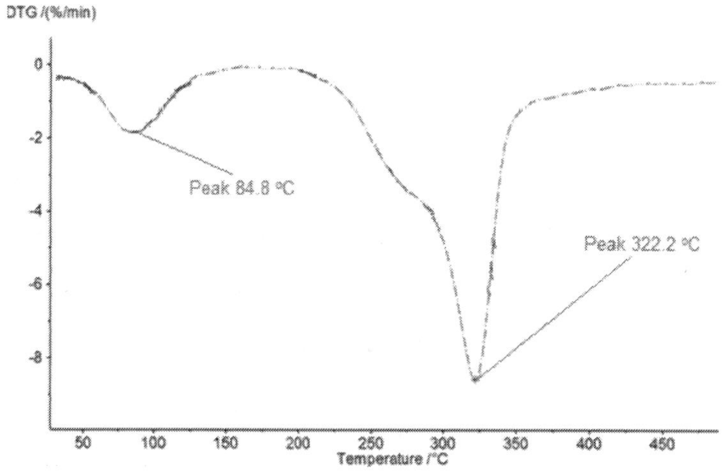

Figure 2. DTG analysis.

The analysis of thermal stability of the areca sheath fiber through TGA, DTA and DTG is presented in the table 1, 2, and 3 respectively.

The thermal stability of fiber was analyzed by using TGA plots and the outcomes are tabulated in table 2 and table 3. The thermal stability of fibers is complemented by the DTG curves and are tabulated in Table 4.

Table 1.TGA analysis of weight loss at various temperature ranges.

1st level of degradation		2nd level of degradation	
Temperature range 1(°C)	Weight loss (%)	Temperature range 2(°C)	Weight loss (%)
241.5 to 297.0	17.13	297.0 to 353.0	29.54

Table 2. Analysis of decomposition temperature and weight loss percentage of Areca sheath fibers.

Decomposition temperature, °C		**Mass change at, %**			**Residual mass at 500 °C (%)**
25% mass change	50% mass change	150 °C	300 °C	500 °C	33.23
289.5	328.5	8.04	21.39	37.32	

Table 3.DTG results of Areca fibers.

Decomposition temperature, °C	
Peak 1	Peak 2
84.8	322.2

As reported by other similar studies on natural fibers [36, 37, 13], areca sheath fibers also exhibited the decomposition of its contents at major two levels. The organic extractives and the moisture got vaporised completely at 241.5 °C and the early weight loss of 10.96% can be witnessed from the TGA curve. From this temperature onwards the areca sheath fiber started losing its thermal stability. In the first level of content decomposition between the temperature range 241.5 °C and 297.0 °C, due to the depolymerization of hemicellulose and pectin through hydrous pyrolysis, dissociation of links of cellulose and lignin, 17.13% of weight loss is observed. In the second level of decomposition between the temperature range 297 °C and 353 °C, due to the decomposing of α-cellulose as well as lignin, weight loss of 29.54% can be noticed [36]

The very complex structure of lignin with various branches of aromatic rings results in its gradual degradation starting from 241.5 °C till the end. Similar trend of lignin degradation can be observed in the curve from 353.0 °C to 500 °C resulting in a 9.16% weight loss. The overall weight loss due to the decomposition of 25% is recorded at 289.5 °C and 50% at 328.5 °C and is presented in Table 3. The weight losses of 8.04%, 21.39% and 37.32% are observed at 150 °C, 300 °C and 500 °C respectively. At 499.4 °C, the final residual weight percentage of areca sheath untreated fiber is 33.23% and contain carbonaceous left outs and un-degraded fillers if any [38].

The DTG curve presented in Figure 2 complements the aforesaid results. The decomposition rate corresponding to the weight loss experienced by the specimen is depicted by DTG curve. The first peak is observed at 84.8 °C and the latter peak at 322.2 °C. The first peak is correlated to the evaporation of the organic extractives and the moisture content present in the untreated areca sheath fiber. Similarly, the latter peak can be correlated to the degradation of cellulose content. The second bigger peak indicates the higher degradation rate of the content took place at 322.2 °C and hence the change in the weight loss.

3.5 XRD analysis

The structural information of the specimen under test is studied by X-ray powder diffraction method. X-ray diffraction is an excellent tool to notice the difference between amorphous and crystalline material present in the natural fiber. Most common general diffraction method employs the diffraction of monochromatic X-rays from a powder specimen under investigation and the intensity of the scattered rays yield the great deal on structural information. The examinations were done by the diffractometer with the intensity of Cu- Kα radiation wavelength of 0.154 nm. The percentage crystallinity and crystallinity index (CI) of the fiber was determined by using following equations [39].

$$\%\mathrm{Cr.} = \frac{\mathrm{I_{crystaline}}}{(\mathrm{I_{crystaline}} + \mathrm{I_{amorphous}})} \times 100 \tag{3}$$

$$\mathrm{C.I.} = \frac{(I_{crystaline} - I_{amorphous})}{I_{crystaline}} \tag{4}$$

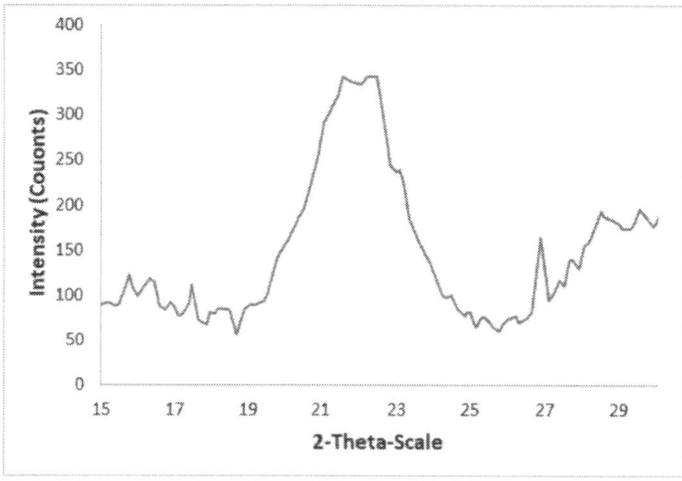

Figure 3. XRD spectrum.

The XRD pattern of the areca sheath is displayed in Fig. 3. The two enlarged diffraction peaks at 18° and 22° are clearly seen in most of the bio-fibers. The peak at 2θ = 18° point out the existence of amorphous constituents in fiber. The peak at 2θ =22° denotes the crystalline content present in the cellulose. The percentage crystallinity and crystallinity index of areca sheath was calculated as 81.33 and 0.77. The crystallinity index 0.77 is higher than that of other bio fibers, such as C. grandis (0.52), F. foetida (0.53), Saharan Aloevera (0.57), S. cordifolia (0.57), C. pangorei (0.41), Acacia leucophloea (0.51), Acacia Arabica (0.52) and almost equal to that of Jute (0.71) and Hemp (0.88). The good crystallinity index indicates the lower amount of amorphous materials like hemicellulose, lignin, and some other non-cellulosic material in the areca sheath fiber.

3.6 Tensile properties of areca sheath single fiber

The areca sheath fiber of 25mm length is investigated for its tensile properties on digital universal testing machine (Mecmesin 2.5 *Xt*) at a cross head speed of 2 mm/min. The test is done in accordance to ASTM D3822-07 standards. Average of 20 specimens under test is reported. Following results are observed: Tensile strength: 59.62±2.20 MPa; Young's modulus: 1.24 ± 0.12 GPa and the Elongation at break: 3.02 ± 0.36%.

3.7 SEM image analysis

SEM micrographs display a general view of cellulosic fibers with a rough surface suitable for a good bond with matrix polymer. The areca sheath fiber under SEM is shown in the figure 4. The fibrils are exhibiting a network structure and interlinked by the hemicellulose and lignin. The fibrils are parallel to each other and oriented in the direction of the fiber. The adhesives, lignin fatty substances are populated in the intercellular spaces and attribute to the firm bonding of the fibrils. Further, it can be clearly seen in the image, there is no cracks, micro voids and helical fibrils on the surface of fiber. So it is revealed that these fibers can be used as reinforcement in polymer composite manufacturing.

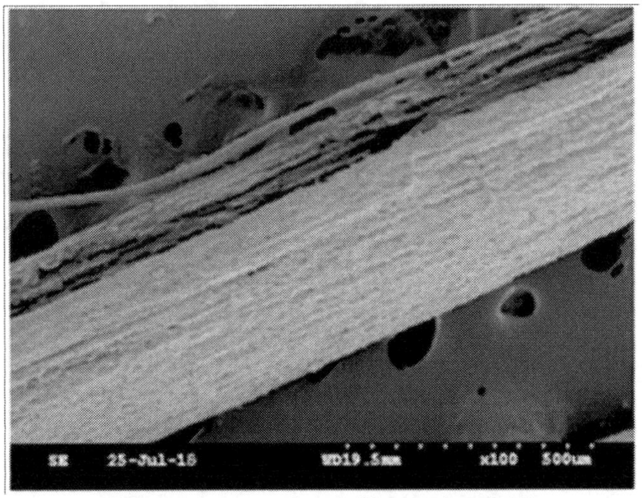

Figure 4. Areca sheath fiber under SEM.

3.8 Water uptake study

The hydrophilic character of areca sheath fiber is studied using desiccators with distilled water. The hot air oven at 60 °C is used to dry the fiber specimens prior to the investigation. It is then weighed using a precise 4-digit balance and placed in the distilled water set up in the humidity chamber. The specimens were taken out after 24 h and wiped with the filter paper and weighed immediately after wiping with a filter paper. The difference in weight is quantified in terms of water absorption by the specimen. The %water uptake is calculated using the following equation 3 and resulted to 60%.

$$\%\,of\ Water\ absorption = \frac{Final\ weight - Initial\ weight}{Initial\ weight} \times 100 \qquad (5)$$

4. Conclusion

The properties of the fiber studied such as density, length and diameter, chemical composition, crystallinity, tensile strength, surface characteristics, thermo-stability and moisture uptake will serve as catalyst in further studies on fiber composites. The density of 0.9 g/cm^3 will show its potential use in light weight load bearing applications. The crystallinity index is 0.77 and the cellulose content of 65.02% is very promising which imparts higher strength to the fiber. The lower lignin content of 8.26 wt.% is contributing the increased tensile strength. The surface study shows smoother surface indicating lower microfibrillar angle which will also supplement to the strength of the fiber. Thermal stability of 322.2 °C and the moisture uptake behaviour (60%) and above mentioned qualities will qualify the fiber as a reinforcement agent in the manufacturing of environment friendly composite materials. The composites may find applications in light weight load bearing applications, automobile panels or interiors, packaging sheets, seat rear supports, etc.

References:

[1] Jawaid M and Abdul Khalil H P S 2011 *Carbohydr. Polym.* **86** 1
[2] Koronis G, Silva A and Fontul M 2013 *Compos. B. Eng.* **44** 120
[3] Manimaran P, Saravanan S P, Sanjay M R, Siengchin S, Jawaid M and Khan A 2019 *J. Mater. Res.* **8** 1952
[4] Sanjay M R 2018 *J. Clean. Prod.* **172** 566
[5] Sanjay M R, Siengchin S, Parameswaranpillai J, Jawaid M, Pruncu C I and Khan A 2019 *Carbohydr. Polym.* **207** 108
[6] Senthamaraikannan P, Saravanakumar S S, Arthanarieswaran V P and Sugumaran P 2016 *Int. J. Polym. Anal. Charact.* **21** 207
[7] Thakur V K and Thakur M K 2014 *Carbohydr. Polym.* **109** 102

[8] Indran S, Edwin Raj R and Sreenivasan V S 2014 *Carbohydr. Polym.* **110** 423
[9] Kathiresan M, Pandiarajan P, Senthamaraikannan P and Saravanakumar S S 2016 *Int. J. Polym. Anal. Charact.* **21** 663
[10] Al-Khanbashi A, Al-Kaabi K and Hammami A 2005 *Polym. Compos.* **26** 486
[11] Mwaikambo L Y and Ansell M P 1999 *Die Angewandte Makromolekulare Chemie* **272** 108
[12] Venkatachalam N, Navaneethakrishnan P and Sathishkumar T P 2016 *J. Ind. Text.* 1528083716682923
[13] Sampathkumar D, Punyamurthy R, Bennehalli B and Venkateshappa S C 2015 *Ciência & Tecnologia dos Materiais* **27** 121
[14] Kabir M, Wang H, Aravinthan T, Cardona F and Lau K-T 2011 *Proceedings of the 1st international postgraduate conference on engineering, designing and developing the built environment for sustainable wellbeing (eddBE2011)* (Queensland University of Technology) p 94
[15] Bledzki A K and Gassan J 1999 *Prog. Polym. Sci.* **24** 221
[16] Bhattacharya D 2006 *Engineering Biopolymers: Homopolymers, Blends, and Composites* (Auckland: Hanser)
[17] Zakaria S and Kok Poh L 2002 *Polym Plast Technol Eng* **41** 951
[18] Vijay R, Singaravelu D L, Vinod A, Paul Raj I D F, Sanjay M R and Siengchin S 2020 *J. Nat. Fibers* **17** 1739.
[19] Belouadah Z, Ati A and Rokbi M 2015 *Carbohydr. Polym.* **134** 429
[20] Fiore V, Scalici T, Valenza A 2014 *Carbohydr. Polym.* **106** 77
[21] Binoj J S, Edwin Raj R and Daniel B S S 2017 *J. Clean. Prod.* **142** 1321
[22] Jayaramudu J, Guduri B R and Varada Rajulu A 2010 *Carbohydr. Polym.* **79** 847
[23] Reddy N, Yang Y 2009 *Carbohydr. Polym.* **77** 898
[24] Maheshwaran M V, Hyness N R J, Senthamaraikannan P, Saravanakumar S S and Sanjay M R 2018 *J. Nat. Fibers* **15** 789
[25] Monteiro S N, Calado V, Rodriguez R J S and Margem F M 2012 *Mater. Sci. Eng. A* **557** 17
[26] Bledzki A K, Mamun A A, Lucka-Gabor M and Gutowski V S 2008 *Express Polym. Lett.* **2** 413
[27] Shinoj S, Visvanathan R, Panigrahi S and Kochubabu M 2011 *Ind Crops Prod* **33** 7
[28] Rowland A P and Roberts J D 1994 *Commun Soil Sci Plant Anal* **25** 269
[29] Loganathan T M, Sultan M T H, Jawaid M, Md Shah A U, Ahsan Q, Mariapan M and Majid M S B A 2020 *J Bionic Eng* **17** 185
[30] Poddar P, Islam M, Sultana S, Nur H and Chowdhury A M S 2016 *J Material Sci Eng* **5** 1000270
[31] Mohammed L, Ansari M N M, Pua G, Jawaid M and Islam M S 2015 *Int J Polym Sci* 243947
[32] Mwaikambo L Y, Tucker N and Clark A J 2007 *Macromol Mater Eng* **292** 993
[33] Alias A H, Tahir P M, Abdan K, Sapuan M, Wahab M S and Saiman 2018 *Sains Malays* **47** 1853
[34] Truong M, Zhong W, Boyko S and Alcock M 2009 *J. Text. Inst.* **100** 525`
[35] Kumar R, Ul Haq M I, Raina A and Anand A 2019 *Int. J. Sustain. Eng.* **12** 212
[36] Albano C, González J, Ichazo M and Kaiser D 1999 *Polym. Degrad. Stab.* **66** 179
[37] Gañan P, Garbizu S, Llano-Ponte R and Mondragon I 2005 *Polym. Compos.* **26** 121
[38] Arbelaiz A, Fernández B, Ramos J A and Mondragon I 2006 *Thermochim Acta* **440** 111
[39] Manimaran P, Senthamaraikannan P, Sanjay M, Marichelvam M and Jawaid 2018 *Carbohydr. Polym.* **181** 650

IOP Conf. Series: Materials Science and Engineering 1065 (2021) 012021 doi:10.1088/1757-899X/1065/1/012021

Compressive properties of 3D braided flax fiber textile fabric reinforced PLA composites

S Kanakannavar[1] and J Pitchaimani[2]

Department of Mechanical Engineering, National Institute of Technology Karnataka, Surathkal, Mangalore, 575 025, India.

[2]**Corresponding Author**: pjeyaemkm@gmail.com

Abstract. In this study, fabric woven using 3D braided flax yarn is used to reinforce Poly Lactic Acid (PLA) composites. Solid braiding technique is used to convert the typical yarns into braided yarn and further these yarns are used to prepare woven fabric. Finally, composites are fabricated through layer stacking method using hot-press hydraulic compression moulding machine. Compression and shrinkage tests are performed on the braided composites. The braided flax fiber composite results are compared with the as received PLA polymer results. Results showed significant enhancement in the compression properties of PLA composites due to the braided fabric reinforcement. These properties are further increased with the increase in filler content respectively. Thermal shrinkage of the braided flax fiber composites is lower than the PLA polymer. This is due to the higher thermal sustainability property associated with the braided flax fiber.

1. Introduction

These days, fiber reinforced composites (FRC's) are generally utilized in car, aviation, military and building applications. Polymers are typically used in composites as a matrix material because of lower cost, simple processability, low density and strong mechanical properties comparable to metals and ceramics. Sadly, these polymers are based on petroleum and are non-biodegradable and non-regenerative polymers. Meanwhile, filler materials normally utilized are synthetic fibers (glass and carbon), which are non-degradable fibers. In this way, environmental-friendly composite material production has been an interesting research field [1]. Bio-composites fabricated with plant-fibers (for example flax, sisal, hemp, coir, bamboo and jute) reinforced thermoplastic biopolymers matrices (PLA, PHA, PCL..) are of great interest [2].

Poly (lactic-acid) (PLA) is a most promising reproducible resource based and bio-degradable thermoplastic matrix [3]. Over the past few decades its low production cost, facile availability and excellent mechanical properties received considerable attention [4,5]. Lactic acid is the basic monomer of the PLA derived by fermentation process from starch. Then lactic acid is polymerised by ring opening polymerisation or poly-condensation to poly (lactic acid) (PLA) [6]. The PLA polymer cost is high in contrast with regular thermoplastic-polymers and for certain applications these matrix materials are weaker. To lessen the matrix price and enhance the material characteristics without altering its "green-polymer" picture, analysts are attempting to get ready characteristic plant fiber strengthened bio-composites [7]. Recent research has examined the production of bio-degradable composites strengthened with natural-fibers for example hemp [8,9], flax [10], kenaf [11], jute [12] and ramie [13] fibers. Some researchers are also used flax nonwoven fabric [14], woven flax fabric, bamboo fabric, cotton fabric, jute fabric [7,15–17] and knitted flax fabric [6] as filler materials for bio-composite fabrication. Natural fiber filled PLA composites mechanical characteristics are based on the volume of fiber in composites, fiber-matrix interfacial bonding and type of matrix, manufacturing conditions and reinforcement materials surface treatment.

Among the plant fibers, flax fibers are most popularly used as fillers for composite applications at present. Generally reinforcement of natural fiber (flax fiber) with polymer matrix enhances the composites mechanical characteristics. Oksman *et al.* [10] manufactured flax fiber filled PLA bio-composites using twin screw extruder with 30 and 40% of fiber content and then composites plates are prepared with compression-molding method. The Flax-PLA composites strength is 50% higher than the flax-PP composites, which are usually used for applications of automotive. Bodros *et al.* [18] showed that the tensile characteristics of the PLLA-flax and PLA-flax composites are high compared to PP-flax composites. Tensile characteristics (strength and modulus) of the PLLA-flax composites are nearer to the glass-fiber filled polyester-composites. Also, the PLLA-flax composites displayed better specific-strength compared to flax-fiber filled polyester composites. Kumar et al. [7] fabricated flax fabric reinforced PLA composites with montmorillonite clay (MMT) addition. Addition of MMT enriched storage modulus and mechanical characteristics of bio-composites. Duc *et al.* [19] conducted experimental study on the unidirectional and twill weave 2/2 flax-fiber filled with epoxy, polypropylene and poly (lactic acid) matrices with 40% of fiber volume and are compared with the glass fiber and carbon fiber filled composites. Damping and stiffness received with twill weave 2/2 flax-fiber filled PLA composites have best compromise.

'In braiding three or more number of yarns are cross-linked along the axis and length of the thread, to obtain a thick and strong thread' [20]. These are used in composites, ropes and biomedical applications. Rajesh and Pitchaimani [21] fabricated basket type woven-fabric with flat-braided yarn filled polyester-composites and they proved that the flat-braided yarn composites mechanical characteristics are greater than short fiber and conventional woven fabric composites. Also, they showed that the braided composites have greater storage modulus compared to knitted fabric and conventional woven fabric filled composites [22]. Kanakannavar and Pitchaimani [23] are produced fabric with solid 3D braided flax yarn filled PLA bio-composites and the buckling at thermal load study revealed that the addition of braided yarn fabric enriched the thermal buckling sustainability and decreased deflection for applied temperature load compared to pure PLA. Also, the addition of woven fabric with flax braided yarn improved the wear resistance quality of PLA [24].

Present research aims to find the effect of 3D braided flax fiber woven fabric reinforcement with PLA on compressive strength and modulus. Braided yarns are prepared by solid braiding method and braided yarn is interwoven to prepare fabric. The prepared fabric is used as filler material with PLA matrix. Composites are fabricated with film stacking and compression-molding methods. On these composites compression and shrinkage tests are performed and then composites results are compared with the pristine PLA.

2. Methodology

2.1. Materials

PLA is procured from Nature Tech India Private Limited, Chennai, India. Simply twisted flax yarn is supplied by The Matrix Enterprises, Tiruppur, India. Dichloromethane (DCM) solvent is purchased from Sri Durga Laboratory Equipment Suppliers, Mangalore, India.

2.2. Preparation of Braided-yarn and woven fabric with plain architecture

Purchased four simply twisted yarns are used to manufacture solid 3D braided flax yarn on a braiding machine. The prepared braided flax yarn is interlaced in warp (longitudinal) direction and weft (transverse) direction to manufacture the plain woven fabric using handloom weaving machine. Figure 1 and Figure 2 represents the sample of 3D braided flax yarn and natural-fiber 3D braided-yarn plain woven-fabric (NFBF).

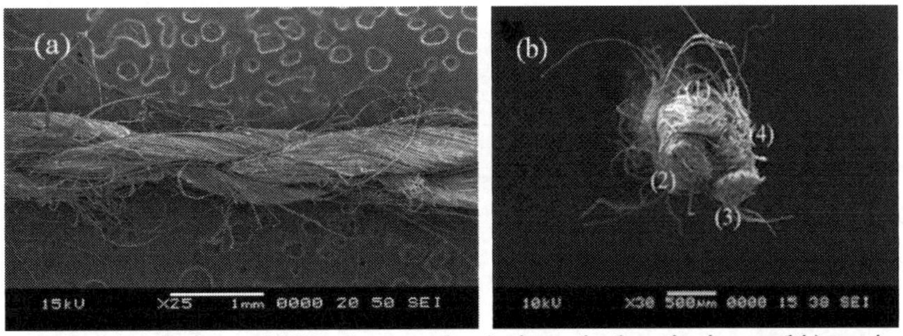

Figure 1 3D braided flax yarn SEM micrographs a) along the length view and b) sectional view

Figure 2 Woven fabric with flax fiber 3D braided yarn (a) photograph of fabric and (b) SEM micrograph of fabric

2.3. Fabrication of composites

In this study, composites are prepared by using two processes. Firstly solution casting is used to manufacture the PLA and flax braided fabric PLA sheets. The solution of PLA/DCM is prepared by dissolving PLA matrix in DCM (dichloromethane). Then this solution is poured in the aluminium square trays, for pure PLA sheet only PLA solution is poured. For braided yarn fabric/PLA (PLA/NFBF) sheet, a known quantity of PLA/DCM solution is poured, on this solution fabric layer is laid by hand layup method. Then another known quantity of PLA/DCM solution is poured above this fabric. To evaporate DCM from pure PLA and braided yarn fabric/PLA trays are kept in room temperature for 48 hours. Secondly the composites are fabricated by film sequencing and hot-press compression method. For this neat PLA and PLA/NFBF sheets are stacked such that PLA sheet comes top and bottom of the PLA/NFBF sheet. Then these laminas are compressed in hydraulic compression molding machine with 5 MPa pressure at 180° C temperature for 15 minutes as shown in Figure 3. After room temperature cooling for 6 hours (or below 40° C temperature), composite laminates are taken out from the machine and coupons are cut as per testing standards. The composites are fabricated with various natural fiber weight fractions (W_F) using following equation.

$$W_F = \frac{W_{NFBF}}{W_{NFBF} + W_{PLA}} \times 100 \tag{1}$$

where, 'W' represents the weight and suffixes '*NFBF*' and '*PLA*' represents the natural fiber braided yarn fabric and poly (lactic acid) matrix respectively.

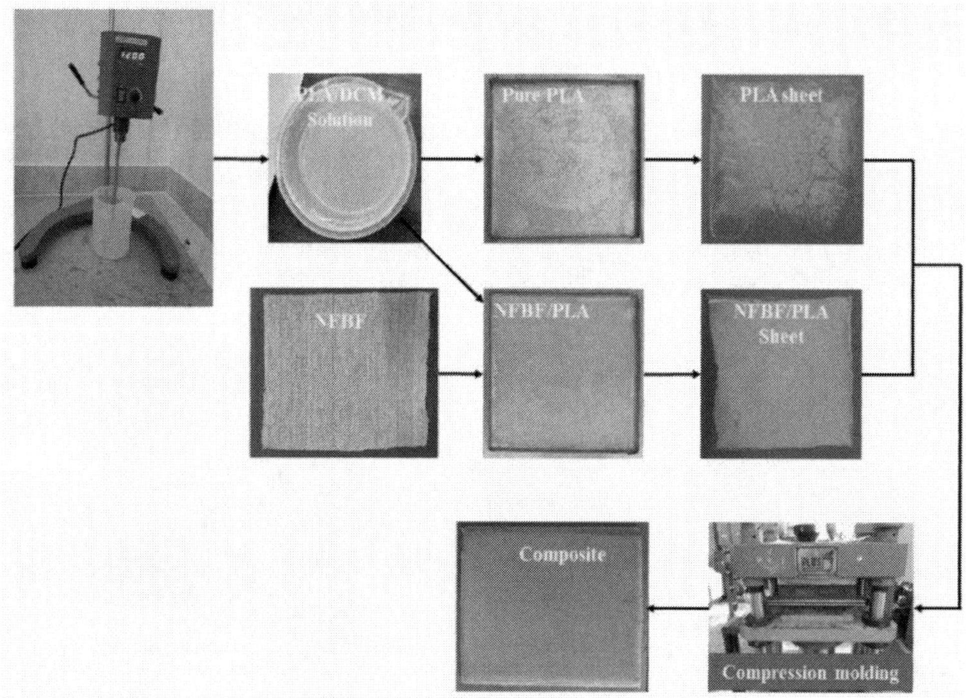

Figure 3 Composites fabrication flow chart

2.4. Compressive test

Compressive test is carried out as out lined in the ASTM D695 standard on a universal testing machine (H75KS, 50KN load cell capacity, Tinius Olsen make, UK) at 1 mm/min cross head velocity. For each fiber loading, five specimens with size of 20 × 20 × 3 mm^3 [25] are used and calculated average results are presented. Braided bio-composites results are compared with the virgin PLA.

2.5. Shrinkage test

Thermal stability of the bio-composites are characterised by placing the PLA/NFBF composite specimens and pure PLA specimens in an vacuum oven at 80° C and the thermal shrinkage of the specimens are measured as the function of time [26]. This study is conducted on the coupons of size 20 × 20 × 3 mm^3. From each composition 3 specimens are used for the testing and average values are calculated. Finally the results of composite are compared with pure PLA.

3. Results and discussion

In the present study, composites are prepared by reinforcing 1, 2 and 3 layer of braided yarn fabric and these are having weight fraction of 11, 22 and 33% respectively. According to their weight percentage these are named as NFBF11, NFBF22 and NFBF33 respectively. Pure PLA is shown as NFBF0.

3.1. Compressive properties

The compressive test is carried out to analyze the effect of NFBF (natural fiber braided yarn woven fabric) reinforcement on the compressive characteristics of PLA composites. Neat PLA and PLA/NFBF composites compressive load and displacement curves are showed in Figure 4. Average maximum compressive load and compressive modulus of neat PLA and PLA/NFBF composites are indicated in Figure 5. From Figure 4 it can be observed that the behavior of the compressive load versus displacement curves are linear for PLA and PLA/NFBF composites up to 0.7 mm displacement. After 0.7 mm displacement curves shows nonlinearity and it represents the crack initiation in the PLA and PLA/NFBF composites. The virgin PLA showed average maximum compressive strength of 71.74 MPa and average compressive modulus of 1.63 GPa. From Figure 5, it can be seen that for braided yarn woven fabric reinforcement, composite showed the increased compressive strength and modulus as compared to neat PLA and these are further enriched with the filler content increase. The fiber content of 33 weight percentage showed the maximum compressive strength of 83.61 MPa and compressive modulus of 1.92 GPa and these values are 16.54% and 17.17% higher than the virgin PLA.

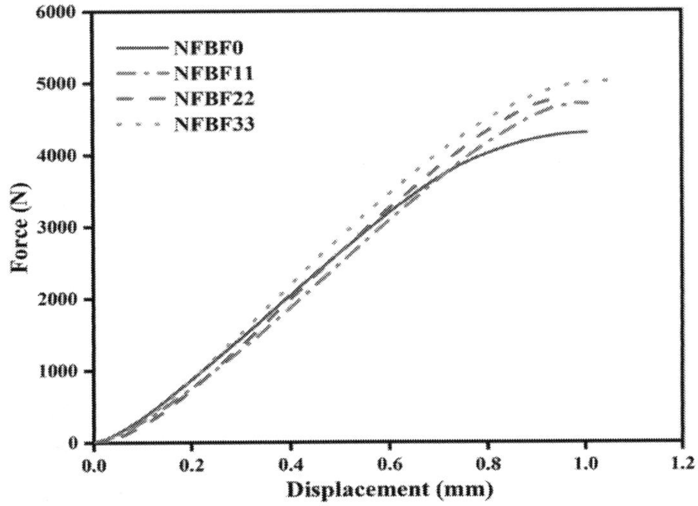

Figure 4 Force versus displacement curves for pure PLA and PLA/NFBF composites

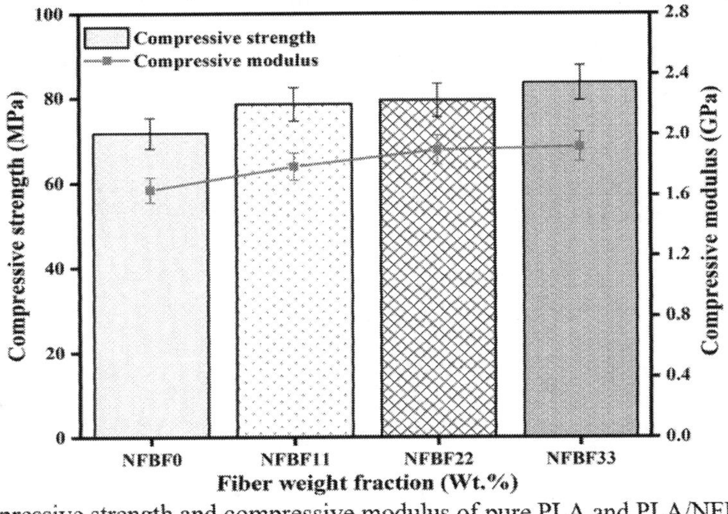

Figure 5 compressive strength and compressive modulus of pure PLA and PLA/NFBF composites

Kain et al. [27] carried out compressive study on 3D printed wood fiber reinforced PLA composites and they noticed that for 25% of fiber reinforcement the compressive strength of the composites is 68 MPa. Chen and Lin [17] fabricated composites reinforcing jute fabric with PLA using hot press compression molding and the compression study of these composites revealed that the maximum compressive strength of the composites is 59.8 MPa. Baba and Ozmen [28] performed a compressive study on long chicken feather fiber filled PLA composites and short chicken feather fiber filled PLA composites. These composites fabricated by extrusion and injection molding methods. Neat PLA sample showed compressive strength of 56.70 MPa. The compressive strength of PLA is diminished with the reinforcement of chicken feather. These results are further decreased with the fiber content increase for both short and long chicken feather fiber reinforcement. Compared to these available literature results the present NFBF reinforced PLA composites results are better.

3.2. Shrinkage properties

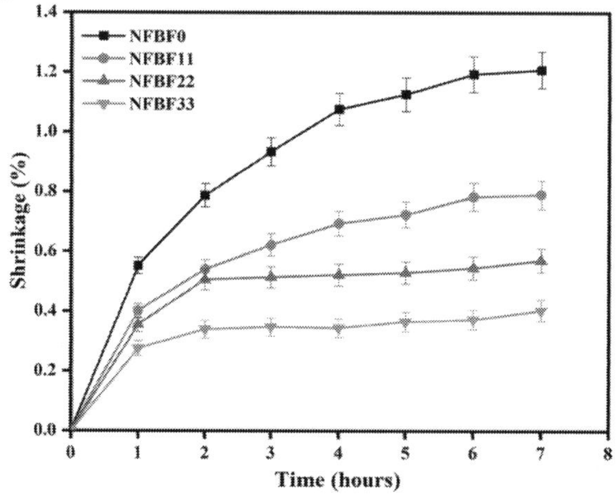

Figure 6 Shrinkage of the neat PLA and PLA/NFBF composites over time at 80° C

Additional to the compressive study of the natural fiber braided yarn strengthened PLA composites thermal stability of the composites are also analyzed experimentally. Figure 6 represents the shrinkage behavior of virgin PLA and PLA/NFBF composites for 80° C of fixed temperature. The percentage shrinkage of virgin PLA and PLA/NFBF composites are increased with time period respectively. From Figure 6, it is observed that the neat PLA sample has higher shrinkage percentage with 1.207%. Reinforcement of the natural fiber thermal shrinkage value of the composites is decreased. Because, natural fibers are porous structures and bad conductors of heat [29]. Hence, filling these fibers with PLA resists the shrinkage of composites. Due to this reason composites shrinkage percentage diminished. Also, Kanakannavar and Pitchaimani [23] thermal deflection study showed that the reinforcement of the natural fiber braided yarn woven fabric with PLA enhances the thermal sustainability of the composites. Due to improved thermal sustainability of these composites required higher temperature to shrink. Further the composites shrinkage percentage is decreased with the fabric content. The composites filled with 33% of fabric showed lowest shrinkage value of 0.403%. Similar trend is noticed by Jang *et al.* [26] for coconut fiber filled PLA composites. This shrinkage study results confirms that the thermal stability of the composites is high compared to virgin PLA.

4. Conclusions

The biodegradable composites are produced by solution casting and hot-press compression molding methods using braided yarn fabric as filler material and PLA as matrix with different fabric weight percentage. The compressive characteristics and thermal shrinkage results of the PLA composites are investigated experimentally. The investigation shows that compressive properties of the composites are enriched with the braided yarn fabric addition compared to virgin PLA and these values are further enhanced with the filler content respectively. Composites with 33 weight percentage addition of fabric with PLA showed highest compressive strength and modulus, which are 16.54% and 17.17% greater than the pristine PLA. The shrinkage study revealed that the reinforcement of the fabric with flax fiber braided yarn enhanced the thermal stability of PLA/NFBF composites.

Acknowledgements

Authors wishes to acknowledge Prof. Narendranath, Material Characterisation Lab, NITK Surathkal, for providing testing facilities and Prof. S M Kulkarni, FIST Composite Centre, NITK, Surathkal, for providing composite fabrication facilities.

References

[1] Chalid M and Prabowo I 2015 The Effects of Alkalization to the Mechanical Properties of the Ijuk Fiber Reinforced PLA Biocomposites Int. J. Chem. Mol. Nucl. Mater. Metall. Eng. 9 342–6

[2] Duigou A Le, Deux J, Davies P and Baley C 2011 PLLA/Flax Mat/Balsa Bio-Sandwich Manufacture and Mechanical Properties Appl. Compos. Mater. 18 421–38
[3] Pickering K L, Sawpan M A, Jayaraman J and Fernyhough A 2011 Influence of loading rate, alkali fibre treatment and crystallinity on fracture toughness of random short hemp fibre reinforced polylactide bio-composites Compos. Part A 42 1148–56
[4] Xia X, Shi X, Liu W, Zhao H and Li H 2017 Effect of flax fiber content on polylactic acid (PLA) crystallization in PLA/flax fiber composites Iran. Polym. J. 26 693–702
[5] Khan G M A, Terano M, Gafur M A and Alam M S 2016 Studies on the mechanical properties of woven jute fabric reinforced poly (L-lactic acid) composites J. King Saud Univ. - Eng. Sci. 28 69–74
[6] Adomavičiūtė E, Baltušnikaitė J, Jonaitienė V and Stanys S 2015 Formation and properties of textile biocomposites with PLA matrix reinforced with Flax and Flax/PLA weft knitted fabrics Fibres Text. East. Eur. 23 45–50
[7] Kumar R, Yakabu M K and Anandjiwala R D 2010 Effect of montmorillonite clay on flax fabric reinforced poly lactic acid composites with amphiphilic additives Compos. Part A Appl. Sci. Manuf. 41 1620–7
[8] Pickering K L and Efendy M G A 2016 Preparation and mechanical properties of novel bio-composite made of dynamically sheet formed discontinuous harakeke and hemp fibre mat reinforced PLA composites for structural applications Ind. Crop. Prod. jo 84 139–50
[9] Islam M S, Pickering K L and Foreman N J 2010 Influence of Hygrothermal Ageing on the Physico-Mechanical Properties of Alkali Treated Industrial Hemp Fibre Reinforced Polylactic Acid Composites J. Polym. Environ. 18 696–704
[10] Oksman K, Skrifvars M and Selin J F 2003 Natural fibres as reinforcement in polylactic acid (PLA) composites Compos. Sci. Technol. 63 1317–24
[11] Ochi S 2008 Mechanical properties of kenaf fibers and kenaf/PLA composites Mech. Mater. 40 446–52
[12] Arao Y, Fujiura T, Itani S and Tanaka T 2015 Strength improvement in injection-molded jute-fiber-reinforced polylactide green-composites Compos. Part B Eng. 68 200–6
[13] Yu T, Jiang N and Li Y 2014 Study on short ramie fiber/poly (lactic acid) composites compatibilized by maleic anhydride Compos. Part A Appl. Sci. Manuf. Part A 64 139–46
[14] Linganiso L Z, Bezerra R, Bhat S, John M, Braeuning R and Anandjiwala R D 2014 Pultrusion of flax/poly (lactic acid) commingled yarns and nonwoven fabrics J. Thermoplast. Compos. Mater. 27 1553–72
[15] Porras A and Maranon A 2012 Development and characterization of a laminate composite material from polylactic acid (PLA) and woven bamboo fabric Compos. Part B Eng. 43 2782–8
[16] Fazita N, Rawi M, Jayaraman K and Bhattacharyya D 2014 Bamboo Fabric Reinforced Polypropylene and Poly (lactic acid) for Packaging Applications: Impact, Thermal, and Physical Properties Polym. Compos. 35 1888–99
[17] Chen J and Lin J-C 2018 Manufacturing and properties of cotton and jute fabrics reinforced epoxy and PLA composites Int. J. Mod. Phys. B 32 1–5
[18] Bodros E, Pillin I, Montrelay N and Baley C 2007 Could biopolymers reinforced by randomly scattered flax fibre be used in structural applications? Compos. Sci. Technol. 67 462–70
[19] Duc F, Bourban P E, Plummer C J G and Månson J E 2014 Damping of thermoset and thermoplastic flax fibre composites Compos. Part A Appl. Sci. Manuf. 64 115–23
[20] Rawal A, Saraswat H and Sibal A 2015 Tensile response of braided structures: A review Text. Res. J. 85 2083–96

[21] Rajesh M and Pitchaimani J 2017 Mechanical Properties of Natural Fiber Braided Yarn Woven Composite: Comparison with Conventional Yarn Woven Composite J. Bionic Eng. 14 141–50
[22] Rajesh M and Pitchaimani J 2016 Dynamic Mechanical and Free Vibration Behavior of Natural Fiber Braided Fabric Composite: Comparison with Conventional and Knitted Fabric Composites Polym. Compos. 39 2479–89
[23] Kanakannavar S and Pitchaimani J 2020 Thermal buckling of braided flax woven polylactic acid composites J. Reinf. Plast. Compos. 1–12
[24] Kanakannavar S, Pitchaimani J and Ramesh M R 2020 Tribological behaviour of natural fibre 3D braided woven fabric reinforced PLA composites J. Eng. Tribol. 1–12
[25] Thiagamani S M K, Krishnasamy S, Muthukumar C, Tengsuthiwat J, Nagarajan R, Siengchin S and Ismail S O 2019 Investigation into mechanical, absorption and swelling behaviour of hemp/sisal fibre reinforced bioepoxy hybrid composites: Effects of stacking sequences Int. J. Biol. Macromol. 140 637–46
[26] Jang Y J, Jeong K T, Oh J H, Youn R J and Song Y S 2012 Thermal stability and flammability of coconut fiber reinforced poly (lactic acid) composites Compos. Part B Eng. 43 2434–8
[27] Kain S, Ecker J V, Haider A, Musso M and Petutschningg A 2020 Effects of the infill pattern on mechanical properties of fused layer modeling (FLM) 3D printed wood/polylactic acid (PLA) composites Eur. J. Wood Wood Prod. 78 65–74
[28] Baba B O and Ozmen U 2015 Preparation and Mechanical Characterization of Chicken Feather/PLA Composites Polym. Compos. 38 837–45
[29] Kymäläinen H and Sjöberg A 2008 Flax and hemp fibres as raw materials for thermal insulations Build. Environ. 43 1261–9

IOP Conf. Series: Materials Science and Engineering 1065 (2021) 012022 doi:10.1088/1757-899X/1065/1/012022

Computational Fluid Dynamics Modelling and Experimental Study on Pressure drop through Vertical Packed Clay Composite Pellets

C R Hiremath, P V Goggal, SV Hiremath, Punit Hiremath

Department of Mechanical Engineering, BLDEA'S V P Dr P. G. Halakatti College of Engg & Technology, Vijayapur, Karanataka, India.

Email: mech.hiremath @gmail.com

Abstract: The importance of computational fluid dynamics in the field of packed bed medium is acceptable due to the high-end processor power availability. The studies of the literature have presented the feasibility of CFD as an advanced practical tool for estimating the pressure drop through the packed porous medium. In this work a CFD model was evolved to simulate the pressure drop through vertical clay pellets packed bed by using CFX analysis. Experiments are conducted to measure the pressure drop across vertical static clay composite pellets bed. The maximum deviation between CFD analysis and experimental results for pressure drop through 500 mm bed height is 13%.

1. Introduction

Packed bed arrangements are commonly used in chemical, absorption, distillation, stripping, catalytic conversion, metallurgy, nuclear industries, thermal storage, building heating and cooling [1]. The important engineering component of the packed bed systems are pressure drop, dispersion, heat and mass transfer. Pressure drop is of prime parameter for the design and configuration of packed bed reactors [2]. Frictional factors during fluid flow and inertia effects are the two causes that results in pressure loss. The drop in pressure along the vertical static bed is dependent on process air and bed parameters. The process air parameters are viscosity, density and velocity. The parameters related to bed are shape and position of particles, void volume, particle periphery roughness and bed to particle diameter ratio. Particle porosity, specific surface area and mean pressure drop across the bed are most indicative factors for the design of packed bed systems [3].

The complex phenomenon of flow of process fluid through packed bed of spheres is parameter dependent. The most convincing way to estimate pressure drip across a packed column is to conduct experiment on actual system. It is costly to build the real system in practice, commonly tests are performed on virtual model of actual arrangement. It involves required geometry generation, transporting fluid through the bed and pressure drop measurement using manometers. This regular mode is also a moderately time intensive and uneconomical.

The flow parameters inside the packed beds are investigated using magnetic resonance imaging (MRI). The other technique is computed tomography (CT). The short coming of these techniques leads to adaptation of other methods for packed beds. Though empirical correlations such as Ergun mathematical statement is employed to find the pressure drop, computational fluid dynamics (CFD) has become a feasible solution to adopted methods. The complexity of packed pressure drop can be better understood by CFD with respect to pressure and velocity field profiles. The simulation data proves to be in agreement with the values estimated from empirical counterpart [4].

CFD, is appropriate for the researchers to understand the pressure drop dynamics of fixed beds. It arranges a numerical tool that can precisely demonstrate the fluid flux over the porous packed beds with the focus on pressure drop prediction. The first part of the present work involves CFD modeling and simulating the pressure drop flow patterns in vertical firm bed comprising spherical shaped clay additive pellets.

Though the CFD models provide virtual design and data but the more development still depends on experimental verification. In the second segment experiments are conducted through vertical packed clay additives composite pellets. Finally, the experimentally measured pressure drop values are compared with CFD pressure drop values.

2. Simulation

The numerical analysis work is carried out in two steps, initially design is done using CATIA V5 and analysis is carried out using Computational Fluid Dynamics (CFD). A drafting of 3D model is generated in CATIA, dimensions are indicated. A complete detail view of pellets is shown with scale of 3:1. A model of packed bed and granules is prepared using CATIA V5 software. A cylinder of length 0.5 m, inside 0.05 m and outside of 0.052 m diameter is created. Inside the cylinder the balls of 0.01 m diameter are created. The virtual models created are presented in Figure 1.

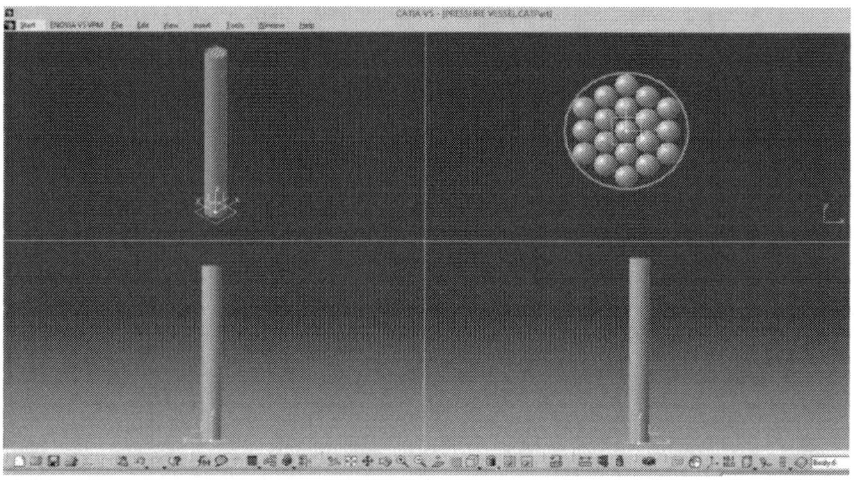

Figure 1. Design of solid three-dimensional models for cylindrical bed and spherical pellets

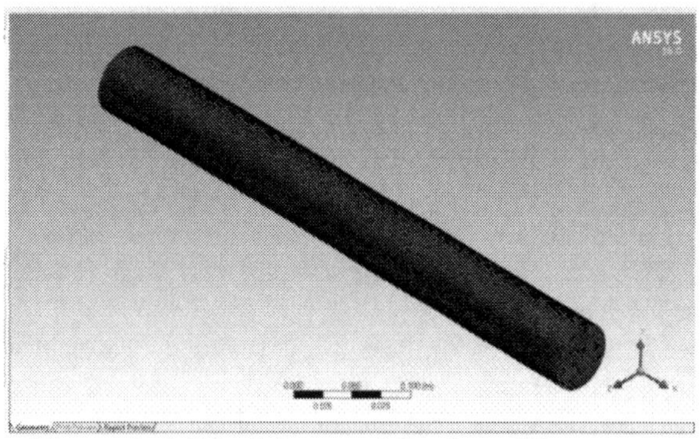

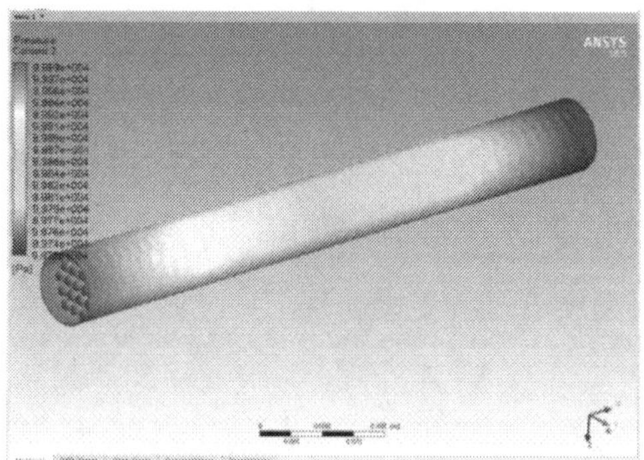

Figure 2. (a)Mesh imported for the solid geometry of packed bed system and (b)Pressure contours generated for packed bed system

To generate fine mesh 388803 nodes and 1851886 elements are considered. Quad meshing elements and triangular meshing type are adopted and the mesh is created for imported model. The mesh produced is presented in Figure 2(a). The pressure distribution presented in Figure 2(b) shows inlet pressure as 99.99 kPa and out let pressure is 99.730 kPa. Medium intensity and eddy viscosity ratio turbulence model is considered for clay bed domain. Subsonic flow regime is considered for the analysis at atmospheric pressure of 1 bar and with air density of 1.145kg/m^3.

3. Experimental work

Experimental work goes in two stages. The first part of the work involves preparation of spherical shaped 0.01 m clay - horse dung pellets. The heat treatment at 500°C produces burnt clay composite particles or pellets. Second part of work experiments are conducted to investigate pressure drop of burnt clay - horse dung of different size and shape. The clay and additive horse dung are collected from a local pot maker. The clay and horse dung are manually cleaned and processed by using a sieve. Horse dung of 20 g per 100 g of clay is mixed and pasty clay composite material is produced. Water is added to produce the pasty and dough clay material. The pasty raw material is then molded manually to a size of 0.01 m spherical shape. The fabricated pellets are shadow dried and then heated in a muffle furnace of 3.5 kW capacities. The prepared samples are heated to a temperature of 500°C for the duration of one hour. The pellets are air cooled to cool in the furnace. Finally, prepared samples are weighed and checked for identical weight and shapes. The second part of experimental work involves setting and arranging the experimental set up. The lab scale arrangement is drawn in Figure 3.

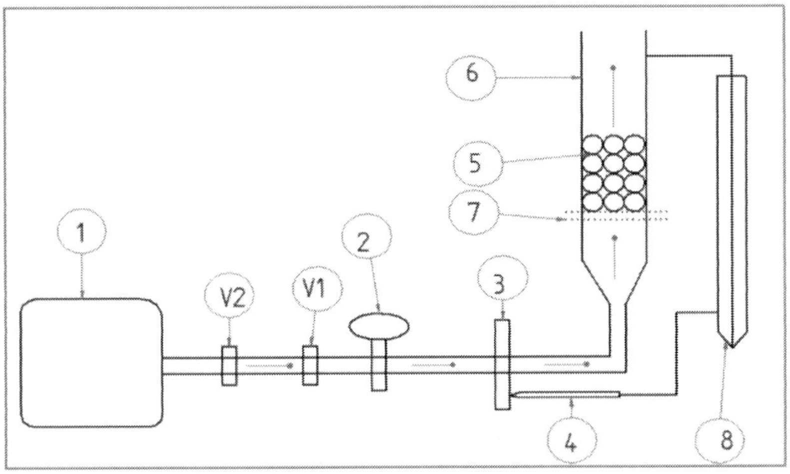

1. Compressor
2. Pressure gauge
3. Orifice meter
4. Hose pipe
5. Clay pellets
6. Vertical column
7. Wire mesh
8. U - Tube manometer, V1 and V2: Gate valves.

Figure 3. Schematic layout of experimental setup

The air from the reciprocating air compressor is made to flow through the galvanized iron pipe of 0.025 m. The atmospheric air drawn by the compressor is stored in air tank. The air from the air tank is let into the packed bed. The gate valves are used to direct and control the flow of air through the pressure line and the bed. A calibrated orifice meter with diameter 0.010 m and with discharge coefficient of 0.68 is used to find the volume flow rate of process air. The estimated bed inlet velocities are from 0.5 to 2 m/s. The vertical column is made of perspex or acrylic material enables visualization of cylindrical vertical packed bed. The height of the column is 0.5 m and the diameter is 0.05 m. The column is connected to the pressure line by a flange. A U-tube manometer connected across the column measures the bed pressure drop.

4. Results and Discussion

The prediction of pressure drop characteristics of static beds of spherical shaped clay composite pellets is carried by CFD. The experimental measurements for drop in pressure through vertical packed

spherical clay pellets having a bed to particle diameter ratio of 5 are recorded. The effect of changing the bed mass and bed inlet velocity on pressure drop across the vertical packed bed is presented.
As presented in Figure 4, for the fixed bed mass of 250 g and 500 g and for the increase in bed inlet velocity results in hhigher drop in pressure. The inertial and viscous effects causes ppressure drop. At lower velocities the inertial effects causes pressure drop, whereas at higher velocities viscous effects are prime to cause the pressure drop through the packed bed. The frictional losses due to obstructions results in pressure loss for the fluid passing through the packed bed. With increase in bed height and for the identical bed inlet velocities the resistance to flow increases. Due to increase in friction and resistance the pressure drop increases.

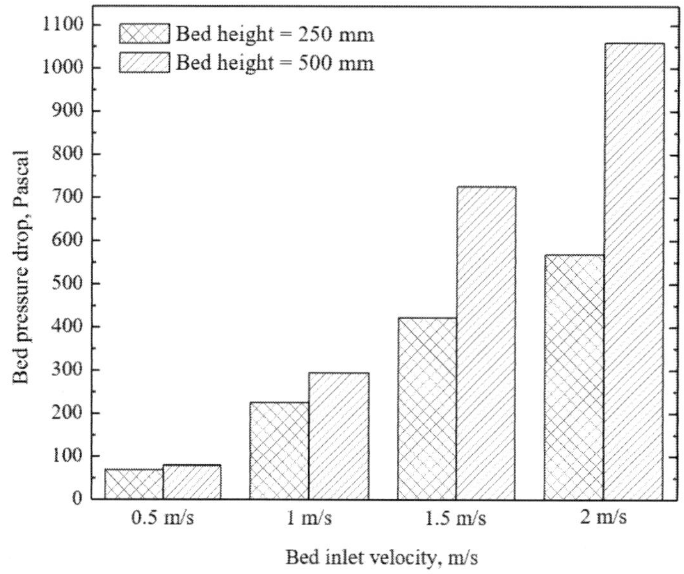

Figure 4. Variation in pressure drop with velocity for 250 mm and 500 mm bed heights.

The comparison of simulated and experimental pressure drop values for bed height of 500 mm and bed inlet velocity of 1 m/s are graphically shown in Figure 5. The comparison shows lower values of pressure drop values as depicted by simulation. The deviation between experimental and CFD results is due to complexity of flow phenomenon through the packed beds.

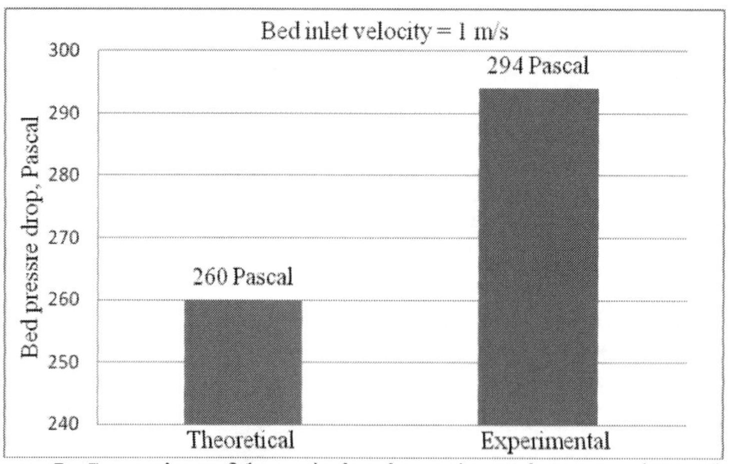

Figure 5. Comparison of theoretical and experimental pressure drop values

5. Conclusion

A CFD model was developed for simulating the pressure drop of process air through vertical packed clay pellets. The observation of experimental measurements shows good agreement with the CFD simulation. The CFD analysis and experimental results for packed bed of 500 mm height with inlet pressure of 1bar yielded the pressure drop of 260 Pa and 294 Pa respectively. The deviation between simulation and experimental values is about 13%. In consequence to this study, it is inferring that using CFD could dispense useful data for the pressure drop through packed bed systems and better predict the pumping power required.

References

[1] Theodoros A and Eugeny Y K 2009 CFD-based analysis of the wall effect on the pressure drop in packed beds with moderate tube/particle diameter ratios in the laminar flow regime J. Chem. Eng 155 404 - 410.

[2] Calis H P A, Nijenhuis J, Paikert B C, Dautzenberg F M. van den Bleek C. M 2001 CFD modelling and experimental validation of pressure drop and flow profile in a novel structured catalytic reactor packing Chemical Engineering Science 56 1713 – 1720.

[3] Mohammed N A 2011 Modeling Of Porosity Equation For Water Flow Through Packed Bed Of Monosize Spherical Packing Journal Of Engineering And Development 15 1813- 7822.

[4] Soroush A, Farhang S 2018 Study of Pressure Drop in Fixed Bed Reactor Using a Computational Fluid Dynamics (CFD) Code Chem Engineering www.mdpi.com/journal/chemengineering.

[5] Ookawara S, Kuroki M, Street D, Ogawaa K 2007 High-fidelity DEM-CFD modeling of packed bed reactors for process intensification";, Proceedings of European Congress of Chemical Engineering (ECCE-6) Copenhagen, 16-20.

[6] Allen K G, Backstrom T W, Kroger D G 2013 Packed bed pressure drop dependence on particle shape, size distribution, packing arrangement and roughness Powder Technology 246 590-600.

[7] Preller A C N 2011 Numerical Modelling of Flow through Packed Beds of Uniform Spheres North-West University: Potchefstroom, South Africa.

[8] Ergun S 1952 Fluid flow through packed columns Chem. Eng. Prog 48 89–94.

[9] Ribeiro A M, Neto P C, Pinho 2010 Mean porosity and pressure drop measurements in packed beds of monosized spheres: side wall effect International review of chemical engineering (I.R.E.CH.E), vol2, No 1.

[10] Melda O C, Emrah O 2008 A simplified correlation for fixed bed pressure drop Powder Technology, 187 94-101.

Fabrication and Characterization of Aluminum 8011 Alloy and nano Zro_2 Metal Matrix Composite

Thippeswamy J C[1] **Dr.Sathisha N** [2]

[1] Assistant Professor, Department of Mechanical Engineering. Shridevi Institute of Engineering and Technology, Tumakuru, Karnataka, India

[2]Professor and Head, Department of Mechanical Engineering. Yenepoya Institute of Technology, Moodabidri, Karnataka, India

thippeswamyjc2004@gmail.com

Abstract: Composites of nano metal matrix produced by stir casting technology have major attractive properties like high ductility, highly conductivity, light weight, and high strength to weight ratio compared to some other approaches. Nano Zro_2 particulates with varying weight percentages (2, 4 and 6%) are reinforced with aluminium 8011 using the process of stir casting. Mechanical properties such as tensile, compression and hardness behavior have been investigated. In the current analysis, scanning electron microscopy (SEM) is carried out to show Nano Zro_2 particulates are successfully incorporated into the matrix of aluminium 8011 with homogeneous distribution. The findings showed that improved mechanical properties such as tensile strength, compression strength and hardness, were enhanced by increasing weight percentage of Nano Zro_2 particulates in the matrix of aluminium 8011.

I. INTRODUCTION

The demand for materials with a high strength to weight ratio in the aerospace and automotive industries is contributing to the production of innovative new materials. The combination of two or more creates this new material with matrix and reinforcement [1]. Aluminum Metal-Matrix Composites (MMCs) are an important class of engineering materials of extreme promise because of their comparatively stronger physical and mechanical properties, aluminium MMCs are potential materials for various applications with the inclusion of sufficient reinforcements [2-3], properties such as wear, hardness, creep, specific strength and fatigue can be increased relative to conventional engineering materials. The alloy of aluminium 8011 has a rare mix of desirable features, such as its low weight, corrosion resistance and simple final product maintenance. The matrix content of aluminium 8011 has a low density, non-toxic content and high thermal conductivity [4-6]. For the past 30 years, nano material-reinforced composites have been widely favored in the automotive and aerospace industries in terms of wear resistance, damping properties and mechanical strength [7], Aluminium reinforced with Nano ZrO_2 Composite was developed effectively by means of the casting process. The nano particles dispersion in aluminium matrix will strengthen the base metal [8].

IOP Conf. Series: Materials Science and Engineering 1065 (2021) 012023 doi:10.1088/1757-899X/1065/1/012023

II. LITERATURE REVIEW

With the assistance of stir casting process, the proposed experimental research on titanium carbide (TiC) and red mud is reinforced with aluminium 7075 at 3, 6, 9 and 12 percent by weight. In which, the mechanical properties are tested and compared with considering yield strength, ultimate tensile strength and micro hardness. It is observed that 48% improvement in the ultimate tensile strength compared to base metal has been investigated in [1]. The authors in [2] reviewed on the stir casting process with considering how base metal is liquefied. The analysis is carried out at different temperature ranges with maintained state, particles condition, mixing time and mixing speed. The mixing speed of final composite material is significantly affected on the electrical resistance furnace than the fired coke furnaces. The electric stir casting process used to fabricate Aluminium/Nano Sic composite [3]. Mechanical properties such as tensile strength, micro-hardness and wear behavior have been studied. The findings revealed that nano-silicon carbide (SIC) powder incorporation increases the wear resistance of nano-aluminum composites and enhances mechanical properties such as stiffness and tensile strength. The Aluminum 7075 alloy reinforced with boron carbide with different weight percentage by stir casting technique reveals in [4] shows increased Yield strength and Ultimate tensile strength by increasing boron carbide weight percentage. It is explicitly observed that the percentage of boron carbide increases, the wear rate decreases.

A study conducted in [5] on latest developments and techniques used in preparing particulate metal matrix composites using friction stir process. This study reveals that friction stir processing has been shown to be an environmentally sustainable solid-state extreme plastic deformation process with reduced carbon footprint, reasonably energy-efficient for fine-grained structure surface and bulk composites processing. The method of stir-casting [6] is used to process the different particle sizes of SiC (63, 76, 89) μm is conjugated with aluminium 8011.The conjugation was made with 2 and 4 weight percentage. The findings show that with the rise in the weight fraction of SiC, the stiffness, yield strength and tensile strength increase, while the resilience and ductility decrease. A review on Metal Matrix Composites Strengthened by Nano-Particles in [7] shows the most important processing methods used for the synthesis of bulk metal matrix nano composites. Impressive results have been obtained in terms of hardness, mechanical efficiency, wear resistance, creep activity and damping properties. Experimental and FEM work on Aluminium 8011 alloy reinforced with Fly-Ash and E-Glass fiber composite prepared by stir casting was discussed in [8]. The results show that the ultimate tensile strength and compression strength of the hybrid composites are improved in comparison with base Aluminum 8011 alloy. The characterization of aluminum 6061 and nano zirconia as a reinforcement was conducted in [9] with different weight percentages. The present work focuses on mechanical characterization through different weight percentages of reinforcement by tensile test, hardness test and compression test. The results reveal that ultimate tensile strength, compression strength and hardness increased with increasing weight percentage of nano zirconia. The Review in [10] summarizes the effect of processing techniques, types of reinforcement on the mechanical properties of aluminium metal matrix composites. The more emphasis was given to tribological properties. Stir casting, high-pressure torsion, and powder metallurgical methods. The study in [11] on impact of temperature on the dry sliding wear behavior of 7075 aluminum with ZrB_2 particulates at 0, 3, 6, 9 and 12 wt.%. The wear resistance of the produced composites is improved with the increased substance of ZrB_2 particulates at all test temperatures. Mechanical Characterization of Ceramic Nano B_4C and Aluminum 2618 Alloy Composites in [12] validate the homogenous distribution of nano B_4C particles in the matrix of aluminum. It was noted that the tensile strength (ultimate strength and yield strength) and compression strength increased with the addition of 6 wt% nano B_4C particles. In the research work [13] characterization and wear Behavior of Aluminum 2014-ZrO_2 Nano- composites are studied. From the investigation, it was observed that the tensile behavior of composites was enhanced due to addition of nano-sized ZrO2 particles in the Aluminum 2014 alloy matrix. The Enhancement of mechanical and tribological properties of SiC and Carbon black reinforced aluminum 7075 hybrid composites [14].The study demonstrates the novel feature that 2.5 times increment in the ultimate tensile strength and significant improvement in wearing resistance.

IOP Conf. Series: Materials Science and Engineering 1065 (2021) 012023 doi:10.1088/1757-899X/1065/1/012023

III. OBJECTIVES

1. Material Selection: Aluminium 8011 alloy was used as the base matrix material and Nano Zro_2 particulates as reinforcement.
2. Fabrication of Aluminium 8011 and Nano Zro_2 composite with different weight percentages (2, 4 and 6 %) by stir casting.
3. To prepare specimens for tensile test, compression test and micro hardness test according to ASTM standards.
4. To conduct scanning electron microscopy to study the distribution of Nano Zro_2 particulates in the matrix of Aluminium 8011.

IV METHODOLOGY

The detailed methodology used for the manufacturing of composite materials as shown in figure below

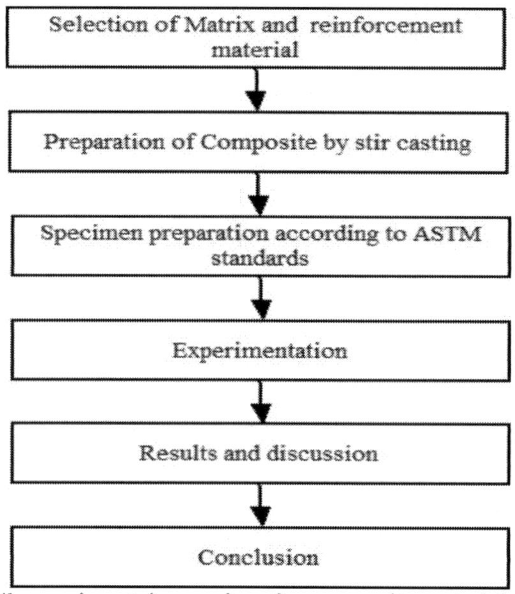

Fig.1: Detail experimental procedure for composite material preparation

V EXPERIMENTAL PROCEDURE

A. Selection of materials

Matrix: Aluminum 8011 was used as a strong machinability matrix material. In both ordinary ambient and marine environments, it exhibits excellent corrosion resistance. Aluminum for aerospace, spacecraft and maritime applications has a high strength-to-low weight ratio.

TABLE 1: Chemical composition of Aluminum 8011 by weight percentage

Chemical Composition	Si	Fe	Cu	Mn	Mg	Cr	Zn	Ti	Al
Aluminum 8011	0.50	0.60	4.35	0.6	1.5	0.1	0.25	0.15	Remaining

Reinforcement: Zirconium oxide nano particles (ZrO2) are a technologically significant substance with strong natural color, high strength, resistance to transformation, high chemical stability, outstanding resistance to corrosion and chemical and microbial resistance. In the form of nano dots, nano fluids and nano particulates, zirconium oxide nano particles (ZrO2) are available.

Zirconium Oxide Nano Powder Specifications

Molecular Formula: ZrO_2

IOP Conf. Series: Materials Science and Engineering 1065 (2021) 012023 doi:10.1088/1757-899X/1065/1/012023

Molecular Density: 5.89 g/cm3
Appearance: White
Particle Size: 30 NM

Table 2: Chemical composition of nano $ZrO_2$by weight percentage

PARAMETERS	SPECIFICATIONS
ZrO_2	99.9%
Al	<0.06%
Fe	<0.02%
Pb	<0.02%

Fig. 2: Aluminum 8011 sheets

Fig. 3: zirconium oxide nano particles (ZrO2)

B. Manufacturing of composite materials

By stir casting technique, composites are prepared. In the electric resistance furnace, the base metal aluminum 8011 is then heated and melted in a crucible at a temperature of 750^{0}c. In the molten metal bath, the stirrer was immersed and the stirrer rotates at a speed of 100 to 300 rpm to create vortexes in the liquid metal. Preheated nano particles of zirconium oxide (ZrO_2) are applied from the side of the vortex to the molten metal. The stirrer was continued for 5 minutes after completion of the addition of nano ZrO_2 and pour molten metal mixture into the necessary mould cavity. After casting, aluminium 8011 metal matrix with zirconium oxide nano particles (ZrO_2) composite materials have been obtained as seen in the figures below.

Fig. 4: Aluminum 8011+ 2% nano zro_2

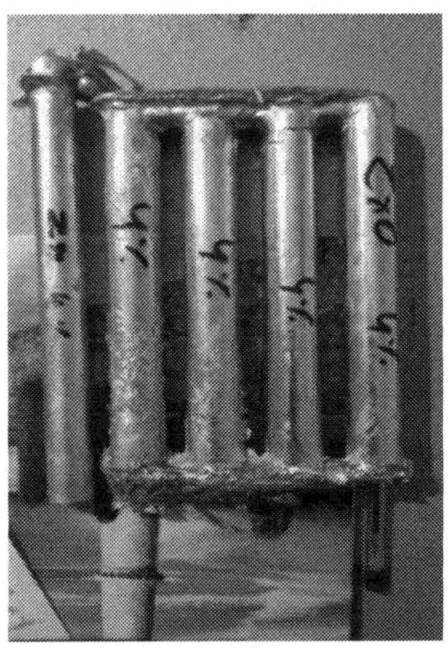

Fig. 5: Aluminum 8011+ 4% nano zro_2

Fig. 6: Aluminum 8011+ 6% nano zro_2

V.RESULTS AND DISCUSSION

A. Tensile test: The tensile strength of the various formulations is measured using a 5 tone size servo-hydraulic universal test unit. Test samples were held in parallel with the applied load. The experiments

IOP Conf. Series: Materials Science and Engineering 1065 (2021) 012023 doi:10.1088/1757-899X/1065/1/012023

carried out were carried out in compliance with the ASTM E8-82 guidelines and the findings obtained from this test are shown in Table 3.

Fig 7: Specimens after the tensile test

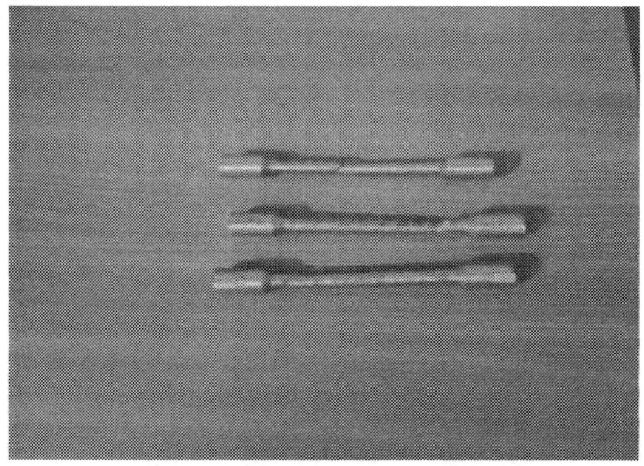
Fig. 8: Specimens after the tensile test

Table 3: Tensile test results

Specimen		**Peak load in KN**	**% of elongation**	**Tensile strength in N/mm²**
Test 1	2%	13.230	14.62	110.981
Test 2	4%	13.350	24.16	115.714
Test 3	6%	12.100	19.82	116.089

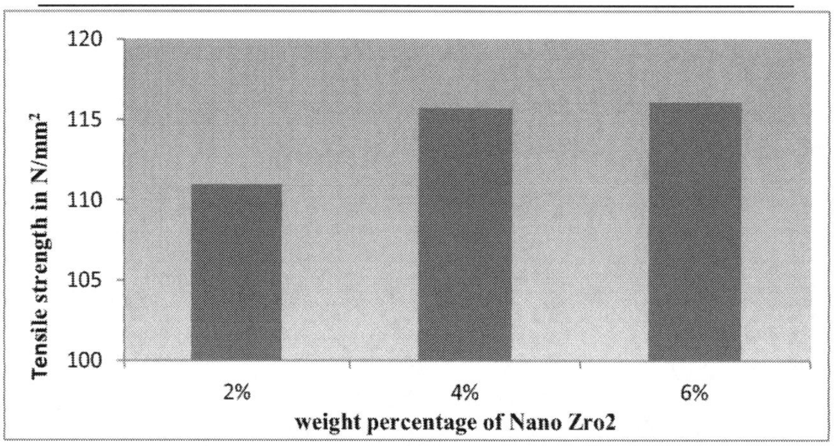

Fig. 9: variations of Tensile strength v/s weight percentage of nano zro_2

B. Compression test: The compression test was carried out using a 5 tonne universal test machine capability and is carried out on 22 mm long machined and 13 mm diameter cast composites. Initially, the loads are added steadily and the resulting strains of the following composites, as seen in Table 4, are determined before the failure of the specimen. The related experiments are carried out at room temperature, according to ASTM E9.

Table 4: Compression test results

SPECIMEN		Load at peak in KN	Elongation at peak in mm	Compression strength in in N/mm^2
Test 1	2%	187.446	17.790	586.945
Test 2	4%	187.728	20.870	598.454
Test 3	6%	210.306	19.650	657.536

Fig 10: specimen prepared for compression test

Fig 11: Specimens after compression test

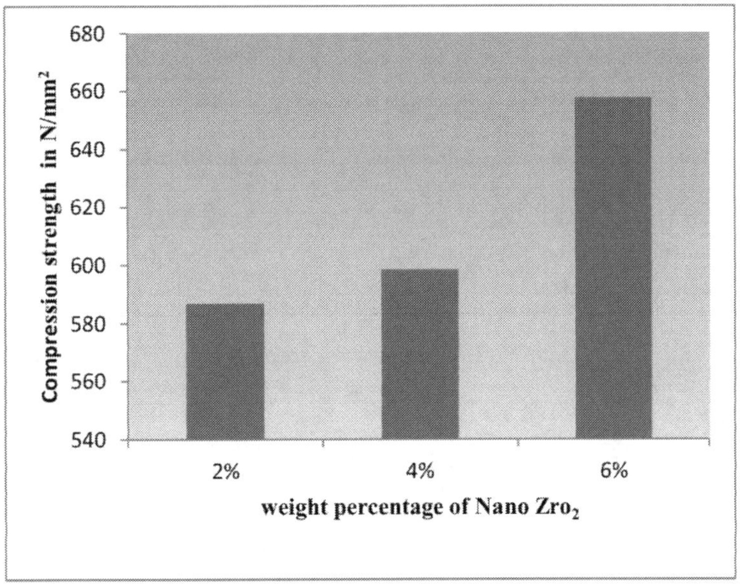

Fig. 12: variations of compression strength v/s weight percentage of nano zro_2

C. Vickers Hardness Test: In this work a hardness test is conducted to detect the tolerance of the plastic deformation of the composites under static or dynamic loads. Micro Vickers Hardness Test is selected to determine the hardness of the specimen. In this test, the indenter has a diameter of 5 mm shall be tested by adding a load of 0.5 kg on each specimen for 20 seconds. The related test parameters are shown in Table 5.

Table 5: Hardness test results

Composition	Ball diameter in mm	Applied load in kg	Average diameter of indentation	VHN
Aluminium 8011-2wt% Nano ZrO_2	5	0.5	2.620	38.6
Aluminium 8011-4wt% Nano ZrO_2	5	0.5	2.750	40.0
Aluminium 80111-6wt% Nano ZrO_2	5	0.5	2.745	42.2

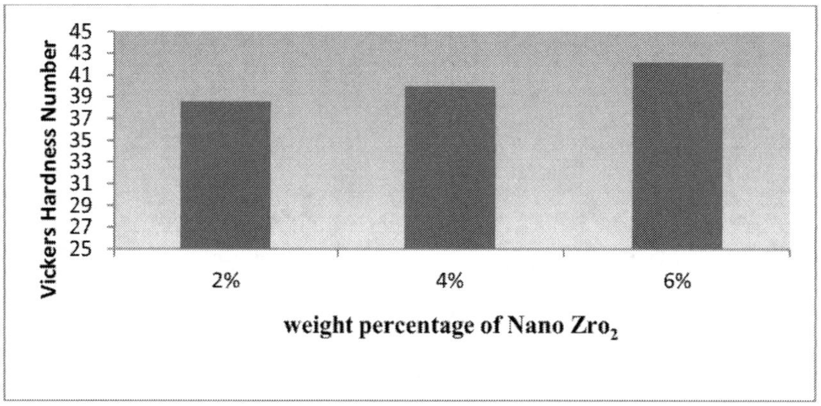

Fig. 13: variations of Hardness number v/s weight percentage of nano zro_2

D. Microscopy: The dispersion of reinforcements in the MMC was measured using optical microscopy and reveals the microstructure of the MMC with 2, 4 and 6 percent percentages of particulate reinforcements. Reinforcement marks are indicated by the arrow marks on the diagram. The SEM images show that the distributions of zirconium oxide nano particles (ZrO2) are even and competitive with Aluminum 8011. The study showed that a much lower porosity was observed in zirconium oxide nanoparticles (ZrO2) with Aluminum 8011.

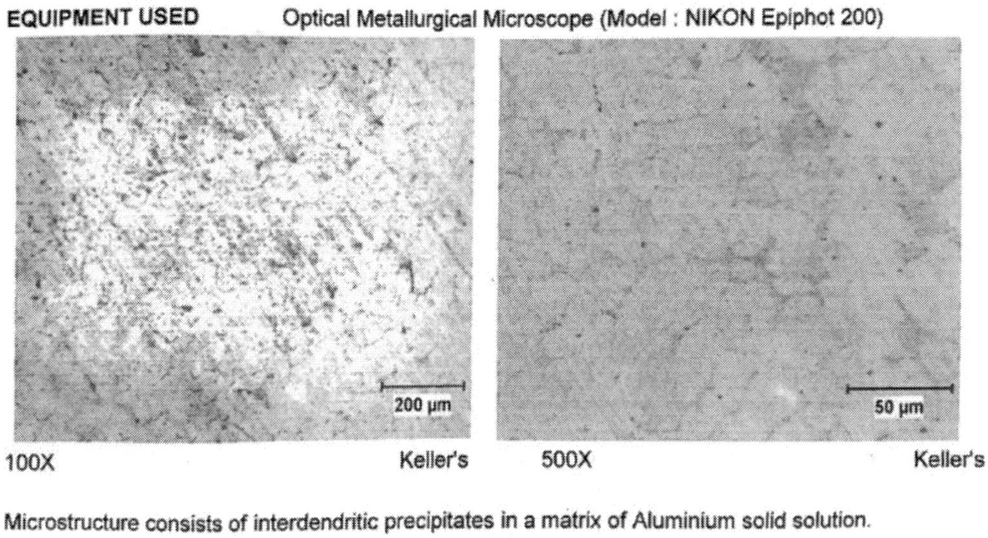

Fig 14: Microstructure of Aluminum 8011 - 2 wt % nano zro_2

EQUIPMENT USED Optical Metallurgical Microscope (Model : NIKON Epiphot 200)

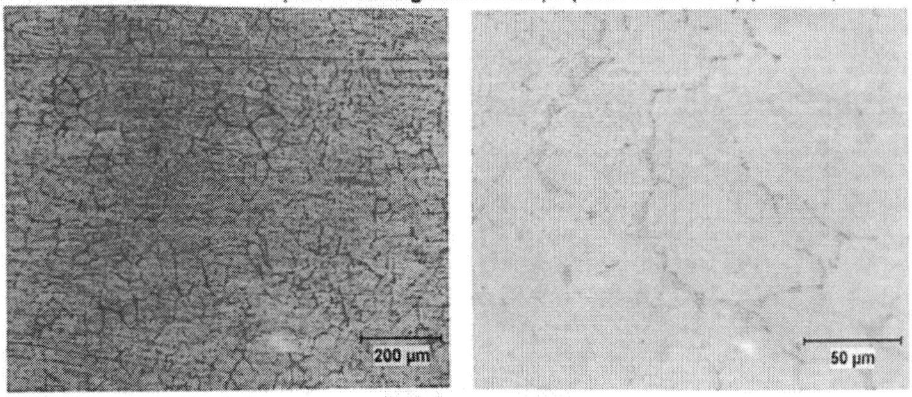

100X Keller's 500X Keller's

Microstructure consists of interdendritic precipitates in a matrix of Aluminium solid solution.

Fig 15: Microstructure of Aluminum 8011 - 4 wt % nano zro_2

EQUIPMENT USED Optical Metallurgical Microscope (Model : NIKON Epiphot 200)

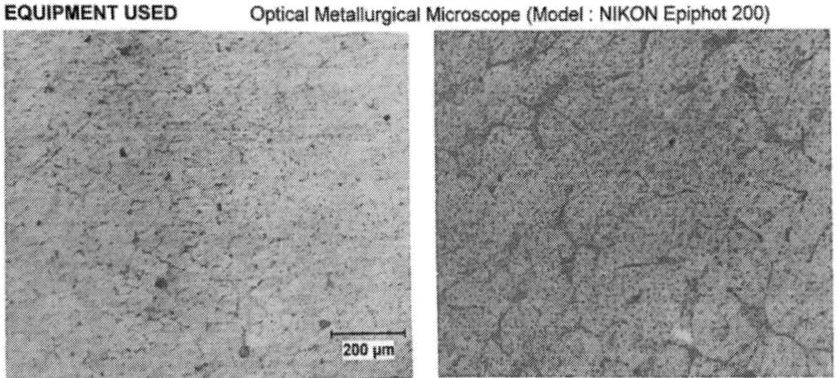

100X Keller's 500X Keller's

Microstructure consists of interdendritic precipitates in a matrix of Aluminium solid solution.

. Fig 16: Microstructure of Aluminum 8011 – 6 wt % nano zro_2

VI. CONCLUSION

Based on the study conducted on aluminium 8011 and Nano ZrO_2 Composite the following conclusions are made:

1) Using stir casting method Nano ZrO_2 particulates can be successfully introduced in the aluminium 8011 alloy matrix to fabricate composite material.

2) The maximum tensile strength was observed with aluminium 8011 + 6 wt% Nano ZrO_2 is 116.089 N/mm^2.

3) Highest compression strength was observed with aluminium 8011 + 6 wt% Nano ZrO_2 is 657.536 N/mm^2..

4) By increasing the reinforcement percentage better hardness properties are obtained.

5) The microscopic images have been revealed that uniform distribution of Nano ZrO_2 in aluminium 8011 with inter dendritic precipitation.

REFERENCES

[1]. Kar, Chinmayee, and B. Surekha. "Characterisation of aluminium metal matrix composites reinforced with titanium carbide and red mud." *Materials Research Innovations* (2020): 1-9.

[2]. Srivastava, Ashish, Amit Rai Dixit, and Sandeep Tiwari. "a review on fabrication and characterization of aluminium metal matrix composite (AMMC)." *International Journal of Advance Research and Innovation* 2, no. 2 (2014): 516-521.

[3]. Suresh, S., G. Harinath Gowd, and MLS Deva Kumar. "Mechanical and wear characterization of Al/nano-SiC NMMCs by liquid state process." *Journal of Bio-and Tribo-Corrosion* 5, no. 2 (2019): 43.

[4]. Kumar, K. Praveen, CP S. Prakash, B. Mallikarjun, and A. Shantharam. "A study on tensile and tribological properties of aluminum 7075 metal matrix with boron carbide reinforced composite material by using computer interface tensile and wear test." In *2017 International Conference on Innovative Mechanisms for Industry Applications (ICIMIA)*, pp. 739-745. IEEE, 2017.

[5]. Bajakke, Padmakumar A., Vinayak R. Malik, and Anand S. Deshpande. "Particulate metal matrix composites and their fabrication via friction stir processing–a review." *Materials and Manufacturing Processes* 34, no. 8 (2019): 833-881.

[6]. Lastnosti, Vpliv Velikosti Delcev Na Mehanske,. "Effect of particles size on the mechanical properties of sic-reinforced aluminium 8011 composites." *Materiali in tehnologije* 51, no. 4 (2017): 667-672.

[7]. Casati, Riccardo, and Maurizio Vedani. "Metal matrix composites reinforced by nano-particles—a review." *Metals* 4, no. 1 (2014): 65-83.

[8]. Kulkarni, Chetan, Akshay Mohan Pujar, and Balappa Hadagali. "Experimental and FEM Analysis on the Mechanical Properties of Al-8011 Alloy Reinforced with Fly-Ash and E-Glass Fibers." In *Materials, Energy and Environment Engineering*, pp. 33-43. Springer, Singapore, 2017.

[9]. Anand, V., M. N. Akshay, S. Abhilash, and G. Deepak. "Recent Advances in the Development of Aluminium Matrix Composites (AMCs)." In *Emerging Trends in Mechanical Engineering*, pp. 619-626. Springer, Singapore, 2020.

[10].Chelladurai, S. J. S., T. Murugesan, T. Rajamani, S. Anand, S. J. P. Asok, and S. Kumaravel. "Investigation on mechanical properties and tribological behaviour of stir cast LM13 aluminium alloy based particulate hybrid composites." *Materialwissenschaft und Werkstofftechnik* 50, no. 7 (2019): 864-874.

[11].Selvam, J. David Raja, I. Dinaharan, Rajakumar S. Rai, and P. M. Mashinini. "Role of zirconium diboride particles on microstructure and wear behaviour of AA7075 in situ aluminium matrix composites at elevated temperature." *Tribol. Mater. Surf. Interfaces* (2019): 1-9.

[12].Nagaral, Madeva, Shivananda Kalgudi, Virupaxi Auradi, and Shivaputrappa Amarappa Kori. "Mechanical Characterization of Ceramic Nano B4C-Al2618 Alloy Composites Synthesized by Semi Solid-State Processing." *Transactions of the Indian Ceramic Society* 77, no. 3 (2018): 146-149.

[13].Nagaral, Madeva, Shivananda Kalgudi, Virupaxi Auradi, and Shivaputrappa Amarappa Kori. "Mechanical Characterization of Ceramic Nano B4C-Al2618 Alloy Composites Synthesized by Semi Solid-State Processing." *Transactions of the Indian Ceramic Society* 77, no. 3 (2018): 146-149.

[14].Roy, Pratip, Subhash Singh, and Kaushik Pal. "Enhancement of mechanical and tribological properties of SiC-and CB-reinforced aluminium 7075 hybrid composites through friction stir processing." *Advanced Composite Materials* 28, no. sup1 (2019): 1-18.

IOP Conf. Series: Materials Science and Engineering 1065 (2021) 012024 doi:10.1088/1757-899X/1065/1/012024

Effect of Fuel Intake Temperature on Performance and Emission of Diesel Engine

Vijay V S[1, a)] and Joseph Gonsalvis[2, b)]

[1] Assistant Professor, St Joseph Engineering College, Mangaluru

[2] Professor, RNS IT, Bangalore

Corresponding author: vijayv@sjec.ac.in

Abstract- Heating diesel fuel before injecting into the engine cylinder is the technique employed in this experiment with the primary objective of improving combustion and decreasing emission. This results in gasification of diesel when injected into the engine cylinder leading to better mixing, evaporation and combustion of diesel fuel. This also results in decrease of physical delay period. Experiments are done at standard injection timing with preheating done in the interval of 100°C, 175°C and 250°C. At each temperature, readings are taken at no load, 25%, 50% and 75% of full load. It is observed that preheating results in decreased emission of CO and HC at all the loads. At 50% and 75% load CO and HC decreased by about 35% and 50% respectively. NOx is also increased showing that combustion of diesel fuel is improved. Maximum brake thermal efficiency (BTE) and minimum brake specific fuel consumption (BSFC) is obtained at preheating temperature of 100°C at 75% load condition. With increase in load condition, performance is found to be better at higher preheating temperature. With increase in preheating temperature, performance and emission is better and efficient. Performance improvement is better at 50% load than at no load and 25% load.

1. Introduction

Automotive sector is experiencing disruption due to the emerging technologies like electric vehicle and fuel cell vehicle resulting in the gradual phase out of engine powered vehicle in the near future. But the amount of research going on in the field of IC engine and also well-established infrastructure in terms of fuel refilling stations makes very difficult to neglect its role in automobile. Also the progress in the reduction of emissions and increase in fuel efficiency of IC engines makes it still a favourable option for passenger vehicle. Therefore in this present experimental work, effort is made to reduce the emission and fuel consumption of a diesel engine by heating the neat diesel fuel before injecting into the combustion chamber. This temperature ranges from 40°C to 300°C which increases the chances of fuel getting into gasified ranges producing a homogeneous mixture of air and fuel which will lead towards improved combustion of the fuel. Therefore it is expected that through a detailed study of injecting fuel at different temperatures and measuring the performance of the engine conclusions may be drawn.

Several literatures studied in this regard shows that preheating of fuel is commonly employed in biodiesel applications to reduce viscosity. Other applications is the recently developing Homogeneous Charge Compression Ignition engines. Vijayeshwar [1] explains about diesel vapor combustion system, where diesel fuel in vapour form is introduced in a petrol engine using hot air vaporization techniques. Here hot air is supplied at the bottom of a diesel sub tank resulting in air bubbles which extract diesel fumes from liquid fuel forming diesel vapors. In this paper, author has not explained about the performance study when using this concept.

Avinash kumar Agarwal et al. [2] studied about exhaust particulates from diesel fuelled Homogenoeus Charge Compression Ignition (HCCI) engine using diesel vaporizer technology. Diesel is less volatile and has more viscosity compared to gasoline. Hence this technology helps in producing more homogeneous mixture of fuel and air. It consists of a chamber where diesel is injected to an air stream which is electrically heated and also uses heat from exhaust gases to achieve diesel vaporization. Temperature of heating was found to be around 160oC. Engine was run with 0%, 15% and 30% EGR in HCCI mode and performance and emission was compared with normal diesel mode. Results showed that performance were equal with normal diesel mode but NOx and PM found to reduce drastically in HCCI mode. HC and CO showed increased trend due to lower in-cylinder temperature due to leaner mixture and also due to trapping of homogeneous air-fuel mixture in crevice volume and dead volumes in combustion chamber. Another work in the reference [3] showed that preheating inlet air was done to improve the performance of diesel engine with ethanol emulsion. Here air is preheated to 30°C, 40°C and 50°C. From the experiment, it is observed that there is reduction in smoke density, particulate matter and exhaust gas temperature with an increase of nitrogen oxides (NOx) and brake thermal efficiency.

In the series of experiments [4, 5, and 6],vegetable oils like Waste cooking oil, raw rapeseed oil etc were blended with diesel in different proportions with preheating up to a temperature of 135°C showed improved performance and reduction in emissions. A patent works [7] shows a diesel pre-heater is developed using heat from engine coolant without any description on the performance results. Also, there are several pre-heaters [8] available in market which claim that by preheating diesel using engine coolant or by some other methods, fuel efficiency increases by 20 – 30%. A fuel preheater for use with internal combustion engines is disclosed [9]. The preheater works on the principle that by preheating the fuel, the fuel is more effectively vaporized, resulting in more efficient combustion. This preheating is accomplished using heat normally wasted via the radiator. The preheater has a housing, through which heated engine coolant on its way from the engine block to the radiator is routed. A coiled steel gas line is routed through the housing, and is connected between the regular fuel line and the engine. Test results show that there is 40% to 50% of fuel saved with preheating compared to without preheating condition in a 2.5 litre diesel engine.

Most important work related to this experiment is known as Flash Boiling Injection concept [10]. It proposes a new method of utilizing the flash boiling injection effect of high temperature diesel fuel to improve the fuel-spray evaporation and mixing in homogenous charge compression ignition (HCCI) diesel engines. The combustion and exhaust emissions of the HCCI diesel engine were tested by increasing the fuel temperature to 160 – 200 °C., and the effect of high-temperature flash boiling was clearly demonstrated by the test results. Benefits includes reduction in droplet size by about 50% at 200oC, reduced BSFC and reduction in NOx. But above 150oC of fuel temperature, BSFC, CO and HC emissions increased due to possibly due to reduced spray penetration and extremely short ignition delay. Also under high load condition and high diesel fuel temperature, it is observed that there is excessive injection retard and relative lack of oxygen leading to deterioration of the late combustion and finally result in increased fuel consumption.

2. EXPERIMENTAL SETUP

The performance characteristics were carried out on variable compression diesel engine. The specifications of the engine are as stated as below. A four-stroke single cylinder, water cooled diesel engine is used for the experiment and specification is provided in table.1

Figure 1 shows the schematic arrangement of a diesel engine test rig used for the experimentation purpose. A rope brake type dynamometer is used for loading the engine. Heating of the fuel is done using a electric heater installed on the fuel line between the fuel pump and injector. Full load of the engine was found to be 16 kg (based on rated power and considering brake drum radius of 0.15m). During experiment engine was loaded with 0%, 25%, 50% and 75% of full load. Due to safety reasons and engine condition, engine load was limited to 75% of full load (12 kg). At each loading, fuel temperature varied from room temperature (normal diesel) to 250°C at the interval of 100°C and 175°C.

Fuel temperature was controlled manually with the help of thermocouple arrangements. Standard injection timing was found to be 23° before TDC and experiments conducted with variable injection timing is not reported here.

Table 1. Specifications

Sl. No	Name	Details	Sl. No	Name	Details
1	No. of cylinder	1	7	Compression ratio	16.5:1
2	No. of stroke	4	8	Brake drum diameter	275mm
3	Fuel	Diesel	9	Rope diameter	19mm
4	Rated power	3.7 kw@1500 rpm	10	C_d of orifice	0.65
5	Cylindrical Diameter	87.5mm	11	Orifice diameter	19mm
6	Stroke length	110mm			

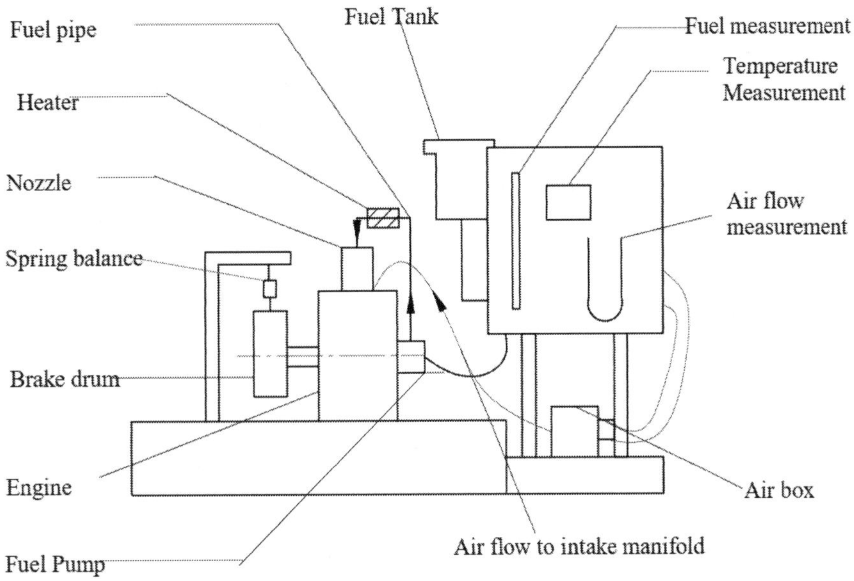

Figure 1. Schematic arrangement of Diesel Engine Test Rig with Pre heater

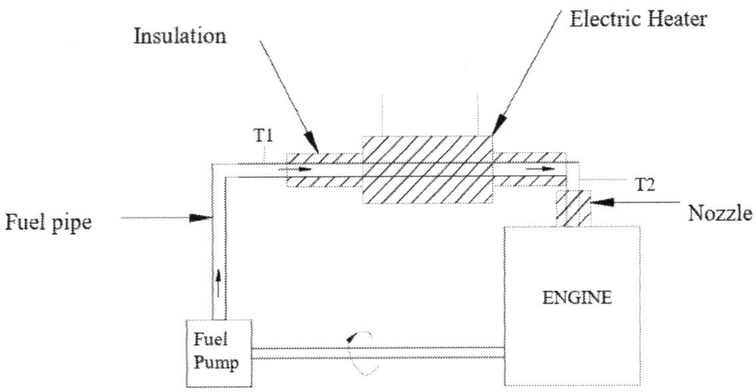

Figure 2. Heater arrangement in the fuel pipe line

Figure 2 shows the heater arrangement in the diesel engine setup. Electric heater of 750 W capacity was used to heat the fuel by mounting it on the fuel pipe line which connects the fuel pump to the injector or nozzle. Since the fuel is flowing in a closed pipeline it can be heated above the flash or fire point of the diesel. Also heating the fuel in the location between pump and nozzle [10], will not disturb working of the pump and pumping capacity.

3. RESULTS AND DISCUSSION

3.1 Performance and Emission at 75% load condition:

Figure 3, 4 and 5 shows variation of Brake Power (BP), BSFC, and BTE for pure diesel at 75% load condition. Compared to room temperature, these parameters remained almost constant at higher temperatures. BSFC reduced by 1.65% at 100°C and brake thermal efficiency increased by 0.34%. As the fuel is heated, vaporization increases and physical delay period should reduce. This will affect the peak pressure rise and heat release rate. Since in this experiment, injection timing is fixed at 23° before Top Dead Centre (TDC), there is no appreciable improvement is observed. It is also found in literature [10], as the fuel temperature is increased, it affects the spray penetration and shortens ignition delay.

Figure 6, 7 and 8 shows NOx, HC, and CO emissions with temperature. It is observed that with increase in temperature HC decreases. NOx and CO initially increases but decreases after 100°C. The minimum value of CO and HC is at temperature of 250°C and is decreased by 37.5% and 56% respectively. It proves that the heating of fuel improves mixing of fuel and air, increases homogeneity and improves combustion. In case of NOx emission, initially it is increased by 42% and then decreases. This shows that mixture is tending to become homogeneous and lean resulting in decrease in NOx at higher temperature. Hence it can be concluded that preheating has a great benefit on reducing emissions.

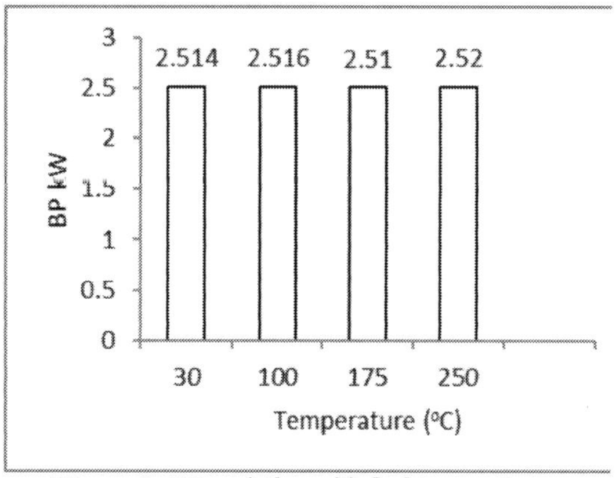

Figure 3. BP variation with fuel temperature

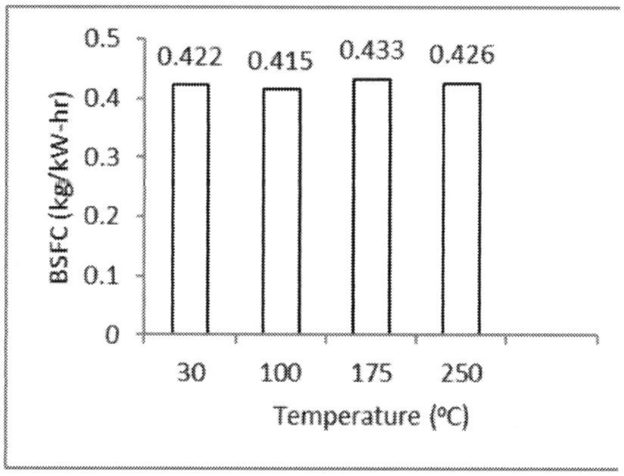

Figure 4. BSFC variation with fuel temperature

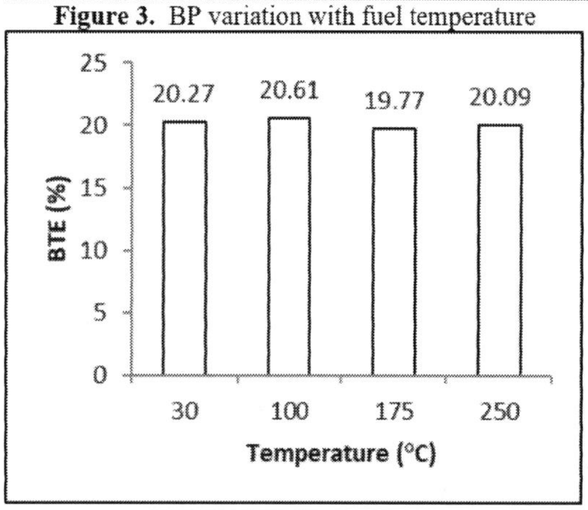

Figure 5. BTE variation with fuel temperature

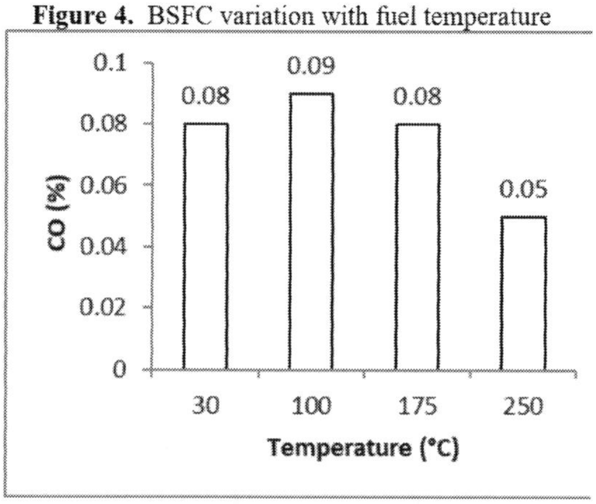

Figure 6. CO variation with fuel temperature

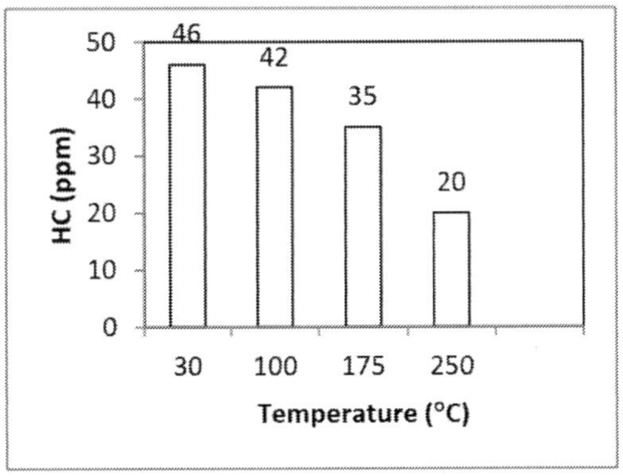

Figure 7. BP variation with fuel temperature

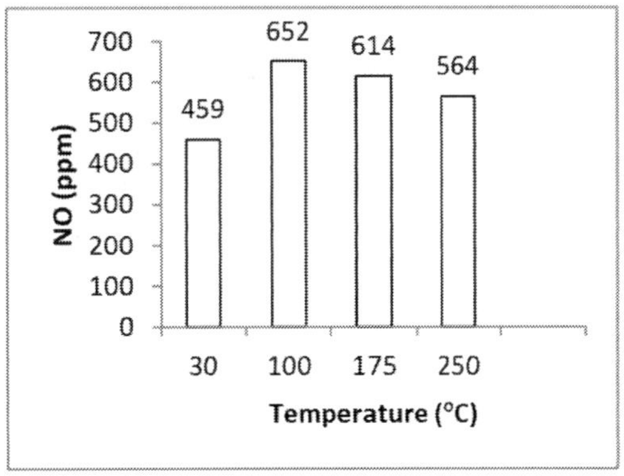

Figure 8. BSFC variation with fuel temperature

3.2 Performance and Emission at 50% load condition:

Figure 9, 10 and 11 shows variation of BP, BSFC, and BTE for diesel at 50% load condition. Compared to 75% load, 50% load shows better performance improvement. The maximum value of BP is at temperature of 100°C which increases by 1.01% and the minimum value of BSFC is at a temperature of 175°C and is found to decrease by 7.41%. Maximum value of brake thermal efficiency is at 175°C which increases by 1.45%. This shows that preheating leads to increase in evaporation and mixing of fuel with air leading to better combustion. It also proves that engine load also affects the mixture formation, spray penetration and combustion ability of fuel air mixture. Also literature shows that, at higher load and fuel temperature, injection is retarded and affects late combustion leading to decrease in performance. Hence performance at 50% load is better than 75% load.

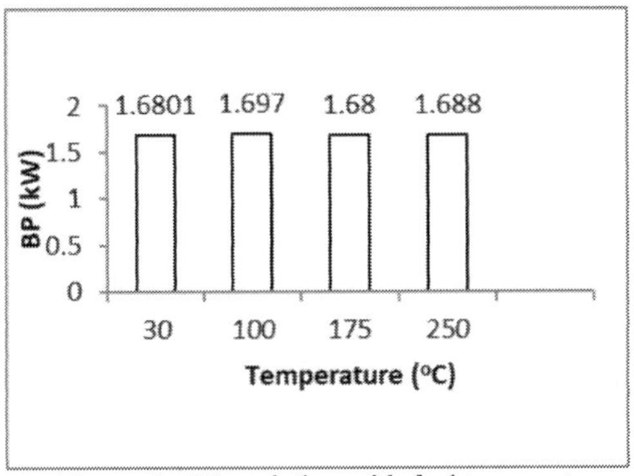

Figure 9. BP variation with fuel temperature

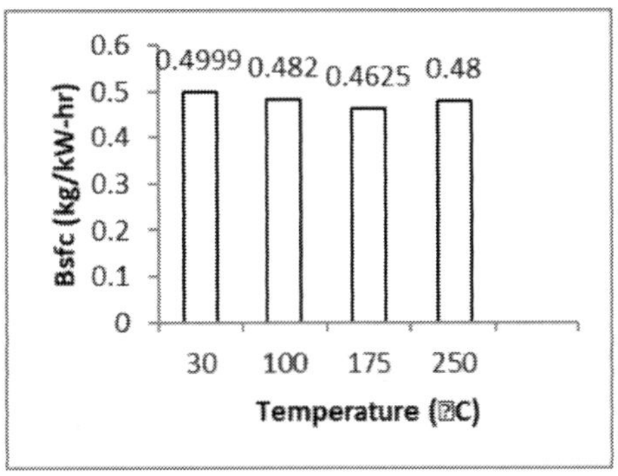

Figure 10. BSFC variation with fuel temperature

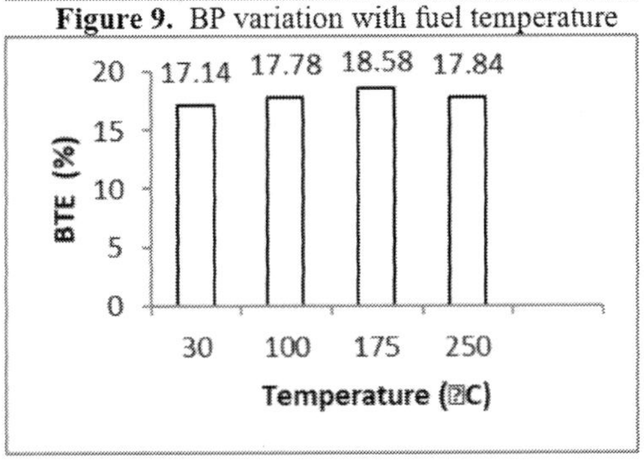

Figure 11. BP variation with fuel temperature

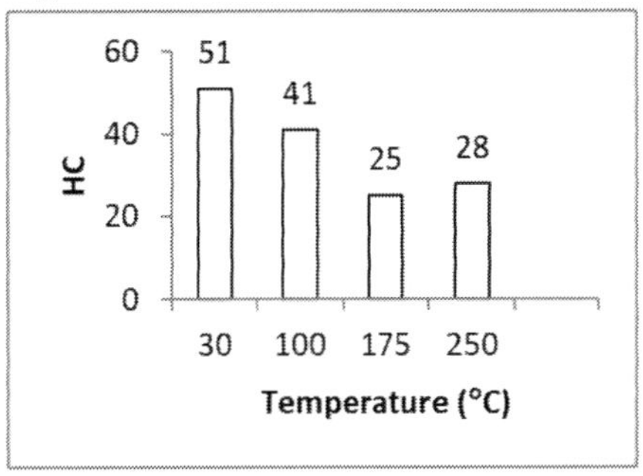

Figure 12. BSFC variation with fuel temperature

Figure 12, 13 and 14 shows HC, NOx and CO emissions with temperature. It is observed that with increase in temperature HC and CO decreases, but NOx increases. At room temperature, CO is 0.13, HC is 51 and NOx is 88. The minimum value of CO is at temperature of 250°C and is decreased by 38% and the minimum value of HC is at a temperature of 175°C which decreases by 51%. NOx is maximum at a temperature of 250°C and is increased by 246.5%. This shows that preheating leads to improved combustion resulting in reduction in CO and HC. This also results in increase in NOx as evidenced in graph.

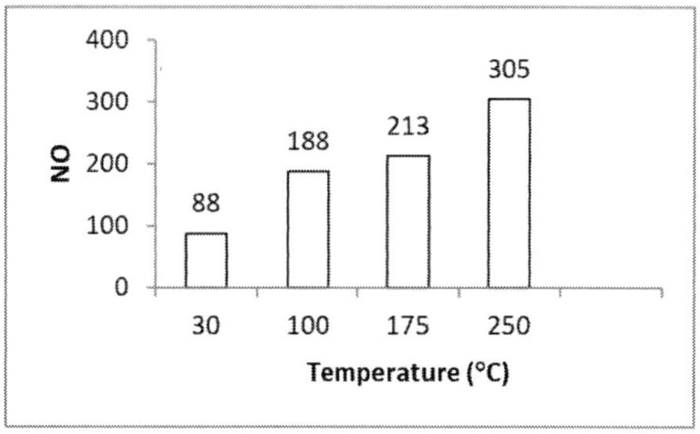

Figure 13. BP variation with fuel temperature

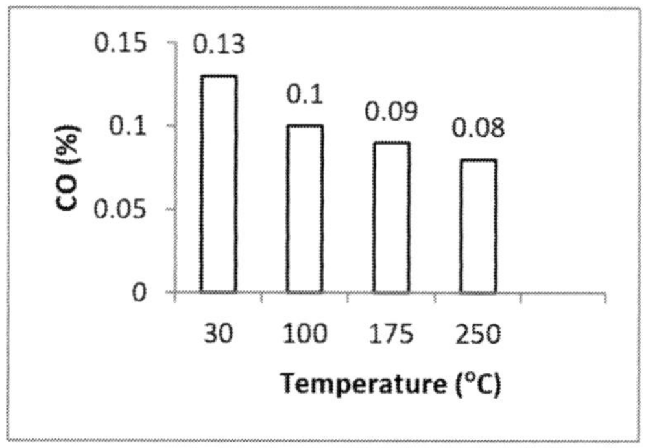

Figure 14. BSFC variation with fuel temperature

4. Conclusion

The experiments done by heating diesel fuel to a high temperature before injecting into the combustion chamber of a diesel engine shows that it has a benefit on improvement of performance and reduction in emission. Experiments were conducted at standard injection timing and results were shown for 50% and 75% of full load at different temperature. Results at 50% load is more encouraging than at 75% load. At 50% load, BSFC decreases by about 7.4%, BTE improves by 1.4%, HC and CO decreases by 51% and 38% respectively. Increases in NOx shows combustion is improving. Thus, it can be concluded that fuel heating before injection has advantages but it is sensitive to injection timing and engine loads. More experimentation is needed to understand the effect of injection timing on performance and emission of diesel engine.

References

[1] Vijayeshwar.B.V "A review: Concept of diesel vapor combustion system" International Journal of Innovative Research in Science, Engineering and Technology, Vol. 5, Issue 4, April 2016

[2] Avinash Kumar Agarwal, Akhilendra Pratap Singh, Jithin Lukose, Tarun Gupta, "Characterization of exhaust particulates from diesel fueled homogenous charge compression ignition combustion engine", Journal of Aerosol Science 58 (2013) 71–85.

[3] Ashok, M. and Saravanan, C., "The Effect of Preheating the Inlet Air to Study the Performance and Combustion Characteristics of Diesel Engine using Ethanol Emulsion," SAE Technical Paper 2007-01-0628, 2007, doi:10.4271/2007-01-0628.

[4] M. Pugazhvadivua, K. Jeyachandran , Investigations on the performance and exhaust emissions of a diesel engine using preheated waste frying oil as fuel, Renewable Energy, Volume 30, Issue 14, November 2005, Pages 2189–2202, Copyright © 2005 Elsevier Ltd

[5] Hanbey Hazar , Hüseyin Aydin, Performance and emission evaluation of a CI engine fueled with preheated raw rapeseed oil (RRO)–diesel blends Applied Energy, Volume 87, Issue 3, March 2010, Pages 786–790, Copyright © 2009 Elsevier Ltd

[6] Lea-Langton, A., Li, H., Andrews, G., and Biller, P., "The Influence of Fuel Pre-Heating on Combustion and Emissions with 100% Rapeseed Oil for a DI Diesel Engine," SAE Technical Paper 2009-01-0486, 2009, doi:10.4271/2009-01-0486.

[7] Clarence R. shepherd, Diesel Engine Fuel Pre-Heating, US patent 4,343,283, Aug. 10, 1982
[8] Daniel. G. Judson, Diesel fuel preheater, United States Patent Number: 4,498,44, Feb. 12, 1985
[9] Paul D. Leonard, Fuel preheating system for internal combustion engines, US Patent Number: 5,218,944, Jun, 15, 1993.
[10] She, J., "Experimental Study on Improvement of Diesel Combustion and Emissions Using Flash Boiling Injection," SAE Technical Paper 2010-01-0341, 2010, https://doi.org/10.4271/2010-01-0341.

IOP Conf. Series: Materials Science and Engineering 1065 (2021) 012025 doi:10.1088/1757-899X/1065/1/012025

Thermodynamic investigation of a SOFC- Steam turbine hybrid system powered by Synthetic gas from Municipal Solid Waste

Venkitesh K S[1], Joseph Daniel[2], Manavalla Sreekanth[3]

[1] Research scholar, School of Mechanical Engineering, VIT University, Chennai.

[2,3] Associate Professors (Sr.), School of Mechanical Engineering, VIT University, Chennai-600127, T.N

Corresponding author email: venkiteshks@gmail.com

Abstract: This study presents the performance evaluation of a system consisting of a Solid oxide fuel cell (SOFC) and steam turbine where the fuel cell is powered by synthetic gas from municipal solid waste (MSW). Analysis is performed based on first and second laws of thermodynamics. The modelling of the system is done in a flow sheeting software Cycle Tempo®. This work mainly focusses on the effective utilization of solid waste by producing the Syngas from gasification. Syngas after cleaning is supplied to the SOFC along with air. The syngas undergoes internal reforming at the fuel cell producing hydrogen which acts as fuel for SOFC. The influence of various operating parameters such as compressor pressure ratio, turbine inlet temperature (TIT), upon thermal efficiency and specific power output are studied. Exergy destruction of various components of the system are also evaluated. The evaluation results are helpful for making suitable modification of the parameters for obtaining optimum configuration of the system.

1. Introduction

The everyday life of a human being is directly or indirectly connected to the energy available in different forms in nature. The average demand for energy rises every day which in the near future may result in a heavy energy crisis. The key focus of the energy sector is on steps to obtain and transform energy from different forms. The key energy sources are those derived from fossil fuels, and they are used in all energy systems. The biggest drawback is that day after day their reserve is depleting and cannot be renewed or reused in any way. As such, we must think about transitioning to non-conventional or renewable energy sources entirely or partially.

Hybridization is a technique which combines conventional sources of energy with renewable sources of energy which results in reduced fuel consumption, decreased emissions of fossil fuels, and improved productivity and output. Fuel cells are devices, where the chemical energy of supplied fuel is converted in to heat and electricity by electrochemical reactions. Because of the direct conversion of power, they ensure constant electricity supply. the anode of the system is fed with fuel and cathode is fed with the oxidant. Fuel undergoes electrochemical oxidation at the anode and the oxidant undergoes reduction at the cathode. Due to its reactivity, which minimizes costly catalysts, the preferred fuel for fuel cells is hydrogen. Hydrocarbon fuels may also be used in fuel cells, but must eventually be converted into hydrogen. This includes a "reformer" or fuel processing system to turn HC fuel into hydrogen [1]. The fuel reformer transforms hydrocarbon fuel into hydrogen-rich gases that are further refined to create pure hydrogen.

Steam turbine combined with SOFC is considered here for hybridization. Syngas derived from municipal solid waste is fed as input to the SOFC where it undergoes internal reforming releasing hydrogen which is the fuel for the fuel cell. SOFC has the ability to withstand higher operating temperatures.

Gaseous fuels have the advantage that they are pure and have high thermal resilience, making burning process ash and soot free. The majority of natural gases have a high content of heat, while those produced from different processes range from moderate to low [2]. High-calorific-value fuels contain less nitrogen, while low-calorific-value fuels contain up to 55% nitrogen [3,4].

Syngas is a blend of fuel mainly composed of carbon monoxide and hydrogen, and most commonly, a little carbon dioxide. It is used as an intermediate in the manufacture of ammonia, hence the name [5]. Syngas is primarily used for the generation of electricity when used with energy conversion systems and is generated by the gasification process. The energy density of syngas is just half of that of natural gas, and can be derived from wood, coal and natural gas. Syngas is an intermediate source for the manufacture of ammonia, hydrogen, methanol and hydrocarbon fuels. Natural gas or liquid hydrocarbons undergoes steam reforming reaction to generate syngas. Syngas is an intermediate source to produce hydrogen during coal gasification and biomass gasification [6].

2. Gasification Municipal solid waste

The globe is looking for solutions for successful solid waste disposal and greenhouse gas reduction. The amount of urban solid waste is rising day by day and can cause significant environmental problems and health risks unless handled properly. Laws and legislation have improved significantly the recycling and reuse of solid waste products in the U.S and European countries. Only $1/4^{th}$ of the overall MSW is recovered in these regions, despite substantial improvements in recycling and energy recovery, leaving the remaining $3/4^{th}$ to be scrapped off in landfills or incinerated.

Energy research personnel concentrate on the integration of alternative energy resources with conventional power systems that can effectively reduce fuel consumption and emissions. Gasification process converts solid wastes in to useful source of energy. Before gasification, the raw solid waste will be processed. Due to the advantage of gas over solid fuels, a part of the heating value of the solid fuel is transferred to a gaseous energy carrier through the process of gasification. [7].

Gases are easy to clean, transport and burn efficiently with less excess air and contains lower levels of some types of contaminants [8]. Numerous studies have been conducted on the gasification process of different biomass feed stocks [9, 10] and coal [8, 11-16]. Gasification appears to be an interesting option for plastic and rubber waste utilization based on high polymer consumption [17]. Although energy recovery and recycling have increased significantly in some zones, only about a quarter of the total solid waste is recovered, and the remaining three-fourths are incinerated or disposed off in landfills [18].

The process of incineration involves burning the wastes with high volume of air which produces carbon monoxide and thus gasification is preferable over incineration. The hot gases which exit from the system are used to produce electricity in energy systems. Gasification transforms MSW into synthetic gas or syngas, which is very much useful. The ash that falls right from the gasifier in the high-temperature gasification process is in a liquid state, which is cooled by quenching. A glassy, non-leachable slag is produced that can be used for producing roofing shingles, cement, and as an asphalt filler.

In this work, effective management of solid waste to produce power is discussed by considering a hybrid energy system consisting of an SOFC integrated with a steam turbine. The fuel is the synthetic gas derived from the gasification of solid waste and is fed directly to the anode side of the solid oxide fuel cell (SOFC) and the air is fed to the cathode side of the SOFC. The air and fuel streams are preheated by the exhaust gases from the combustor or after burner, before entry in to the fuel cell. The fuel undergoes internal reforming reaction at the fuel cell producing hydrogen which acts a fuel for the fuel cell. Heat and electricity are produced within the fuel cell.

IOP Conf. Series: Materials Science and Engineering 1065 (2021) 012025 doi:10.1088/1757-899X/1065/1/012025

The unburned fuel and the gases are further burned in an after burner increasing the temperature further. These hot gases are used in a boiler to produce high temperature steam which runs a steam turbine producing additional power.

Nomenclature

P	Pressure	bar
T	Temperature	K
W	Work	kJ
$\dot{m}$	Mass flow rate	kg/s
h	Enthalpy	kJ/kg
E	Reversible cell voltage	Volt
E^0	Ideal cell voltage at standard conditions	
$\bar{R}$	Universal gas constant	(8.314 kJ/kg. K)
n_e	Number of electrons	-
N	No of cell in SOFC stack	
F	Faraday's constant	96485 (C/mole)
I	Current	Ampere
i	Current density	A/m^2
A	Area	m^2
V	Voltage	Volt
$\dot{Q}$	Heat transfer rate	kJ
ex	Exergy	kJ
y	Mole fraction	-
LHV	Lower heating value of fuel	kJ/kg

Greek symbols

η	Efficiency
γ	Ratio of specific heats (kJ/kg. K)
ϕ_w	Exergy associated with work
ϵ	Effectiveness of heat exchanger

Subscripts/Superscripts

C	Compressor
Act	Activation
Ohm	Ohmic
Conc	Concentration
DC	Direct current
AC	Alternating current
SOFC	Soli Oxide Fuel Cell
Inv	Inverter
In	Inlet
Out	Outlet
AB	After Burner
ST	Steam turbine
CW	Cooling water
P	Pump
k	k^{th} component
D	Destruction
Ph	Physical
Ch	Chemical
T^0	Ambient Temperature

IOP Conf. Series: Materials Science and Engineering 1065 (2021) 012025 doi:10.1088/1757-899X/1065/1/012025

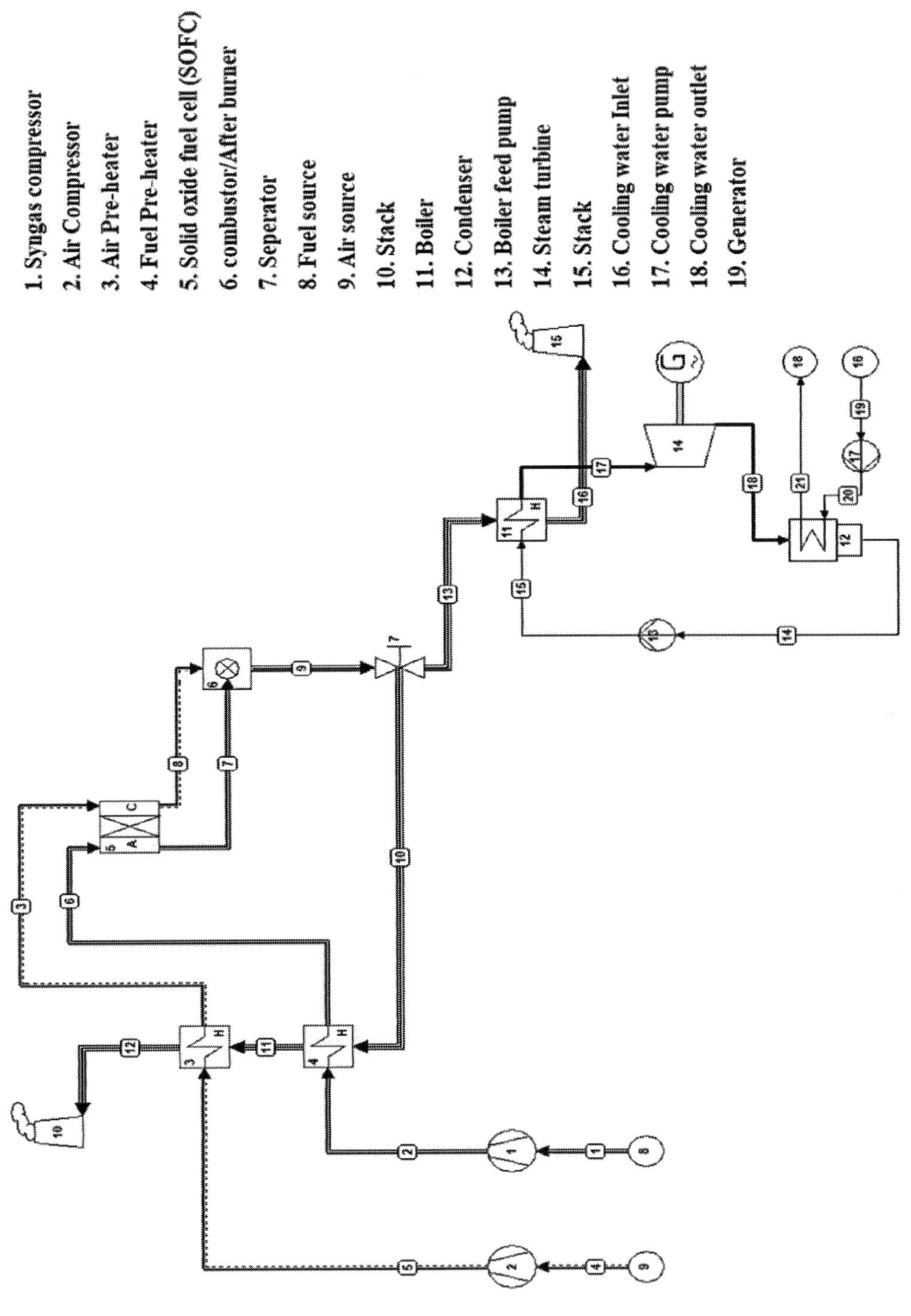

Figure. 1 Layout of Hybrid-SOFC- Steam turbine system powered by syngas from MSW.

3. Description of the system.

The system consists of 2 compressors one for compressing inlet air and other compressing the syngas fuel. They are preheated using the exhaust gases from the combustor before entry in to the solid oxide fuel cell. Preheated air enters the cathode and syngas fuel enters the anode side of the SOFC. Hydrogen is produced as a result of electrochemical reaction inside the fuel cell. The unburned gas and residual fuel are further burned in an after burner which further increases the temperature. The hot gases are used to produce high temperature steam in a boiler and the steam expands in a steam turbine producing additional power.

4. Modelling of the system

Modelling of the proposed hybrid system is done with the flow sheeting software Cycle Tempo® developed by the Delft university of Technology. The program has got inbuilt models of all the components used in energy systems and it calculated the performance parameter based on the inputs given. The analysis of the system is based on laws of thermodynamics and it calculated the energy and exergy efficiency of the entire system, destruction of exergy in individual components of the system. By changing the parameters, the performance of the system can be assessed thermodynamically.
The system routine considers the real state of various components without any assumption of steady state.

4.1. Air and Fuel Compressors

The compression process is assumed to be adiabatic. From the knowledge of isentropic efficiency and pressure ratio of the compressor, the outlet temperature and work done can be determined.

$$\left(\frac{T_{5s}}{T_4}\right)=\left(\frac{P_5}{P_4}\right)^{\left(\frac{\gamma-1}{\gamma}\right)} \tag{1}$$

$$\eta_C=\frac{W_{isentropic}}{W_{actual}}=\frac{h_{5s}-h_4}{h_5 \quad h_4} \tag{2}$$

$$W_C=\overset{g}{m}_4\times\left(h_5-h_4\right) \tag{3}$$

Similar equations can be written for the fuel compressor.

4.2. SOFC

The fuel cell considered here is the direct internal reforming type where the heat released by the electrochemical reactions is used for the reforming reaction which is endothermic in nature. The fuel is the syngas derived from the gasification of municipal solid waste. The reforming, shifting and electrochemical reactions occurring within the SOFC are given below [19-21].

Reforming reaction:

$$CH_4+H_2O\rightarrow CO+3H_2 ----------(4)$$

Water gas Shifting reaction:

$$CO+H_2O\rightarrow H_2+CO_2 -----------(5)$$

Electrochemical Reaction:

$$H_2+0.5O_2\rightarrow H_2O -------------(6)$$

The amount products of reforming and shifting reactions which are equilibrium reactions, depends on the final temperature and initial composition of the reaction.

The reversible cell voltage produced by the SOFC stack is given by Nernst equation given below [22]

$$E = E^0 = \frac{\overline{R}T}{n_e F}\ln\left(\frac{\left(P_{H_2O}\right)}{P_{H_2}\left(P_{O_2}\right)^{0.5}}\right) ------(7)$$

Where E^0 is the ideal cell voltage at standard conditions (298K and 1 bar) and R is the universal gas constant, F represents the Faraday's constant (96485 C/mole), n is the number of electrons and T is the temperature of the stack.

Calculation of the actual cell voltage involves determination of various losses such as activation loss (V_{act}), ohmic loss (V_{ohm}) and concentration loss (V_{conc}) associated with the fuel cell. The actual cell voltage is given by the following equation.

$$V_{cell} = E - \left(V_{act} + V_{ohm} + V_{conc}\right) = \text{E-}\left(\Delta V_{loss}\right) -------(8)$$

The amount of current and power generated by the fuel cell can be calculated from the following set of equations from (9) through (12).

$$I_{cell} = i \times A_{cell} ----------------------(9)$$

$$I_{total} = N \times I_{cell} ----------------------(10)$$

$$(\dot{W})_{DC_{SOFC}} = V_{cell} \times I_{total} ----------(11)$$

$$(\dot{W})_{AC_{SOFC}} = (\dot{W})_{DC_{SOFC}} \times \eta_{inv_{SOFC}} ----------(12)$$

Where $\eta_{inv_{SOFC}}$ is the efficiency of DC-AC inverter.

4.3. Afterburner

The unreacted fuel and gases from the SOFC stack are further burned in an afterburner for further increasing the temperature. The reaction at the afterburner is completely exothermic and no external fuel is supplied. The energy balance can be written as follows

$$\sum m_{in}h_{in} - m_{out}h_{out} - (\dot{Q})_{loss.AB} = 0 -----------(13)$$

$$Q_{loss.AB} = (\dot{m})_{fuel} \times \left(1 - U_f\right) \times \left(1 - \eta_{AB}\right) \times LHV ---------(14)$$

4.4. Air and fuel preheaters and steam generator

The exhaust gases from the afterburner is split in to two sections, one going to preheat the air and fuel streams from respective compressors and the other one going directly in to the steam generator. The outlet temperatures of the preheaters and the steam generator are calculated by the following energy balance equations.

$$\varepsilon = \frac{h_6 - h_2}{h_{10} - h_2} -------------(15)$$

$$\dot{m}_2\left(h_6 - h_2\right) = \dot{m}_{10}\left(h_{10} - h_{11}\right) -------(16)$$

The equations for fuel preheater and steam generator are similar to the above ones.

4.5. Steam turbine

The hot steam generated at the steam generator is expanded in the steam turbine producing power and is discharged to the condenser.

$$\frac{T_{18s}}{T_{17}} = \left(\frac{P_{18}}{P_{17}}\right)^{\frac{(\gamma-1)}{\gamma}} -----------(17)$$

$$\eta_{ST} = \frac{W_{ST.actual}}{W_{ST.isentropic}} = \frac{h_{17} - h_{18}}{h_{17} - h_{18s}} ------(18)$$

$$W_{ST} = \dot{m}_{17}\left(h_{17} - h_{18}\right) --------(19)$$

4.6. Condenser

The exit steam from the turbine is entering the condenser to which cooling water is supplied externally. The following energy balance equation can be written for the condenser.

$$\dot{m}_{18} h_{18} + \dot{m}_{20} h_{20} = \dot{m}_{14} h_{14} + \dot{m}_{21} h_{21} ---(20)$$

4.7. Feedwater and cooling water pumps

The energy balance equations for the pumps used in this system are as follows

$$W_{CW.pump(isentropic)} = \dot{m}_{19} v_{19}\left(P_{20} - P_{19}\right) ----(21)$$

$$\eta_{pump} = \frac{W_{isentropic}}{W_{actual}} -------(22)$$

4.8. Hybrid system

The efficiency of the entire hybrid system is obtained from the following equation

$$\eta = \frac{\dot{W}_{net}}{m_{fuel} \times LHV_{fuel}} --------(23)$$

$$W_{net} = W_{SOFC} + W_{ST} - \Sigma W_P - \Sigma W_C ---(24)$$

4.9. Exergy equations

The performance of the system is assessed based second law of thermodynamics through exergetic analysis. In helps in identifying the processes which are thermodynamically undesirable. The irreversibility associated with the entire system are computed from exergy loss and exergy destruction associated with various components. The mass, energy and exergy balance equations are considered in the exergetic assessment of the system. Each part of the system is considered as a control volume. The exergy balance equation is written in general as follows.

$$\Sigma\left[1 - \left(\frac{T_0}{T}\right)\right] Q_k + \Sigma ex_{in} = \Sigma \varphi_w + \Sigma ex_{out} + Ex_D ---(25)$$

The first term in the above equation represents the exergy rate due to heat transfer, ϕ_w is the exergy rate due to work transfer, Ex_{in} and Ex_{out} are the exergy input and output of the control volume and Ex_D is the exergy destruction. The exergy rate of entire system is obtained by summing up the physical and components of exergies as follows

$$\dot{E}_x = \dot{E}_{x\,ph} + \dot{E}_{x\,ch} \qquad (26)$$

IOP Conf. Series: Materials Science and Engineering 1065 (2021) 012025 doi:10.1088/1757-899X/1065/1/012025

$$E_{xph} = \overset{g}{m}\left[\left(h - h_0\right) - T_0\left(s - s_0\right)\right] \tag{27}$$

$$E_{xch} = \overset{g}{m}\left[\sum y_i \times ex_{ch.in} + \bar{R} \times T_0 \sum_i y_i \ln y_i\right] \tag{28}$$

The exergy efficiency is given by the following expression

$$\eta_{ex}^{cycle} = \frac{W_{net}}{\sum E_{x.in}} \tag{29}$$

Table 1. Input parameters for the Hybrid System

Parameter	Value Input
Ambient temperature	25^0C
Ambient Pressure	1 bar
Syngas Temperature	800^0C
Calorific value of Syngas	5 MJ/kg
SOFC Temperature	900^0C
Fuel utilization factor	0.85
Cathode and anode pressure drop	4%
Current density	3000 A/m^2
Resistance of the cell	5×10^{-5} ohm m^2
DC-AC Inverter efficiency	89%
Compressor isentropic efficiency	0.81
Turbine isentropic efficiency	0.85
Generator isentropic efficiency	0.95

5. Results and discussions

The proposed system is modelled and simulated by varying the parameters and the system performance was determined. The main parameters that affect the performance of the plant are the fuel cell current density, Cell pressure etc. The influence of these parameters on energy and exergy efficiencies, cell voltage is determined. The exergy destruction for various components is determined. This will help in finding ways for minimizing exergy loss and improving the performance of the entire system.

5.1. Influence of current density

Figure 2 shows the impact of current density on energy efficiency at fuel utilization factor of 0.85, cell temperature of 1173K and cell pressure of 3 bar. The energy efficiency has a decreasing trend with respect to the current density. As the current density increases, the power developed from the SOFC stack decreases and thus the efficiency.

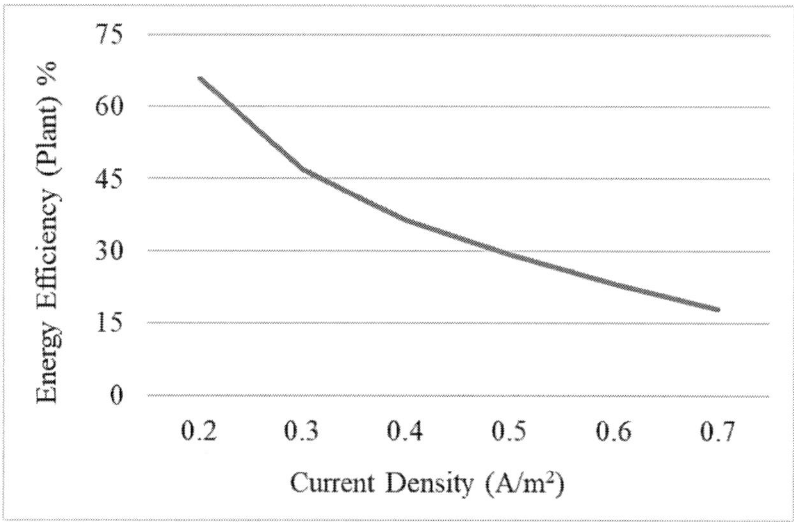

Figure. 2 Influence of current density on energy efficiency

Exergy efficiency as shown in figure 3 also has a similar trend as that with energy efficiency with current density. The exergy destruction at cell temperature of 1173K and 3 bar pressure is shown in figure 4. The exergy efficiency decreases while exergy loss increases with increasing SOFC current density. At higher current densities, there is a decrease in the cell voltage and the requirement for average number of cells is high which leads to low output power from the fuel cell.

5.2. Influence of Cell pressure

Figure 5 shows the impact of cell pressure on the cell voltage. Pressure of the cell is varied from 1 to 6 bar. The cell voltage increases as the Nernst voltage increases. As a result of this, cell voltage increases. When cell pressure increases, the overpotentials decreases (Akkaya et al. 2007) which also result in increased cell voltage.

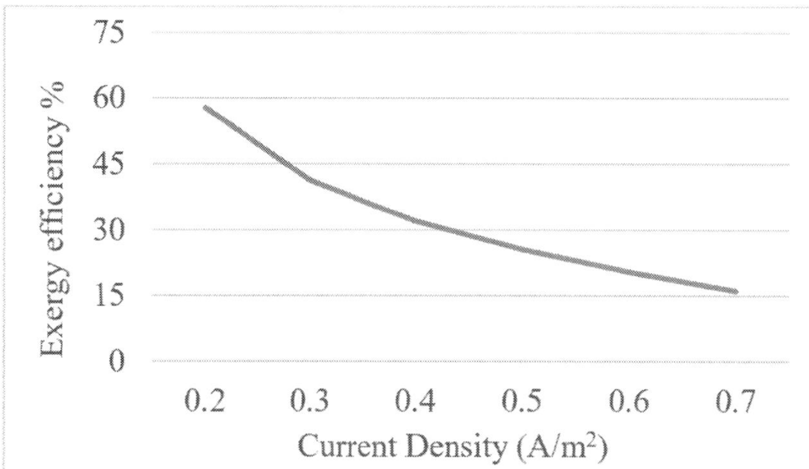

Figure. 3 Influence of current density on exergy efficiency

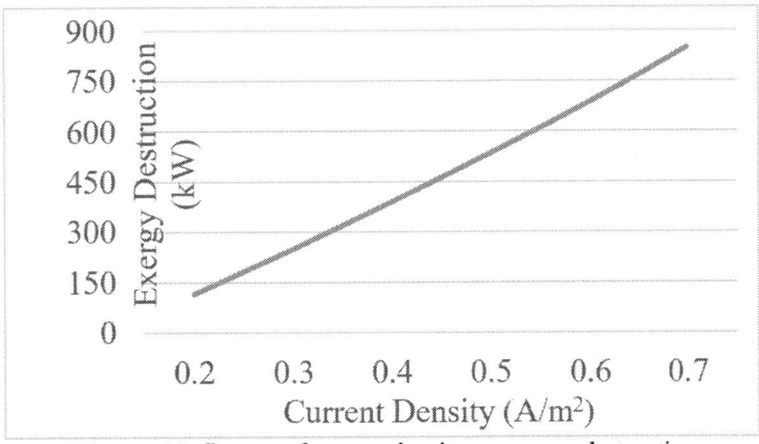

Figure. 4 Influence of current density on exergy destruction

Figure 6 shows the destruction of exergy in different components of the system. Exergy destruction is maximum at the combustor followed by the heat exchanger and fuel cell. This is because of the highest irreversibility associated with the heat transfer occurring in these components.

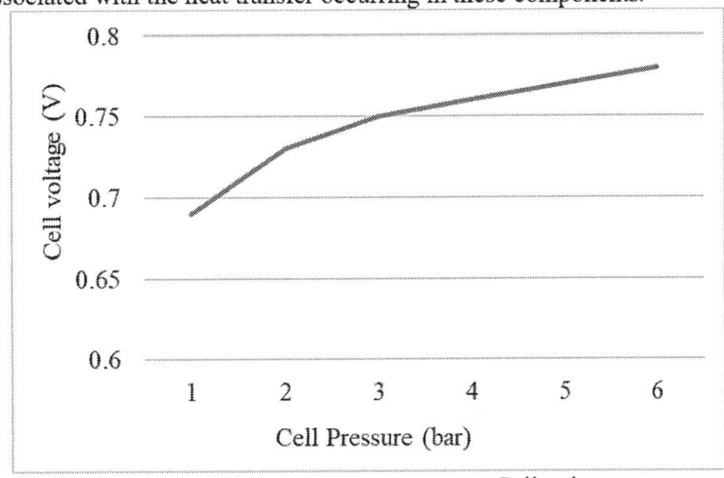

Figure. 5 Effect SOFC pressure on Cell voltage

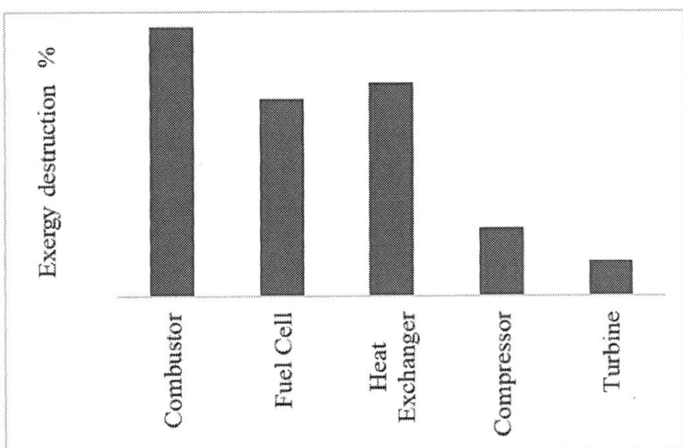

Figure. 6 Exergy destruction in the system

6. Conclusion

The proposed hybrid system was modelled and analyzed thermodynamically. The exergy analysis gives the destruction of exergy in various components. The exergy destruction is maximum in the combustor due to the higher heat transfer rate. Maximum efficiency of the entire system is 46% and the exergy efficiency is 41%. The net output power of the system is 180kW. The system can be used as a domestic CHP (combined heat and power) to power up restaurants, small offices and shopping centres.

References

[1]. Jack Brouwer. *Hybrid Gas Turbine Fuel Cell Systems*, National Fuel Cell Research Center, University of California.

[2]. Arthur H. Lefebvre, 1983 *Gas turbine Combustion*, (Hemisphere Publishing corporation)

[3]. Y. B. Zeldovich, P.Y Sadovnikov and D. A. Frank-Kamenetskii 1947 Oxidation of Nitrogen in Combustion, *Academy of Sciences, U.S.S.R.*, Moscow-Leningrad.

[4]. D. N. Anderson 1975 Effects of equivalence ratio and dwell time on exhaust emissions from an experimental premixing pre-vaporizing burner, *ASME Paper* 75-GT-69.

[5]. Beychok, M.R May 1975 Process and environmental technology for producing SNG and liquid fuels, *U.S. EPA report* EPA-660/2-75-011.

[6]. Beychok, M.R. September 1974 Coal gasification and the Phenosolvan process, *American Chemical Society 168th National Meeting*, Atlantic City,

[7]. Van Swaaij W. P. M., Dec 1981 Gasification - the process and the technology. *Resour. Conserv.* 337-49

[8]. Comolli A. G, Lee L.K and Pradhan. V 1997 Alternate fuels from the co-liquefaction of coal, oil, and waste plastics. *Prepr. Pap. - Am. Chem. Soc., Div. Fuel Chem.* **42** (2), 736-740.

[9]. Wang Y. Yoshikawa K. Namioka T and Hashimoto Y. 2007 Performance optimization of two-staged gasification system for woody biomass. *Fuel Process. Technol.*, **88** (3), 243-250.

[10]. Butterman H. C. Castaldi M. J 2007 Influence of CO_2 Injection on biomass gasification. *Ind. Eng. Chem. Res.* **46** (26), 8875-8886.

[11]. Braekman-Danheux C, D'Haeyere A., Fontana A and Laurent P 1997 Upgrading of waste derived solid fuel by steam gasification. *Fuel* **77** (1/2), 55-59.

[12]. Kajitani S. Har, S. Matsuda H 2002 Gasification rate analysis of coal char with a pressurized drop tube furnace. *Fuel* **81** (5), 539-546

[13]. Luo C, Watanabe T, Nakamura M, Uemiya S. and Kojima T 2001 Gasification kinetics of coal chars carbonized under rapid and slow heating conditions at elevated temperatures. *J. Energy Resour. Technol.* **123** (1), 21-26.

[14]. Lee J. G, Kim J. H, Lee H. J, Park T. J and Kim S.D 1996 Characteristics of entrained flow coal gasification in a drop tube reactor. *Fuel* **75** (9), 1035-1042

[15]. Shin Y, Choi S and Ahn D.H. 2000 Pressurized drop tube furnace tests of global coal gasification characteristics. *Int. J. Energy Res.* **24** (9), 749-758.

[16]. Yamamoto T, Sato H, Matsukura Y, Ujisawa Y, Ishida H, Sasaki S and Hata Y 2004 Gasification and smelting system using oxygen blowing for plastic waste including polyvinyl chloride. *J. Mater. Cycles Waste Manage.* **6** (1), 6-12

[17]. Eilhann Kwon, Kelly J. Westby and Marco J. Castaldi 2009 An Investigation into the Syngas production from municipal solid waste (MSW) gasification under various pressures and CO_2 concentration atmospheres. *Proceedings of the Annual North American Waste-to-Energy Conference NAWTEC17*, 17th.

[18]. http://www.gasification-syngas.org/applications/waste-to-energy-gasification/

[19]. Ozcan H, I Dincer 2014 Performance evaluation of an SOFC based trigeneration system using various gaseous fuels from biomass gasification. *Int. J Hydrogen Energy*. **40** (24):7798–7807.

[20]. Ma S, Wang J, Yan Z, Dai Y and Lu B 2011 Thermodynamic analysis of a new combined cooling, heat and power system driven by solid oxide fuel cell based on ammonia–water mixture. *J Power Sources*. **196** (20):8463–8471.
[21]. Chan SH, Ho HK, Tian Y 2002 Modelling of simple hybrid solid oxide fuel cell and gas turbine power plant. *J Power Sources*. **109** (1):111–120
[22]. *Fuel cell handbook.* 7th ed 2004 (Morgantown, WV: EG&G Services Parsons, Inc)

IOP Conf. Series: Materials Science and Engineering 1065 (2021) 012026 doi:10.1088/1757-899X/1065/1/012026

MECHANICAL CHARACTERIZATION OF MMT NANO CLAY/EPOXY/ BASALT FIBER COMPOSITES

Susilendra Mutalikdesai, Vani R, Arpan Bhandari, Aishwarya Jain Y

Department of Mechanical Engineering, Yenepoya Institute of Technology, Moodbidri, India

arpanbhandary75@gmail.com

Abstract. The importance of composite fibers have been rapidly increasing since the inception of its discovery, in almost every engineering field. They possess beneficial properties such as low weight, high strength, highly resistive towards corrosion which makes them unique. The objective of present study is to fabricate composite using epoxy as polymer and basalt fiber as reinforcement and to investigate the influence of dispersing MMT Nano clay (Montmorillonite Nano-clay) in polymer. Conventional hand lay-up technique was used for the fabrication of composite laminate. The bi-directional basalt fiber was used as reinforcement and epoxy was taken as matrix material. Mechanical Characterization of MMT Nano clay/epoxy/basalt fiber composites was carried out for tests such as tensile, flexural and impact test, the results showed that the laminate had comprehensible properties, concluding usage of composite material is more beneficial and cost effective than conventional materials in many ways.

Keywords. Hand lay-up technique, Tensile Test, Flexural Test, Impact Test

1. Introduction

Two or more dissimilar material combined together forms a composite material which has the properties superior to its parent material. In the finished structure, the blended materials remain separate and distinct, differentiating composites from solid solutions and mixtures [1]. Researchers have started to include computation, sensing, communication and actuation into composite material known as Robotic Materials. Wattle and Daub is one of the oldest composite synthetic materials, more than 6000 years old. [2]. There are also greener ways of acquiring fibers from the nature which are eco-friendly and do not leave any traces behind at end life. Natural fibers can be obtained from plants, animals and many other sources as well. The world's second most plentiful natural polymer is Chitin, whereas collagen is first. It is a "linear polysaccharide of β-(1-4)-2-acetamido-2-deoxy-D-glucose" [3].

A notable change in the use of natural fibers was reported for the manufacture of composites all over the world [4-6]. Natural fibers, moreover, have the potential to replace synthetic fibers, such as glass, spandex, and nylon, in many ecological factors but not in respect of mechanical strength.

Basalt fibers are extruded continuously from selected basalt stones (volcanic, effusive rocks with 45-25% silica) of high temperature melt (around 1500°C) [7]. The chemical structure of basalt fiber is similar to that of glass, with slightly higher density (0.26 g/cm3 than glass) [8]. The more important factor is that basalt fibers are chemically more stable when compared to glass fibers, especially in an acidic environment [9]. This allows basalt fibers for more effective binding to sizing agents, such as organ silanes, resulting in reduced usage of chemicals which has to be used along glass fibers [10]. Also, basalt fibers can be used in a wide range of temperatures, from 200°C to 600°C [11].

Synthetic resins are viscous substances that are industrially manufactured and processed into rigid/solid polymers through the curing process. Resins typically contain reactive end groups in order to undergo curing [12], such as acrylates or epoxides. Some synthetic resins may have identical qualities to natural plant resins, too [13]. Epoxy resin, made by polycondensation or polymerization-polyaddition reactions, is the classic variety used as a thermoset polymer for composites and adhesives [14]. Epoxy resin, seamless and waterproof, is two times stronger than concrete. Therefore, it has been primarily in use since the 1960s for commercial flooring purposes. However, since 2000, epoxy and polyurethane resins have also been used in interior designs, particularly in Western Europe.

In present work Nano composite is fabricated by dispersing MMT Nano clay with varying weight ratio 0%, 2%, 4%, 6% with epoxy resin (LY556) and basalt fiber as re-enforcement. The present work also focuses on mechanical characterization of MMT Nano clay (Montmorillonite Nano-clay) /epoxy/basalt fiber Nano composites.

2. Material selection, Fabrication and Material characterization

2.1. Material Selection

The details of the processing of the composites and the experimental procedures carried out for their characterization and tests to which the composite specimens are subjected are described in material selection and manufacture. In this job, the raw materials used are:

- Basalt fiber
- Nano clay-15A
- Synthetic Epoxy text.

2.1.1. Basalt fiber. A recent rise can be seen globally in the use of eco-friendly, natural fibres as reinforcement for the manufacture of lightweight, low-cost polymer composites. Basalt fibre, which provides outstanding properties over glass fibres, is one such material of interest currently being used extensively. High basic mechano-physio-chemical properties, synthetic degradability, and non-abrasive qualities, to name a few, are the prominent advantages of these composites. This paper provides a brief analysis of basalt fibres used as a composite reinforcement material and explores them as an alternative to the use of glass fibres. The paper also addresses the basics of and classification of basalt chemistry. In addition, the growing trend in research publications and activities in the field of basalt fibres is also discussed in an attempt to show. The improvement in mechanical, thermal and chemical resistant properties achieved for applications in specific industries is discussed in further sections.

2.1.2. Nano Clay. Nano clays are clay minerals designed for use in multi-functional structures with many property improvements aimed at unique applications in Nano composites. Polymer clay- Nano composites are a particularly well-researched class of materials of this kind. Nano clays are a wide class of naturally occurring inorganic minerals, the most widely used material applications of which are plaques such as montmorillonites. Nanaoclay-15A was selected for the present job.

2.1.3. Epoxy LY 556. The chemically synthesised high molecular polymer is synthetic resin, short for resin. Therefore, after the name of the synthetic resin it is made from, various kinds of plastic can be named. Synthetic resin makes up 30 % to ~60 % or more of its composition, the essential raw material of plastic. It has the feature of agglutination, not only binding together but also binding strongly together the other materials. The physical and mechanical properties of plastic often change as the form, property, and amount of synthetic resin change. The key characteristics of plastic are therefore based on the synthetic resin from which it is made. LY 556 is a medium viscosity epoxy resin based on bisphenol-A with excellent mechanical properties which can be modified by using different hardeners as well as fillers.

2.2. Fabrication. The composite laminate fabrications are carried out using traditional hand lay-up techniques. The bi-directional basalt fibre is used as a reinforcement and the matrix content taken is epoxy LY556. Basalt fibre is manufactured by Nickunje Exhimp Entp P Ltd. Zenith Polymers India Ltd supplied with epoxy resin and the accompanying hardener. The synthetic epoxy resin and hardener are combined by weight percentage at a ratio of 10:1. Composites of varying compositions with three distinct filler content percentage. The most commonly used and simple process for producing fibre-reinforced polymer composites is the hand lay-up technique. Hand Layup technique was adopted for fabrication of laminates. Number of fibre layers and weight of resin mixture was taken according for calculation. Apply resin layer by layer by using hand lay-up method. For curing, the laminate is kept for 24 hours of recovery at room temperature. After curing specimens of standard sizes are cut from laminates for different mechanical testing.

2.3. Material Characterization. On prepared laminates, the mechanical properties of basalt and Nano clay composites have been determined. In compliance with ASTM standards ASTMD 638 and ASTMD 790, tensile and flexural tests were performed on the universal test unit. The tensile test is the most widely used mechanical test in which the sample is steadily loaded until it breaks or fails. One of the main mechanical properties of any material is flexural strength. Flexural strength is the ability of the composite material to withstand bending forces applied perpendicular to its longitudinal axis before the three-point flexural test technique breaks or begins to yield permanently. Impact strength is the material's ability to withstand a suddenly applied load, or it can also be described as the material's ability to absorb mechanical energy in the deformation and fracture process under impact loading. Impact experiments were carried out on the impact measuring unit, according to ASTMD 179EU.

3. Result

3.1. Tensile test. The specimen has been prepared in compliance with the ASTMD 638 requirements and it can be inferred by examining tensile tests for various composites that the dispersion of filler content in synthetic epoxy resin results in a decrease in tensile strength. Synthetic epoxy basalt composite with 0 % filler content having 388MPa has demonstrated material tensile strength.

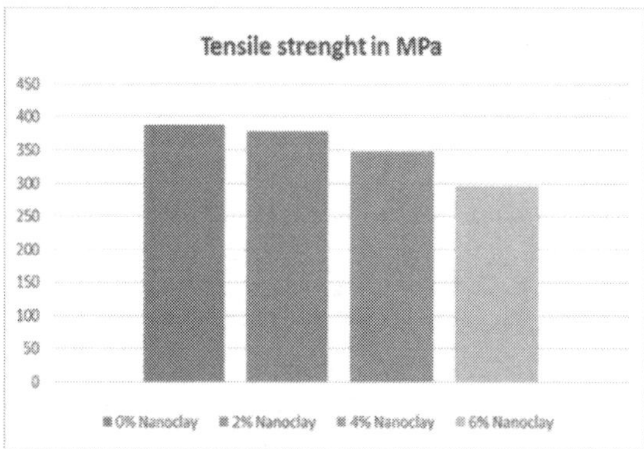

Figure 1. Tensile strength of different laminates

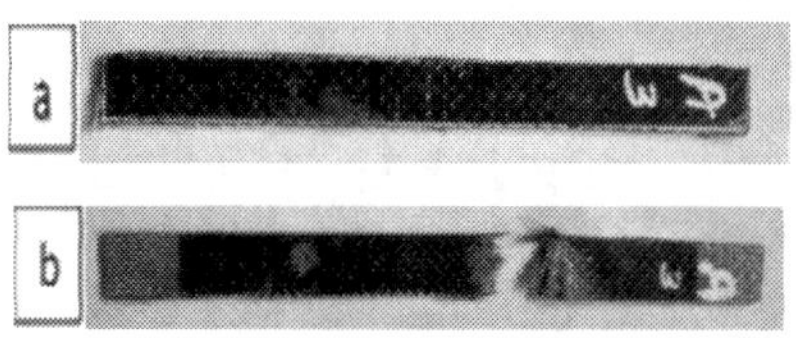

Figure 2. Tensile test specimens (a) before (b) after

3.2. Flexural Test. According to the ASTMD 790 standards the specimen has prepared and tested. From graph analysis and test it is obtained that composite Nano clay as the filler material 2% has exhibit good flexural strength 387MPa compared to others.

IOP Conf. Series: Materials Science and Engineering 1065 (2021) 012026 doi:10.1088/1757-899X/1065/1/012026

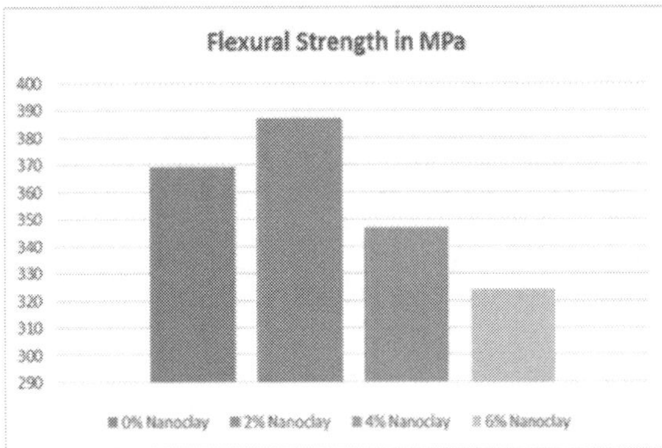

Figure 3. Flexural strength of different laminates

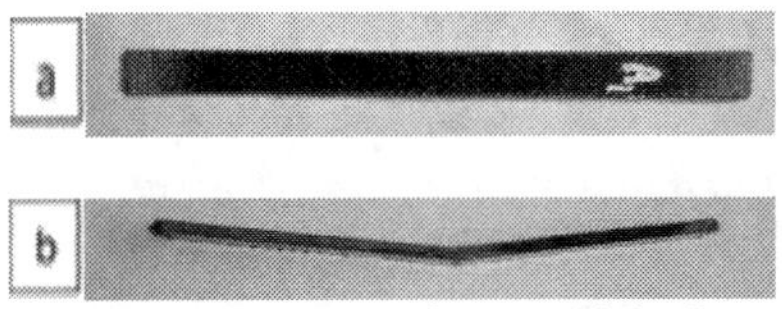

Figure 4. Flexural test specimens (a) before (b) after

3.3. Impact Test. According to the ASTMD 179EU standards the specimen has prepared and tested. From graph analysis and test we concluded that 6% Nano clay has good impact strength 153.34kJ/m2 compared other.

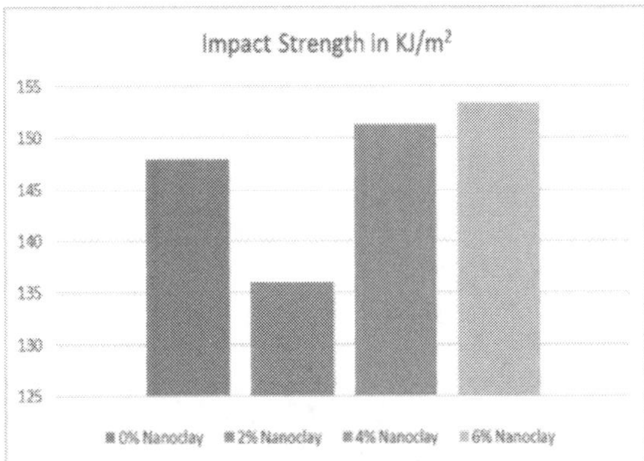

Figure 5. Impact strength of different laminates

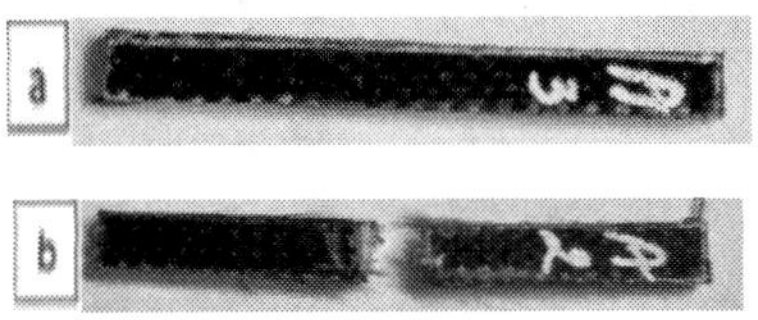

Figure 6. Impact test specimens (a) before (b) after

4. Conclusion

Experimental tests were conducted on basalt fibre reinforced plastic laminates with Nano clay dispersed by 0%, 2%, 4%, and 6%. Mechanical experiments have been carried out, such as tensile tests, bending tests and impact tests.

Findings on examination of materials and conclusions, based on the experimental results and analysis, the following conclusions were arrived at,

- Tensile measurements also helped assess the tensile strengths of various percentages of fillers.

- 388MPa, 377MPa, 347MPa and 295MPa respectively are the tensile strength of laminates with 0 percent, 2 percent, 4 percent and 6 percent Nano clay fillers. The findings show that the tensile strength of the material is not quite appreciable with the addition of filler to the composites.
- The overall obtained flexural strength was 2 percent Nano clay 387MPa. Followed by 0 percent Nano clay 369MPa, 4 percent Nano clay has 347MPa and 6 percent Nano clay has 324MPa. Improves the flexural strength of composites by specifically showing the inclusion of filler.
- With the addition of fillers, impact strength has also increased. The impact strength of 153.43kj/m2 is improved by 6% Nano clay.

References

[1] Fazeli, Mahyar; Florez, Jennifer Paola; Simão and Renata Antoun (April 2019). *Improvement in adhesion of cellulose fibers to the thermoplastic starch matrix by plasma treatment modification*. Composites Part B: Engineering. 163: 207–216.
[2] Shaffer and Gary D. (Spring 1993). *An Archaeomagnetic Study of a Wattle and Daub Building Collapse*. Journal of Field Archaeology. 20 (1): 59–75.
[3] Meyers, M.A.; Chen and P.Y. (2014). *Biological Materials Science*. United Kingdom: Cambridge University Press.
[4] Maya Jacob John and Sabu Thomas, *Biofibers and biocomposites*. Carbohydr. Polym 2008; 71(3):343-64.
[5] Uma Devi L, Bhagawan S S and Sabu Thomas, *Mechanical properties of pineapple leaf fiber-reinforced polyester composites*. J. Appl. Polym. Sci 1997; 64(9):1739-48
[6] Murali Mohan Rao K, Mohana Rao K and Ratna Prasad A V, *Fabrication and testing of natural fiber composites: vakka, sisal, bamboo and banana*. Mater. Des2010; 31(1): 508-813.
[7] Czigány T and Vad JPK. *Basalt fiber as reinforcement of polymer composites*. Period Polytech Mech Eng 2005; 49:3–14
[8] Susilendra Mutalikdesai, G. Sujaykumar, Amal Raju, C. J. Moses, Jithin Jose and Vishak Lakshmanan, *Mechanical Characterization of Epoxy/ Basalt Fiber/ Flax Fiber Hybrid Composites*. American Journal of Materials Science 2017, 7(4): 91-94 DOI: 10.5923/j.materials.20170704.04.
[9] Wei B, Cao H and Song S. *Environmental resistance and mechanical performance of basalt and glass fibers*. Mater Sci Eng A 2010; 527:4708–15.
[10] Ivashchenko E. *Sizing and finishing agents for basalt and glass fibers*. Theor Found Chem Eng 2009; 43:511–6.
[11] Deak T and Czigany T. *Chemical composition and mechanical properties of basalt and glass fibers: a comparison*. Text Res J 2009; 79:645–51.
[12] Chemistry, International Union of Pure and Applied. *IUPAC Compendium of Chemical Terminology*. iupac.org. IUPAC. doi:10.1351/goldbook.RT07166
[13] Collin and Gerd; et al. (2005). "*Resins, Synthetic*". Ullmann's Encyclopedia of Industrial Chemistry. Weinheim: Wiley-VCH. doi:10.1002/14356007.a23_089. ISBN 3527306730.
[14] Cripps and David. "*Epoxy Resins*". NetComposites. Retrieved 28 September 2018.

Effect of Mean Stress on Fatigue Behavior of Glass Fiber-Reinforced with Epoxy Composites

Hemanth Kumar C[1], R. P. Swamy[2]

[1] Department of Mechanical Engineering, B N M Institute of Technology, Bengaluru

[2] Department of Mechanical Engineering, U B D T College of Engineering, Davangere

Email: hemanthkumar.sh17@gmail.com

Abstract. In this study, fatigue tests have been conducted on the glass fiber reinforced with epoxy (GFRE) composites under completely reversed loading. Test specimens were prepared for fiber orientation of $[\pm 45°]_2$. Further, experimental fatigue data obtained from the S-N curve at stress ratio, R = -1 are analyzed using the Goodman approach for different mean stress levels. It was noted that stress ratios such as 0.5, 2, and 10 had a detrimental effect on the fatigue life of GFRE composites. Whereas, stress ratios such as -1, -2, and -0.5 had beneficial effect on increasing the fatigue life of GFRE composites under cyclic loading.

Keywords: Fatigue Test, Glass Fiber, S-N Curve, Mean Stress

1. Introduction

Composite materials refer to a combination of two or more different materials that are combined at various proportions to attain the relative advantages of their constituents while retaining their individual distinctiveness. Major constituents of composites are particles, fibers, laminae, fillers, flakes, and matrix materials. Composites are usually given bulk form by the matrix materials. The reinforcements are embedded within the matrix and are protected against environmental effects/damages. The structure of the composites is determined by the type of fibers, laminae, fillers, flakes used. Fiber-reinforced polymer (FRP) matrix-based composites are progressively being employed in making structural members such as ship hulls, wind turbines, airframes and in electric/solar vehicles due to its high strength to weight ratio and stiffness properties. In addition to this, these materials also possess good damage tolerance characteristics [1]. Failure of composite materials can be viewed from macroscopic and microscopic levels. Macroscopic damage in composites is due to manufacturing inaccuracies associated with the fabrication process/methods. These defects are present in composites before the application of loads. Microscopic level damage usually starts with the application of external loads. With the continuous application of the external loads, these microscopic defects grow at the macroscopic level and result in the failure of engineering structures. Fatigue life prediction of fiber-reinforced polymeric composites under repeated loads has gained considerable importance in recent years due to their wide range of applications in spacecraft, automobile, aerospace, and aircraft industries. For FRP composites, cyclic strength is relatively greater in tension, in contrast to metallic materials such as steel or aluminum alloys. For ferrous materials like steel, a definite endurance limit will exist. Whereas for non-ferrous and anisotropic materials like aluminum alloys, copper, brass and composites, a definite endurance limit does not exist. Therefore special care has to be taken by the designers while designing the structural machine components using these materials to avoid unpredictable failures.

1.1. Literature Review
Several studies in the literature have addressed the modelling and characterization of fatigue behaviour for composite materials. Agarwal and James [2] examined the influence of stress ratio on the fatigue behavior of GFRP composites under low cycle fatigue. It was revealed by the authors that the fatigue life of GFRP composites was negatively affected by stress ratio and matrix failure was observed prior to the final failure of the composites. Abd-Allah et al. [3] studied the effect of fiber volume fraction on the fatigue life of composites, and they concluded that the fatigue strength was proportional to the fiber volume fraction. Mandell et al. [4] conducted a fatigue study on the effect of fiber and matrix and fiber, and they determined that structural integrity and strength were largely dependent on a stable matrix system for wind turbine blades. Piggott [5] stated that the temperature rise in fiber composites was insensitive at test frequencies below 30 Hz. Also, several other researchers in the literature have evaluated the fatigue strength of fiber-reinforced epoxy composites for different stacking sequences, fiber orientation, stress ratios, and test frequencies using glass fibers and epoxy matrix [6-11]. From these studies, it has been observed that the fatigue strength of the fiber composites is largely influenced by the type and orientation of fibers. Composites with fiber orientation in off-axis directions were found to be weaker than unidirectional laminates. Stiffness degradation was witnessed at high-stress levels leading to catastrophic failure of fibers. Matrix degradation was observed owing to large plastic deformation at high-stress levels and test frequencies. In this study, the tension-compression bending fatigue behavior of glass fiber reinforced with an epoxy matrix is characterized. Then, the S–N curve and modified Goodman diagram are drawn to show the effect of mean stress on the fatigue life of the developed composite specimens under cyclic loading.

2. Materials and Method
Glass fibers and epoxy matrix (GFRE) were used as reinforcement and matrix to prepare the test specimens. The properties of the constituent materials are given in Table 1. Test specimens were prepared for the dimensions as follows:
a) External diameter = 13mm;
b) Thickness = 2.5 mm and
c) Gauge Length = 60mm.

Table 1. Properties of constituent materials

Properties	Glass Fibre	Epoxy
Tensile Strength [N/mm^2]	2700-3500	300-320
Elastic Modulus [N/mm^2]	70000-75000	300-4000
Elongation (%)	4 to 5	5 to 7
Density (kg/m^3)	2570	1100
Thermal Expansion coefficient (* 10^{-6}/°C)	4.9 to 5.4	45-65

Fatigue tests were carried out at room temperature for zero mean stress at a frequency of 10 Hz using a rotating bending fatigue tester [12]. Five stress levels were selected based on the tensile strength of the GFRE composites. The selected stress levels fall in the range of 40-80% of the static tensile strength of GFRE composites as shown in Table 4.

3. Results and Discussion
3.1. Density of GFRE composites
The theoretical density of the GFRE composite specimens is determined by burning off the epoxy matrix. The volume fraction rate of the epoxy matrix from the GFRE composites was calculated based on the mass of glass fiber and epoxy matrix to determine the density of GFRE composite specimens using the rule of mixture. The obtained density of the GFRE specimens is tabulated in Table 2.

Table 2. Density of composites

Sl. No	Composite Code	Volume Rate		Density (kg/m³)
		$V_{Epoxy\ Matrix}$	$V_{Glass\ Fibres}$	
1	GFRE Composites $[\pm45°]_2$	0.47	0.53	1791

3.2. Tensile results of GFRE composites

The tensile test was carried out by applying load in a uniaxial direction. The strain rate chosen for this study is 0.03mm/sec. Figure 1 shows the stress-strain plot of GFRE Composites. The tensile properties like elastic modulus, tensile strength, proof stress, and strain at failure are tabulated in Table 3.

Table 3. Tensile properties of GFRE composites

Sl. No	Properties	Particulars
1	Elastic Modulus (N/mm²)	152
2	Proof Stress, (N/mm²)	9
3	Tensile Strength, (N/mm²)	14
4	Fracture Strain	0.12

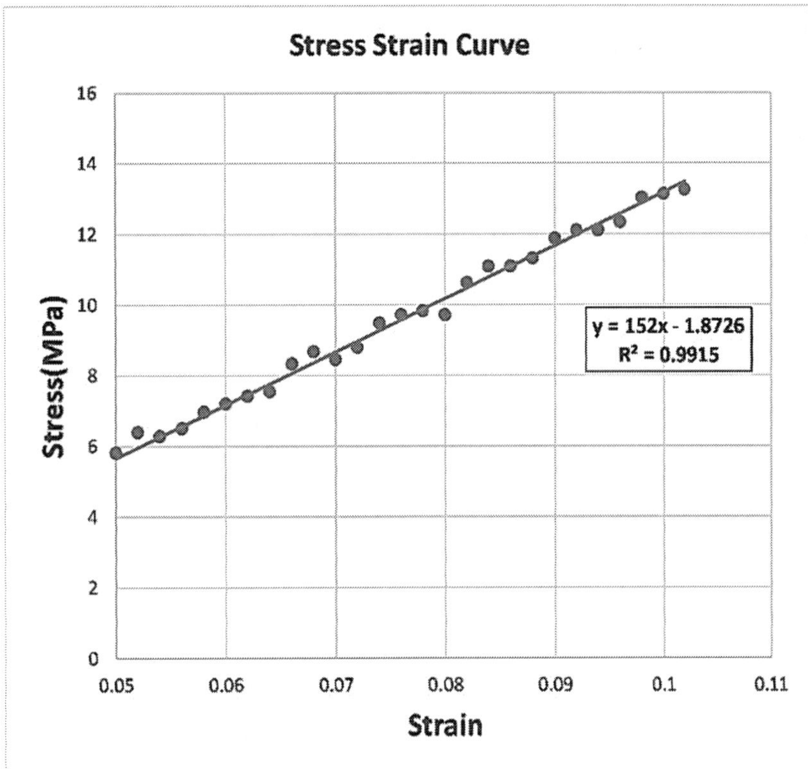

Figure 1. Stress-strain curve of GFRE Composite

3.3. Fatigue results of GFRE composites

Constant amplitude type fatigue testing was carried out for developed composites specimens R. R Moore type fatigue tester. During the test, as the specimens were rotated fully, it underwent through a complete reversed type of tension and compression loading cycles. Figure 2 shows the S-N curve for GFRE composites. The required bending stress was calculated by using equation (1).

$$\sigma = \frac{M_b}{Z}\ N/mm^2 \tag{1}$$

where M_b= Applied Bending Moment; Z= Section Modulus

Table 4. Fatigue Stress Levels and Cycles to Failure

Sl.No.	Selected Stress Level	Number of cycles to failure [a]
1	($0.4\sigma_{ult}$)	293464
2	($0.5\sigma_{ult}$)	122858
3	($0.6\sigma_{ult}$)	53748
4	($0.7\sigma_{ult}$)	360
5	($0.8\sigma_{ult}$)	65

a = average of five specimens

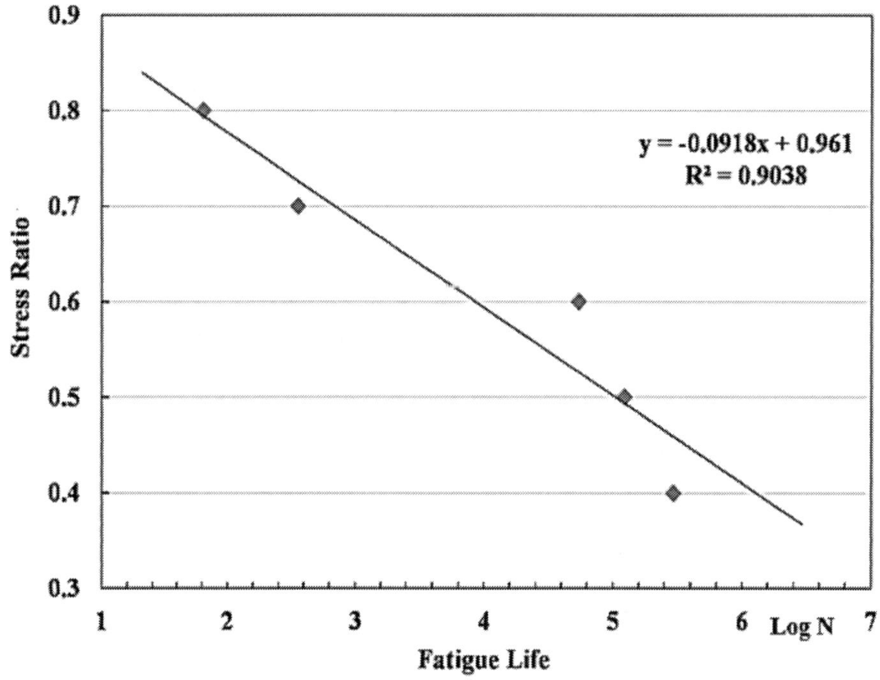

Figure 2. S-N Curve for Developed GFRE Composites

3.4. Effect of Mean Stress

The magnitude of mean stress is the average value of the maximum and minimum stress cycles acting on the machine component. Hence an attempt is made in this work, to assess the fatigue behavior of the developed GFRE composite specimens for a wide range of alternating and mean stress using the simplest method called the "Goodman approach". This method helps us to understand the mean stress effects and material properties collectively on estimating the fatigue life cycle of the machine components under consideration. "R-Ratio" is mathematically given as follows.

$$R = \frac{\sigma_{min}}{\sigma_{max}} \tag{2}$$

For stress cycles with the mean stress σ_m and stress amplitude value σ_a, the corresponding equivalent stress value σ_{eq} at R= −1 is stated as follows:

$$\sigma_{eq} = \frac{\sigma_a \, x \, UTS}{UTS - \sigma_m} \quad \text{for } \sigma_m > 0, \tag{3}$$

$$\sigma_{eq} = \frac{\sigma_a \, x \, UCS}{UCS - \sigma_m} \quad \text{for } \sigma_m > 0, \tag{4}$$

The above-described model was applied to the available experimental data attained from the fatigue tests. The value of mean stress σm and alternating stress σ_a were estimated using equations (3) and (4). The obtained results are presented in Figure 3.

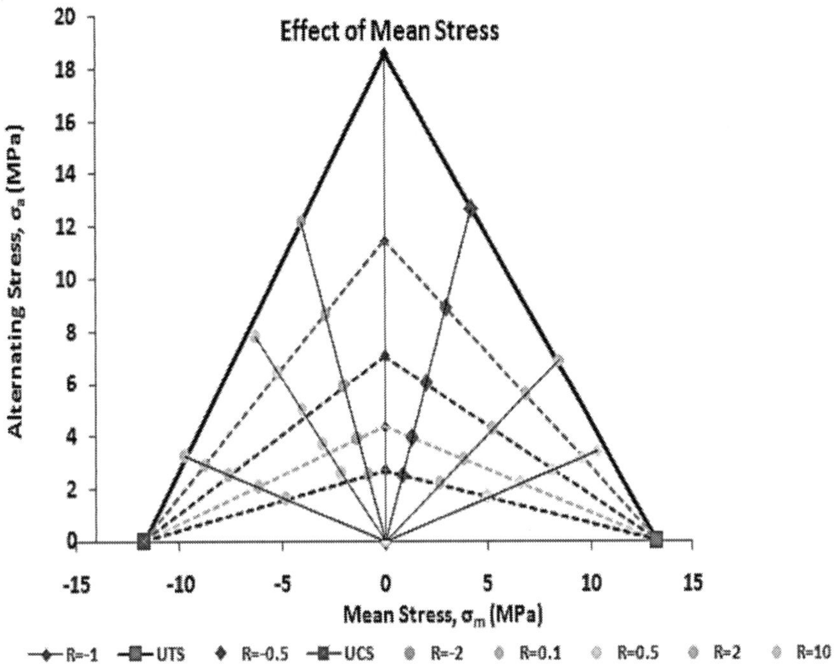

Figure 3. Constant life diagram of GFRE composites

The corresponding equivalent value of stress σ_{eq} after stress amplitude can be found by using the relations given in equations (3) and (4). This will give us the value of several cycles to failure from the selected S-N curve at desired stress ratio R. The different stress ratios R considered are 2, 10, -2, -1, -0.5, and 0.5 as shown in Figure 3. The obtained results are depicted in the form of an S-N curve as shown in Figure 4. It was observed that, for the developed GFRE composite specimens, stress ratios such as 0.5, 2, and 10 have detrimental effects under fatigue loading leading to shorter fatigue lives. Whereas stress ratios such as -1, -2, and -0.5 will not affect the fatigue life of the machine component significantly.

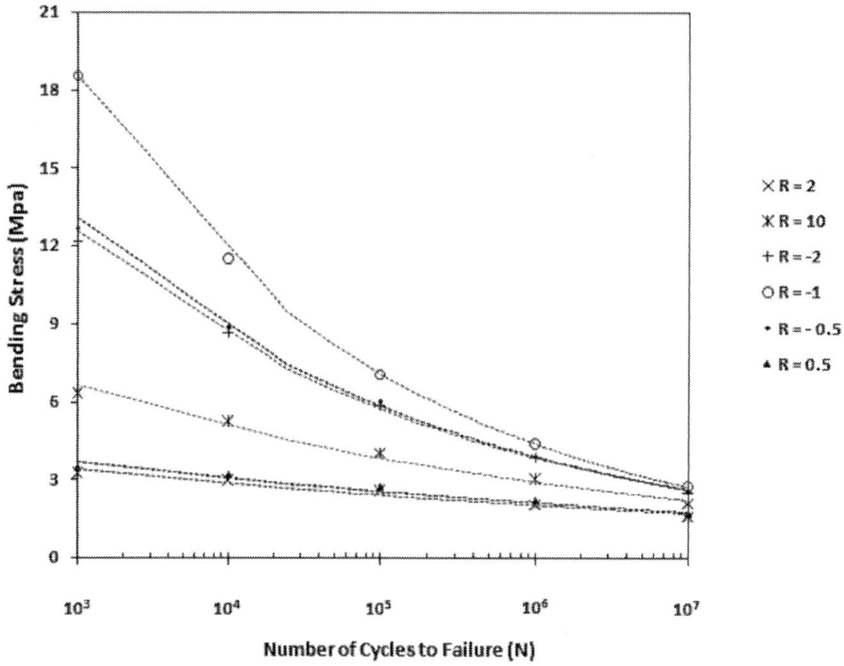

Figure 4. Estimated S-N curves for different stress ratios

4. CONCLUSION

In this work, an experimental and analytical method was adopted to illustrate the effect of mean stress on the fatigue life of GFRE composites. Non-linear behaviour was observed from tensile test results due to the orientation of fibers in off-axis directions. The S–N curve was drawn for completely reversed loading (R= -1) with a coefficient of determination (R^2) being greater than 0.9. It was observed that, for the developed GFRE composite specimens, stress ratios such as 0.5, 2, and 10 have detrimental effects under fatigue loading leading to shorter fatigue lives. Whereas, stress ratios such as -1, -2, and -0.5 have beneficial effects under cyclic loading. Hence, for the proposed GFRE composites, the stress ratios should be around -1 for the conservative design of machine components under fatigue loading.

Reference

[1] Hemanth Kumar C, Swamy R P. 2015 "Analysis of Metallic and Composite Tail Rotor Drive Shaft for Ballistic Impact," International Journal of Mechanical Engineering and Robotics Research, 4 (1) 455-462.

[2] Agarwal, B.D. and James, W.D. (1975). Prediction of Low-cycle Fatigue Behavior of GFRP: An Experimental Approach, Journal of Materials Science, 10(2): 193–199.

[3] Abd-Allah, M.H., Abdin, E.M., Selmy, A.I. and Khashaba, U.A. (1996). Effect of Fiber Volume Fraction on the Fatigue Behavior of GFRP Pultruded Rod Composites, Composites Science and Technology, 56(1): 23–29.

[4] Mandell, J.F., Samborsky, D.D. and Scott, M.E. (2003). DOE-MSU Wind Turbine Blade Composite Material Fatigue Database, Department of Chemical Engineering, Montana State University at Bozman, Montana, USA.

[5] Piggott, M.R. (1980). Load-Bearing Fiber Composites, Pergamon Press, New York, NY.

[6] G. Belingardi, M.P. Cavatorta. 2006 "Bending fatigue stiffness and strength degradation in carbon–glass/epoxy hybrid laminates: Cross-ply vs. angle-ply specimen". International Journal of Fatigue 28 815–825.

[7] A. Bernasconi, P. Davoli, A. Basile, A. Filippi. 2007 "Effect of fibre orientation on the fatigue behaviour of a short glass fibre reinforced polyamide-6". International Journal of Fatigue. 29 199–208.
[8] N.K. Kar, Y. Hu, E. Barjasteh, S.R. Nutt. 2012 "Tension–tension fatigue of hybrid composite rods" Composites: Part B 43 2115–2124.
[9] P.K. Mallick, Yuanxin Zhou. 2004 "Effect of mean stress on the stress-controlled fatigue of a short E-glass fiber reinforced polyamide-6, 6" International Journal of Fatigue 26 941–946.
[10] S. Mortazavian and A. Fatemi 2015 "Effects of mean stress and stress concentration on fatigue behavior of short fiber reinforced polymer composites". Wiley Publishing Ltd. Fatigue Fract. Engg. Mater. Struct. 001–18.
[11] Bernasconi Robb M. Kulin. 2015 "Effect of frequency upon fatigue strength of a short glass fiber reinforced polyamide 6" © Wiley Publishing Ltd. Fatigue Fract. Engg. Mater. Struct. 154–161.
[12] Hemanth Kumar C & R.P. Swamy. "Experimental Investigation on the Fatigue Behavior of Glass Fiber-Reinforced Epoxy Composites Under Rotating Bending Loads". J Fail. Anal. and Preven. https://doi.org/10.1007/s11668-020-01105-3

Acknowledgments

The authors wish to express gratitude to late Dr. C. S. Venkatesha, Professor, Department of Mechanical Engineering, U.B.D.T.C.E, Davangere for his valuable guidance and support.

Production of biodiesel and experimental investigation on IC engine: A review

Hemanth Kumar B[1*] and Manjunath H[1]

[1] Siddaganga Institute of Technology, Tumakuru, Karnataka, India

E-mail: hemanthkumarb909@gmail.com

Abstract. Due to the depletion and increase in the price of petroleum derived fuel, it is necessary to find out an alternative fuel. Biodiesel is an important alternative fuel source from the non edible oil to fulfil the energy demand. The common way to produce biodiesel from non edible oil is through transesterification. This paper reviews the production of biodiesel using catalysed esterification and factors affecting the biodiesel yield are discussed. This article reported that combustion characteristics of the diesel and biodiesel are similar. Shortened ignition delay, higher ignition pressure, higher ignition temperature were found in the biodiesel blends. The power output and thermal efficiency of the engine fuelled with biodiesel was found to be equivalent to diesel fuel. Emission characteristics of biodiesel found that higher NO_x emissions and lower CO and HC emissions due to complete combustion.

Keywords: Biodiesel, Transesterification, Combustion, Performance, Emissions

1. Introduction

Petroleum reserves play a very important role in the development of transportation, industrial growth, agricultural sector and to meet other basic human needs. These fuels are limited and depleting as the consumption is increasing rapidly, it is necessary to find out an alternative fuel. Biodiesel is an important alternative fuel source to fulfil the energy demand. Biodiesel, renewable energy, can be produced from animal fats, vegetable oils or waste cooking oil. Biodiesel can be produced from both edible and non edible oils. Edible oil use may create shortage of food and oil, to overcome this disadvantage biodiesel from non edible oils were preferred by the researchers. Nonedible oil which is reacted with alcohol to form biodiesel and glycerol [1]. Some physicochemical properties of different oils and their yield shown in Table 1. Biodiesel cannot entirely replace petroleum derived diesel fuel, there are several advantages of biodiesel over diesel fuel. Biodiesel has higher cetane number, combustion efficiency than diesel fuel and it is biodegradable. Biodiesel reduces monoxide and hydrocarbon emissions except NO_x. Biodiesel can be produced from different techniques such as micro-emulsions, transesterification, thermal cracking, direct blending with raw oil etc. The common way to produce biodiesel from non edible oil is through transesterification. The main aim of transesterification is to reduce the viscosity of oil [2]. The main objective of this paper is to give an overview on the advancement in the production of biodiesel through catalysed transesterification, factors affecting the biodiesel yield, combustion, performance and emission characteristics.

IOP Conf. Series: Materials Science and Engineering 1065 (2021) 012028 doi:10.1088/1757-899X/1065/1/012028

Table 1: Physicochemical properties of oil for biodiesel production

Type of oil	Density(g/cm³)	Kinematic Viscosity(cst, at 40°C)	Heating value(MJ/kg)	References
Used cooking oil	0.90	44.7	-	[2]
Tea seed oil	0.907	27.0	37.14	[3]
Honne oil	0.910	32.47	39.10	[4]
Cotton seed oil	0.910	34	36.80	[4]
Honge oil	0.924	45.23	39.44	[4]
Rice bran oil	0.901	42.55	38.95	[4]

2. Production of biodiesel using catalyzed transesterification

Edible, non edible and animal fats are esters of saturated and unsaturated monocarboxylic acids with the trihydric alcohol glyceride, these are called as triglycerides. Triglycerides can react with alcohol in the presence of a catalyst to form biodiesel and glycerol, a process known as transesterification. Simarouba, Mahua and waste cooking oils were used in the production of their methyl esters. Biodiesel production consists of two steps: acid esterification and alkaline esterification. A single step transesterification for waste cooking oil and two step transesterification for simarouba and mahua oils were used to convert the oils to their ester [5]. Biodiesel production from a non edible oil from honne oil. A three stage process pre-treatment, alkali-catalysed transesterification and post treatment were adopted for production [6].

Most of the biodiesel production is by the alkali-catalyzed transesterification process. Flow chart of the alkali-catalyzed process is shown in Figure 1. The maximum amount of FFA acceptable in an alkali-catalyzed process is below 2.0 wt.%. If the FFA of the oil is above 2.0 wt.%, a pre-treatment step is necessarily adopted before transesterification process [5]. In pre-treatment process the oils were first converted in to esters using sulphuric acid as catalyst. Experiments were conducted at 60°C, below the boiling point of methanol and reaction is allowed for 30 to 60 min [5]. The oil having FFA below 2.0 wt.% was used for the alkaline transesterification process.

In alkali-catalyzed transesterification process most commonly used catalysts are sodium hydroxide and potassium hydroxide. The most commonly used alcohols in the process include methanol, ethanol and propanol [7]. Ester from the pre-treated process or the oil needs transesterification. Sodium methoxide (methanol+sodium hydroxide) is included in pre-treated ester or in the oil and heated up to 60°C, just below the boiling point of methanol. Constant temperature is allowed for 1h with continuous bending and then allowed for normal cooling. Glycerol is separated from methyl esters and the biodiesel is moved for water wash [8]. Physicochemical properties and yield of biodiesel from different oil source is showed in Table 2.

Factors affecting the biodiesel yield were molar ratio of alcohol to oil, reaction time, reaction temperature, catalyst concentration, the molar ratio for reaction requires 3 mol of alcohol for 1 mol of oil to produce 3 mol of ester and 1 mol of glycerol. The biodiesel yield is increased when molar ratio of alcohol to oil ration increased beyond 3. Yield reaches maximum at a reaction time less than 90min and remains constant beyond 90min. The transesterification reaction temperature must be less than boiling point of alcohol, to ensure that alcohol will not leak out by vaporization. Optimal temperature ranges from 50°C to 60°C. Biodiesel yield can be affected by catalyst concentration, increase in the catalyst concentration increases the yield of biodiesel from oil. Optimal value of yield reaches when the concentration reaches 1.5wt.% and then yield decreases with slightly increase in catalyst concentration [9].

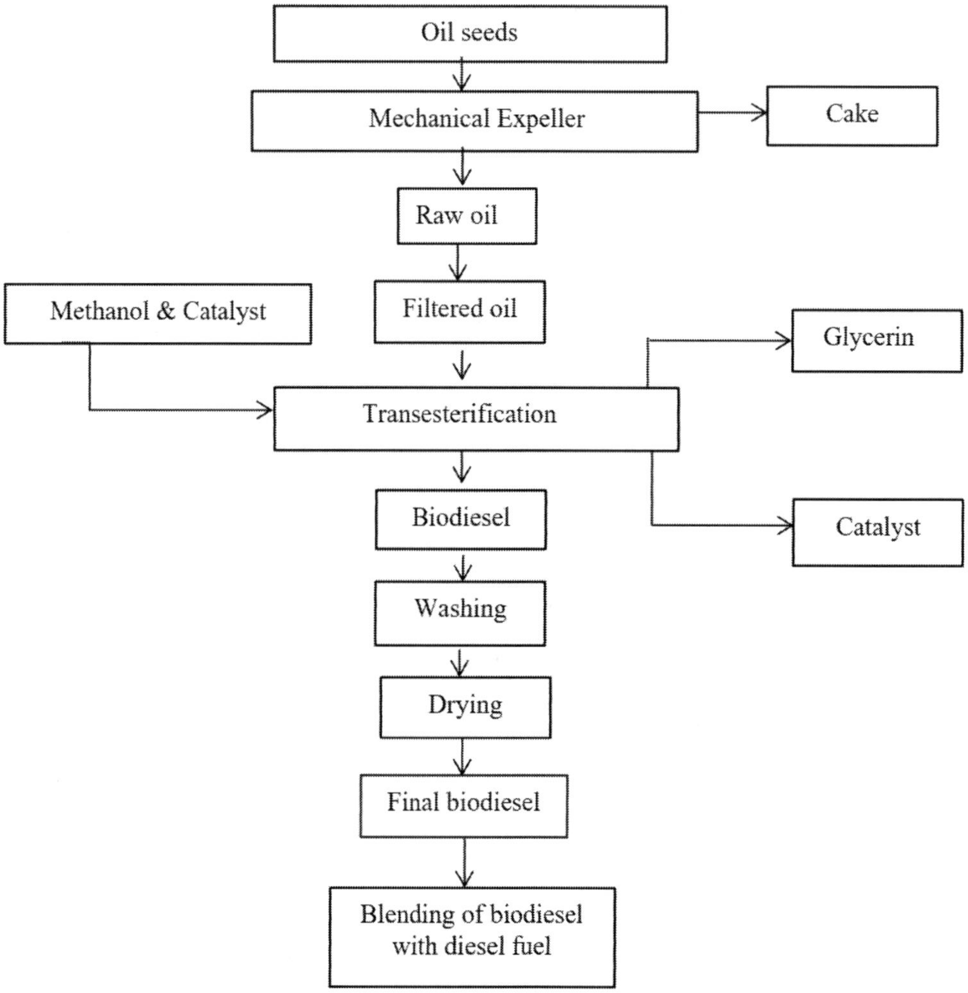

Figure 1: Flow chart of the alkali-catalyzed process

Table 2: Physicochemical properties and yield of biodiesel from different oil source

Biodiesel	Kinematic viscosity (cst, at 40°C)	Density (g/cm^3)	Heating value (MJ/kg)	Yield (%)	Reference
Simarouba	5.09	0.879	40.62	95.14	[5]
Mahua	5.45	0.874	38.21	94.21	[5]
Used cooking oil	4.72	0.882	38.91	96.31	[5]
Jatropa curcas	4.78	0.863	41.12	98.0	[2]
Honne	4.82	0.881	39.02	96.5	[6]

Properties are compared with ASTM D6751 and EN 14214 for using in the diesel engines

3. Combustion characteristics of biodiesel

The combustion characteristics such as ignition temperature, ignition delay period, combustion duration, maximum rate of pressure rise and heat release rate of different biodiesel had reviewed in a detailed manner in the following paragraphs.

[4] Venkanna and Venkataraman Reddy investigated the use of blends of honne oil and diesel in a single cylinder direct injection diesel engine. They found that combustion characteristics of fuel blend 20/80 are found to be close to diesel fuel, exhibits lower maximum net heat release rate, slightly higher ignition delay period, lower maximum rate of pressure rise and higher combustion duration compared to diesel. [10] Kannan et al. investigated the effect of waste cooking oil methyl ester in a diesel engine. Compared to diesel at same load condition, waste cooking oil methyl ester exhibited shorter ignition delay period, lower heat release rate and slightly longer combustion duration.

[11] Selman et al. investigated the effects of thermal barrier coating on the combustion characteristics of a diesel engine fuelled with biodiesel produced from cotton seed oil and ultra-low sulphur diesel. They found that shortened ignition delay period results in the elimination of engine knocking and lowered the engine noise values. Coated engine exhibits higher cylinder pressure values, the reason for this is to be higher temperature. The values of heat release in coated and uncoated engine were found quite similar, the heat is released slightly earlier in coated engine. [12] Ahmed et al. investigated the combustion characteristics of a diesel engine fuelled with Jatropa, ethanol and n-butanol as co-solvent. They found that maximum rate of pressure rise and heat release rate increased by adding ethanol. The ignition delay period is increased with the addition of ethanol and the combustion duration is almost same as diesel. Addition of n-butanol as co-solvent enhanced the stability of the mixture of Jatropa oil methyl ester with ethanol. [13] Manjunath Channappagoudra investigated the combustion characteristics of a diesel engine fuelled with dairy scum biodiesel and copper oxide as nanoadditive. He found that blends of nanoadditive have shown better combustion characteristics such as heat release rate, peak pressure, ignition delay period, combustion duration and rise of pressure compared to 20/80 dairy scum oil methyl ester fuel blend. [17] Karthikeyan et al. investigated the effect of cerium oxide additive on the combustion characteristics of a compression ignition engine operated with blends of rice bran biodiesel. With the addition of cerium oxide nanoparticles leads to early combustion, reduced peak pressure rise and heat release rate increased.

4. Performance characteristics of biodiesel

The performance characteristics such as specific fuel consumption, power output, brake thermal efficiency and exhaust gas temperature of different biofuels had been reviewed in the following paragraphs.

[14] Amarnath and Prabhakaran studied the thermal performance of a variable compression ratio diesel engine fuelled with Karanja biodiesel. They reported that engine gives better performance at higher compression ratio and the exhaust gas temperature of Karanja biodiesel is slightly less compared to that of normal diesel. [15] Venkata Subbaiah and Raja Gopal investigated the performance characteristics of a diesel engine fuelled with Rice bran biodiesel an blends of ethanol. They found that addition of ethanol reduces the exhaust gas temperature of biodiesel and the maximum brake thermal efficiency and minimum brake specific fuel consumption was obtained with 2.5% ethanol blended with Rice bran biodiesel at full load condition. [16] Vedarman et al. investigated the effect of various additives on performance characteristics and NO_x emission reduction in 20/80 Palm oil biodiesel blend. They reported that with increase in blend percentage, thermal efficiency of the engine reduces.

Performance and emission characteristics of biodiesel had investigated by Venkanna and Venkataraman [4]. The engine performance characteristics of a fuel blend 20/80 are found to be close to normal diesel fuel, with increase in the blend ratios performance found to be inferior to normal diesel fuel. [11] Selman et al. investigated the effects of thermal barrier coating on the performance characteristics of a diesel engine fuelled with biodiesel produced from cotton seed oil and ultra-low sulphur diesel. At full load condition the thermal efficiency increased for a coated engine operation. [13] Manjunath Channappagoudra investigated the performance characteristics of a diesel engine fuelled with dairy scum biodiesel and copper oxide as nanoadditive. Addition of copper oxide nanoadditive with 20/80 dairy scum biodiesel blend enhance the brake thermal efficiency and slightly lower brake specific fuel consumption. [17] Karthikeyan et al. investigated the effect of cerium oxide additive on the performance characteristics of a compression ignition engine operated with blends of rice bran biodiesel. Enhancement in brake thermal efficiency was found with cerium oxide blends compared with 20/80 blends at full load condition. Specific fuel consumption for 20/80 blend is higher than that of cerium oxide present in the blends.

[18] Nithin et al. investigated the performance characteristics of diesel engine fuelled with Jatropa biodiesel with alumina nanoparticles. Performance of Jatropa methyl ester is comparatively low than that of diesel fuel. But the alumina nanoparticle blended fuel showed better performance than Jatropa biodiesel and normal diesel fuel. [8] Karthik et al. analysed the performance characteristics of tamanu oil diesel blends in compression ignition engine. They reported that slightly higher exhaust gas temperature and lower specific fuel consumption in fuel blends. [19] Aparna et al. reported that diesel engine fuelled with Jatropa biodiesel blends, increase in power output, shorter ignition delay and increase in brake thermal efficiency is observed with increase in load.

5. Emission characteristics of biodiesel

The different emission characteristics such as carbon, monoxide, carbon dioxide, sulphur oxides, oxides of nitrogen and smoke had briefly reviewed in the subsequent paragraphs.

Performance and emission characteristics of biodiesel had investigated in a diesel engine fuelled with honne oil blends by Venkanna and Venkataraman [4]. The engine emission characteristics CO, CO_2, HC of a fuel blend 20/80 are found to be close to normal diesel fuel. And the values shows that O_2 emissions are low in honne oil compared to normal diesel fuel. [10] Kannan et al. investigated the effect of waste cooking oil methyl ester in a diesel engine. Compared to diesel at same load condition, waste cooking oil methyl ester exhibited decreased 43.3% CO, 52.7% HC, 23% NO and 15.5% smoke emissions compared to normal diesel fuel. [12] Ahmed et al. investigated the combustion and emission characteristics of a diesel engine fuelled with Jatropa, ethanol and n-butanol as co-solvent. They found that CO, HC, NO_x emissions are reduced by 40% with adding ethanol. [13] Manjunath Channappagoudra investigated the performance and emission characteristics of a diesel engine fuelled with dairy scum biodiesel and copper oxide as nanoadditive. He found that better HC and CO emission characteristics compared to other nanoadditive fuel blends.

[14] Amarnath and Prabhakaran studied the thermal performance and emissions of a variable compression ratio diesel engine fuelled with Karanja biodiesel. They reported that for blend ratio 20/80, 66% decrease, 50% decrease and 39.2% increase in CO, HC, NO_x emissions respectively compared to diesel fuel. [15] Venkata Subbaiah and Raja Gopal investigated the emission characteristics of a diesel engine fuelled with Rice bran biodiesel an blends of ethanol. They found that addition of 2.5% ethanol reduces CO, HC and unused oxygen emissions. And the smoke emissions reduced by 27.47% with ethanol blending. [16] Vedarman et al. investigated the effect of various additives on NO_x emission reduction in 20/80 Palm oil biodiesel blend. They found that additives with 20/80 ratio blend with 2% distilled water shows lowest NO_x emission, due to lower combustion temperature. [18] Nithin et al. investigated the performance and emission characteristics of diesel engine fuelled with Jatropa biodiesel with alumina nanoparticles. The emission characteristics CO, HC, smoke emissions were reduced to a great extent, and NO_x emission of biodiesel was slightly reduced with alumina nanoparticles. [8] Karthik et al. analysed the emission characteristics of tamanu oil diesel blends in compression ignition engine. They reported that increase in NO_x and decrease in HC emissions were observed for all blends [8]. [20] Ganaram et al. investigated the effect of carbon nanotube additive blended with neem biodiesel fuelled in diesel engine. They found that addition of carbon nanotubes reduces NO_x, HC, CO, smoke emissions by 9.2%, 6.7%, 5.9% and 7.8% at all load conditions.

6. Conclusion

Biodiesel is a renewable, alternative diesel fuel. Various methods are available for biodiesel production, the most commonly used method is alkali-catalyzed transesterification. The factors affecting the yield of biodiesel were molar ratio of alcohol to oil, reaction temperature, reaction time and catalyst concentration. Longer combustion duration, shorter ignition delay period and slightly lower heat release rate were observed in biodiesel compared to diesel fuel. Brake specific fuel consumption slightly increases and thermal efficiency of biodiesel is equivalent to diesel fuel. The decrease in CO, HC and smoke emissions and increase in the NO_x emissions was observed in biodiesel and biodiesel fuel blends as compared to diesel fuel.

References

[1] Ozcanli M, A Keskin and K Aydın 2011 *J. Green Energy* **8** 518–28
[2] Dennis Y C Leung, Xuan Wu and M K H Leung 2010. J. Appl. Energy **87** 1083–95
[3] Hasan Serin, Mustafa Ozcanli, Metin Kemal Gokce and Gokhan Tuccar 2013. *J. Green Energy* **10** 370–77
[4] B K Venkanna and Venkataraman reddy 2015 *J. Green Energy* **12** 728–736
[5] Manjunath H, Omprakash Hebbal and Hemachandra Reddy K 2015 *J. Green Energy* **12** 424–430
[6] B K Venkanna and Venkataraman reddy 2009 *J. Biores. Technol.* **100** 5122–5
[7] Rashid U and Anwar F 2008 *J. Fuel* **87** 265–73
[8] K Karthik, T Ramesh, P Sankarganesh and I Vinoth Kanna 2020 *J. Amb. Energy* DOI: 10.1080/01430750.2020.1783353.
[9] Leung DYC, Guo Y 2006 *J. Fuel Process* **87** 883–9
[10] G R Kannan, K R Balasubramanian, S P Sivapirakasam and R Anand 2011 *J. Ambient Energy* **32** 179–93
[11] Selman Aydın, Cenk Sayın, Şehmus Altun and Hüseyin Aydın 2016 *J. Green Energy* **13** 1102–90
[12] Ahmed I El-Seesy, Hamdy Hassan, Latif Ibraheem, Zhixia He and Manzoore Elahi M Soudagar 2020 *J. Green Energy* **17** 793–804
[13] Manjunath Channappagoudra 2018 *J. Amb. Energy* 01430750.2018.1557553
[14] H K Amarnath and P Prabhakaran 2012 *J. Green Energy* **9** 841–63
[15] G Venkata Subbaiah and K Raja Gopal 2011 *J. Green Energy* **8** 197–208
[16] N Vedaraman, Sukumar Puhan, G Nagarajan and K C Velappan 2011 *J. Green Energy* **8** 383–97
[17] S Karthikeyan, A Elango and A Prathima 2014 *J. Green Energy* **13** 267–73
[18] Nitin Shrivastava, Devanshu Shrivastava and Vipin Shrivastava 2018 *J. Green Energy* **15** 136–43
[19] Aparna Singh, Shailendra Sinha, Akhilesh Kumar Choudhary and H Chelladurai 2020 *J. Amb. Energy* DOI: 10.1080/01430750.2020.1789741
[20] Ganaram Ramakrishnan, Purushothaman Krishnan, Sivasubramanian Rathinam, Thiyagu R and Yuvarajan Devarajan 2019 *J. Green Energy* **16** 435–441

INHIBITION EFFECTS OF BANANA AND ORANGE PEEL EXTRACT ON THE CORROSION OF BRIGHT STEEL IN ACIDIC MEDIA

Vani R, Arpan Bhandari, Aishwarya Jain Y

Yenepoya Institute of Technology, Moodbidri, India.

arpanbhandary75@gmail.com

Abstract. From the beginning of its invention, steel has become the most significant engineering alloy used by man. However, it possesses a serious problem of corroding when exposed to normal climatic conditions. Using organic produce from nature has become a trend in tackling such corrosion of Steel. The main focus of this study is to produce green corrosion inhibitor from banana and orange peel. The gravimetric method was used to determine the corrosion of bright steel in acidic media (HCl) in which equal proportions of BPE and OPE were added as a corrosion inhibitor. The result of this study showed that as the concentration of inhibitor increases the rate of corrosion decreases. Maximum efficiency of 76.76% was achieved for 1200 PPM in 0.1N HCl and 79.19% was achieved for 1200 PPM in 0.3N HCl. The result shows that extracts of Banana and Orange peel can be used as corrosion inhibitors more widely when compared to their counterpart that is, conventional toxic inhibitors which are harmful to the environment.

Keywords. Orange Peel Extract (OPE), Banana Peel Extract (BPE), Bright Steel, Corrosion, Inhibitor Efficiency, Concentration.

1. INTRODUCTION

Steel has had a huge effect on our lives, the car we drive, the homes in which we live, and many other things in between. It is the most important, multifunctional, and most adaptive material by far [1]. Human growth would have been unlikely, except for steel. The strength and inherent use of steel have been the foundation of industrialized economies. But it is vulnerable to a condition called corrosion when exposed to normal climatic conditions. In metallic structures, corrosion occurs in different ways, such as uniform corrosion, pitting corrosion, crevice corrosion, galvanic corrosion, intergranular corrosion, cracking stress corrosion, and others Acidic solutions are commonly used in industrial processes such as acid washing, pickling, descaling, and drilling operations in oil and gas production, and iron and steel vessels or surfaces used in these conditions are vulnerable to corrosion [2]. And such undesired activity leads to damage of greater magnitude creating a risk to the environment. Therefore, a more impactful step towards controlling such an undesired effect of corrosion on the metallic structure is to produce green corrosion inhibitors. They not only reduce the corrosion rate but also have minimal or zero effect on the environment as far as the threat to life is concerned. The inhibition efficiency of more than 75% in most of the cases can be achieved using green corrosion inhibitors. And can be produced at a very low cost when compared to conventional inhibitors. Moreover, the mechanical properties of the metallic structures can be retained without a thick expenditure. Synthetic organic inhibitors have also been commonly used, but their use is now marred by their toxicity and elevated development costs. This has spurred researchers to explore other fields to develop green corrosion inhibitors that are eco-friendly, inexpensive, and biodegradable to replace inorganic and synthetic organic inhibitors. Efficient corrosion inhibitors have been identified as natural products such as plant extract, amino acids, proteins, and biopolymers [3]. The performance of these compounds depends primarily on their ability to adsorb the polar groups acting as the adsorptive hubs on the metal surface and block the active sites so that the corrosion rate is reduced [4-5]. Global corrosion costs have been measured using a rational approach and it is seen that India's contribution is just 1% of the global cost of corrosion [6]. The cost of corrosion in Australia, Great Britain, and Japan is about 3 to 4 percent of GDP [7]. Hence, using a green corrosion inhibitor proves not only to be environment-friendly but also to ameliorate a nation's economy. Due to corrosion, a lot of money is lost every year. Industries and governments lose around S276 dollars annually or 3.1 percent of GDP in the United States [8].

2. METHODOLOGY

2.1. Preparation of specimen.

The bright steel used for the investigation consisted of the following composition as shown in the table 1. The test specimen was prepared according to ASTM Standards as shown in fig 1.

Table 1. Composition of bright steel.

Element	C	Si	Mn	P	S	Cu	Ni	Cr
Percentage	0.32-0.38	0.15-0.35	0.60-0.90	<0.030	<0.035	<0.30	<0.20	<0.20

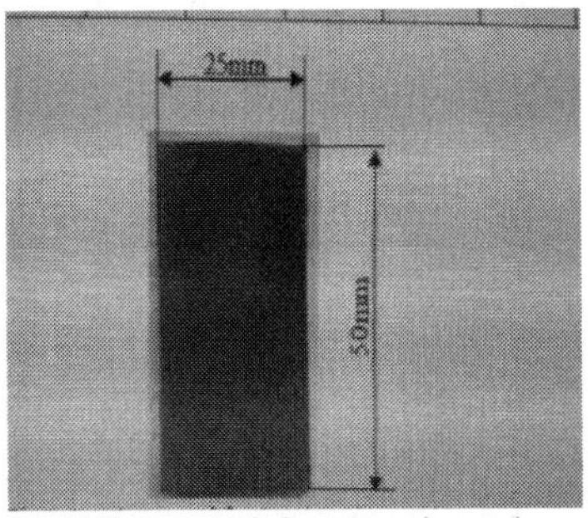

Fig 1. Specimen used for Gravimetric analysis.

The surface of the specimen was finely machined and ground with emery paper.

2.2. Preparation of inhibitor (Banana and Orange peel extract).

The banana and orange peels were dried for about a week under sunlight. And it was heated at 100°C in a microwave oven. The dried peels were then finely powdered separately in a blender and passed through a 53-micron grit.

2.3. Methods.

The experiments were conducted on bright steel samples of standard specimen that were washed with acetone and heated prior to the experiment. 0.1N HCl and 0.3N HCl solutions were prepared for conducting tests. Equal quantities of prepared banana and orange peel powder were added at varying concentrations for a variety of tests, such as 500 PPM, 800 PPM, 1200 PPM, 1600 PPM and 2000 PPM. The initial weight of all the test specimen was noted down after cleaning with acetone. Initially the bright steel specimen were dipped in a 40 ml blank solution of 0.1N HCl and 0.3N HCl in a beaker as shown in fig 2.

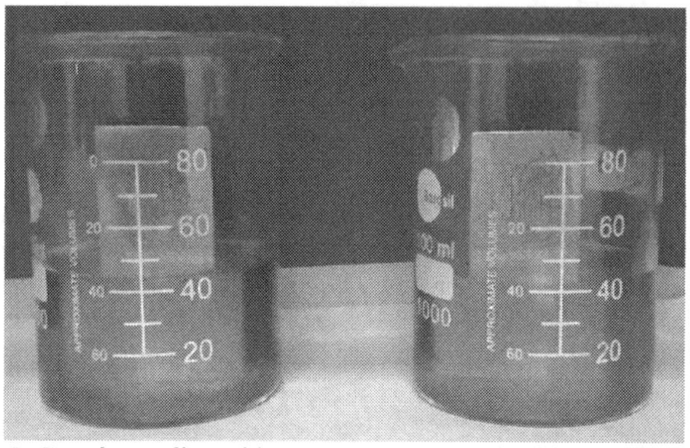

Fig 2. Specimen dipped in 0.1N & 0.3N HCl without inhibitor.

The specimen were taken out after 24 hours of immersion and weighed to understand the loss of material due to corrosion. The same procedure was followed for the following specimen immersed in HCl solutions with different concentration of the inhibitors (BPE 50% + OPE 50%).

2.4. SEM analysis.

In order to create a magnified image for study, scanning electron microscopy is a test method that scans a sample with an electron beam. During research, the signals generated produce a two-dimensional image and disclose sample information, including surface topography, chemical composition, when used with the EDS function, and alignment of the sample materials.

3. Results and Discussion

3.1. Weight Loss analysis.

The steel specimens were suspended for 24 hrs at room temperature in 40 mL of HCl solution with and without the inhibitor. The cleaned specimens were weighed before and after immersion into the corrosive solution. The specimens were removed and dried. The weight loss of the steel samples was measured by an analytical balance with an accuracy of 1 mg. The percentage inhibition efficiency (IE) was obtained from the following equation:

$$IE = \frac{W_o - W_1}{W_o} \times 100,$$

Where W_o and W_1 are the weight of the bright steel in the absence and in the presence of inhibitor. The initial and final weights of the steel sample immersed in 0.1N & 0.3N HCl without the presence of inhibitor is shown in table 2.

Table 2. Variation in the weight of samples.

	0.1N HCl	0.3N HCl
Initial Weight	58.459 grams	57.485 grams
Final Weight	58.289 grams	57.211 grams
Weight Loss (W_o)	0.17 grams	0.274 grams

Fig 3. Specimen taken out from HCl in the absence of inhibitor.

The initial and final weights of steel samples immersed in 0.1N HCl solution in the presence of inhibitor is shown in table 3.

Table 3. Weight loss of specimen and efficiency of inhibitor in 0.1N HCl.

Concentration of inhibitor in PPM	Loss in weight in grams (W_1)	Inhibitor efficiency $\frac{W_o - W_1}{W_o} \times 100$ %
500	0.1320	22.35
800	0.0635	62.64
1200	0.0525	69.11
1600	0.0460	72.94
2000	0.0395	76.76

The table 3. Validates the significant increase in the inhibitor efficiency with increase in concentration of inhibitor when tested on bright steel specimen at room temperature. The linear increase in inhibitor efficiency can be clearly analysed in fig 4.

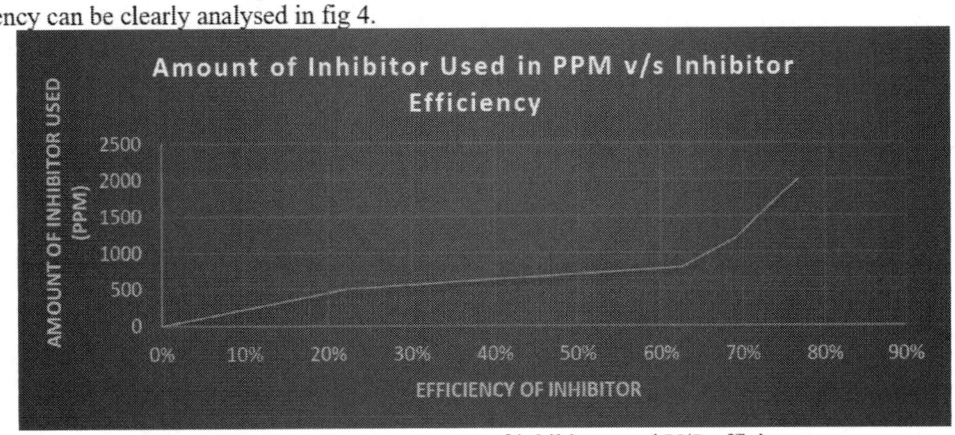

Fig 4. Graph indicating amount of inhibitor used V/S efficiency.

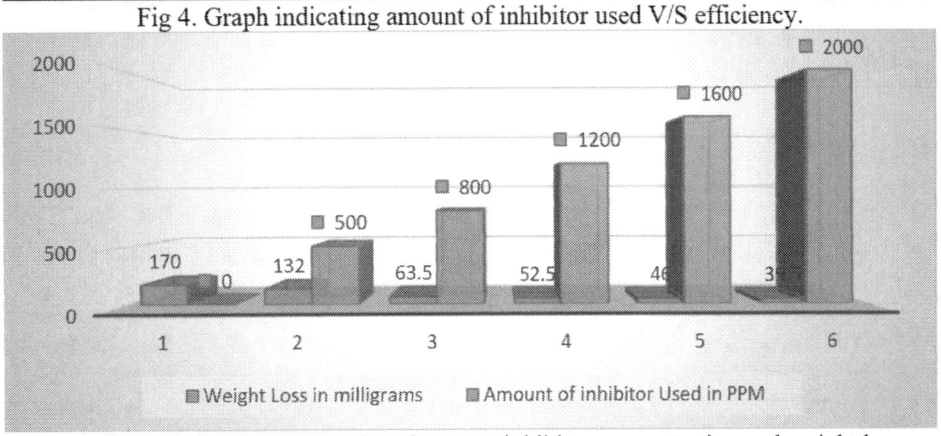

Fig 5. Graph indicating comparison between inhibitor concentration and weight loss.

Fig 5. Shows the decrease in loss of weight in steel specimen immersed in acidic media in the presence of different concentration of banana and orange peel extract corrosion inhibitor.

The initial and final weights of steel samples immersed in 0.3N HCl solution in the presence of inhibitor is shown in table 4.

Table 4. Weight loss of specimen and efficiency of inhibitor in 0.3N HCl.

Concentration of inhibitor in PPM	Loss in weight in grams (W_1)	Inhibitor efficiency $\frac{W_o - W_1}{W_o} \times 100\ \%$
500	0.02630	4.01
800	0.2520	8.02
1200	0.0980	64.23
1600	0.0395	70.80
2000	0.0570	79.19

Unlike the results obtained for steel specimen immersed in 0.1N HCl solution the steel samples in 0.3N HCl showed some interesting characteristics. The inhibitor efficiency did not show any drastic increase until 1200 PPM. But as soon as 1200 PPM of inhibitor was employed the efficiency leaped from 8% to 64%. Despite being a stronger corrosive medium compared to 0.1N HCl the inhibitor exhibited improved efficiency as shown in table 4.
The increase in efficiency with an increase in inhibitor concentration in a corrosive environment can be visualized in g 6.

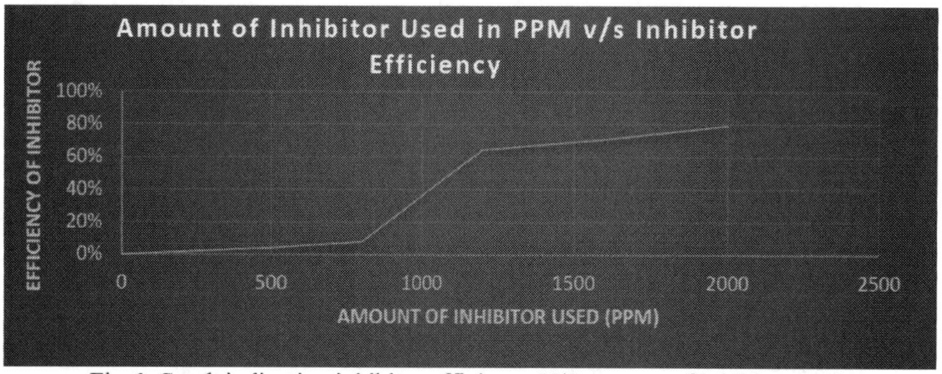

Fig 6. Graph indicating inhibitor efficiency V/S amount of inhibitor used.

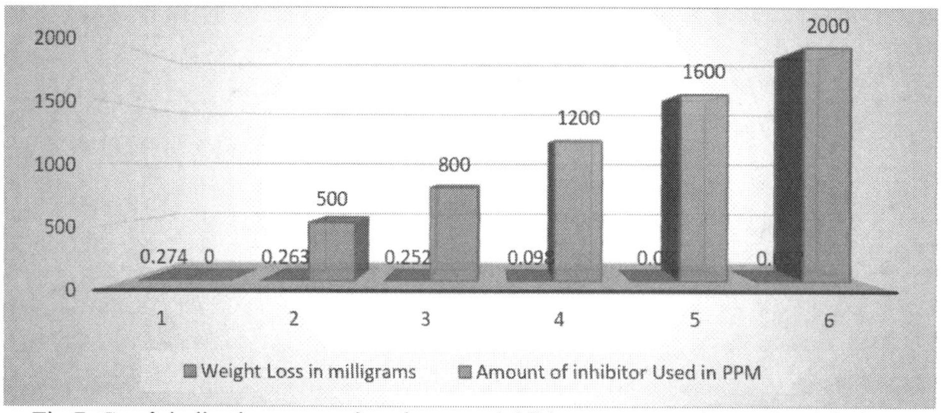

Fig 7. Graph indicating comparison between inhibitor concentration and weight loss.

Fig 7. Shows the decrease in loss of weight in steel specimen immersed in 0.3N HCl solution for 24 hrs in the presence of a different concentration of corrosion inhibitor of banana and orange peel extract.
Some of the rich sources of antioxidant compounds in banana and orange peels are delphinine, cyaniding, polyphenols, carotenoids and vitamins C and E, and anthocyanin. It has been shown that phenolic compounds, particularly anthocyanin, possess significant antioxidant activity, which is mainly based on their structural characteristics. Abundant sources of pectin, which is a structural heteropolysaccharide, are also found in banana and orange peels [9-14]. Banana peel is also made up

IOP Conf. Series: Materials Science and Engineering 1065 (2021) 012029 doi:10.1088/1757-899X/1065/1/012029

of a complex polyphenol compound called Tannin, which is adsorbed to form a thin film on the metal surface. This layer is not visible to naked eyes, but the corrosion process in metal is halted [15].

3.2. Surface Analysis.

The surface morphology of bright steel samples immersed in 0.1 N and 0.3N HCl was analysed in the absence and presence of 2000 PPM of inhibitor at room temperature for 24 hours, with a scanning electron microscope (SEM) working at a voltage of 5.00 kV.

Fig.8 and Fig.9. Shows the SEM micrographs of the metal surface after exposure to acidic solution in the absence and presence of an inhibitor. After immersion in a corrosive solution, Fig.8 displays the abraded bright steel surface and exposes the SEM image of a bright steel surface immersed in 0.1 N HCl solution in the absence of an inhibitor. In comparison, Fig.9 displays the SEM image of the specimen immersed in a 0.1N HCl solution with an inhibitor of 2000 PPM. It is seen from Fig.8 that the metal surface has been heavily corroded due to the presence of a corrosive solution. However, after the incorporation of the inhibitor, the appearance of the bright steel surface is different.

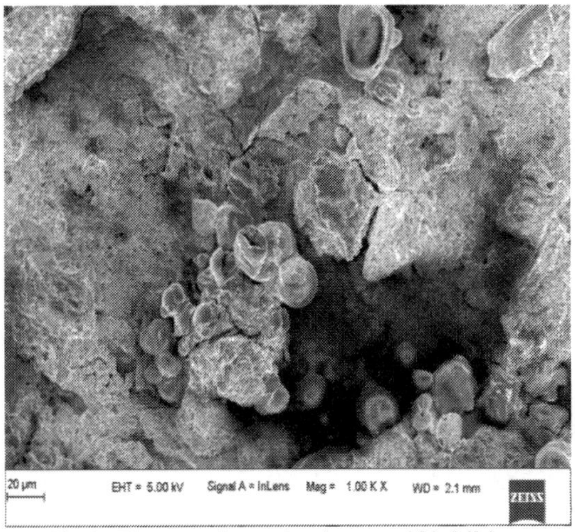

Fig.8. The specimen in absence if inhibitor

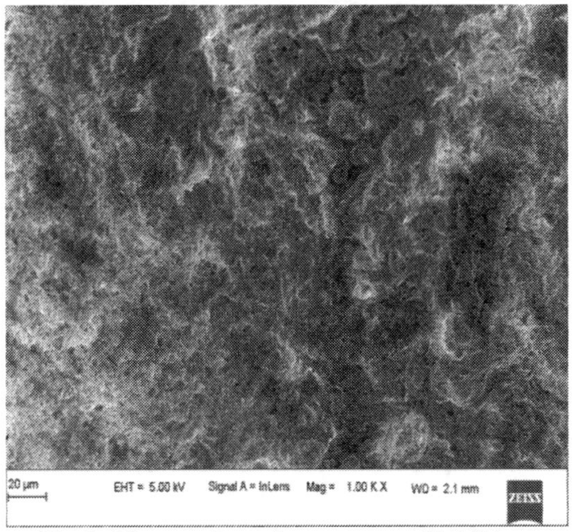

Fig.9. The specimen in presence of inhibitor

Fig.10 and Fig.11 shows the SEM micrographs of the metal surface after exposure to acidic solution in the absence and presence of an inhibitor. After immersion in a corrosive solution, Fig.10 displays the abraded bright steel surface and exposes the SEM image of a bright steel surface immersed in 0.3N HCl solution in the absence of an inhibitor. In comparison, g. 11 displays the SEM image of the specimen immersed in a 0.3N HCl solution with an inhibitor of 2000 PPM. It is seen from Fig.10 that the metal surface has been heavily corroded due to the presence of a corrosive solution. However, after the incorporation of the inhibitor, the appearance of the bright steel surface is different. In comparison with the 0.1N HCl solution, the inhibitor provides more conclusive evidence in inhibiting the corrosive action when mixed with 0.3N HCl by forming an anti-corrosive lm on metal surface which can be observed precisely in Fig.11.

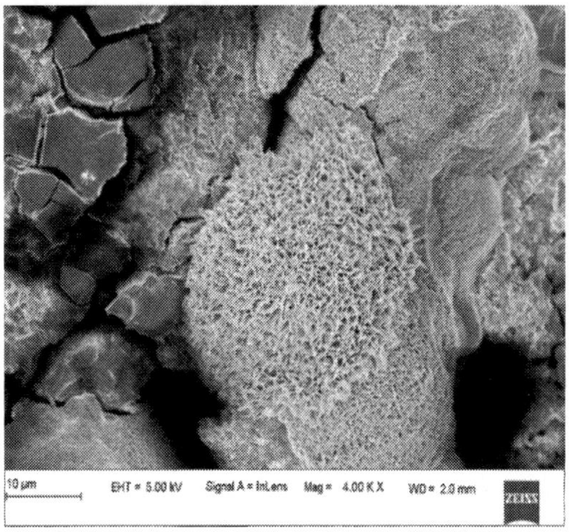

Fig 10. The specimen in absence if inhibitor.

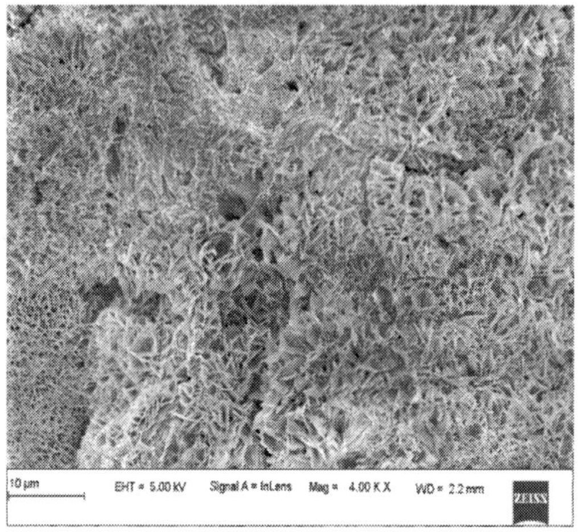

Fig 11. The specimen in presence of inhibitor.

4. Conclusion

The Banana and Orange peel extracts in HCl solution serve as a green inhibitor for the prevention of corrosion in bright steel. The efficiency of inhibition increases with the rise in inhibitor concentration. The SEM analysis also shows conclusive evidence supporting the efficiency of the green inhibitor against corrosive action.

The usage of such inhibitors to prevent corrosion of bright steel proves to be more economical and eco-friendly in comparison with toxic inhibitors such as 2-5 zinc phosphates, 6 polyphosphates, lead oxide, etc.

The accessibility of banana and orange peel throughout the year makes it more beneficial and cheap to be prominently used as green corrosion inhibitors.

References

[1] Higgin R.A., Krieger R.E., Malaber F.L. (1990): Engineering metallurgy, 5th edition, Krieger, Melabar, FL, 1983.

[2] M. V. Fiori-Bimbi, P. E. Alvarez, H. Vaca, and C. A. Gervasi, "Corrosion inhibition of bright steel in HCLsolution by pectin," Corrosion Science, vol. 92, pp. 192–199, 2015.

[3] G. Sığırcık, T. Tüken, and M. Erbil, "Assessment of the inhibition efficiency of 3, 4-diaminobenzonitrile against the corrosion of steel," Corrosion Science, vol. 102, pp. 437–445, 2016.

[4] S. Deng, X. Li, H. Fu, "Acid violet 6B as a novel corrosion inhibitor for cold rolled steel in hydrochloric acid solution," Corros. Sci., vol. 53 pp. 760-768, 2011.

[5] D.B. Hmamou, R. Salghi, A. Zarrouk, H. Zarrouk, M. Errami, B. Hammouti, L. Afia, L. Bazzi, "Adsorption and corrosion inhibition of bright steel in hydrochloric acid solution by verbena essential oil," Res. Chem. Intermed., vol. 39, pp. 973-989, 2014.

[6] Bhaskaran, R. & Palaniswamy, N. & Rengaswamy, N. (2009). COST OF CORROSION AND CORROSION AUDITING IN INDIAN INDUSTRIES. Corrosion Reviews. 27. 1-35. 10.1515/CORRREV.2009.27.S1.1.

[7] Kruger J. and Uhligs W. (2000): Corrosion Handbook, 2nd edition, Wiley, (pp 165 – 170.)

[8] Gerhard H., Michael K., Brongers P., Neil G. (2002): Corrosion costs and preventive strategies in the united states – Publication no FHWA – RD – 01- 156.

[9] Ribeiro SMR, Barbosa LCA, Queiroz JH, Knodler M and Schieber A. Phenolic compounds and antioxidant capacity of Brazilian mango (Mangifera indica L.) varieties. Food Chemistry. 2008; 110(3):620-626. foodchem.2008.02.067.

[10] Ajila CM, Naidu KA, Bhat SG and Prasada Rao UJS. Bioactive compounds and antioxidant potential of mango peel extract. Food Chemistry. 2007; 105(3):982-988.2007.04.052.

[11] Anagnostopoulou MA, Kefalas P, Papageorgiou VP, Assimopoulou AN and Boskou D. Radical scavenging activity of various extracts and fractions of sweet orange peel (Citrus sinensis). Food Chemistry. 2006; 94(1):19-25.
[12] Wang YC, Chuang YC and Hsu HW. The flavonoid, carotenoid and pectin content in peels of citrus cultivated in Taiwan. Food Chemistry. 2008; 106(1):277-284.2007.05.086.
[13] Berardini N, Knodler M, Schieber A and Carle R. Utilization of mango peels as a source of pectin and polyphenolics. Innovative Food Science & Emerging Technologies. 2005; 6(4):442-452.
[14] Janaina Cardozo da Rochaa , José Antônio da Cunha Ponciano Gomesa , Eliane D'Eliab, Aqueous Extracts of Mango and Orange Peel as Green Inhibitors for Carbon Steel in Hydrochloric Acid Solution, Materials Research. 2014; 17(6): 1581-1587.
[15] Komalasari, S P Utami, M I Fermi, Y Aziz and R S Irianti, Corrosion control of carbon steel using inhibitor of banana peel extract in acid diluted solutions, IOP Conf. Series: Materials Science and Engineering 345 (2018) 012030 doi:10.1088/1757-899X/345/1/012030.

IOP Conf. Series: Materials Science and Engineering 1065 (2021) 012030 doi:10.1088/1757-899X/1065/1/012030

A Review on Thermal Properties of Hybrid Polymer Matrix Composites

Gurushanth B Vaggar[1*], S C Kamate[2], Rithesh Shetty[3].

[1]Research Scholar, Alva's Institute of Engineering and Technology, Mijar, Moodbidri, Visvesvaraya Technological University, Belagavi, Karnataka state, India.

[2]Principal Hirasugar Institute of Technology, Nidasoshi, Visvesvaraya Technological University, Belagavi, Karnataka state, India.

[3]UG student, Alva's Institute of Engineering and Technology, Mijar, Moodbidri, Visvesvaraya Technological University, Belagavi, Karnataka state, India

E-mail: gvgr.aiet@gmail.com

Abstract. The requirement and demand of hybrid polymer composites increasing in every filed of applications day by day due to low weight high strength. Since the polymers have found low thermal properties and low strength to high temperature conditions, similarly glass fibre reinforced epoxy resin composites are high strength and stiffness but poor in thermal stability and easily degrades at high temperatures. Which encourages to enhance thermal properties to increase the thermal stability of glass fibre reinforced epoxy resin composites without compromising the strength and stiffness, solution of that is adding high thermal conductivity filler particles to glass fibre reinforced epoxy resin composites, increases the thermal stability and thermal resistivity.

1. Introduction
Most of the polymers are low thermal conductivity, low thermal and heat resistant and high coefficient of expansion, to improve their thermal stability and thermal properties, fibres and fillers are used in neat polymers to develop hybrid polymer composites. The complete review on hybrid polymer composites concentrated on enhancement of thermal property and thermal stability.

1.1. Polymers
Polymer matrix are of two types as given below.

a) Thermoplastics: Materials made by intermolecular linear polymer structures are called thermoplastics. Thermoplastics are excellent elastic property, poor strength to high load. Melting of thermoplastics gives the restructure to intermolecular polymer structures. The linear structure chemical bonds loses their strength to high temperature exposure. Thermoplastics soften or fuse when heated, harden and become rigid after cooling. Unlike thermosets, thermoplastics can be modified or reused upon the need. Thermoplastics have longer shelf life and higher fracture toughness than thermoset resins. Thermoplastic resins have high viscosity and less creep resistance when compared to thermosets. Examples of thermoplastics are Polyethylene, Polystyrene, Polyamides and Nylons.

IOP Conf. Series: Materials Science and Engineering 1065 (2021) 012030 doi:10.1088/1757-899X/1065/1/012030

b) Thermosets: Materials with more number of crosslinked structure made by polymer matrix with highly chemical bonds are called thermoset materials. Due to high crosslinked structure thermosets are stronger than thermoplastics and poor in elastic property. Once the thermosets are transferred from liquid to solid state during the solidification, can't bring back to liquid form due to non-ability of recycle. Examples of thermosets are Epoxy, Polyester and Phenolic polyamide resins. Epoxy resin has excellent adhesion property and mechanical properties compared to other resins. Epoxies are expensive compared to polyester and less resistant to moisture. Polyester has the advantages of low cost, ease of handling, good chemical resistance with reasonable mechanical properties. Polyester and epoxy makes approximately 85% of the fibre reinforced polymer composites.

Polymer matrix composites have advantages, high strength in tensile, Stiffness more, Fracture due to toughness is high, Resistance due to abrasion is good, Resistance due to puncture is good, Good resistance against corrosion. Polymer matrix composites also have certain disadvantages, Resistance against thermal conditions is low, Thermal expansion coefficient is high, Methods of developing composite materials are slow in nature.

1.2. Resins
Thermoplastic materials are formed by addition polymerization. Thermoplastics soften or fuse when heated, harden and become rigid after cooling. Unlike thermosets, thermoplastics can be modified or reused upon the need. Thermoplastics have longer shelf life and higher fracture toughness than thermoset resins. Thermoplastic resins have high viscosity and less creep resistance when compared to thermosets. Epoxy resin has excellent adhesion property compared to other resins and have excellent mechanical properties. Epoxies are expensive compared to polyester and less resistant to moisture. Polyester has the advantages of low cost, ease of handling, good chemical resistance with reasonable mechanical properties. Polyester and epoxy makes approximately 85% of the fibre reinforced polymer composites. Vinyl ester offers good mechanical properties and excellent corrosion resistance. The bonding ability of vinyl ester is good to glass fibre but less efficient with carbon or Kevlar. Phenolic resin cures through condensation reaction, which produces water during the reaction.

1.3. Additives
Additive materials are primarily used to modify and tailor material properties of the composite. By introducing additives into resin system it enhances the processability or durability of the composite. Additives help in increasing the performance or a specific property as well it increases the overall cost of the product. Various additives widely used to boost the thermal and electrical conductivity of the resin are graphite powder, chopped carbon fibre, carbon nano-tubes, SiC nanoparticles, Cu nanoparticles etc.

2. Composites by polymer matrix
Composites made by polymer matrix is the material comprising of a polymer matrix mixed with a fibres dispersed by reinforcing method. Simple methods of fabrication with economical cost made polymer matrix composites more popular. Polymer matrix composites possess lightweight, have good mechanical strength and properties, reinforcement occur in wet well which leads to good adhesion. Polymer reinforced composites are low-density materials.

Matrix is the continuous phase and surrounds the reinforcements. The role of matrix is to transfer equal amount of stresses to fibres and fillers to protect them from surroundings and environment.

Reinforcement is the distributed fibres which enhances the matrix properties. In polymer matrix composite (PMC) materials the basic constituent (the matrix) contribute at least 45% by weight. The materials used as a matrix in PMCs are usually Epoxy resin, Polyester, Phenolic polyamide resins. Glass Fibre, Carbon Fibre, Aramid are used reinforcement materials in PMCs.

The widely used thermoset polymers are epoxy resin due to low shrinkage, less moisture absorption, excellent interface, quick curing, high strength stiffness, good mechanical properties and easy to process at RTP (room temperature and pleasure). Epoxy resins are strong strength in bonding at low pressure curing [1].

The matrix used in FRPCs are continuous phase, may be metal, ceramic or polymer. Ceramics are high strength and brittle, polymers have low strength, metals matrices have moderate strength and ductile in nature. Glass fibre with chopped strand mat (CSM) are suitable for FRPCs due to intact held with matrix [2].

3. Polymer matrix composites manufacturing methods

Polymer composites are manufactured by various methods as shown below.

3.1. Hand Layup Method

It is one of oldest and simplest method to manufacture polymer composites. Polymer matrix, fibres are used in desired shape of moulding box, laid down the one by one layer of fibres into the mould box by applying the required amount polymer matrix on every layer of fibres. To avoid stickiness a non-stick material is applied on to the surface of mould box before placing the fibres and polymer matrix. Curing takes more time to settle down for solidification, accuracy in dimensions and shape lies on skill of worker.

3.2. Compression Moulding Method

This method of manufacturing composites is used for similar kind bulk composite materials by compression moulding technique. Compression moulding machine is designed in such a way that it consists of two parts one is male and another female, fibres and polymer matrix together placed in a female part of a mould box and male part is placed above the female part. Then apply the desired amount of pressure by using hydraulic press and allow cure at normal temperature and pressure. After curing remove the composite material from mould box. This method is more suitable for automobile applications.

3.3. Resin Transfer Moulding (TRM) Method

This manufacturing method uses a mould with an inlet to introduce resin/catalyst mixture and vent ports to allow air to escape. In this method, resin and catalysts are mixed proportionally in an injection head and then pumped into the mould. Dry reinforcement is placed inside the mould and is closed. Required amount of resin is pumped to the mould and allowed to cure under atmospheric condition. After composites are removed from the mould. RTM has been further modified by using vacuum to suck resin catalyst mixture inside. In this case, vacuum bags are used along with moulds enhances resin flow and reduces void fraction. RTM process is cleaner with less emissions of volatiles. The applications of this process includes manufacturing of auto body panels, swim pool panels, sandwich panels etc.

3.4. Pultrusion Method

Pultrusion is a continuous and highly automated moulding process used for manufacture of similar kind of composites in mass production. Reinforcement materials are arranged in such a pattern so that they match the profile of the die. The fibres are pulled through a resin bath and then to the heated metal pultrusion die. The die is maintained at a précised temperature so that it can transfer heat to polymer matrix and fibres. The heating process is used to carry on the polymerization of the resin to matrix. The cooled solid part is then pulled from the die and cut to desired length. Applications of this process are in the fields like construction, transportation, electrical etc.

A detailed method of manufacturing hybrid polymer composites shown in figure 1.

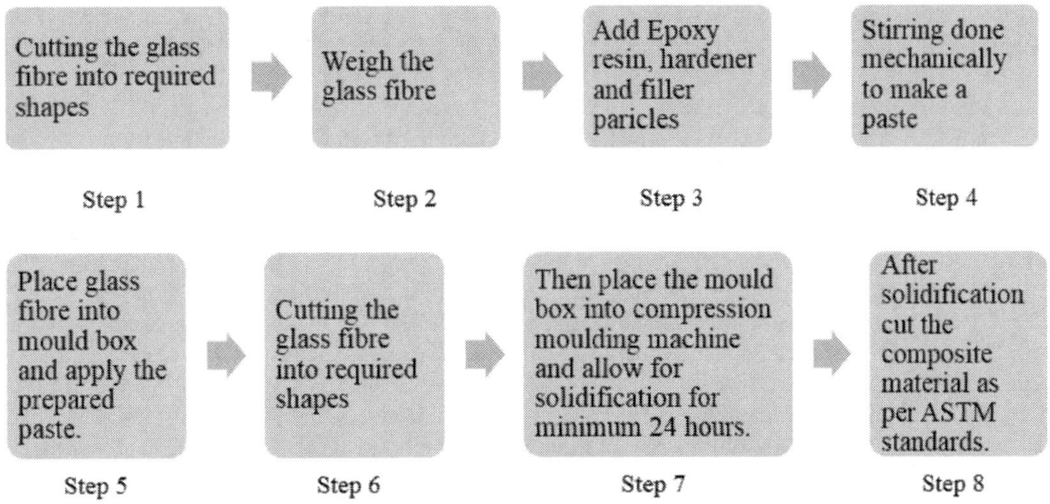

Figure 1. Steps involved in manufacturing hybrid polymer composites.

4. Thermal property of hybrid polymer composite materials

4.1 Thermal Conductivity

For various applications designing and developing of hybrid composites, a study of thermal behaviour needs to be consider under varying conditions. The thermal property variations are nonlinear in hybrid composites, it is very difficult to identify and predict the temperature behaviour of these materials with unknown thermal conductivity. However, thermal conductivity measurement of hybrid composite materials is not an easy task due to temperature variations complexity [3]. Addition of Silicon carbide filler particles to the neat epoxy polymer composites increases thermal conductivity as the volume fraction percentage increases; 20% and 30% SiC filler particles have shown 0.41 and 0.51 W/m K thermal conductivity respectively. Thermal conductivity of silicon carbide alone is very high as compared to hybrid composites, silicon carbide epoxy resin composites exhibit low thermal conductivity due to size of filler particles are smaller. However, they are not useful in generating conductive paths [4]. Hence, electronic devices are going to fail because of inadequate cooling. Durability and performance of electronic devices depends on the limit of specified operating temperature range. The high performance of semiconductors depends on the accurate thermal management [5].

The advantage of using HPCs depends upon the filler materials which give highest thermal conductivity. Hybrid nano-filler modified polymer matrices have created wide opportunities in research to achieve excellent thermal and mechanical properties for advances in many fields of applications. The commonly used nanoparticle-fillers are carbon nanoparticle-filler, carbon nanotubes, nano clay and silicon carbide fibres [6]. Since polymers have low thermal conductivity, an effort has been made to increase thermal conductivity and electrical resistivity of polymers by adding boron nitride, silicon carbides, glass Fibre, mica, aluminum nitride, alumina and zinc oxide fillers [7]. Improved Fire resistance and more thermal stability is found in silica hybrid composites, due to high strength of interfacial bonding of silica particles and epoxy resin [8]. The addition of graphene nano-plates (1% weight) increased thermal conductivity and the enhancement was significant due to homogenous dispersion of graphene nano-plates in the epoxy polymer matrix and internal bonding. Thermal conductivity improvement is 33.3% for 1% graphene nano-plates compared to plain epoxy resin [9].

On comparison of experimental thermal conductivity for the particulate filled polymer composites with theoretical mathematical models, such as parallel thermal conductivity model, maxwell thermal conductivity model and rule of mixture thermal conductivity model, all the three mathematical models

IOP Conf. Series: Materials Science and Engineering 1065 (2021) 012030 doi:10.1088/1757-899X/1065/1/012030

and experimental results have been reported to be close to each other at low % fibre content. They observed that experimental thermal conductivity and theoretically calculated values increase with increase of fibre content; the errors associated with Maxwell model, Parallel model and Rule of mixture model compared to experimental values lie in the range of 31 to 75.2%, 28.5 to 73.2% and 33.25 to 79.2% respectively [10]. Further, thermal conductivity and thermal stability of polymers composites can be enriched by making them hybrid polymer composites by adding fillers such as CNT, graphene, copper powder, copper mesh in carbon fibre reinforced polymer (CFRP) [11].

4.2 Coefficient Thermal Expansion.

Coefficient of thermal expansion improves for 5% weight multi walled nanotubes added in polyurethane composites and thermal degradation temperature improves from 409°C to 421°C [12]. Hybrid composites prepared with epoxy resin, glass fibre, fly ash, stone powder and silicon carbide in a 50:50 matrix fibre and 5% & 20% fillers in weight proportionate. From thermal experiments it is observed that glass fibre epoxy resin (GFER), GFER + silicon carbide 5% and GFER + silicon carbide 20% shows high thermal conductivities compared to other composites, but GFER + fly ash 5%, GFER + silicon carbide 20% and GFER + stone powder 20% exhibits low CTE when compared with other composites [13]. Use of hybrid fillers (Aluminium nitride, Silicon carbide, Boron nitride) in polymer composites was found more effective in decreasing the coefficient of thermal expansion [14].

4.3 Thermogravimetric Analysis.

The composites made by polyester resin, glass fibre and jute fibre undergone thermogravimetric analysis and found that the composite PO77-JU23-V10 (polyester 77% + Jute 23% + glass fibre 0%) without glass fibre lost more mass in function of temperat1re and the composite with more percentage of glass fiberPO56-JU21-VI23 (polyester 56% + Jute 23% + glass fibre 23%) lost less weight with increasing temperature[15]. The TGA comparison of glass fibre composite, Carbon fibre composite and hybrid polymer composites, the thermal stability of hybrid composites found better [16]. The TGA of PMMA toughened glass-epoxy composite shows higher mass loss whereas SiC addition retains the temperature stability [17]. Several graphene-based nanomaterials including graphene oxide (GO), reduced graphene oxide (rGO), graphene nanoplatelets (GNPs) and multi-walled carbon nanotubes (MWCNTs) were used to modify the epoxy matrix and the surface of glass fibres. The results indicate that the introduction of GNPs, GO, rGO, and MWCNTs enhance thermal conductivity. The thermal conductivity of two-phase epoxy/nanoparticle composites was increased by up to 6.0% for 1.2 wt. % of GNPs in GFER composites [18].

4.4 Effect of filler materials on glass transition temperature of polymers.

Micro and Nano filler particles using in epoxy resin glass fibre (ERGF) composites are the reason in glass transition temperature (GTT) rise of hybrid polymer matrix composites. As the GTT of polymers lies in the range of 90°C to 120°C, and thermogravimetric analysis tests are conducted at a temperature range from 200°C to 300°C or more and is much higher than the epoxy resin glass transition temperature (GTT), filler particles thermal conductivity values are higher compared to glass fibre and epoxy resin, which causes the polymer hybrid composites bear higher temperature than that of GTT of neat polymers, individually no effect on GTT of polymer matrix in addition of percentage volume fraction of filler particles and fibres but as a whole polymer hybrid composites GTT improves in comparison with neat polymer matrix.

Adding higher thermal conductivity value fillers to epoxy resin glass fibre (ERGF) leads to delink the covalent bonds of polymer matrix and allows fillers to disperse evenly in epoxy resin Fibre reinforced composites, which in turn increases the thermal conductivity and thermal stability. Fillers are added in small percentages either in volume fraction or weight fraction 5%, 10%, 15%, 20%, 25% and 30% so on. The shape of fillers may be in spherical, flakes, tubes, and whiskers. Minimum size of fillers added in microns and advanced study of composites shows Nanoparticle fillers more effective in hybrid polymer composites. The effect of thermal properties varies with respect to the use of fibres and filling materials. The orientation of fibres, additives, size of fillers and type of resins will affect the Thermal conductivity, Coefficient of thermal expansion and thermal stability with respect to high temperature. Some composite are also tested for low temperature for structural stability.

Table 1. Thermal Characteristics of glass fibre epoxy resin reinforced hybrid composites

Matrix	Type of fibre	Powder filler	Powder filler Vol.%	Mixing method	Technology	Improved properties	Reference
Epoxy / PMMA	E-Glass	SiC	0.5% and 1.0%	Mechanical stirred with epoxy	Hand layup	Thermal stability increases	[17]
Epoxy / LY556	E-Glass	Al_2O_3, $Mg(OH_2)$, SiC	10% and 15%	Mechanical stirred with epoxy	Hand layup	Thermal conductivity, Coefficient of thermal expansion and fire resistance	[18]
Epoxy / Hardener	E-Glass	MWCNTs, GO, rGO, GNPs.	1.1% to 5.5%	Mechanical Stirring at 400 rpm	Vacuum Assisted Resin Transfer Moulding (VARTM)	Thermal conductivity and thermal stability	[19]
Epoxy resin(L12) / Hardener	E-Glass	Fly ash, Stone powder, SiC.	5% and 20%	Mechanical stirred with epoxy	Hand layup	Thermal properties (Thermal conductivity, CTE, Specific heat)	[13]
polypropylene	---	CNT, SND, BN, Cu.	4%	Mechanical stirred with polypropylene	Compression moulding using mould box	Overall thermal properties	[20]
HDPE	---	Silicon particulates	5%, 10%, 15% and 20%	Mechanical stirred with HDPE	Compression moulding using mould box	Thermal conductivity, CTE, TGA	[21]
E-51 Matrix / MeHHPA / DMP-30	---	GNPs,		Mechanical Stirring	Compression moulding using mould box	Thermal conductivity.	[22]

Conclusions

Thermal property characterization of polymer hybrid materials provides the information about withstanding and sustainability of polymer hybrid composite materials under steady, transient temperature conditions it may be low or high temperature exposure conditions. Thermal properties and thermal characterization study of hybrid polymer composites with various fillers volume fractions navigates importance of decomposition temperature, increase or decrease of thermal conductivity value and CTE variations. Without altering the basic strength of composites, improve or enhance the thermal properties of composite materials by adding fillers to make them hybrid composites. The failure analysis of advanced composites are having more complexity due to various reasons, failure may takes along or across the fibre, failure may occur micro level cracks in Fibre or matrix, fibre debonding with matrix due to environmental and high range of temperature variant conditions makes difficulty of identifying and finding the solution to micro level failure of advanced composites, encourage to work more in improving the thermal stability and thermal properties of composite materials to reduce the failure percentage.

References

[1] M. K. Gupta, *American Journal of Polymer Science & Engineering,* (2018) ISSN 2572-5734, Volume 6, Issue 1.
[2] R Ambigai and S Prabhu, *Materials Science and Engineering* 402 (2018) 012136
[3] Dilek Kumlutas et al, *Composites Science and Technology* 63 (2003) 113–117, Elsevier.
[4] Krishnamachar Srinivas and Mysore Siddalingappa Bhagyashekar, *Journal of Minerals and Materials Characterization and Engineering*, 2015, 3, 76-84.
[5] Th. Schubert, et al, *Advanced Materials Research* Vol. 59 (2009) pp 169-172.
[6] Naveen Jesuarockiam et al, *Polymers* 2019, 11, 1085.
[7] Watthanaphon Cheewawuttipong et al, *Energy Procedia* 34 (2013) 808 – 817.
[8] Dongho Kim et al, *Fibers and Polymers* 2013, Vol.14, No.12, 2141-2147.
[9] Minh-Tai Le and Shyh-Chour Huang, *Materials* 2015, 8, 5526-5536.
[10] Daniel J. Thomas, *J Fail. Anal. and Preven.* (2016) 16:513–514.
[11] Rakesh S et al, International Journal of Advance Research, *Ideas and Innovations in Technology,* Volume 4, Issue 5, ISSN: 2454-132X, 2018.
[12] Dinesh Kumar and Prashant Jindal, *Materials Research Express* 6 (2019) 105336.
[13] Ravi Kumar B N et al. *International Journal of Mechanical Engineering Research & Applications.* ISSN: 2347-1719, Vol. 1 Issue 4, September – 2013.
[14] Geon-Woong Lee et al, *Composites: Part A* 37 (2006) 727–734, Applied Science & Manufacturing, Elsevier.
[15] Braga, R. A et al. *Int. Journal of Engineering Research and Applications,* ISSN: 2248-9622, Vol. 5, Issue5, (Part -2) May 2015, pp.113-118.
[16] Bhasker Bommara, et al, *International Journal of Mechanical Engineering and Technology,* International Journal of Mechanical Engineering and Technology, ISSN Print: 0976-6340 and ISSN Online: 0976-6359.
[17] V Ravi Raj and B Vijaya Ramnath, *Springer Nature B.V. 2020,* doi.org/10.1007/s12633-020-00580-5.
[18] K.Devendra and T. Rangaswamy, *International Journal of Computational Engineering Research (ijceronline.com)* Vol. 2 Issue.5, ISSN Online: 2250-3005. (2012).
[19] M. Rafiee et al, *Composites Part B (2018),* https://doi.org/10.1016/j.compositesb.2018.11.051.
[20] MS Nurul and M Mariatti, *Journal of Thermoplastic Composite Materials,* 26(5) 627–639, DOI: 10.1177/0892705711427345.
[21] T.K. Dey and M. Tripathi, *Thermochimica Acta* 502 (2010) 35–42, doi:10.1016/j.tca.2010.02.002.
[22] Junwei Gu et al, *International Journal of Heat and Mass Transfer* 92 (2016) 15–22

IOP Conf. Series: Materials Science and Engineering 1065 (2021) 012031 doi:10.1088/1757-899X/1065/1/012031

Design and Analysis of Wind Amplifying Rotor Platform for Wind Turbine

Dr. B.J.Patil

Principal, Basav Engineering School of Technology, Zalaki, Indi, Vijayapur Karnataka, India-586204

bjpatil5@gmail.com

Abstract. Generally, annual wind speed available in India is close 6.5 m/s. Design of optimum, cost effective wind turbine for maximum annual energy extraction is a challenging task. The present work carried out the Design, Analysis and evaluation of Wind amplifying rotor platform module for cost effectiveness and suitability for Indian conditions. The degree of amplification depends on module and system configuration and is typically over 1.5 to 1.8 times of free stream velocity. Two micro wind turbines have been designed and analysed with respect to twist, thickness distribution, blade profile and aerodynamic characteristics using available airfoil data at National Aerospace Laboratories (NAL), Bangalore, India.

1. INTRODUCTION

The World watch Institute has issued a fact stating that the wind turbine industry is growing faster (about 39% annually) than the personal computer industry and almost as fast as the cellular phone market. Wind energy has become the most economic and environmentally benign source of energy from the existing renewable energy technologies. It has the ability to be used as an electrical power supplier from the electrical utility scale (wind farms) to a single-use power supplier. Present day large bladed wind systems (to be referred to as “POP” -- propeller-on-pole) and the 4th generation in wind energy technology -- WARP (Wind Amplified Rotor Platform). Present day wind energy technology (POPs) have continued to evolve through the use of ever larger and larger blades and gearboxes or large, complex customized generators. This continuous increase in size has subsequently increased the costs and complexity of manufacturing, transporting, constructing and maintaining these systems. As with any system, the more complex it is, the greater the chance for system failures (e.g. historically blade and gearbox failures).

The general public has recently voiced concerns regarding the use of POPs. In the forefront of their concerns are the large footprints that wind farms employ (i.e. land sprawl of thousands of acres), the low frequency noise pollution, horizon pollution with its disco effect and the “flicker” interference with TV and telecom transmission, as well as the perception that they are prone to be bird killers. These drawbacks are in addition to the fact that they are only capable of capturing the winds near the surface (that is why they have, in part, gone to taller towers and larger blades -- attempting to reach the greater wind speeds at higher elevations). Finally, the turbine industry has built only about 20,000 turbines, yet no standardization has been achieved. Each larger blade design requires the industry to basically engineer anddesign it from scratch. Each wind site with its unique characteristics requires another round

IOP Conf. Series: Materials Science and Engineering 1065 (2021) 012031 doi:10.1088/1757-899X/1065/1/012031

of re- engineering and design or compromise performance. These changes also required extensive retooling at great cost to the industry. The larger the blades, the larger the gearbox typically required. Each increase in size subsequently increased the risks of blade and gearbox failures – increasing the risk of reliability, durability and sustainability of the life of system.

2. OBJECTIVES:

The main objective is to design a system for extracting wind energy from low wind speed regions where commercial large wind turbines are not viable. The work is also focussed on the following aspects.

a) 1.Three dimensional CFD analysis of isolated single module of WARP tower
b) Design of micro wind turbine using data like twist, thickness distribution, blade profile and aerodynamics characteristics of aerofoil which are available at NAL
c) Performance evaluation of micro-wind turbine using commercially available computer code

3. COMPUTATIONAL METHODOLOGY

The methodology adopted for the study is described under the following sections

Geometrical details and Grid generation details of WARP module: CFD- GEOM was used as pre-processor for modelling the WARP tower single module and for generating structured grid. The 2D & 3D drawings of the WARP module shown in the figure 1 and figure 2 respectively

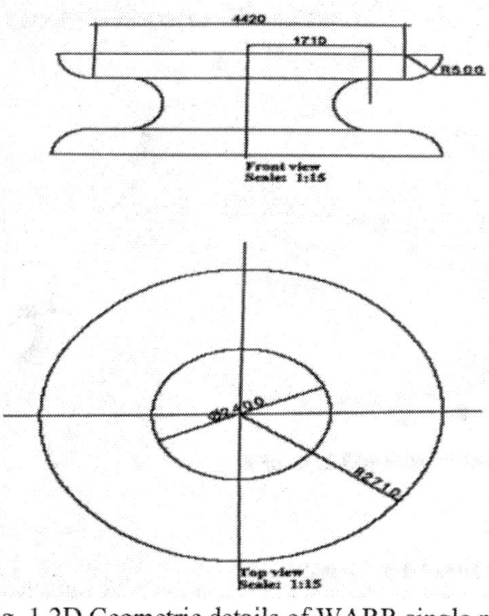

Fig. 1 2D Geometric details of WARP-single module

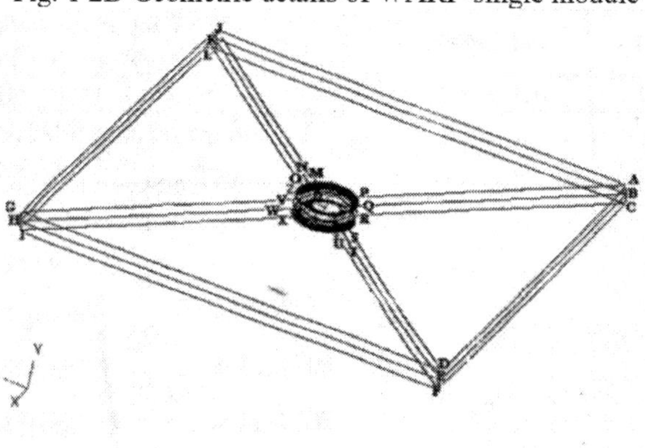

Fig. 2 3DView of computational domain

B Grid Generation Details: The details of grid generation and meshing of central wall in the CFD software is shown in the figure 3, figure 4 and table 1

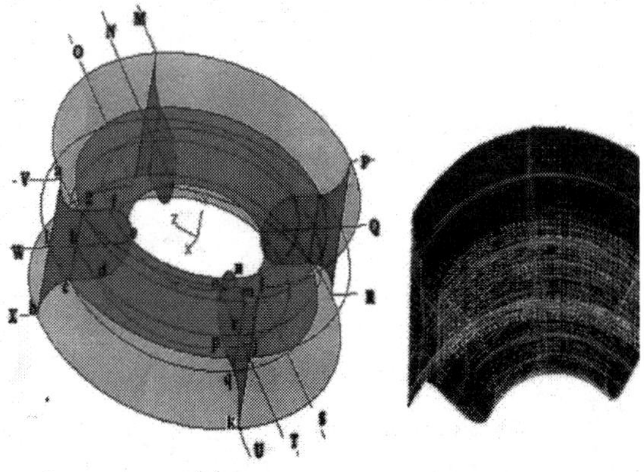

Fig. 3 Details of module grid generation

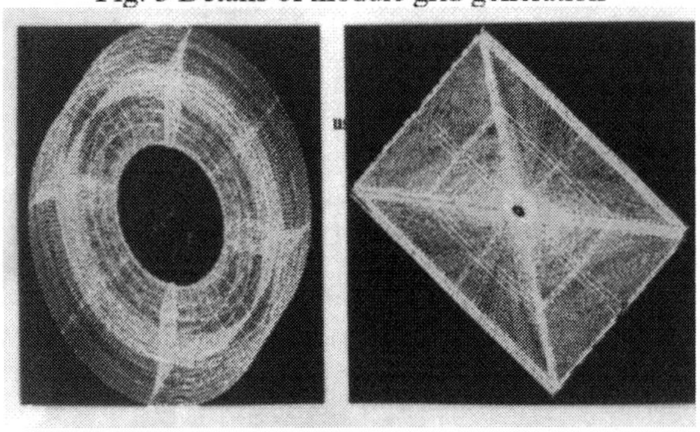

Fig. 4 Details of meshing of central wall and total module

TABLE 1 Grid generation details

Sl. No.	**Curves**	**No. of Points**	**Distribution Type**	**Power**	**Type**
1	AB,AD,BC,DC	50	Power law	1	Forward
2	AF,BG,CH,DE	50	Power law	1	Forward
3	ai,ib,lj,gh,hc,fe,ed,mr ,rq,no,op	15	Power law	1	Forward
4	bc,cd,gf,ag,he,ih	10	Power law	1	Forward
5	Al,bk,ij,gm,hr,cq,fn,e o,dp		Power law	1	Forward

Grid Summary:
Total number of nodes = 386512 Number of quad faces = 1121708 Total number of faces = 1121708
Total number of cells = 367696 Total number of domains = 16

4. RESULTS AND DISCUSSION

The results of CFD analysis of WARP modules are discussed and graphical representation of performance evaluation of micro wind turbine.

A. *Result of CFD Analysis of WARP Modulr.*

The velocity magnitude contour results of inlet free stream wind conditions of V=8m/s, P=101325N/m^2, T =288.15K is show in figure 5 and figure 6

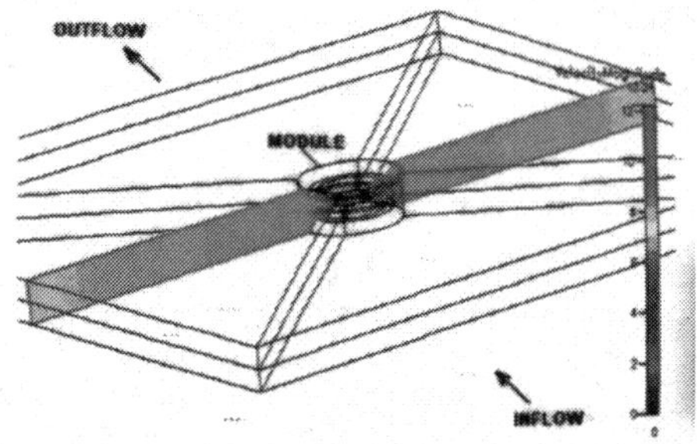

Fig. 5 Velocity magnitude contour (X-cut) for the inlet velocity 8m/s

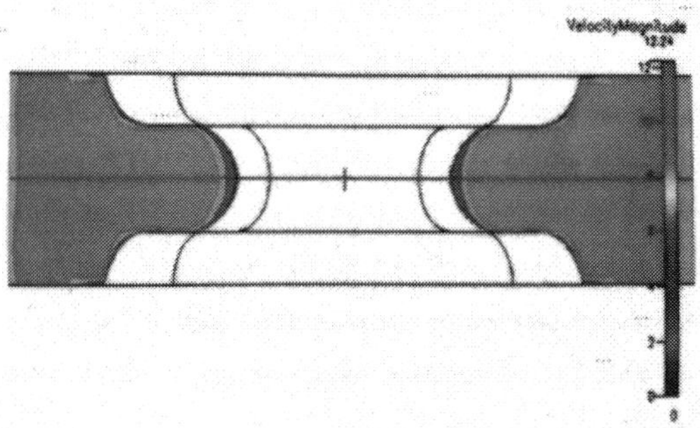

Fig.6 Enlarged view of the velocity contour for inlet velocity 8m/s

Fig. 5 to 12 show vector as well as contours of wind flows over a WARP module. Fig 5 shows the velocity magnitude at the expected location of the micro wind turbine. The flow direction is as shown in the figure. Two Micro wind turbines would be mounted perpendicular to the flow on either side of the WARP module. It can be observed from the figure 5 that the flow velocity increases as it approach the module channel and the flow velocity gets decreased as you move out of the module channel or far-field. The amplification of flow at the turbine location is in the range of 1.3 to 1.5 times the inlet flow. Figure shows that the velocity is zero at the wall boundary as expected within the flow module. figure 6 shows enlarged view of the flow near the module.

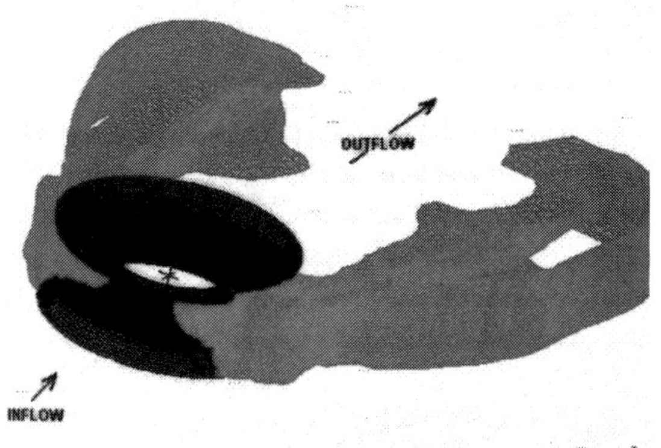

Fig.7 Free stream wind velocity amplified zone at module flow channel.

Figure 7 shows Iso-metric view of the amplified flow pattern around the module in the down stream. A large wake has been created at the downstream of the flow module due to WARP model. It could be observed from the figure 8 that the magnitude of ISO velocity is zero close to the WARP module flow channel and the magnitude increases as one move outwards up to certain distances and then approaches the free stream velocity lines. This is as expected from the WARP module. Even though certain patch of velocity magnitude is as high as 1.5 times the free stream, average velocity amplification is close to 1.3. This results in 120% of increased power in comparison to free stream velocity.

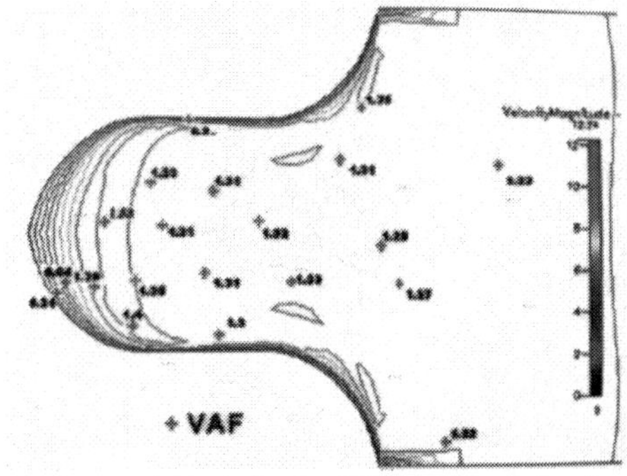

Fig. 8 ISO-'Velocity contours and Local VAF in the WARP module channel free stream wind speed 8m/s

Figure 9 and Figure 10 shows the flow pattern plotted at different planes. It can be observed that flow decelerates as it approaches the WARP module and flow gets split near the module and then again it's starts accelerating and reaches its maximum near the location of turbine. However as earlier as large wake could be seen after the WARP module

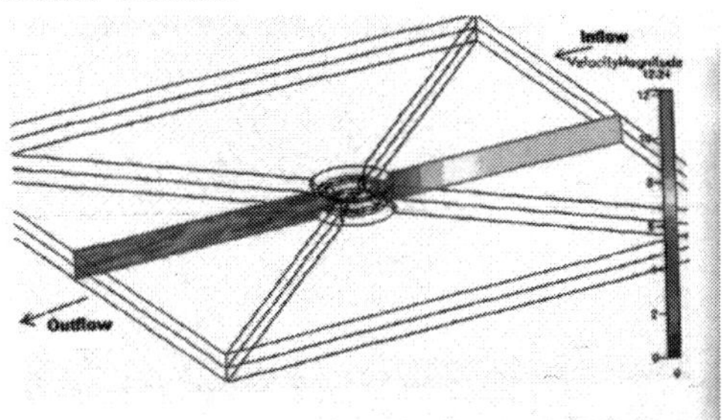

Fig. 9 Velocity magnitude contour (Z-cat) for inlet velocity 8m/s

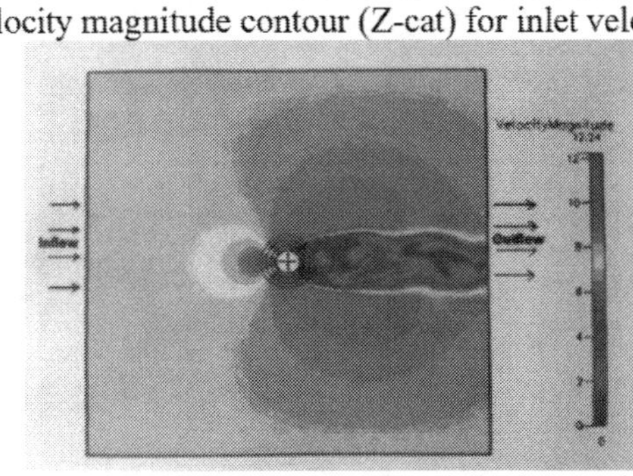

Fig 10 velocity magnitude contour (Y-cut) for inlet velocity 8m/s

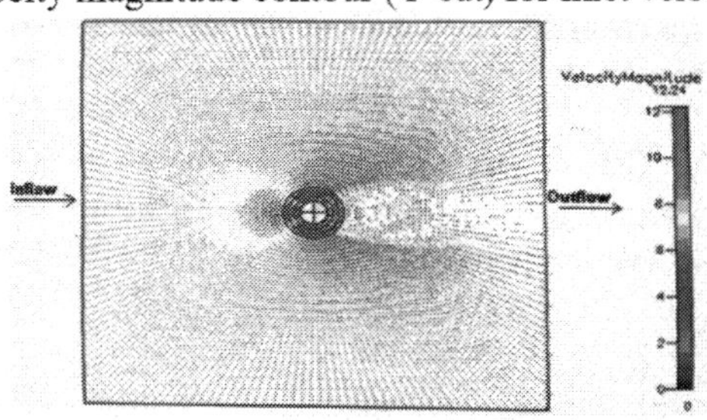

Fig. 11 Vector plot of velocity contour for inlet velocity 8m/s

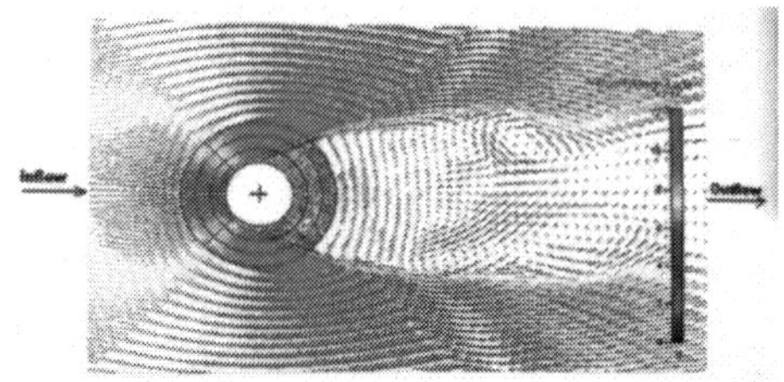

Fig.12 Enlarged view of velocity vector plot

Figure 11 and Figure 12 Shows the vector plot of the velocity plot for the free stream wind speed of 8m/s. In this plots the flow direction at every grid point can be observed. It can be observed from the plot that how the flow stream splits near the WARP module in the upstream and guided through the module channel along the downstream. Figure 3.8 shows the enlarged view of vector plot, which shows high intensity of separation with reverse flow at downstream of the WARP module. With in the wake region you can observe number of high intensity local whirl formed due to the WARP module. At the same time the effect of increased wind speed near to the location of wind turbine could also be seen.

A figure 13 shows the static pressure contour at wind speed of 8m/s, though there is no the applicable change in static pressure, the small change in the static pressure could be observed from the figure

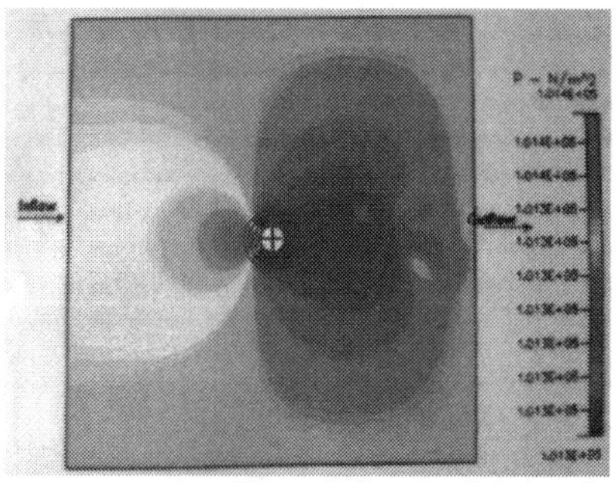

Fig.13 Static pressure plot (Y-cut) for the inlet velocity 8m/s

TABLE 2 Computed values of amplified wind speed and average amplified wind speed

Air free stream Velocity rn/s	Amplified Velocity at Tower Module, m/s	VAF (Velocit y Amplifi cation Factor)	Average Available velocity at WARP module m/s	Avg. VAF
2	3.081	1.54	2.6	1.30
3	4.602	1.53	3.96	1.32
4	6.108	1.53	5.2	1.30
5	7.695	1.54	6.56	1.31
6	9.324	1.55	7.85	1.31
7	10.87	1.55	9.22	1.32
8	12.24	1.53	10.54	1.32
9	13.77	1.53	11.83	1.31
10	15.50	1.55	13.18	1.3
11	17.16	1.56	14.45	1.3
12	18.48	1.54	15.8	1.32

Figure 14 and Table 2 shows the variation of free stream wind velocity v/s increased wind velocity at the WARP module flow channel. Figure 15 shows the velocity amplification factor for free stream wind speed. Though the amplification factor ranges fromto 1.5 around the flow channel, the average amplification factor is close to 1.32. It is observed from the graph that for a given free stream wind velocity, the wind velocity near the WARP module increases and it is ranging from 1.3 to 1.5 times the free stream wind velocity. This results in higher wind power, since the wind power is proportional to the cube of the wind velocity. The design of flow channel requires more focused distribution of amplified wind source to achieve more power at higher amplification factor.

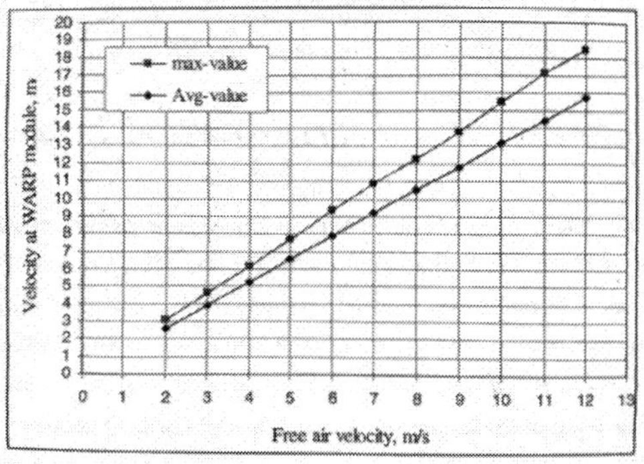

Fig. 14 Increment in free stream wind velocity at WARP module

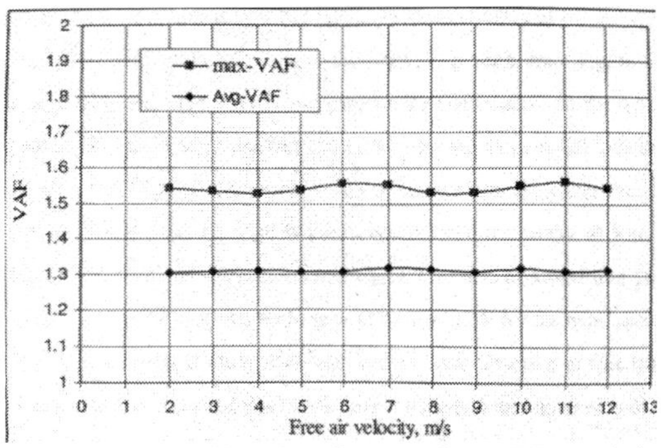

Fig 15 Wind amplification factor of WARP module

5. CONCLUSION

1. There is close agreement between flow analysis of WARP module using CFD –ACE for present configuration and results obtained from published paper.
2. An average increase of 30% wind speed results in 120% increase in power output of turbine, which is enough to encourage concept in power generation.

References

[1] Shikha, Bhatti T.S., Kothari D.P., (2004) "Wind Energy in India: Shifting Paradigms and Challenges Ahead", J. of Energy Engg. Vol.130 No.3, Dec.1, 2004, 67-79Another reference

[2] Alfred L. Weisbrich, P.E. Fuel Cell Augmented Offshore Warp Wind Power: A Proposed set to a Hydrogen Economy, www.warp- enco.com/documents/Power GenE200/Paper.PDE

[3] Stover. D., The Best of What's New-Science & Technology-Energy: Wind Power Towers, Popular Science, December 1996

[4] IEEEE TRANSACTIONS ON INDUSTRY APPLICATION A publication of IEEE Industry Application Society July/Aug 1996 Vol.32 No.4 ITIACR (ISSN 0093-9994)

[5] Brown S., Here Come the Print-Size Power Plants-The Windmill's Narrow Niche; FORTUNE: (Industry Management & Technology issues) April 1, 1996

[6] Duffy. R. E. (Rensselaer Polytechnic Institute), Weisbrich, A. L. (ENECO), Liebowitz, B. Wind

Power Needle Application of the TARP Wind Energy Conversion System, 3rd ASME Wind Energy Symposium, New Orleans, L A, Feb. 12, 1984.

[7] Dr.Ostrow, S.L., Padalino, J.(Raytheon), Weisbrich. A.L., COE Projections for The Modular WARPTM Wind Power Systems for Wind Farms & Electric Utility Power Transmission; American Power Conf., Chicago, IL, 4/18/1995.

[8] Dr. Rainy, D.L. (Rensselaer Polytechnic Institute/Hartford Graduate Center), Weisbrich, A. L. (ENECO), WARPTM Technology for Low Cost & Environmentally Friendly Marine Based Wind Power Plants; British Wind Energy Association Conf.(BWEA-20), Cradiff, Wales, UK, Sept.2-4,1998

[9] Dr. Rainy, D.L. (Hartford Graduate Center), Weisbrich, A. L. (ENECO), mass Customization of WARPTM wind Power Plant Design & Constructions; Proceedings of the American Power Conference, Vol.59,1997.

[10] Prof. Dr. Pucher, K. (Technical University, Graz, Austria), Weisbrich, A. L. (ENECO), Computational Fluid Dynamic Assessment of WARPTM Wind Power System, American Wind Energy Association Conference, Denver, CO, 6/23/1996.

Comparative Study of Carbide Tools in Turning of High-Chrome White Cast Iron using Hard Turning Methods

Ravi A.M[1], S.M. Murigendrappa[2],

[1,2] Department of Mechanical Engineering.,

[1] Department of Collegiate and Technical Education, Karnataka, India,

[2] National Institute of Technology Karnataka, Surathkal, Mangalore, India

E-mail: amravi.**nitk@gmail.com**

Abstract: Thermally enhanced machining (TEM) is a new practice in hard turning of hard materials like White cast iron, Inconel 718 steels etc. Addition of preheating before shear zone improves the elasticity of the material. Before this, Industries invested more capital on toolings and robust machineries/equipments, also it demamed high cost diamond, Cubic boran nitride (CBN) and polyCBN (pCBN) tools. This research paper mainly focusses on the comparative study of machinability characteristics like machining forces and surface roughness of a high hard High chrome white cast iron (HCWCI) when turning with and without the addition preheat using multilayer hard coated carbide (TiC/TiCN/Al_2O_3). Taguchi technique were used to carryout the experimentational work. The experimental and theoretical investigations are done in different phases in order to derive some of the useful process responses like machining forces, surface roughness etc, and to prove Oxy-LPG aided TEM most appropriate methodology to cut HCWCI. Furthermore derived the suitable conditions at which the coated carbide tools perform better. Hence comparative studies of each process responses are required to arrive to the reasonable conclusions.

Keywords: High Chrome White Cast Iron (HCWCI), Thermally enhanced machining (TEM), Taguchi method, Multi coated Carbide insert

1.Introduction

The demand for hard materials like Chromium white cast irons, Tool steels, High manganese steels, Inconel 718 steels, Nickel chromium steels etc has increased significantly in the field of Mining, Agricultural, Aerospace, Defense and Process Industries [1-4]. Machining of such materials has always been a tough job, so Industries opt for casting process to shaping of the product. However, drastic developments in the tooling bring back metal cutting methods like high-speed machining [2] and thermally enhanced machining (TEM) [1]. In hard turning, frictional heat softens the work material by increasing the cutting speed or adding external heat to the tool-work interface [4,5]. Interestingly TEM method overrides the conventional high-speed machining and brings economical reprieve to the Industries [1,4].

In their research work, Ravi AM et. al (2014) conducted series of experimentations to optimize the machinability characteristics of high hard HCWCI with and without addition of preheat using CBN and multilayer coated carbide tools. And stated that, the favourable properties of CBN always be a better choice in hard turning of HCWCI, however the introduction of coating techniques brings back the hard carbide tools in-line with the CBN tools.

Since, the local heat added to the tool-work interface causes severe tool wear on the flank and rake faces of the tool, but, setting up of optimum cutting conditions made TEM an effective tool in hard turning of HCWCI [1-4,7,8]. The main objective of this research paper is to compare the performances of carbide tools at different cutting conditions in order to optimize the process parameters for the benefit of the Industry in manufacturing components for various applications.

2. Experimental setup

The material chosen for this experimentation was sand blasted air quenched HCWCI made in the form round bars having dimensions; 50mm diameter and 400mm long. The work material was an abrasion wear resistant hard material, and its chemical compositions are; 2.7%C, 0.7%Si, 0.66%Mn, 0.068%S, 0.7%P, 0.5%Cu, 27.93%Cr and 0.5%Mo. Some of the important properties HCWCI are;1260°C Melting point, 7.7gm/cc Density, 13W/mk Thermal conductivity, 541 BHN hardness. The tooling's used are; Multilayer hard coated carbide (TiC /TiCN/Al_2O_3) inserts, and indexable tool holder (Mitsubishi make, PCLNR/L 2525MO9 type, insert seating geometry angles; normal rake angle -6°, back rake angle -6°, clearance angle 6° and angle of lap 95°) to mount the insert[1,3,6,10].

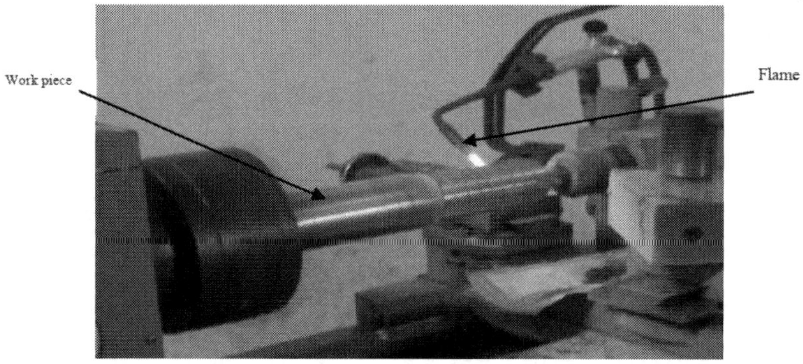

Figure 1. Experimental set up for TEM process.

Here Oxy-acetylene gas was used as the heat source to generate the flame, and the mixture was controllled with the help of flow control valves. A non-contact type thermal sensor was used to set surface temperatures. Figure 1 illustrate the set up for hot machining process where preheat (Flame) added to the work before metal cutting process.

Table 1. Control factors selected for hard turning and TEM process [1].

Sl. No.	Control factors	Levels			Units
		I	II	II	
a	Cutting speed, S	55	88	132	m/min
b	Depth of cut, d	0.1	0.2	0.3	mm
c	Feed rate, f	0.096	0.124	0.179	mm/rev
d	Surface Temperature, T_s	250	350	450	°C

3. Results and discussions

Adopted Taguchi's procedures to carryout experimentations on HCWCI using carbide inserts with and without the addition of preheat [1,3,10]. Table 1 shows the range of cutting parameters selected for both the experimentations. For each experimental trial a new cutting edge was used, and the experiments were repeated thrice at each condition to keep the experimental error to a minimum value

[1]. During experimentation a special care was taken to identify the crack formation on the surface of the work after machining using flaw detector (Orion model, Type-115D). Cutting forces (F_t, F_c and F_f) were measured using Piezo-electric lathe tool dynamometer (Kistler model 9712B500). The surface roughness of the turned part was measured using Surtronic-10 Taylor Hubson surface roughness measuring instrument. The cutting forces and surface measurements were recorded separately for hard turning and TEM process, and the percentage differences are tabulated in Tables 2 and 3.

Table 2. Comparison of cutting forces of the machining processes with and without the aid of flame heat- Using Carbide tools.

Trail Nos	Control Parameters				Percentage differences		
	S, m/min	**d,mm**	**f, mm/rev**	**T_S, °C**	**F_t, N**	**F_c, N**	**F_f, N**
1	55	0.1	0.096	200	-29	-26	29
2	55	0.1	0.124	325	-73	-19	34
3	55	0.1	0.179	450	-55	-28	30
4	55	0.2	0.096	325	28	25	63
5	55	0.2	0.124	450	30	25	63
6	55	0.2	0.179	200	29	27	68
7	55	0.3	0.096	450	-53	27	71
8	55	0.3	0.124	200	-70	31	74
9	55	0.3	0.179	325	-82	28	71
10	88	0.1	0.096	200	-40	45	-38
11	88	0.1	0.124	325	-61	39	-60
12	88	0.1	0.179	450	-37	38	-57
13	88	0.2	0.096	325	-50	37	53
14	88	0.2	0.124	450	-34	37	48
15	88	0.2	0.179	200	-53	34	48
16	88	0.3	0.096	450	-41	-20	-8
17	88	0.3	0.124	200	-33	-14	-8
18	88	0.3	0.179	325	-30	-11	6
19	132	0.1	0.096	200	-36	48	35
20	132	0.1	0.124	325	-9	53	33
21	132	0.1	0.179	450	-62	44	28
22	132	0.2	0.096	325	-37	48	-15
23	132	0.2	0.124	450	-46	49	-5
24	132	0.2	0.179	200	-67	46	-8
25	132	0.3	0.096	450	-66	24	28
26	132	0.3	0.124	200	-59	24	23
27	132	0.3	0.179	325	-29	30	16
				Total	-1064	642	620
				Average	-39	24	23

3.1. Comparison of cutting forces

It is known that the cutting forces generated during machining of HCWCI are; thrust force, F_t, main cutting force, F_c, and feed force F_f. In this research work all the experiments were conducted keeping the angle of lap as 75° for better angle of contact against the workpiece [1,3,6].

3.1.1. Thrust force analysis

Thrust force is one of the components of the machining force acting along the radial direction of the workpiece during machining. Figure 2 shows the comparison of Thrust force at different cutting conditions during machining of HCWCI using coated carbide tools. The research analysis reveals, thrust force measures high, especially in hard turning [2], but behaved differently in TEM process[1,2,4]. The reason might be due to angle of lap, tool nose radius and cutting parameters[1,4]. Also the contribution of each control factors on the thrust force, in which depth of cut has significant contribution to the thrust force followed by surface temperature. An increase in depth of cut increases the thrust force due to increased tool-work contact length, hence more frictional resistance to shear the metal [1,4]. The broad nose radius of the tool having contact only at the radius rather than cutting edge, especially at the lower depths of cut. Obviously it generates more amount of thrust force [8,11]. Kainth et. al (1990) identified the same during their experiments using CBN tools. Also, the formation of the flank and crater wear has more influences on the cutting forces, predominantly the thrust force [7].

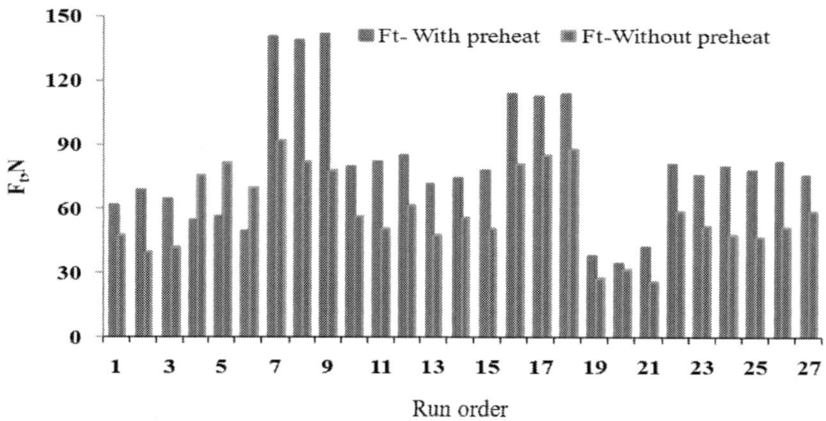

Figure 2. Comparison of Thrust force at different cutting conditions during machining of HCWCI using coated carbide tools.

3.1.2. Main cutting force analysis

The study reveals, in hard turning the main cutting force is the primary force acting along the z-direction (downward). However, its magnitude depends on the hardness of the workpiece, tool geometry and cutting parameters [1,4,6]. Also cutting speed has significant contribution to the cutting force followed by surface temperature. It was observed during the experimentations that, the main cutting force increases as the feed rate or depth of cut increases, on the other hand, decreases as the workpiece surface temperature increases due to softening of the workpiece [1,4]. At a higher shear zone temperature the shear strength of the workpiece declines and improves the volume of metal removal [4]. The increase in feed rate or depth of cut induces large volume of cut metal in a same unit of time. As and when the tool advances, the tool-work contact length increases which forms a longer shear front bonds around the primary shear zone (both plastic and elastic compression). The influences of these may improve the heat conduction at the shear zone [11,12]. When the temperature available at the shear zone is higher, the primary carbides along with eutectic carbides in a matrix of austenite transformed to free secondary carbides (M_7C_3) as a result the carbon content of the matrix decreases [4]. Further increase of the surface temperature decreases the bulk hardness and microhardness of the workpiece [1], hence, lower cutting forces in TEM of HCWCI can be understood. This phenomenon looks similar to the hard turning experimental studies.

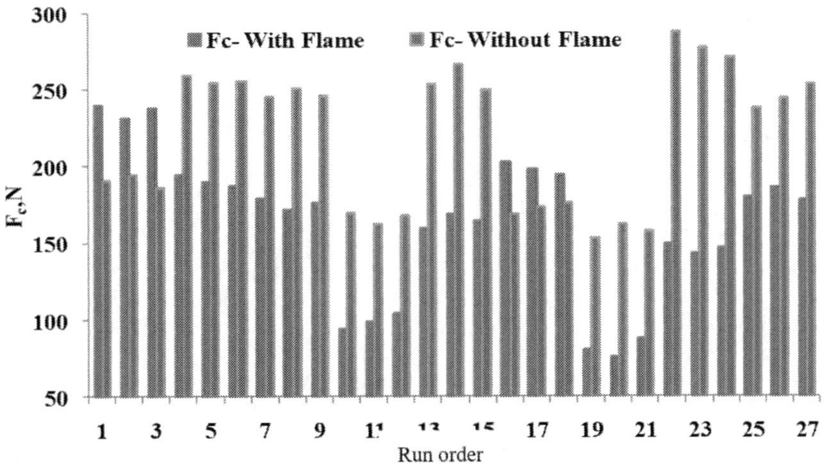

Figure 3. Comparison of Main cutting force at different cutting conditions during machining of HCWCI using coated carbide tools.

Figure 3 shows the comparison of Main cutting force at different cutting conditions during machining of HCWCI using coated carbide tools. The main cutting forces have lesser value in heat assisted trails compared to hard turning for all run over trails.

For the set values of feed rate and surface temperature, the cutting force increases as the depth of cut increases and decrease as the cutting speed increases. This indicates the sufficient amount heat generation at the primary shear zone to soften the metal [1,4]. Interestingly, as the feed rate increases main cutting force decreases, this is due to sufficient amount of heat transfer to the base metal during machining [4].

3.1.3. Feed force analysis

The feed force is the another important cutting force identified in hard turning and TEM process, but it is smaller in value compared to other forces. Figure 4 shows the comparative study of feed force at different cutting conditions during machining of HCWCI using coated carbide tools. It is understood from the characterstics curve that the feed force obtained during turning with or without the preheat have almost similar values. The experimental observations reveal that increase of feed force was due to insufficient heat transfer to the work base metal, and larger tool-work contact length at the shear zone. In their remarks Ravi AM et. al (2014) arrived to the same conclusion when doing experiments with coated carbides. In TEM process, the feed force decreases when surface temperature increaes for constant feed rate and depth of cut. In an another case, for the set values of surface temperature and feed rate, the feed force increases as the cutting speed increases, interestingly, depth of cut has a considerable influence on the feed force.

3.2. Surface roughness

It is interesting to watch the performance of hard-coated carbide against CBN tool, since the tool newly introduced to turn the HCWCI. Table 3 shows the comparison of surface roughness of HCWCI at different cutting parameters and surface temperatures. The characteristic curves of the surface roughness shown in Figure 5 describe the behavior of the workpiece at different cutting conditions, and their trend looks similar. Among them, the curve at 315°C has the lowest surface roughness compared to the curves at 200 and 450°C. At higher surface temperatures, there might be chances of formation of BUE on the tool rake face due to which the cutting area enlarges which results in improper shearing of the metal [4,6].

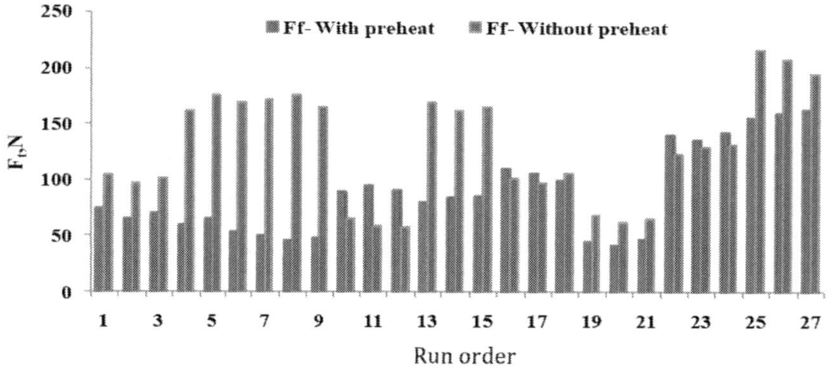

Figure 4. Comparison of feed force at different cutting conditions; during machining of HCWCI using coated carbide tools.

Also the results confirm that, addition of preheat reduces the surface roughness up to 33.65%, except the roughness obtained at low depth of cut (d=0.1mm) with higher cutting speed (S=132m/min). The reason might be due to excess of heat absorbed by the top layer of the workpiece when exposed to the flame, also, the depth of heat penetration decreases as the depth of cut increases [4]. In addition, the broad tool nose radius of the tool has highest contact with the work at its nose radius for lower depths of cut. Hence, roughness of the workpiece is high at these conditions [1,7,8]. At higher cutting speeds, the time for heat conduction to the required depths is not available and so machined surface tends to approach that without heat assistance.

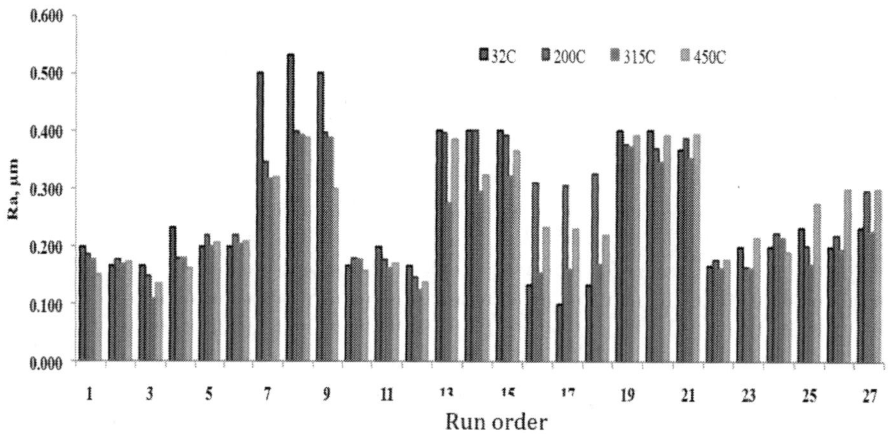

Figure 5. Comparisons of surface roughness of the workpiece at different surface temperature when turning HCWCI using coated carbide tool.

Table 3. Comparison of surface roughness of HCWCI at different surface temperatures.

Trail Nos	Control Parameters			R_a (µm) at different T_s (°C)			
	S, m/min	*d*,mm	*f*, mm/rev	32 °C	200 °C	315 °C	450 °C
1	55	0.1	0.096	0.192	0.186	0.178	0.150
2	55	0.1	0.124	0.172	0.178	0.169	0.173
3	55	0.1	0.179	0.152	0.149	0.108	0.134
4	55	0.2	0.096	0.183	0.179	0.179	0.161
5	55	0.2	0.124	0.202	0.219	0.200	0.205
6	55	0.2	0.179	0.209	0.219	0.204	0.209
7	55	0.3	0.096	0.421	0.346	0.317	0.319
8	55	0.3	0.124	0.425	0.399	0.393	0.387
9	55	0.3	0.179	0.420	0.396	0.387	0.299
10	88	0.1	0.096	0.170	0.178	0.176	0.158
11	88	0.1	0.124	0.181	0.176	0.161	0.170
12	88	0.1	0.179	0.151	0.146	0.125	0.137
13	88	0.2	0.096	0.401	0.396	0.274	0.385
14	88	0.2	0.124	0.400	0.400	0.295	0.323
15	88	0.2	0.179	0.396	0.392	0.321	0.367
16	88	0.3	0.096	0.148	0.311	0.152	0.233
17	88	0.3	0.124	0.151	0.306	0.160	0.230
18	88	0.3	0.179	0.162	0.326	0.169	0.220
19	132	0.1	0.096	0.381	0.378	0.371	0.392
20	132	0.1	0.124	0.374	0.370	0.346	0.392
21	132	0.1	0.179	0.358	0.388	0.353	0.394
22	132	0.2	0.096	0.163	0.176	0.161	0.177
23	132	0.2	0.124	0.210	0.163	0.163	0.214
24	132	0.2	0.179	0.193	0.224	0.214	0.190
25	132	0.3	0.096	0.290	0.201	0.167	0.294
26	132	0.3	0.124	0.198	0.220	0.195	0.299
27	132	0.3	0.179	0.230	0.297	0.226	0.299

4. Conclusion

For the set control factors, thrust forces generated in preheat condition appears to be more than that of no preheat conditions. However, there is a considerable decrement in the main cutting force (23.77% (average of the run order)) and feed forces (22.96%). In both the cases, the work metal receives adequate amount of cutting temperature for the set control parameters. The surface roughness of the workpiece at 315°C surface temperature has the lowest surface roughness compared to 200 and 450°C temperatures. At higher surface temperatures, chances of formation of BUE on the tool rake face are more due to which the cutting area enlarges as a results improper shearing of the metal. Finally, the comparative study reveals that thermally enhanced machining is better than hard turning where no pre heat added to the process.

5. References

[1] Ravi AM, Murigendrappa SM, Mukunda PG, Experimental investigation on Thermally enhanced machining of high-chrome white cast iron and to study its machinability characteristics using Taguchi method and artificial-neural network, Int. J. of Adv. Manuf. Technology, Vol. 72(2014), p1439–1454.

[2] Chryssolouris G, Anifantis N, Karagiannis S, Laser assisted machining: an overview. J. of ManufactSci & Engg, Transactions of the ASME, 119 Iss.4(B) (1997),p766-769.

[3] Ravi AM, Murigendrappa. SM, Wear studies in hard turning of High-chrome white cast iron using ANOVA and RSM techniques, AIP Conference Proceedings 2247, 050002 (2020); https://doi.org/10.1063/5.0004412.

[4] Masood. SH, Armitage K, Brandt M, An experimental study of laser-assisted machining of hard-to-wear white cast iron. Int. J. of Mach. Tools and Manuf. 51, (2011), p450-456

[5] Tönshoff H, Arendt. C, Ben Amor R, Cutting of Hardened Steel, Annals of the CIRP, Keynote Paper 49 (2) (2000), p547-565.

[6] Ravi. AM., Murigendrappa. SM., Mukunda P. G., Experimental investigation of influence of tool temperature on cutting forces in the Thermally enhanced machining of High Chrome White Cast Iron, Elsevier-Procedia Matls. Sci. 5, (2014).

[7] Thamizhmanii S, Hasan S, Measurement of surface roughness and flank wear on hard martensitic stainless steel by CBN and PCBN cutting tools, J. of Achievements in Mat. and Manuf. Engg., 31(2), (2008), p 415-421..

[8] Kainth GS, Dey BK, Experimental investigation in to tool life and temperature during hot machining of EN-24 steels. Bulletin Cercle Etudesdes Me'taux 14(11) (1998), p22.1-22.

[9] Tsao CC, Prediction of thrust force drill in drilling composite material by Taguchi method and radial basis function network, Int. J. of Adv Manuf Technol, 36: (2008),p11-18.

[10] Mantegomery D C, Design and Analysis of Experiments, 7th Edition. Wiley India Edition. India(2000).

[11] Ravi AM, Murigendrappa SM, Mukunda PG, Machinability investigations on High chrome white cast Iron using multi coated hard Carbide Tools, Transactions of Indian Institute Materials-Springer publications, Vol. 67(2014), p485–502.

[12] Ravi AM, Murigendrappa SM, "Development of thermal model to analyze thermal flux distribution in Thermally enhanced machining of High chrome white cast iron", AIP Conference Proceedings 1943 (1), 020004 (2018).

Effect of addition of bismuth on the vibration analysis of Mg-Al/Mg_2Si composites processed using stir casting technique

D E Umesh[1], D Ramesh Rao[2], D Shivalingappa[3]

[1] Department of Mechanical Engineering, Bapuji Institute of Engineering and Technology, Davangere, India
[2] Department of Mechanical Engineering, M.S. Ramaiah Institute of Technology, Bangaluru, India
[3] Dept of Mechanical Engineering, B.N.M Institute of Technology, Bangaluru, India

Email: umeshde1209@gmail.com, umeshdoddamane@gmail.com

Abstract. In this work, the experimental and finite element simulation solutions of vibration characteristics of the Mg-Al/Mg2Si composite has been carried out. The influence of the geometrical structures, percentage of bismuth and material properties on the natural frequency and the material damping of the Mg-Al/Mg2Si composites has been studied. The fast fourier transform (FFT) analyzer has been utilized to find the first three natural frequencies and the modal damping experimentally of the said composite. Finite element simulation method also has been adopted to find the natural frequencies and their mode shapes for the Mg-Al/Mg2Si composites specimens. From the results of the experimental and analytical method it is identified that the composite material exhibits the set of natural frequencies and the modal damping. It is also observed that the modal damping increases with increase in the addition of bismuth.

Keywords: Mg-Al/Mg2Si composites, Vibration analysis, damping, natural frequencies.

1. Introduction

Magnesium alloys are having low mechanical properties, due to which its application in aerospace and automobile industries is limited. From the study it is observed that magnesium alloy composite with different particulate reinforcements shows the enhanced mechanical properties. Now a days, it is observed that magnesium alloy composites with Mg2Si particles has shown higher potential due to its low density and higher hardness values, so Mg2Si reinforced Mg metal matrix composites have wide opportunities in automotive aerospace industries [1-2].
The processing of Mg metal matrix composites includes ex-situ and in-situ method [3]. Ex-situ methods include conventional methods such as powder metallurgy, stir casting, squeeze casting etc. In-situ method, reinforcement are synthesized internally in the matrix during composite fabrication. Conventional casting produce large size Mg2Si particles and they are hardened brittle, it will reduce the mechanical as well as tribological properties [4-5]. Modification is a simple and effective technique for refining the microstructure and improving the mechanical, tribological properties [6] of composites. Recently, La, Y, B, Na, Sb, Gd, Sr, P and some other materials have been used to modify Mg2Si in the

IOP Conf. Series: Materials Science and Engineering 1065 (2021) 012033 doi:10.1088/1757-899X/1065/1/012033

Al and Mg metal matrix composites for obtaining fine microstructure and improved mechanical properties.

Schaller and Mayencourt [7] worked on the vibration behaviour of the Mg-2 wt% Si alloys reinforced with long carbon fibers exhibited higher damping capacities. Zhu and Peng [8] studied the metal matrix composites of nano-sized dispersoids for the damping properties and found the enhanced damping properties.

Vibration analysis is a non-destructive technique [9-10] which helps early detection of machine problems by measuring/ evaluating vibration. Using the careful analysis of vibration, the vibrational characteristics of the equipment can be carried out which reduces the failure of the system [11-12]. Before, vibration investigation required dialing an instrument through the full range to recognize frequencies at which vibration was noticeable. The most recent age of vibration analyzers [13-15] has a larger number of capacities and robotized capacities than their archetypes had. Numerous units show the full vibration range [16] of three tomahawks at the same time giving a depiction of what is new with a specific machine. Vibration analysis can also be used in determining some of the material properties such as damping [17] factors, natural frequency etc.

The main objective of this work is to carryout modal analysis of Mg-Al/Mg2-Si Alloy for different composition of varying Bismuth. The dynamic behavior of MgAl/Mg2Si is investigated by characterizing its properties under different modes of vibration, both analytically and experimentally. The FFT analyzer in combination with PC is used as the data acquisition unit. The input force and output response is measured by force transducer and accelerometer respectively. Analytical analysis is done using ANSYS software. The modal parameters obtained from the modal analysis can be used to analyze the system behaviour under the assumed operating conditions.

2. Materials and Processing

2.1 Raw materials

Commercially pure magnesium ingot (99.3% pure), aluminum ingot (99.2% pure) and Si powder (99.95% purity) are utilized as the basic materials to prepare the Mg-Al/Mg2Si composites. Mg ingots and Al ingots brought from Jagada Industries, Virudhunagar and cut into smaller pieces with the help of power hacksaw in order to keep ingot inside the crucible properly. Si powder brought from Jedee Enterprise, Mumbai powder having the size of 23 microns.

2.2 Modification material

The Mg-Al/Mg_2Si composites were modified with bismuth (Bi). The Bi powder brought from Jedee Enterprise, Mumbai and the powder having the size of 44 microns. Bismuth (Bi) is an element with low melting point (271°C) and it as self-lubricating property during friction process.

2.3Processing of Mg-Al/Mg2Si composites

Commercially pure Mg ingot and Si powder were used as the starting materials to prepare the Mg-Al/Mg2Si composites. The melting process was carried out in a steel crucible location in a 2kW electric resistance furnace under argon gas protective atmosphere. The processing of Mg metal matrix composites were carried out in inert gas atmosphere only because of flammable property of magnesium with oxygen. The furnace containing bottom pouring arrangement ant it is provided with inert gas atmosphere.

About 88.3wt.%Mg and 9wt.%Al were melted and superheated to 760^0C, and 2.7wt.%Si powder preheated 200^0C and packed in aluminium foil was added in to the Mg-Al melt. The melt was held at 760^0C for 10min and stirred about 7min at a speed of 600rpm for the complete dissolution of silicon. After that, different amount of bismuth powder (0.7wt.%, 1.4wt.% and 2wt.%) preheated at 200^0C to remove moisture and gases from the surface of the particulates. Finally, the composite melt was poured into a steel mould (size of 3mm×5mm×120mm) preheated at 400^0C.

3. Experimentation

The experimental setup for conducting modal analysis is shown in Fig 1. The Mg-Al/Mg2Si test specimen is fixed. A point is marked at the middle of the Mg-Al/Mg2Si test specimen for mounting accelerometer to measure acceleration of vibration. Mg-Al/Mg2Si test specimen can be excited for free vibration at various points and response is observed using OROS software for coherence. Once coherence is observed, the results are stored and the point of excitation is changed. The photograph of the test setup is shown in Fig 1.

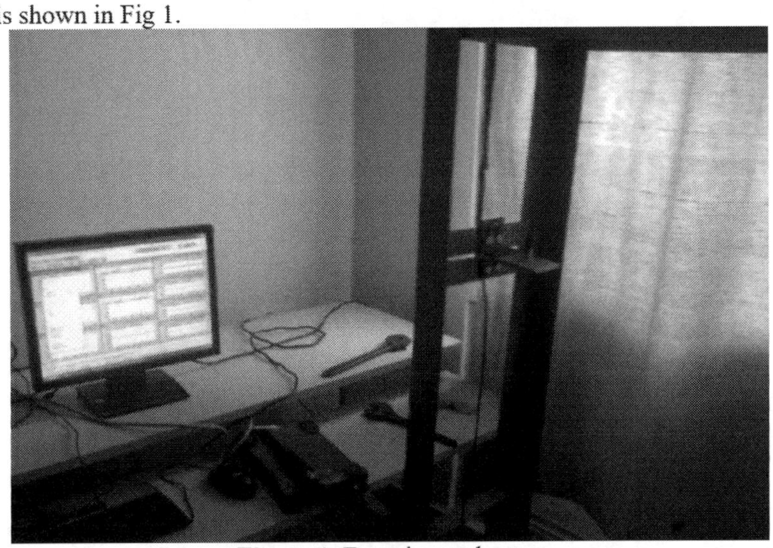

Figure 1. Experimental setup

The response is measured at different points as shown. There are two channels used in this setup. Channel 1 is connected to hammer which is used for exciting the Mg-Al/Mg2Si test specimen for free vibrations. Channel 2 is (shown in Fig 2) connected to the accelerometer which is used for measuring response of the Mg-Al/Mg2Si test specimen.

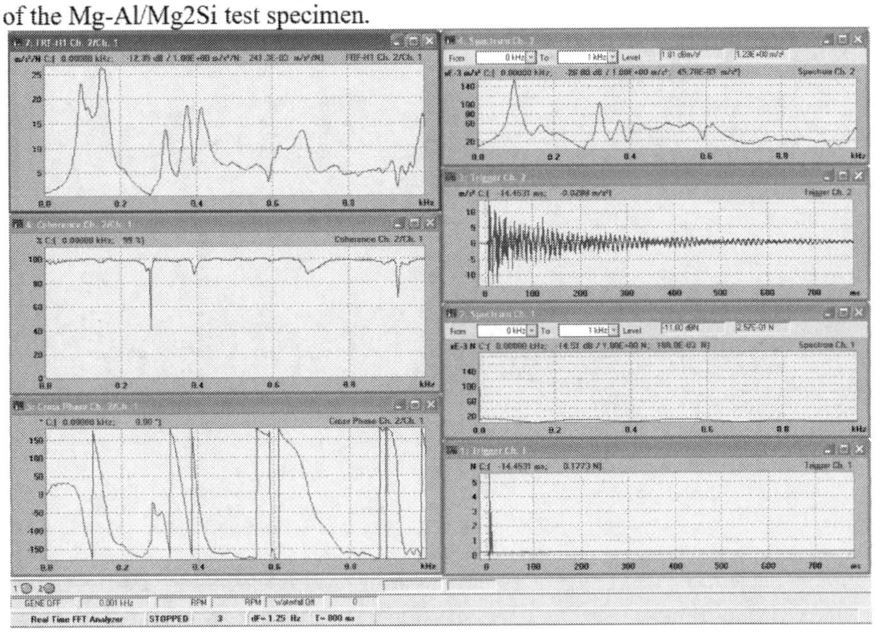

Figure 2 Result obtained through the FFT analyzer showing Spectrum and Coherence.

4. Results and Discussion

4.1 Experimental Results

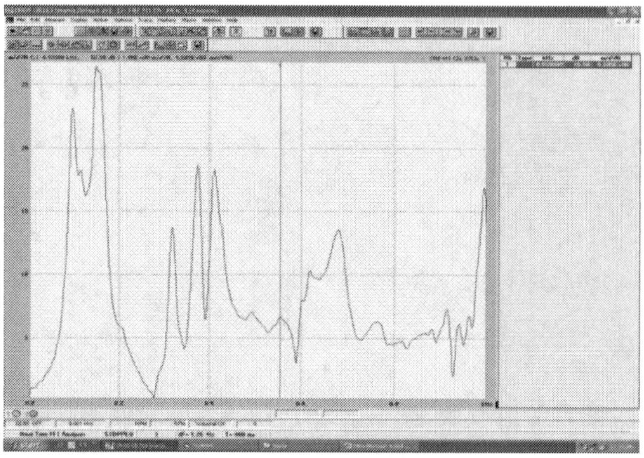

Figure 3 Spectrum showing natural frequency of Mg-Al/Mg2Si+0.0 wt.% Bi.

Table 1 Experimental Model Analysis results of Natural Frequencies of Mg-Al/Mg2Si

Sl No	Natural Frequencies of (Mg-Al/Mg2Si+0.0 wt.% Bi) in Hz	Natural Frequencies of (Mg-Al/Mg2Si+0.7 wt.% Bi) in Hz
1	555	546
2	3415	3349
3	10017	9849

4.2 Modal Damping

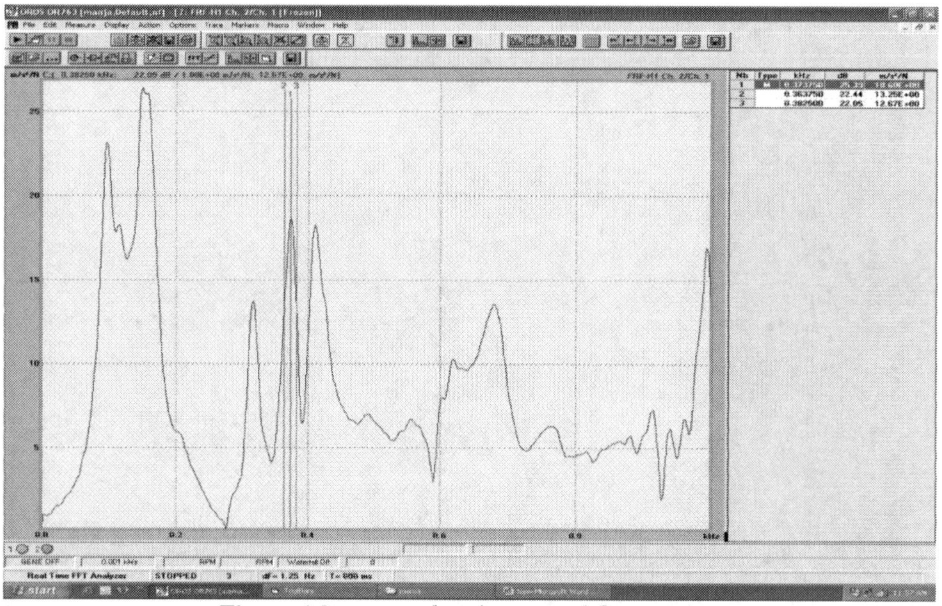

Figure 4 Spectrum showing natural frequencies

Table 2(a) Model damping of Mg-Al/Mg2Si+0.0 wt.% Bi

SL No	ω in Hz	ω1 in Hz	ω2 in Hz	Modal damping
1	373	382	363	0.025
2	515	526	503	0.019
3	666	679	625	0.011

Table 2(b) Model damping of Mg-Al/Mg2Si+0.7 wt.% Bi

SL No	ω in Hz	ω1 in Hz	ω2 in Hz	Modal damping
1	118	127	107	0.084
2	147	158	135	0.078
3	179	194	168	0.072

4.3 Modal Analysis using FEM

The commercially available finite element simulation software ANSYS has been utilized to the natural frequencies of the said composite. ANSYS is capable of solving the many structural, dynamic, fluid, heat transfer etc kind of the practical problems.

The three-dimensional model is created using the 8-noded element called SOLID45. This element has three degrees of freedom at all the 8 nodes at X, Y, and Z directions. A general description of element input is given in Element Input.

The geometric model is created in ANSYS. The finite element model is built using SOLID45 element, for the modal analysis of the Mg-Al/Mg2Si alloy. Modal analysis was carried out and result of the analysis is listed in Table 3.

Table 3 Analytical Modal Analysis results of Mg-Al/Mg2Si

Mode No	Natural Frequencies of Mg-Al/Mg2Si+0.0 wt.% Bi	Natural Frequencies of Mg-Al/Mg2Si+0.7 wt.% Bi
1	570	568
2	3463	3560
3	8284	10836

4.4 Mode Shapes

Mg-Al/Mg2Si+0.0 wt.% Bi

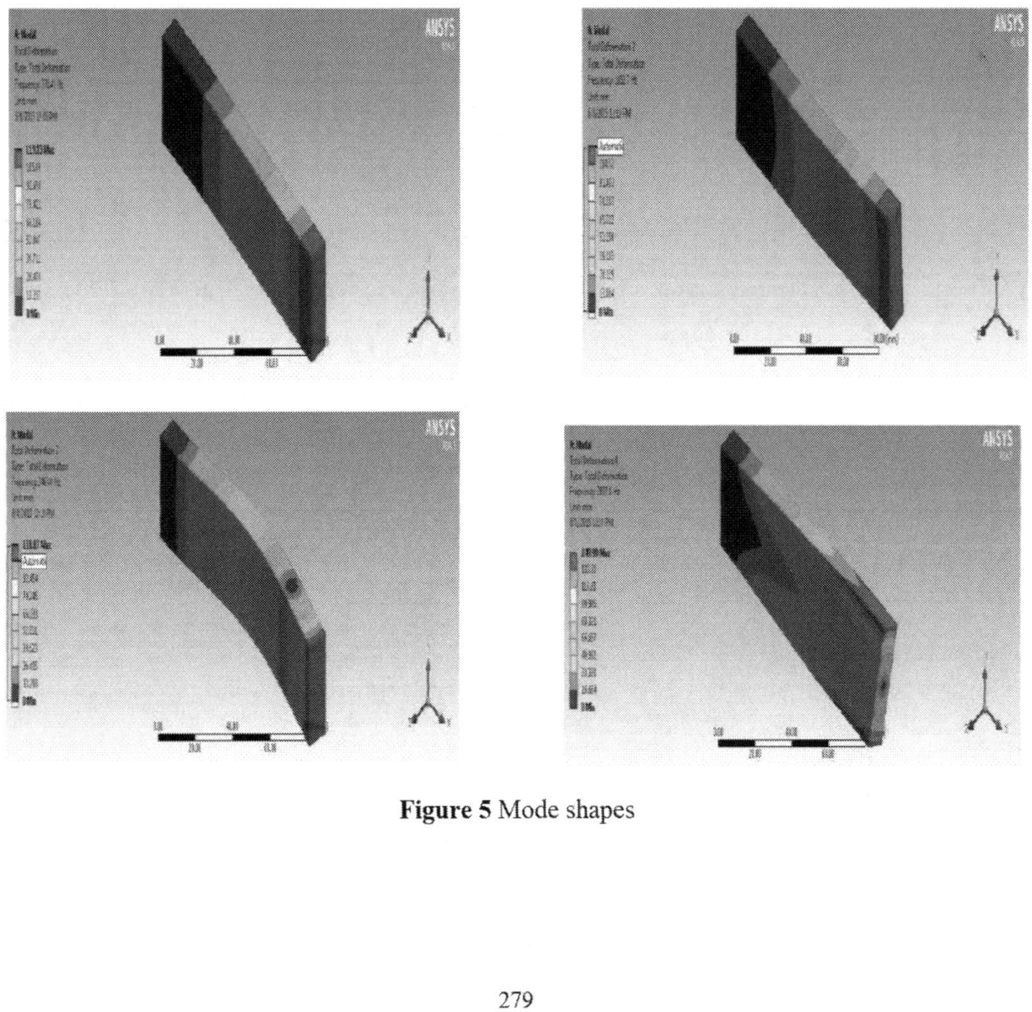

Figure 5 Mode shapes

The data collected from both experimental and FEM analyses are shown in table below.

Table 4 Comparison of FEM and Experimental results of Mg-Al/Mg2Si 0.0 wt % Bi

No.	Experimental results in Hz	ANSYS results in Hz	% Deviation
1	555	570	2.60
2	3273	3463	5.47
3	7546	8284	8.9

5. Conclusion

From the present study the following conclusions are made:

- The processing of Mg-Al/Mg2Si composites can be successfully synthesized by in-situ method.
- From the Comparison of FEM and Experimental. The results obtained by both the methods agree with each other with a deviation of about 2% -9% Damping. Thus, the better agreement in the experimental and finite element simulation results has been observed.
- From this work it is concluded that by the use aluminium and silicon with magnesium, it gives a good damping property and high density.

References

[1] J. Zhang, Z. Fan, Y.Q. Wang, and B.L. Zhou, Microstructural Refinement in Al-Mg2Si in-situ composites, J. Mater. Sci. Lett.,1999, 18, p 783-784.

[2] Q.D. Quin, Y.G. Zhao, and P.J. Cong, Effect of phosphorus on microstructure and Growth manner og primary Mg2Si crystal in Mg2Si/Al composite", Mater. Sci. Eng., 2007, A447, p 186-191.

[3] R. Khorshidi, A. Honarbakhsh Raouf, M. Emamy, and J. Campbell, The Study of Li Effect on the microstructure and Tensile Properties of Cast Al- Mg2Si Metal Matrix Composite, J Alloys Compd., 2011, 509, p 9026-9033.

[4] Srinivasa R. Bakshi, Di Wang, Timothy Price, Deen Zhang, Anup K. Keshri, Yao Chen, D. Graham McCartney, Philip H. Shipway, Arvind Agarwal, Microstructure and wear properties of aluminum/aluminum–silicon composite coatings prepared by cold spraying, Surface and Coatings Technology,204, 4,2009, pp503-510,https://doi.org/10.1016/j.surfcoat.2009.08.018.

[5] Wu, XF., Zhang, GG. & Wu, FF. Microstructure and dry sliding wear behavior of cast Al–Mg_2Si in-situ metal matrix composite modified by Nd. *Rare Met.* **32,** 284–289 (2013). https://doi.org/10.1007/s12598-013-0030-4

[6] Yoshida, I.; Yokosuka, M.; Monma, D.; Ono, T.; Sakurai, M. Damping properties of metal-piezoelectric composites. *J. Alloys Compd.* 2003, *355*, 136-141.

[7] Mayencourt, C.; Schaller, R. Development of a high-damping composite: Mg2Si/Mg. *Phys. Status Solidi A* 1997, *163*, 357-368.

[8] Peng, L.M.; Zhu, S.J. Creep of metal matrix composites reinforced by combining nano-sized dispersoids with micro-sized ceramic particulates or whiskers (review). *Int. J. Mater. Prod.Technol.* 2003, *18*, 215-254.

[9] Jerzy W. Wekezer (1989). Vibrational Analysis of Thin-Walled Bars with Open`Cross Sections. J. Struct. Eng., 115(12), 2965–2978. Volume 115, Issue 12 (December 1989)

[10] Sharayu U. Ratnaparkhi,,S.S. Sarnobat Vibration Analysis of Composite Plate, Department of Production Engineering, D.Y. Patil College of Engineering, Pune, India, (Vol.3, Issue.1, Jan-Feb. 2013 pp-377-380.

[11] Mielczarek A., Trojanova Z., Riehemann W., Lukac P. Influence of mechanical cycling on damping behaviour of short fibre-reinforced magnesium alloy QE22. Mater. Sci. Eng. A, Struct. Mater. 2006;442:484–487.

[12] Rohatgi P.K., Nath D., Singh S.S., Keshavaram B.N. Factors affecting the damping capacity of cast aluminum-matrix composites. J. Mater. Sci. 1994;29:5975–5984.

[13] Elomari S., Boukhili R., Skibo M.D., Masounave J. Dynamic-mechanical analysis of prestrained Al2O3/Al metal-matrix composite. J. Mater. Sci. 1995;30:3037–3044.

[14] Ludwigson M.N., Lakes R.S., Swan C.C. Damping and stiffness of particulate SiC-InSn composite. J. Compos Mater. 2002;36:2245–2254
[15] Zhang J.M., Perez R.J., Wong C.R., Lavernia E.J. Effects of secondary phases on the damping behavior of metals, alloys and metal-matrix composites. Mater. Sci. Eng. R Rep. 1994; 13: 325–389.
[16] M.S. Misra and P. D. Legreca,"Damping behavior of MMCs', Vibration Damping-1984 (Workshop Proceedings), AFWAL-TR-84-3064, V-l, November 1984.
[17] Amreen Taj, Saleemsab Doddamani, T.N Vijaykumar, Vibrational Analysis of Aluminium Graphite Metal Matrix Composite, International Journal of Engineering Research & Technology (IJERT), 6 (4), (2017) pp.1072-1078. http://dx.doi.org/10.17577/IJERTV6IS040720

IOP Conf. Series: Materials Science and Engineering 1065 (2021) 012034 doi:10.1088/1757-899X/1065/1/012034

TRANSIENT HEAT TRANSFER ANALYSIS OF DIMPLED ROD

Abdul Rajak Kaldgi[1*], Abdul Rajak Buradi[2], Vishwanath K C[3], N. Sreenivasalu Reddy[3], Rayid Muneer[1], Mohammed Avvad[1], Chandrasekhar A[4], Asif Afzal[1] and Ahamed Saleel C[5*]

[1]P. A. College of Engineering (Affiliated to Visvesvaraya Technological University), Mangalore, India

[2]NITTE Meenakshi Institute of Technology (NMIT), Bangalore

[3]Raja Rajeswari College of Engineering, Bangalore, India

[4] Bangalore Institute of technology, Bangalore, India

[5]Department of Mechanical Engineering, College of Engineering, King Khalid University, Abha, Saudi Arabia

Corresponding author: ahamedsaleel@gmail.com, arkmech9@gmail.com

Abstract. The considerable heat transfer increase that the dimple causes has led to an interest in its technology recently, with pressure drop penalties lesser than other heat augmentation types. The increase in heat transfer using dimples is based on the fact that the scrubbing of cooling fluid occurs within the dimple and intensifies the delay of flow separation over the surface. The researchers have used different dimple-shaped geometries from all the research studied, such as triangular, ellipsoidal, circular, square, from which the ellipsoidal shape offers better results than other shapes because of prior vortex development. The present work mainly aims to study the dimpled specimen's transient properties and determine their influence on heat transfer.

1. INTRODUCTION

In several fields of application, such as microelectronic cooling, gas turbine internal airfoil cooling, nuclear power plant fuel components, and biomedical equipment, there is a relentless attempt to find ways to increase heat transfer. Two methods are generally adapted for enhancing the heat transfer: Active techniques and Passive techniques. In passive methods, the heat transfer rate is increased by producing surface modifications such as protrusions, dimples, and pin fins. The dimples (surface indentations) are considered necessary because the material is significantly reduced in dimple production, while excessive material is used in pin-fin or rib tabulators, which increases the equipment's weight and cost. A variety of Computational and experimental work has been carried out to enhance the transfer of heat. In 1971, the first person to try using dimples on the surface for heat transfer enhancement was Kuethe[1]. According to him, the dimples would help promote vortexes' creation, resulting in heat transfer improvement [1]. V.N Afnasyev[2] performed an experimental analysis on surfaces formed by spherical cavity networks and observed that, relative to the plane surface, heat transfer improved by 150 percent. Nikolai Kornev[3] investigated vortex structure and heat transfer enhancement in turbulent flow over the staggered dimple array in the narrow channel using Large Eddy Simulation. The impact of the dimple aspect ratio, temperature ratio, Reynolds number, and flow structure in the dimple channel was studied by Mahmood and Ligrani [4]. A numerical simulation of laminar channel flow over the dimpled surface was conducted by Wang [5], and within a single dimple, an asymmetric 3D horseshoe vortex was found. Experimental research was performed by S.L.Borse and I.H Patel[6] on the influence of dimples on heat transfer under forced convection over a flat surface. They claimed that surface dimples contributed to a lower pressure drop in heat transfer enhancement and indicated that heat transfer enhancement in staggered arrangements is more efficient than inline arrangements. The channel height effect on the heat flow over the dimpled surfaces was studied by Moon [7].

The heat transfer coefficient and friction factors were computationally explored in rectangular channels with dimples on one wall. Pisal and Ranaware[8] have experimented to determine whether dimples on a heat sink fin can enhance heat transfer for laminar airflow. It was carried out using two distinct types of circular dimples (spherical) and oval dimples (elliptical). These dimples were located on both sides of the copper plate; it had a relative pitch of S/D=1.20, and a relative depth of δ/D=0.2. The Nusselt number and overall heat transfer coefficient are calculated for these configurations. Between the 600 and 2000 scale, heat transfer improvements were observed. Katkhaw et al. [9] analyzed the flat surface with an outer flow ellipsoidal dimple with ten types of different arrangements and dimple intervals. For such a dimpled surface, the heat transfer was measured and contrasted with a smooth surface's heat transfer. Results revealed that for staggered arrangement, the heat transfer coefficients were 15.8% more than that for a smooth surface and 21.7% more for inline arrangement. By the transient wide band liquid crystal process, an eight-by-eight jet array will affect a staggering array of dimples at Reynolds number 11,500 as investigated by Kanokjaruvijit Martinez-Botas[10]. Two dimple geometries were tested for hemispherical and cusped elliptical shapes. The study of dimple geometry's effect showed that hemispherical and cusped elliptical dimples did not behave differently significantly. However, compared to the economy, production, and loss of pressure, the hemispheric type should be preferred. Various similar studies can be found in [11-29].

From the literature, it is clear that dimples play a significant role in heat transfer enhancement. However, the unsteady heat transfer analysis for different dimple orientation and material is scarce; this is the primary motivation behind this paper. The main objectives were to find the cooling curve, rate of convective cooling at any instant, and the total amount of heat transfer at a given time. Additionally, the effect of dimple orientation on the heat transfer rate, the cooling medium's influence on the heat transfer rate, and the heat transfer coefficients for different cooling mediums and different dimple orientations were also discussed.).

2. EXPERIMENTAL METHODOLOGY

2.1. Experimental setup

The apparatus is a large insulated water bath with a volume of approximately 30 liters (Fig 4.1). A 3 kW electric heater powered by a thermostat to achieve a constant bath temperature is at the bath's bottom. A rotary switch located in the front of the bath regulates the water temperature. The water bath cover assembly is designed to allow the test specimen to be quickly placed into the bath while maintaining the flow conditions. The test sample is connected to the carrier assembly, and for uniform heat transfer through the water bath, the water bath is stirred continuously. The thermocouple labeled T1 indicates the brass temperature, and the thermocouple labeled T2 indicates the water bath temperature.

2.2. Preparation of test specimen

Brass is the material considered for testing, and the dimples have been machined with the assistance of CNC end milling on the surface of the material. The dimples were machined with a diameter of 4mm and a total depth of 4mm in which 2mm depth has a cylindrical profile, and the remaining 2mm has a concave profile. Two types of orientation were considered to study the influence of dimple orientation on the transient properties, which are inline configuration and circumferentially Staggered orientation. The inline configuration has the dimples located at 90° apart with a spacing of 10cm along the length (as shown in Fig 2). Circumferentially staggered orientation has the dimples located 90° apart but are provided with an offset along each corner's length (Fig 4.3).

Fig 1 Unsteady state test rig

Fig 2. Inline configuration

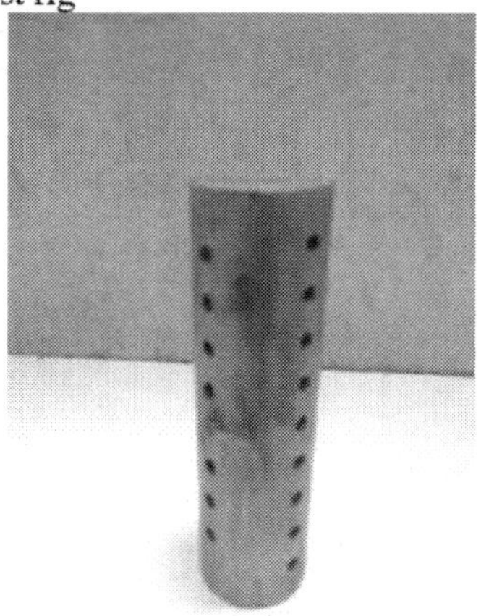

Fig 3. Circumferentially Staggered

3. RESULTS AND DISCUSSION

The study presents the experimental results of a cylindrical brass specimen with dimples machined with two configurations: 1) Inline configuration and 2) Circumferentially staggered configuration. The study presents the cooling curve, rate of convective cooling at any instant of time, the total amount of heat transfer at a given time. It also provides the effect of dimple orientation on the heat transfer rate, the cooling medium's influence on the heat transfer rate, and the heat transfer coefficients for different cooling mediums and different dimple orientation.

The following graph (Fig. 4(a)and 4(b)) illustrates the decay rate in temperature and heat transfer coefficient with time. The decay rate is higher in the dimpled rod than the ordinary rod in air cooling. We can also notice that the better results are obtained for inline when compared to the random arrangement.

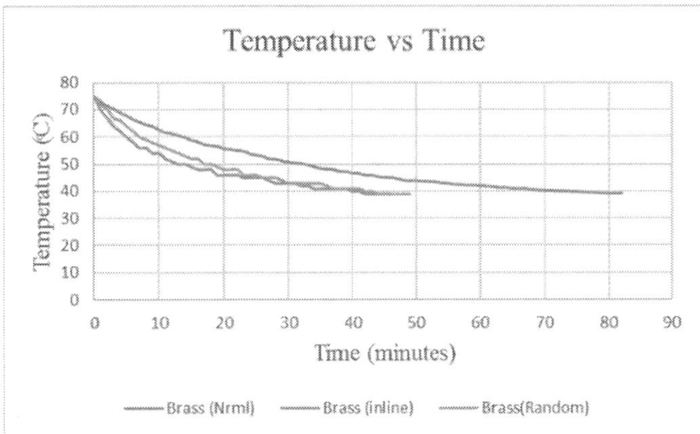

Fig 4 a Temperature vs. Time plot for Brass with air cooling

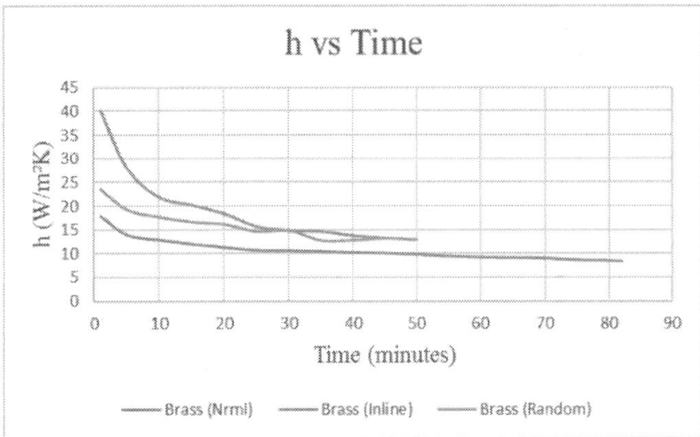

Fig 4 b. Heat transfer coefficient vs. Time plot for Brass with air cooling

Figures 5(a) and 5(b) illustrate the change in temperature and heat transfer coefficient when the cooling medium is water. It can be seen that the decay in temperature is almost the same for both the dimpled orientations. However, significant heat transfer enhancement can be seen in the dimpled surface compared to the ordinary rod.

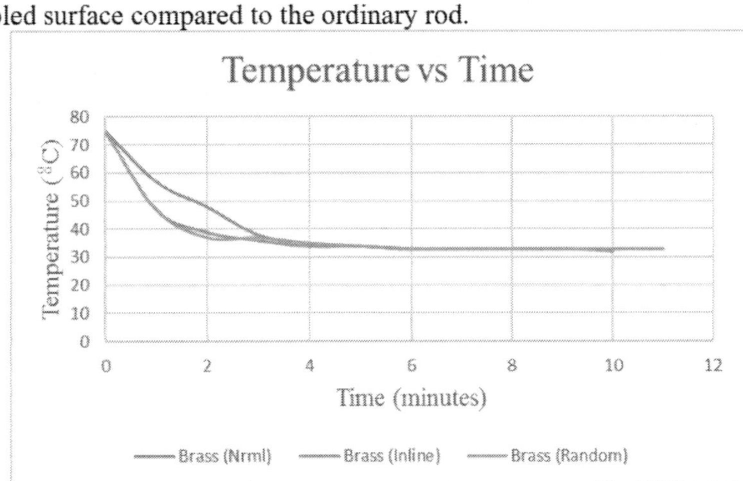

Figure 5 a. Temperature vs. Time plot for Brass with water cooling

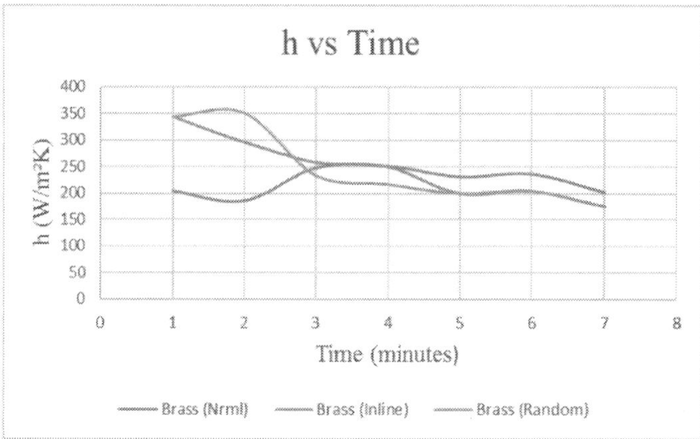

Figure 5 b Heat transfer coefficient vs. Time plot for Brass with water cooling

The following graph (Fig. 6(a) and 6(b)) shows the variation of temperature and heat transfer coefficient with time for a radiator coolant. It illustrates that there was no significant heat transfer enhancement when the cooling medium was a radiator coolant. It is because of the reason that a coolant circulation system is required to remove heat effectively.

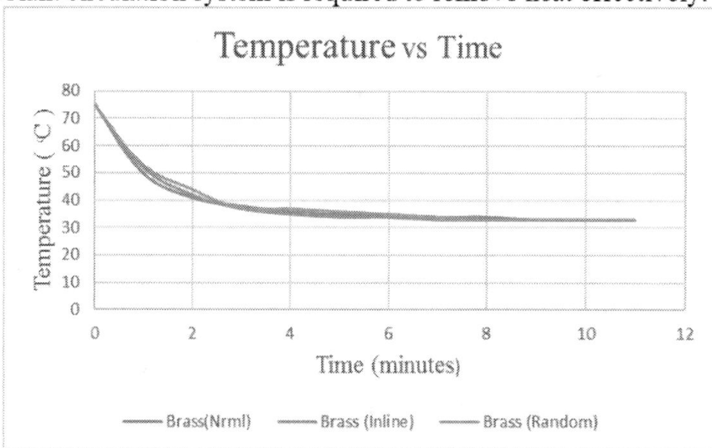

Fig 6 a Temperature vs. Time plot for Brass with coolant cooling

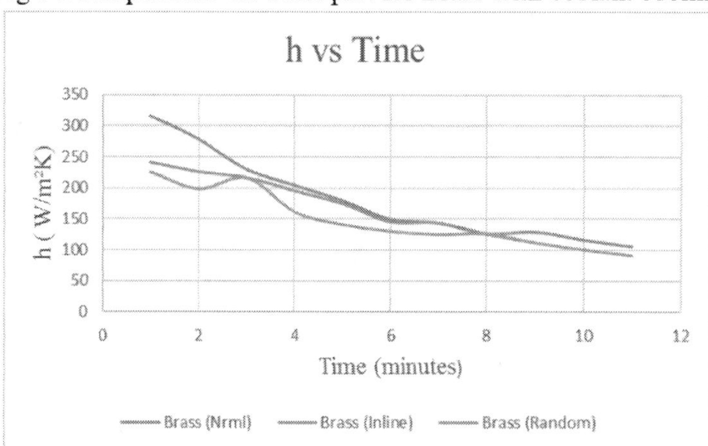

Fig 6 b Heat transfer coefficient vs. Time plot for Brass with coolant cooling

4. CONCLUSIONS

From the results obtained, we can conclude that there was considerable heat transfer enhancement with dimples' introduction. Heat transfer enhancement concerning dimples' inline configuration was more than the staggered arrangement of dimples. The convective cooling and total heat transfer rate also increased after the introduction of dimples for air cooling. When the cooling medium was water, there was no distinction between the heat transfer from specimens with inline and staggered configuration. Surprisingly, when a coolant was used, there was probably no heat transfer enhancement seen.

References

[1] Kuethe A. M., Boundary Layer Control of Flow Separation and Heat Exchange. US Patent No. 1191,1970.

[2] Afanasyev, V.N., Chudnovsky, Y.P., Leontiev, A.I., and Roganov, P.S, Turbulent flow friction and heat transfer characteristics for spherical cavities on a flat plate. Experimental Thermal Fluid Science, Vol.7, Issue 1, pp. 1–8, 1993.

[3] Chyu, M.K., Yu, Y., Ding, H., Downs, J.P., and Soechting, F.O., Concavity enhanced heat transfer in an internal cooling passage In Orlando international Gs Turbine & Aero engine Congress & Exhibition, Proceedings of the 1997(ASME paper 97-GT-437), 1997.

[4] Mahmood, G.I., Hill, M.L., Nelson, D.L., Ligrani, P.M., Moon, H.K. and Glezer.B., Local heat transfer and flow structure on and above a dimpled surface in a channel. J Turbomach, Vol.123, Issue 1, pp: 115–23, 2001.

[5] Dewan, A., Mahanta, P., Raju,K.S., and Kumar.P.S.,Review of passive heat transfer augmentation techniques. Proc. Instn Mech. Engrs, Part A: J. Power and Energy, Vol. 218, pp. 509–527, 2004.

[6] Liu S, and Sakr ,M.A., comprehensive review on passive heat transfer enhancements in pipe exchangers. Renewable and Sustainable Energy Reviews, Vol.19 pp. 64–81, 2013.

[7] Hasibur Rahman Sardar, Abdul Razak Kaladgi. Forced Convection Heat Transfer Analysis through Dimpled Surfaces with Different Arrangements. American Journal of Energy Engineering. Vol. 3, No. 3, 2015, pp. 37-45.

[8] Hasibur Rahman Sardar, Abdul Razak Kaladgi. Forced Convection Heat Transfer Analysis through Dimpled Surfaces with Different Arrangements. American Journal of Energy Engineering. Vol. 3, No. 3, 2015, pp. 37-45.

[9] Katkhaw, N., Vorayos, N., Kiatsiriroat, T., Khunatorn, Y., Bunturat, D. and Nuntaphan, A., 2014. Heat transfer behavior of flat plate having 45 ellipsoidal dimpled surfaces. Case Studies in Thermal Engineering, 2, pp.67-74.

[10] Kanokjaruvijit, K. and Martinez-botas, R.F., 2005. Jet impingement on a dimpled surface with different crossflow schemes. International journal of heat and mass transfer, 48(1), pp.161-170.

[11] Afzal A, Mohammed Samee AD, Abdul Razak RK, Ramis MK. Thermal management of modern electric vehicle battery systems (MEVBS). Journal of Thermal Analysis and Calorimetry. 2020 Apr 3:1-5.

[12] Afzal, A., Mohammed Samee, A. M., Razak, R. A., & Ramis, M. K. (2019). Effect of spacing on thermal performance characteristics of Li-ion battery cells. Journal of Thermal Analysis and Calorimetry, 135(3), 1797-1811.

[13] Razak RA, Afzal A, Samee AM, Ramis MK. ''Effect of cladding on thermal behaviour of nuclear fuel element with non-uniform heat generation''. Progress in Nuclear Energy. 2019 Mar 1; 111:1-14.

[14] Afzal A, Samee AM, Razak RA, Ramis MK. 'Steady and Transient State Analyses on Conjugate Laminar Forced Convection Heat Transfer''. Archives of Computational Methods in Engineering (Springer, Impact factor: 7.605). ISSN: 1134-3060,

26.1,2019.10.1007/s11831-018-09303-x.

[15] Afzal, Asif, AD Mohammed Samee, RK Abdul Razak, Sher Afghan Khan, and Hurmathulla Khan. "Optimum spacing between grooved tubes: An experimental study." Journal of Mechanical Science and Technology 34, no. 1 (2020): 469-475.

[16] Samee, A.M., Afzal, A., Ramis, M.K. and Razak, R.A., 2019. Optimal spacing in heat generating parallel plate channel: A conjugate approach. International Journal of Thermal Sciences, 136, pp.267-277.

[17] Kareemulla, Mohammed, K. M. Chethan, Abdul R. Kaladgi, and K. M. Yashawantha. "Heat Transfer Analysis of Shell and Tube Heat Exchanger Cooled Using Nanofluids." Recent Patents on Mechanical Engineering 12, no. 4 (2019): 350-356.

[18] Afzal, Asif, Mohammed AD Samee, and Abdul RK Razak. "Experimental Investigation of Thermal Performance of Engine Coolant Oil and Water in Helical Coil Heat Exchanger." JOURNAL OF ENGINEERING RESEARCH 7.2 (2019): 333-351.

[19] Kaladgi, Abdul Razak, Asif Afzal and M. K. Ramis. "Investigation of dimensionless parameters and geometry effects on heat-transfer characteristics of liquid sodium flowing over a flat plate." Heat Transfer—Asian Research (2018).

[20] Razak, RK Abdul, Samee Mohammed, and M. K. Ramis. "Numerical investigation of Prandtl number effect on heat transfer and fluid flow characteristics of a nuclear fuel element." Nuclear Energy and Technology 3.2 (2017): 81-91.

[21] Kaladgi, Abdul Razak, AD Mohammed Samee, and M. K. Ramis. "Influence of Prandtl number on heat transfer of a flat vertical plate." International Journal of Nuclear Energy Science and Technology 11.3 (2017): 272-290.

[22] Kaladgi, Abdul Razak, AD Mohammed Samee, and M. K. Ramis. "Thermal performance characteristics of liquid sodium as a coolant flowing past a rectangular nuclear fuel element." International Journal of Nuclear Energy Science and Technology, 10.1 (2016): 11-27.

[23] Mohammad Samee A.D., Afzal A., Kaladgi Razak A.R., Ramis M.K. (2019). Temperature and location of hot spots variation with spacing in a vertical parallel plate channel: conjugate view, International Journal of Heat and Technology, Vol. 37, No. 1, pp. 153-160.

[24] Samee AD, Mohammad, Asif Afzal, and Abdul Razak RK. "Effect of Prandtl number on the average exit temperature of coolant in a heat-generating vertical parallel plate channel: A conjugate analysis." Heat Transfer—Asian Research 47.4 (2018): 603-619.

[25] Afzal, Asif, A. D. Mohammed Samee and Abdul Razak. "Comparative thermal performance analysis of water, engine coolant oil and MWCNT-W nanofluid in a radiator." (2018).

[26] Afzal, A., Samee, A. D., Razak, R. K., & Ramis, M. K. (2018). Heat transfer characteristics of MWCNT nanofluid in rectangular mini channels. International Journal of Heat and Technology, 36(1), 222-228.

[27] Abdul Razak. R.K, Mohammed Samee, and Ramis M. K, "A Study on Critical Radius and Crossover Radius of Insulation for Various Heat Transfer Problems," American Journal of Heat Transfer, Vol. 1 No. 3, pp. 149-156, 2014.

[28] Faheem Akhtar, Abdul Razak R Kaladgi and Mohammed Samee, Heat transfer enhancement using dimple surfaces under natural convection—an experimental study, Int. J.Mech. Eng. & Rob. Res. Vol 4,Issue 1 ,pp 173-175,2015.

[29] Faheem Akhtar, Abdul Razak R Kaladgi and Mohammed Samee, Heat transfer augmentation using dimples in forced convection -an experimental approach. Int. J. Mech. Eng. &Rob. Res. Vol 4,Issue 1 ,pp 150-153,2015.

Experimental investigation of mechanical properties of Acrylonitrile Butadiene Styrene (ABS) based polymer for Submersible pumps

Geetanjali S. Guggari[1]*, S Shivakumar[1], R Nikhil [1], G A Manjunath [1], Asif Afzal[2], C. Ahamed Saleel[3]*

[1] KLS Gogte Institute of Technology, Belagavi, 590008, India.
[2] Department of Mechanical Engineering, P. A. College of Engineering (Affiliated to Visvesvaraya Technological University, Belagavi), Mangaluru 574153, India.
[3] College of Engineering, King Khalid University, Abha, 61421, Kingdom of Saudi Arabia.

E-mail: ahamedsaleel@gmail.com, gsguggari@git.edu

Abstract. Now a day's many companies manufacturing the most advanced submersible pumps with Acrylonitrile Butadiene Styrene (ABS) based polymer, which is available in the market for use in domestic, commercial and even research activities. ABS based pumps are widely used in mining, construction and chemical industries. Acrylonitrile Butadiene Styrenehasmuch application in the temperature range from −20 to 80°C. The ABS based polymer composites are used in submergible pumps due their important properties such as corrosion resistance, light weight, erosion resistance, etc. In the current work, the technique used for the fabrication of GFRP- ABS laminateis compression molding, as per the ASTM standard. The various tests are conducted as per Design of Experiments. Analysis of experimental data was performed using Minitab software and tests were conducted for Flexural strength, Inter Laminar Shear Stress, Tensile strength. Three factors areconsidered for the Design of Experiments namely, fibre-to-resin ratio, fibre orientation and period of immersion in artificial sea water. The outcomepoints towards Flexural strength; ILSS and Tensile Strength areat highest along 40:60 fibre-to-resin ratio with 0°/90°fiber orientation, longer the contact to sea water more is the fall in the strength of the composite material and also due to sea water diffusing into the composite material specimen, there is an increase in its weight.

1. Introduction

Research of ABS Plastic has gained importance from several decades for its significant existence in practical applications and has increased demand nowadays. Adding glass fibre-reinforced composites has much higher strength than ABS material. Several researchers explored explanations for improving the properties of mechanical. Along with clay, the inclusion of glass increases the strength of ABS composites [1]. Better tensile and flexural properties are provided by the use of ABS/PC blend; these properties can be further improved by the addition of Kevlar fibres as reinforcement [2].
The tensile properties of ABS can also be strengthened in the form of short glass fibre/calcite fibre hybrid composites reinforced by ABS[3]. In the case of particle-reinforced ABS composites, the particle size influences the mechanical properties; in a study by L Jiang et al.[4], micron-sized calcium carbonate strengthened the module but decreased the tensile and impact strength, whereas nano-sized

calcium carbonate displayed an improvement in impact properties along with the modulus. Addition of short glass fibre to ABS lowers the fracture toughness, whereas, addition of short glass fibres to calcite-filled ABS improves the fracture toughness [5]. Addition of a very small content 1 wt% of multi-walled carbon nano tubes improves the yield strength and flexural modulus of polyamide6/ABS blend [6]. Impact strength can be improved by sandwich structural composites; a study showed improved impact properties in CFRP/ABS/CFRP (carbon fibre-reinforced epoxy polymer) composites as compared to ABS [7]. Adding sawdust as reinforcement to ABS improved its modulus with a decrease in strength [8]. Waste textile fibres can also be put to use as a reinforcing material for ABS, this improves flexural properties of ABS [9]. Addition of thermal exfoliated graphite as reinforcement enhances electrical conductivity of ABS [10]. Research of ABS Plastic has gained importance from several decades for its significant existence in practical applications and has increased demand nowadays. Adding glass fibre-reinforced composites has much higher strength than ABS material. Many researchers investigated reasons for improving mechanical properties. Addition of glass along with clay increases the toughness of ABS composites [1]. Using ABS/PC blend gives better tensile and flexural properties; these properties can be further enhanced by the addition of Kevlar fibres as reinforcement [2]. The tensile properties of ABS can also improved in the form of hybrid composites of short glass fibre/calcite fibre reinforced be ABS [3]. In case of particle reinforced ABS composites, size of the particles affects the mechanical properties; micron-sized calcium carbonate improved the modulus but lowered tensile and impact strength, whereas, nano-sized calcium carbonate showed increase in impact properties along with modulus in a study by L Jiang et al. [4]. Addition of short glass fibre to ABS lowers the fracture toughness, whereas, addition of short glass fibres to calcite-filled ABS improves the fracture toughness [5]. Addition of a very small content 1 wt% of multi-walled carbon nano tubes improves the yield strength and flexural modulus of polyamide6/ABS blend [6]. Impact strength can be improved by sandwich structural composites; a study showed improved impact properties in CFRP/ABS/CFRP (carbon fibre-reinforced epoxy polymer) composites as compared to ABS [7]. Adding sawdust as reinforcement to ABS improved its modulus with a decrease in strength [8]. Waste textile fibres can also be put to use as a reinforcing material for ABS, this improves flexural properties of ABS [9]. Addition of thermal exfoliated graphite as reinforcement enhances electrical conductivity of ABS [10-38].

2. Methodology

The Figure 1 Describes the methodology followed for mechanical and abrasive studies respectively to complete the research work, and various tools that were employed at various stages of research.

Fabrication of test specimen

The sheet-shaped test specimens are cut according to the ASTM (ASTM-D638), ILSS (ASTM-D2344), impact test (ASTM-D256), flexural test (ASTM-D790) and flexural fatigue test (ASTM D671-93) specifications, respectively.

Preparing artificial sea water

In a temperature-controlled tank, artificial sea water is formulated according to the ASTM D1141 standard. In order to treat artificial sea water, 250 litres of purified water was used. The amounts of sodium chloride, calcium chloride, magnesium chloride, sodium bicarbonate and sodium carbonate weighted precisely were applied to the purified water in the quantities referred to in Table 1.The artificial sea water was prepared in temperature controlled sea water as shown in Figure 2.

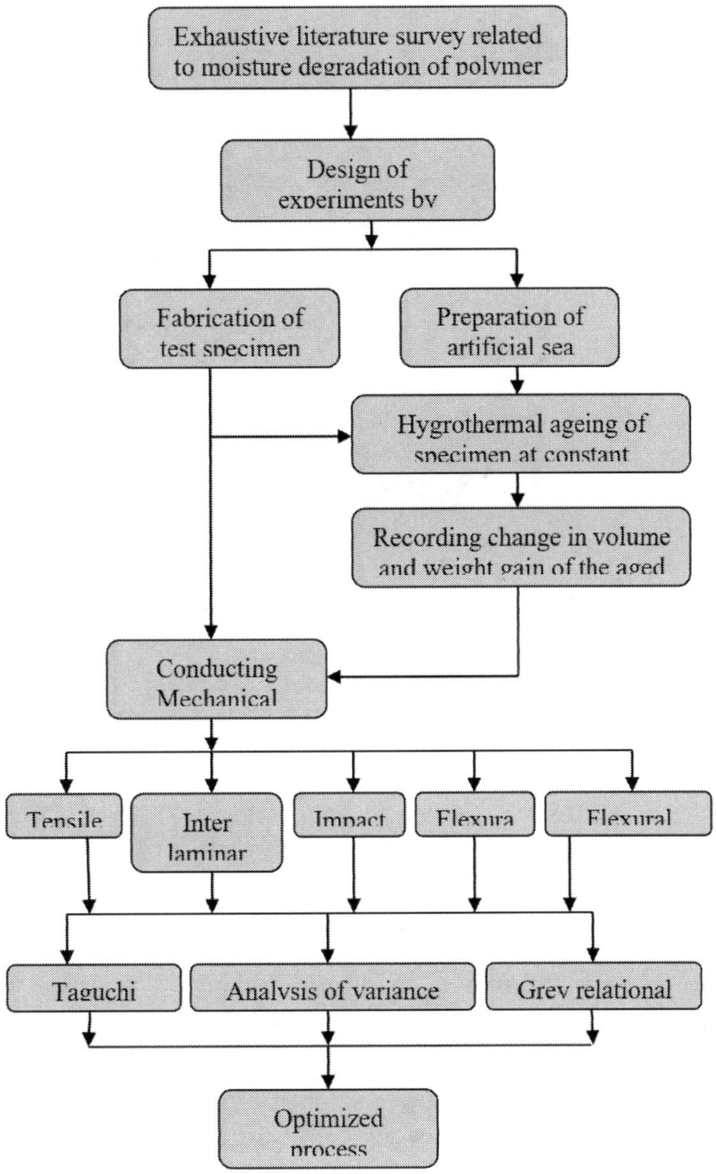

Figure 1. Flowchart for the project methodology (Mechanical properties)

Design of Experiments (DoE)

Minitab V16 software is use to define the DoE with mixed level taguchi orthogonal array. The array was developed using three factors viz.. Period of immersion in artificial sea water with 6 levels (15, 30, 45, 60, 75 and 90 days), fibre-to-resin ratio with 3 levels (60:40, 50:50 and 40:60) and fibre orientation with 3 levels (0o/90o, $45^0/45^0$ and $30^0/60^0$).

Figure 2. Temperature controlled tank

Table1. Composition of artificial **sea** water

Sl. No.	Constituent	Quantity (grams)
1	NaCl	7597.59
2	$CaCl_2$	401.362
3	$MgCl_2$	2907.346
4	$NaHCO_3$	43.68
5	$Na2CO_3$	5.512

3. Results and Discussion

The developed specimens were fabricated for various testing as per their applications. The various testing conducted as per design of experiments. The obtained results are given in this section. Analysis of experimental data was performed using Minitab software and two tests were performed for each factor, i.e. Taguchi analysis and Analysis of Variance (ANOVA) analysis. The factors analysed were, Flexural strength, ILSS, Tensile strength, Impact strength, Swelling and % wt gain and abrasive resistance of the ABS and its composites. The functional properties such as damping, XRD, fracture surface, phase transformation, FTIR and flammability test were presented.

Flexural strength

3-pt bending strength values from the flexural tests were considered for this analysis.

Taguchi analysis

The mean of means and mean of signal-to-noise ratio plots, as shown in, point out highest strength for sample with 40:60 fibre-to-resin ratios with 0o/90o fibre orientation. In addition, longer the contact with artificial sea water further is the lessening in the strength of the composite.

Figure 3 shows weighted mean plots for multiple flexural strength variables. It can be found that the flexural intensity represents a continuously declining tendency as specimen immersion time increases in sea water. For the rising fibre-to-resin ratio, a lessening pattern is seen for flexure strength. And the overall flexural power of the fibre orientation of 0o/90o is followed by 30o/60o & 45o/45o.

The Figure 4 displays the mean of signal to noise ratio plots for diver separameters for flexural. It is noted that the flexure indicates a persistent decreasing pattern with a rise in the time of composite specimen immersion in seawater artificial. A decreasing tendency for the rising fiber-to-resin ratio of flexural intensity is observed. And the 0o/90o fiber orientation displays optimum flexural power, followed by 30o/60o and 45o/45o

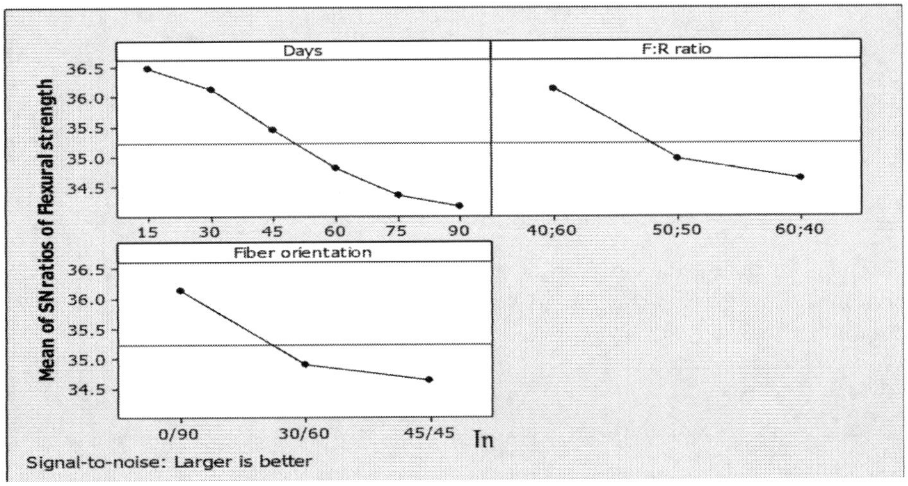

Figure 3. Mean of Signal to noise ratio for Flexural strength

ANOVA for Flexural strength

As seen in Figure 4, residual plots from the ANOVA for flexural intensity are seen. ANOVA is carried out on the basis of a 95% degree of trust.

In Table 2, it is seen that-ratio is highest for fiber orientation with a involvement of 23%, making it the most significant factor followed by fiber to resin ratio and period of immersion in artificial sea water.

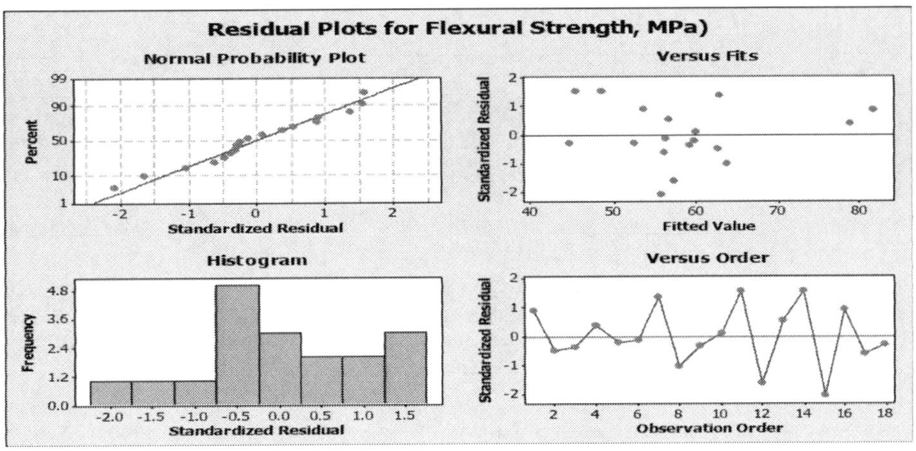

Figure 4. Residual plots for Flexural strength

Table2. Analysis of variance for Flexural strength

Source	DF	Seq SS	Adj SS	Adj MS	F	P	% C
Days	5	678.06	678.06	135.61	2.03	0.178	32.1
F:R ratio	2	412.92	412.92	206.46	3.09	0.101	19.5
Fibre orientation , deg	2	488.83	488.83	244.41	3.66	0.074	23.1
Error	8	534.07	534.07	66.76			25.3
Total	17	2113.88					

Interactions plot for the Flexural strength is as shown in the Figure 5. The curves are crossing each other and it indicates a good interaction between the 3 factors, i.e. period of immersion, fibre-to-resin ratio and fibre orientation.

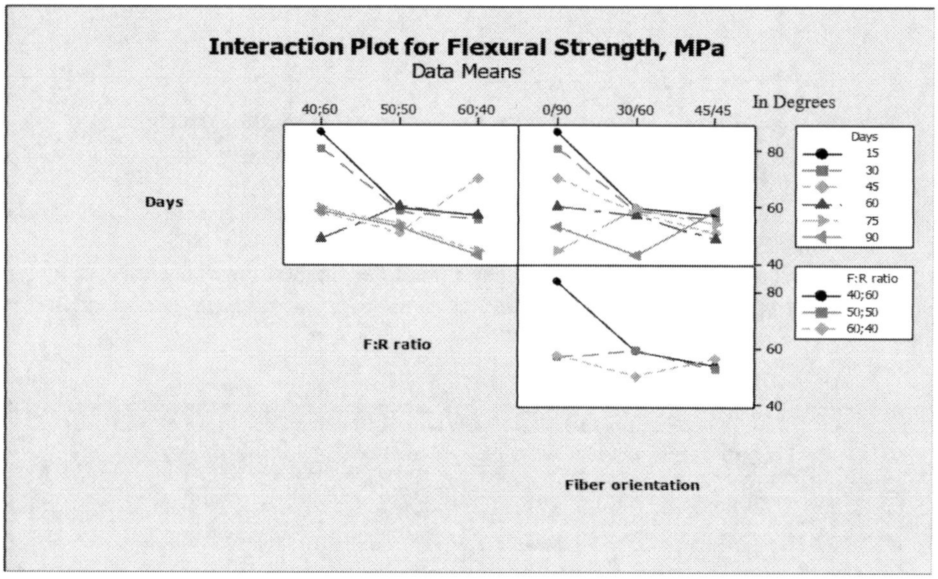

Figure 5. Interactions plot for flexural strength

Inter-laminar shear strength (ILSS) property

Inter-laminar shear strength is the factor under analysis in this test. This test determines the shear stress bearing capability of the composites between the laminas.

Taguchi analysis

Looking at the following mean of means and mean of signal-to-ratio plots, as shown in Figure 5 and Figure 6 respectively, it can be said that maximum strength for composites with least period of exposure, i.e. 15 days and 40:60 fibre to resin ratio with a fibre orientation of $0^0/90^0$.

Figure 6 stand for mean of means plots for different factors for ILSS. It is observed that ILSS shows a continuously decreasing trend with increase in the period of immersion of composite specimen in the artificial sea water. Proper trend cannot be concluded for the increasing fibre-to-resin ratio from the above plot. The fibre orientation of 0o/90o exhibit highest ILSS followed by $30^0/60^0$ and $45^0/45^0$.

Figure 7 shows the different factors for ILSS, reflect mean of means plots. It is found that ILSS displays a steadily declining pattern with a rise in the immersion time of the composite specimen in the seawater artificial. For the increasing fibre-to-resin ratio of the above plot, a proper pattern cannot be inferred. The $0^0/90^0$ fiber orientation displays the largest ILSS, followed by $30^0/60^0$ and $45^0/45^0$.

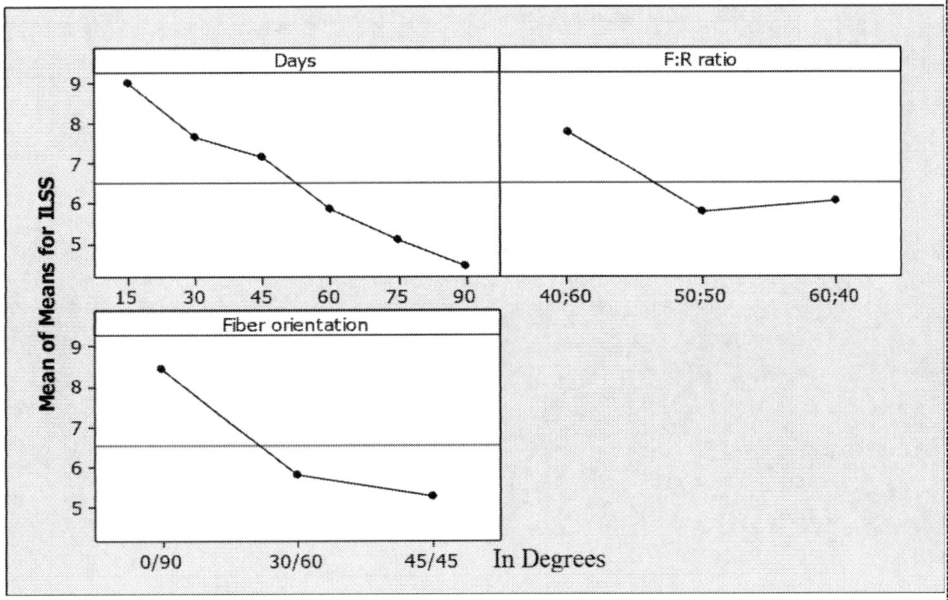

Figure 6. Mean of means for inter laminar shear stress

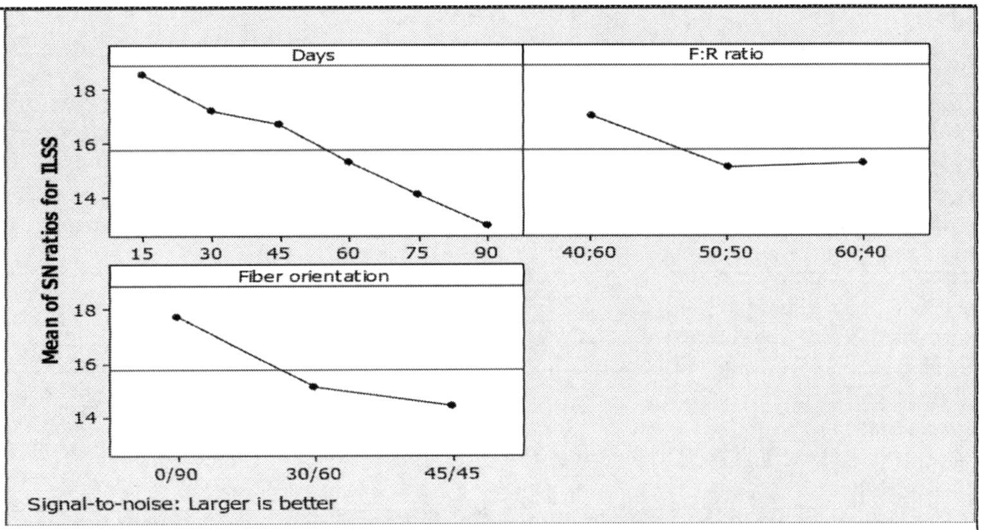

Figure 7. Mean of signal-to-noise ratio for inter laminar shear stress

IOP Conf. Series: Materials Science and Engineering 1065 (2021) 012035 doi:10.1088/1757-899X/1065/1/012035

ANOVA for ILSS

Residual plots from the ANOVA for ILSS are as shown in Fig. ANOVA is performed based on 95% level of confidence.

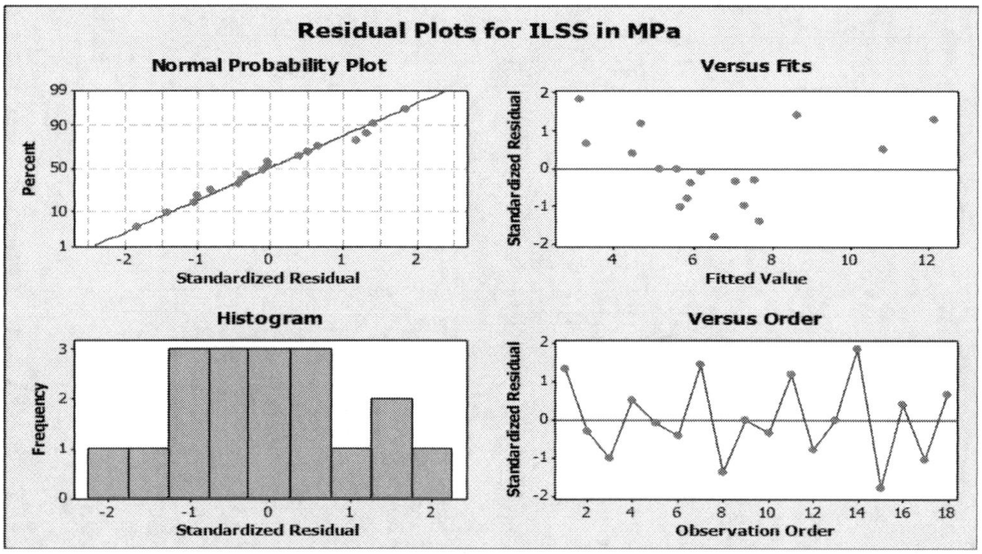

Figure 8. Residual plots for inter laminar shear stress

In Table 3 F-ratio is maximum for fibre orientation with a contribution of 30%, making it the most significant factor fol lowed by time of immersion in sea water and fibre to resin ratio.

Table3. Analysis of variance for inter laminar shear stress

Source	DF	Seq SS	Adj SS	Adj MS	F	P	% C
Days	5	44.142	44.142	8.828	3.12	0.074	38.1
F:R ratio	2	13.831	13.831	6.915	2.44	0.148	11.9
Fibre orientation , degree	2	35.360	35.360	17.680	6.25	0.023	30.5
Error	8	22.638	22.638	2.830			19.5
Total	17	115.972					

Interactions plot between different factors is as shown in Figure 9. The curves are crossing each other and it indicates good interaction between the 3 factors, i.e. period of immersion, fibre-to-resin ratio and fibre orientation.

Tensile strength

The ultimate tensile strength (UTS) values from tensile tests are analysed in this section.

Taguchi analysis

As shown in Figure, the following mean of means and mean of signal-to-noise ratio plots show that optimum intensity is shown by composite specimens with a fiber-to-resin ratio of 40:60, a fiber orientation of $0^0/90^0$ and 15 days of interaction with sea water.

IOP Conf. Series: Materials Science and Engineering 1065 (2021) 012035 doi:10.1088/1757-899X/1065/1/012035

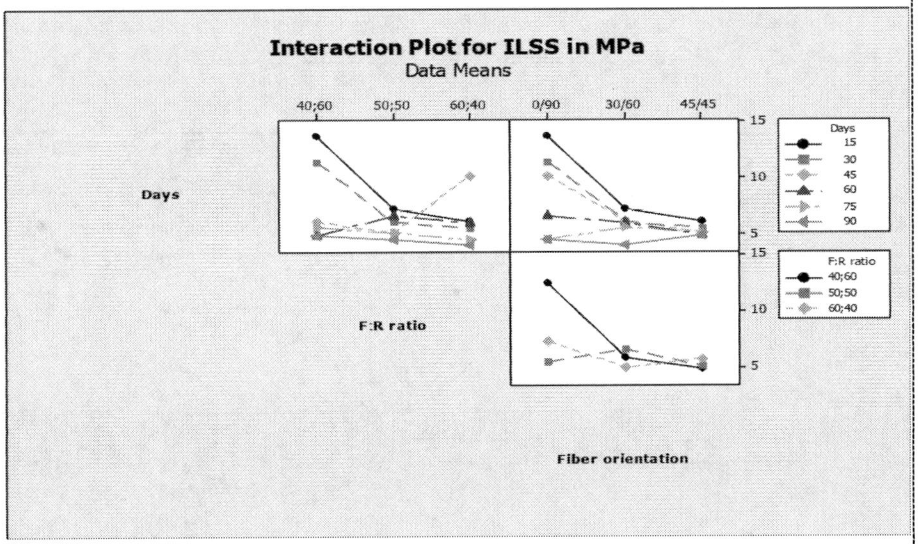

Figure 9. Interactions plot for inter laminar shear stress

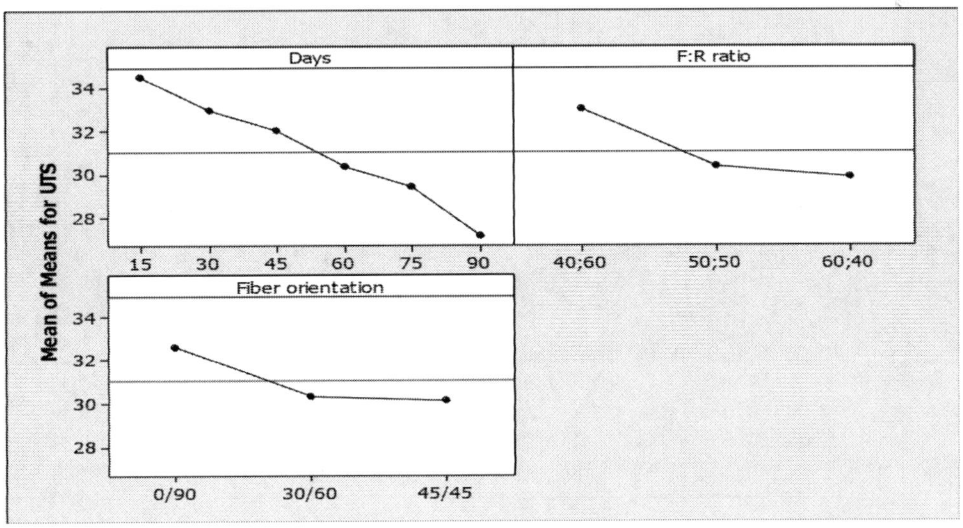

Figure 10. Mean of means for Ultimate tensile strength

The Figure 10 shows average mean graphs for various Ultimate Tensile strength variables. Tensile intensity is found to demonstrate a steadily decreasing pattern with a rise in the time of sample immersion in the sea water. A downward trend from the above plot for the rising fibre-to-resin ratio can be inferred. The $0^0/90^0$ fiber orientation displays the strongest UTS, followed by $30^0/60^0$ and $45^0/45^0$.

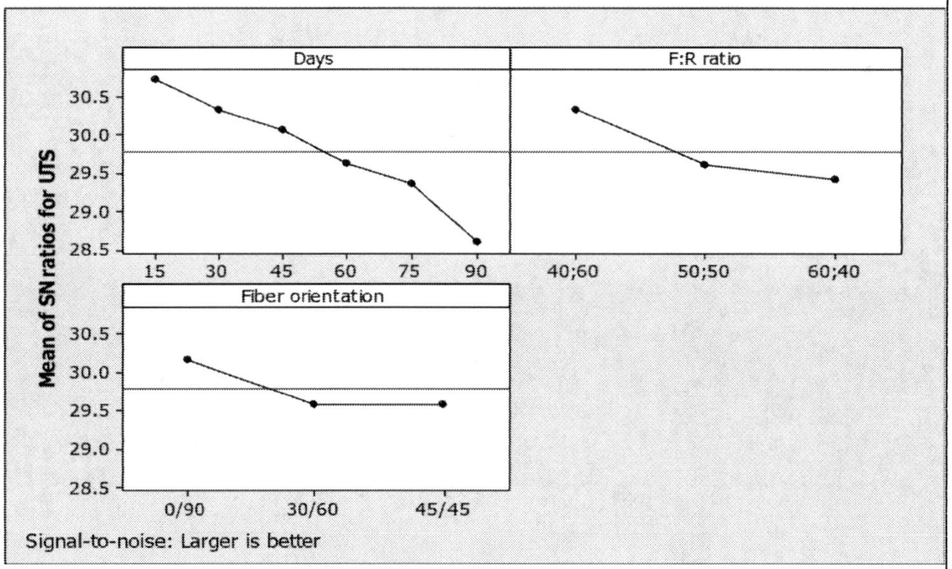

Figure 11. Mean of signal-to-noise ratio for ultimate tensile strength

Table4. Grey-Taguchi analysis

Process parameters				Grey Relational co-efficient								
Expt. No.	**Days**	**F:R ratio**	**Fibre orientation**	**Flexural strength**	**ILSS**	**Ultimate tensile strength**	**Impact strength**	**Swelling**	**Weight**	**Fatigue**	**Grey relational grade**	**Rank**
1	15	40:60	0/90	1.000	1.000	1.000	1.000	1.000	1.000	1.000	1.000	1
2	15	50:50	30/60	0.449	0.427	0.634	0.506	0.778	0.841	0.437	0.582	4
3	15	60:40	45/45	0.424	0.391	0.546	0.438	0.498	0.586	0.479	0.480	9
4	30	40:60	0/90	0.796	0.681	0.807	0.585	0.906	0.806	0.406	0.712	2
5	30	50:50	30/60	0.436	0.390	0.546	0.485	0.647	0.624	0.634	0.537	5
6	30	60:40	45/45	0.412	0.371	0.500	0.430	0.435	0.513	0.528	0.456	11
7	45	60:40	0/90	0.572	0.583	0.807	0.547	0.794	0.667	0.533	0.643	3
8	45	40:60	30/60	0.433	0.389	0.504	0.430	0.526	0.532	0.648	0.495	8
9	45	50:50	45/45	0.378	0.362	0.449	0.379	0.387	0.430	0.531	0.416	16
10	60	50:50	0/90	0.454	0.407	0.592	0.518	0.703	0.630	0.347	0.522	6
11	60	60:40	30/60	0.423	0.386	0.461	0.425	0.489	0.446	0.479	0.444	12
12	60	40:60	45/45	0.363	0.356	0.436	0.374	0.362	0.411	0.629	0.419	15
13	75	40:60	30/60	0.446	0.374	0.577	0.438	0.550	0.586	0.506	0.497	7
14	75	50:50	45/45	0.400	0.362	0.436	0.383	0.446	0.433	0.481	0.420	14
15	75	60:40	0/90	0.340	0.345	0.399	0.374	0.343	0.356	0.787	0.421	13
16	90	40:60	45/45	0.437	0.355	0.550	0.428	0.515	0.492	0.508	0.469	10
17	90	50:50	0/90	0.392	0.344	0.372	0.378	0.417	0.369	0.417	0.384	17
18	90	60:40	30/60	0.333	0.333	0.333	0.333	0.333	0.333	0.385	0.341	18

4. Conclusion

The highest grade of Grey relational rank shows that the results obtained experimentally are closer to idyllic value. In other words, the larger the Grey relational rank, the better will be the multiple performance characteristics. Therefore, experiment 1 in the Table 5hshows the highest Grey relational grading indicating that the parameter of 15 days hygrothermal ageing, 40:60 fibre to resin ratio and $0^0/90^0$ fibre orientation in the orthogonal array has the best multi performance characteristics among 18 experiments.

5. References

[1] Basurto F C, García-López D, Villarreal-Bastardo N, Merino J C and Pastor J M 2013 Composites and nanocomposites of ABS: Synergy between glass fibre and nano-sepiolite,

Composites **50** 42-47.

[2] SarawutRimdusit, ParkpoomLorjia, KuljiraSujirote and SunanTiptipakorn Physical and Mechanical Characteristics of Kevlar Fibre-reinforced PC/ABS Composites 2012 *Engineering Journal* **16** 57-66.

[3] Shao-Yun Fu and Bernd Lauke 1998 Characterization of tensile behaviou of hybrid short glass fibre/calcite particle/ABS composites *Composites* **29A** 575-583.

[4] Jiang L, Lama Y C, Tam K C, Chua T H, Sim G W and Ang L S 2005 Strengthening acrylonitrile-butadiene-styrene(ABS) with nano-sized and micron-sized calcium carbonate *Polymer* **46** 243-252.

[5] Shao-Yun Fu and Bernd Lauke 1998 Fracture resistance of unfilled and calcite particle filled ABS composites reinforced by short glass fibres (SGF) under impact load *Composites: Part A* **29A** 631-641.

[6] Xi-Qiang Liu, Wei Yang , Bang-Hu Xie, Ming-Bo Yang 2012 Influence of multiwall carbon nanotubes on the morphology, melting, crystallization and mechanical properties of polyamide 6/acrylonitrile-butadiene-styrene blends *Materials and Design* **34**355-362.

[7] NaoyaTsuchikura, Michael C Faudree and Yoshitake Nishi 2013 Charpy Impact value of sandwich structural (CFRP/ABS/CFRP) composites constructed with carbon fibre reinforced epoxy polymer (CFRP) and acrylonitrile butadiene styrene (ABS) sheets separately irradiated by electron beam prior to lamination *Materials Transactions* **54(3)** 371-379.

[8] Chotirat L, Chaochanchaikul K and Sombatsompop N 2007 On adhesion mechanisms and interfacial strength in acrylonitrile butadiene styrene/wood sawdust composites *International Journal of Adhesion & Adhesives* **27** 669-678.

[9] Johnny N Martins, Tobias G Klohn, Otavio Bianchi, RudineiFiorio and EstevaoFreire 2010 Dynamic mechanical, thermal and morphological study of ABS/textile fibre composites *Polym Bull* **64** 497-510.

[10] BhardwajNeha, Manjula K S, Srinivasulu B and Shit C. Subhas 2012 Synthesis of exfoliated graphite/ABS composites *Open Journal of Organic Polymer Mterials* **2** 74-78.

[11] Jaymin R Desai, Shit S C, Shah M D and Jain S K 2013 Preparation and characterizations of acrylonitrile butadiene styrene (ABS)-cenosphere composites *Journal of information, knowledge and research in mechanical engineering* **2 (2)** 411-416.

[12] Chi-Yuan Huang, Wen-Wei Mo and Ming-Lih Roan 2004 The influence of heat treatment on electroless-nickel coated fibre (ENCF) on the mechanical properties and EMI shielding of ENCF reinforced ABS polymeric composites *Surface coating technology* **184** 123-132.

[13] Mohammed H Al-Saleh, Walaa H. Saadeh and UttandaramanSundararaj 2013 EMI shielding effectiveness of carbon based nanostructured polymeric materials: A comparative study *Carbon* **60** 146-156.

[14] Haiyun Ma, ZhongbinXu, Lifang Tong, AiguanGu and Zhengping Fang 2006 Studies of ABS-graft-maleic anhydride/clay nanocomposites:Morphologies, thermal stability and flammability properties *Polymer Degradation and Stability* **91** 2951-5959.

[15] Haiyun Ma, Jun Wang and Zhengping Fang 2012 Cross-linking of a novel reactive polymeric intumescent flame retardant to ABS copolymer and its flame retardancy properties *Polymer degradation and stability* **97** 1596-1605.

[16] Sudeepan J, Kumar K, Barman T K and Sahoo P 2013 Study of friction and wear properties of ABS/Kaolin polymer composites using grey relational *Polymer degradation and stability* **97** 1596-1605.

[17] Pinto R, Afzal A, Souza L D, Ansari Z and Mohammed Samee A D 2017 Computational Fluid Dynamics in Turbomachinery: A Review of State of the Art Arch. *Comput. Methods Eng* **24(3)** 467–479 doi: https://doi.org/10.1007/s1183.

[18] Afzal A, Ansari Z, Faizabadi A and Ramis M 2017 Parallelization strategies for computational fluid dynamics software: state of the art review *Arch. Comput. Methods Eng* **24(2)** 337–363 doi: https://doi.org/10.1007/s11831-016-9165-4.

[19] Afzal A, Nawfal I, Mahbubul I M and Kumbar S S 2019 An overview on the effect of ultrasonication duration on different properties of nanofluids *J. Therm. Anal. Calorim* **(135)** 393–418 doi: https://doi.org/10.1007/s10973-018-7144-8.

[20] Kumar R, Sesha V, Butt M M, Ahmed N, Khan S A and Afzal A 2019 Thermo-mechanical analysis and estimation of turbine blade tip clearance of a small gas turbine engine under transient operating conditions *Appl. Therm. Eng* **(179)** 115700 doi: 10.1016/j.applthermaleng.2020.115700.

[21] Afzal A, Imran Mokashi, Khan S A and Muhammad Azami bin H 2020 Optimization and analysis of maximum temperature in a battery pack affected by low to high Prandtl number coolants using response surface methodology and particle swarm optimization algorithm," *Numer. Heat Transf. Part A Appl* doi: https://doi.org/10.1080/10407782.2020.1845560.

[22] Mokashi I, Afghan S, Nur A, Abdullah, Hanafi B, Muhammad Azami, and Afzal A 2020 Maximum temperature analysis in a Li-ion battery pack cooled by different fluids *J. Therm. Anal. Calorim* **32** 1–17 doi: 10.1007/s10973-020-10063-9.

[23] Afzal A, Khan S A, Islam T, Jilte R D, Khan A, and Soudagar M E M 2020 Investigation and back-propagation modeling of base pressure at sonic and supersonic Mach numbers," *Physcics of Fluids* **32** 096109 doi: 10.1063/5.0022015.

[24] Afzal A 2020 Response surface analysis , clustering , and random forest regression of pressure in suddenly expanded high-speed aerodynamic flows *Aerosp. Sci. Technol* **107** 106318 doi: 10.1016/j.ast.2020.106318.

[25] R. Pinto, A. Afzal, L. D'Souza, Z. Ansari, and A. D. Mohammed Samee, "Computational Fluid Dynamics in Turbomachinery: A Review of State of the Art," Arch. Comput. Methods Eng., vol. 24, no. 3, pp. 467–479, 2017, doi: https://doi.org/10.1007/s1183.

[26] Afzal, Z. Ansari, A. Faizabadi, and M. Ramis, "Parallelization strategies for computational fluid dynamics software: state of the art review," Arch. Comput. Methods Eng., vol. 24, no. 2, pp. 337–363, 2017, doi: https://doi.org/10.1007/s11831-016-9165-4.

[27] R. Kumar, V. Sesha, M. M. Butt, N. Ahmed, S. A. Khan, and A. Afzal, "Thermo-mechanical analysis and estimation of turbine blade tip clearance of a small gas turbine engine under transient operating conditions," Appl. Therm. Eng., vol. 179, no. March, p. 115700, 2020, doi: 10.1016/j.applthermaleng.2020.115700.

[28] Afzal, I. Nawfal, I. M. Mahbubul, and S. S. Kumbar, "An overview on the effect of ultrasonication duration on different properties of nanofluids," J. Therm. Anal. Calorim., vol. 135, pp. 393–418, 2019, doi: https://doi.org/10.1007/s10973-018-7144-8.

[29] N. A. A. Asif Afzal, Imran Mokashi, S. A. Khan and H. Muhammad Azami bin, "Optimization and analysis of maximum temperature in a battery pack affected by low to high Prandtl number coolants using response surface methodology and particle swarm optimization algorithm," Numer. Heat Transf. Part A Appl., 2020, doi: https://doi.org/10.1080/10407782.2020.1845560.

[30] Mokashi, S. Afghan, A. Nur, Abdullah, B. Hanafi, Muhammad Azami, and A. Afzal, "Maximum temperature analysis in a Li ion battery pack cooled by different fluids," J. Therm. Anal. Calorim., pp. 1–17, 2020, doi: 10.1007/s10973-020-10063-9.

[31] Afzal et al., "Response surface analysis, clustering, and random forest regression of pressure in suddenly expanded high-speed aerodynamic flows," Aerosp. Sci. Technol., vol. 107, p. 106318, 2020, doi: 10.1016/j.ast.2020.106318.

[32] Afzal, S. A. Khan, T. Islam, R. D. Jilte, A. Khan, and M. E. M. Soudagar, "Investigation and back-propagation modeling of base pressure at sonic and supersonic Mach numbers," Physcics of Fluids, vol. 32, no. July, p. 096109, 2020, doi: 10.1063/5.0022015.

[33] A. Afzal, I. Nawfal, I. M. Mahbubul, and S. S. Kumbar, "An overview on the effect of ultrasonication duration on different properties of nanofluids," J. Therm. Anal. Calorim., vol. 135, pp. 393–418, 2019, doi: https://doi.org/10.1007/s10973-018-7144-8.

[34] Asif Afzal, S. A. Khan, and C. A. Salee, "Role of ultrasonication duration and surfactant on

characteristics of ZnO and CuO nanofluids," Mater. Res. Express, vol. 6, no. 11, p. 1150d8, 2019, doi: 10.1088/2053-1591/ab5013.

[35] A. Afzal, M. Samee A. D, A. Javad, A. Shafvan S, A. P V, and A. Kabeer K. M, "Heat transfer analysis of plain and dimpled tubes with different spacings," Heat Transf. Res., vol. 47, no. 3, pp. 556–568, 2018, doi: http://dx.doi.org/10.1002/htj.21318.

[36] O. David, M. O. Okwu, O. J. Oyejide, E. Taghinezhad, A. Asif, and M. Kaveh, "Optimizing biodiesel production from abundant waste oils through empirical method and grey wolf optimizer," Fuel, vol. 281, no. May, p. 118701, 2020, doi: 10.1016/j.fuel.2020.118701.

[37] M. Kareemullah, A. Afzal, and K. F. Rehman, "Performance and emission analysis of compression ignition engine using biodiesels from Acid oil , Mahua oil , and Castor oil," Heat Transf. - Asian Res., no. August, 2019, doi: 10.1002/htj.21642.

[38] A. Afzal, A. D. M. Samee, R. K. A. Razak, and M. K. Ramis, "Steady and Transient State Analyses on Conjugate Laminar Forced Convection Heat Transfer," Arch. Comput. Methods Eng., vol. 27, pp. 135–170, 2020, doi: https://doi.org/10.1007/s11831-018-09303-x.

IOP Conf. Series: Materials Science and Engineering 1065 (2021) 012036 doi:10.1088/1757-899X/1065/1/012036

Finite Element Simulation of Fracture Toughness of Al6061 Reinforced with Silicon Carbide

Hareesha G[1], Chikkanna N[2], Saleemsab Doddamani[3]

[1] Department of Mechanical Engineering, Government Engineering College, HuvinaHadagali, India

[2] Department of Aerospace Propulsion Technology, VTU, VIAT, Muddenahalli, Chickballapur, India

[3] Department of Mechanical Engineering, Jain Institute of Technology, Davangere, India

Email: harishssb@gmail.com

Abstract. The main objective of the work is to study the fracture behaviour of AA6061-Silicon Carbide material using finite element method. Fracture toughness analysis of AA6061-Silicon Carbide composite has been carried out using finite element methods. Modelling of compact tension specimens was utilized to analyze the fracture toughness using the ANSYS tool. From the results of the comparison, it is recommended that AA6061-9 wt% silicon carbide particulate composite material can be potentially utilized as a replacement of AA6061 in the aerospace application such as helicopter rotor blades.

Keywords: Aluminium alloy, Metal matrix composite, AA6061-Silicon Carbide, FEM, Fracture toughness, compact tension specimens.

1. Introduction

Many two dimensional, three dimensional tutorials ware also available to evaluate the fracture toughness [1-4] using finite element methods, most of them ware used ANSYS tool for the same. While simulating the fracture related issues, it requires the use of PLANE182, which will fit the crack tip section. PLANE182, as shown in figure 1, is a higher order form of the 2-D, eight-node component. It gives more exact outcomes to blended (quadrilateral-triangular) automatic meshes and can hold up under sporadic shapes without as much loss of accuracy. The 8-node components have ideal displacement shapes and are appropriate to model curved boundaries.

The 8- node component is characterized by eight nodes having two degrees of freedom at every node: interpretations in the nodal x and y directions. The component might be utilized as a plane component or as an axis-symmetric component. The component has stress stiffening, large deflection, plasticity, creep, swelling, and extensive strain abilities. The stress intensity factors at a crack for a linear elastic fracture mechanics analysis may be computed. The analysis uses a fit of the nodal displacements in the vicinity of the crack.

Literature shows that failure of helicopter occurred is at the threaded hole region of rotor blade. Many helicopter rotor blades are failed due to the fatigue loading, corrosion of threaded holes [7-8].

IOP Conf. Series: Materials Science and Engineering 1065 (2021) 012036 doi:10.1088/1757-899X/1065/1/012036

Literatures also shows that many helicopter failures occurred in air i.e in the operational conditions [7-9]. Crack developed and propagated, in the rotor blade fittings i.e., bolt hole [8,10], edge spar [9] without giving any prior indication. Many researchers reported that main rotor blades are made of aluminum 6061-T6 alloy [7-11].

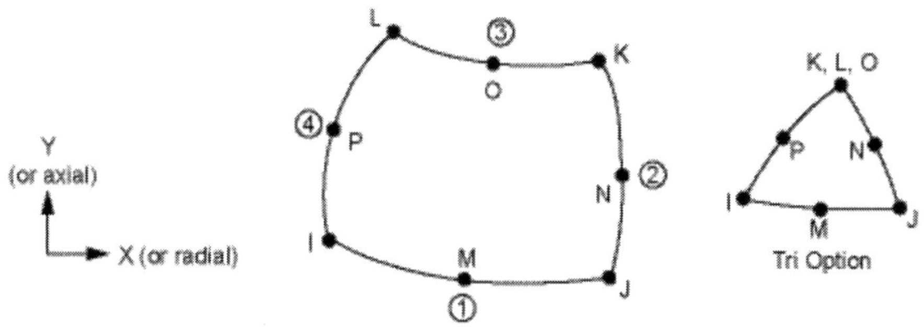

Figure 1. PLANE182 Geometry

Yasmin Begum et al [12], Saleemsab Doddamani et al [13] worked on aluminum matrix reinforced with graphite particles has been studied for the mechanical and fracture characteristics of the composite respectively. ASTM has prescribed compact tension specimen for the fracture toughness testing [14]. Saleemsab Doddamani et al [15] uses compact tension specimen to study the fracture toughness of Al6061-graphite MMC at varied weight fractions. S Doddamani et al [13] utilizes ANSYS software to study the fracture toughness of Al6061-graphite. K.K. Alaneme et al [16] studied the fracture toughness of Al6063-SiC particulate MMC. It is observed from the results of the fracture toughness testing of Al6061-graphite [15] and Al6063-SiC [16] that Al6061-graphite composite has more fracture toughness values.

From the literature it is identified the helicopter rotor blade, made of AA6061, was failed at the threaded portion of the rotor blade. The main objective of the work is to study the fracture behavior of AA6061-SiC material using finite element method [18-20]. The finite element model is built for the AA6061 and AA6061-SiC composite. Through this work an attempt has been made to compare the existing use of material with the composite material to prepare the helicopter rotor blade so as to enhance its fatigue crack resistance property.

2. Materials Selection

In the present work aluminium 6061 is utilized as a matrix and silicon carbide particles are used as reinforcement. A precipitation-hardened aluminium (AA6061) alloy has its main alloying elements as magnesium (0.81 wt%) and silicon (0.70 wt%). Hardness of 95 BHN, E=68.9 GPa, σ_y = 275 MPa, σ_{ut} = 315 MPa and elongation 17% were the some of its mechanical properties.

Silicon carbide (SiC), a form of carborundum, is a ceramic material. It is a combination of silicon and carbon. SiC is one of exceptional abrasive materials used to manufacture abrasive wheels. Now a day, the SiC available is of high quality technical grade ceramic with excellent physical properties. Some of the properties of the SiC are Density = 3.1g/cc, melting point – 2730°C, Appearance –Black in color, Hardness = 45.8 GPa.

3. Model Generation

Using ANSYS, the geometric model was created. A finite element model was constructed utilizing PLANE182 and SOLID185 components shown in figure 2. The nodes around the crack tip appear in figure 3. The finite element model was built using PLANE182 element, for the crack analysis of the aluminum-graphite alloy.

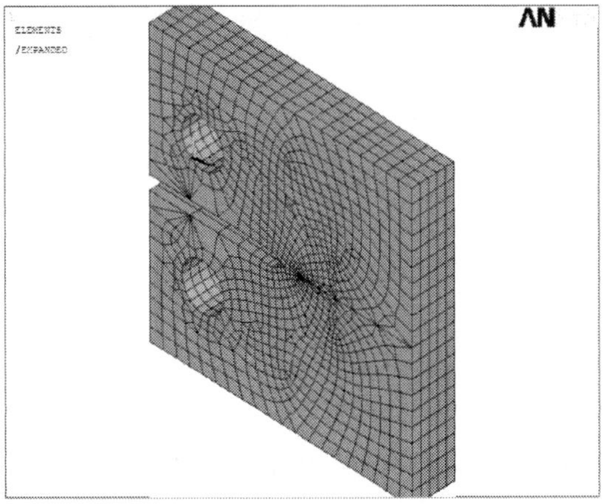

Figure 2. Finite Element Model

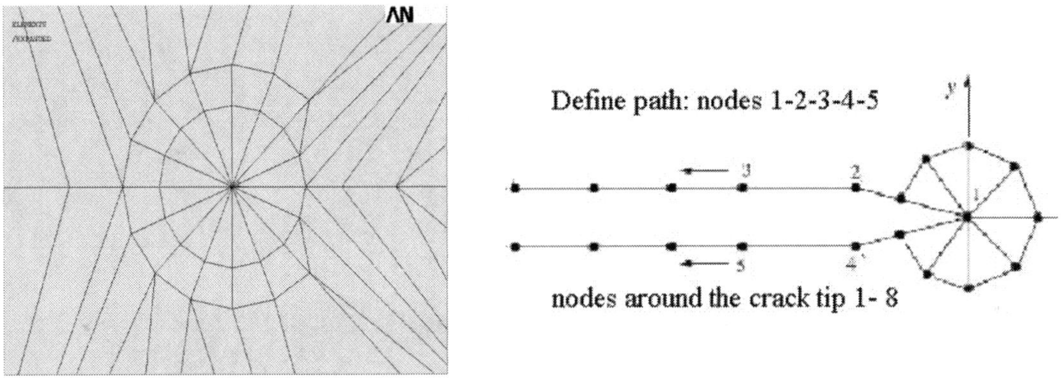

(a)Full Crack Model (b) Nodes around the crack tip

Figure 3. Crack tip region

4. Results and Discussion

The crack analysis was carried out and the resulting analysis is shown in Table 1.

Table 1 Analytical Fracture Toughness results of AA6061 & AA6061-SiC Composites

Sl No	Composite	Fracture load (P_f) kN	K_{Ic} MPa √m
1	AA6061	2.183	09.36
2	AA6061-6%SiC	2.260	10.04
3	AA6061-9%SiC	2.367	10.51
4	AA6061-12%SiC	2.148	09.54

From the fracture toughness analysis using ANSYS software, it is clear that fracture toughness increases with the increase of SiC content. For the AA6061 value of fracture toughness is 09.36MPa√m, whereas for AA6061-9%SiC it is 10.51 MPa√m, which is slightly higher than the 6wt% and 12 wt% compositions and also with the base metal. From the fracture toughness analysis using ANSYS software, it is clear that fracture toughness increases with the increase of SiC content. Up to 9 wt% and at 12 wt% it decreases. The fracture toughness values were additionally observed to be in close concurrence with data got from standard K_{Ic} testing (as shown in figure 4).

IOP Conf. Series: Materials Science and Engineering 1065 (2021) 012036 doi:10.1088/1757-899X/1065/1/012036

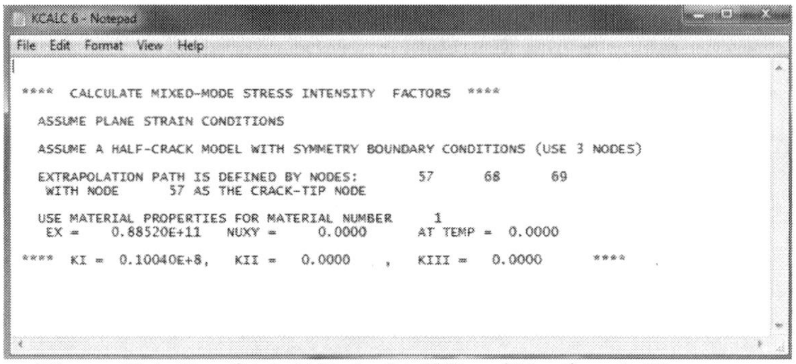

Figure 4(a). K_{Ic} of AA6061-6wt% SiC – a/W ratio = 0.45

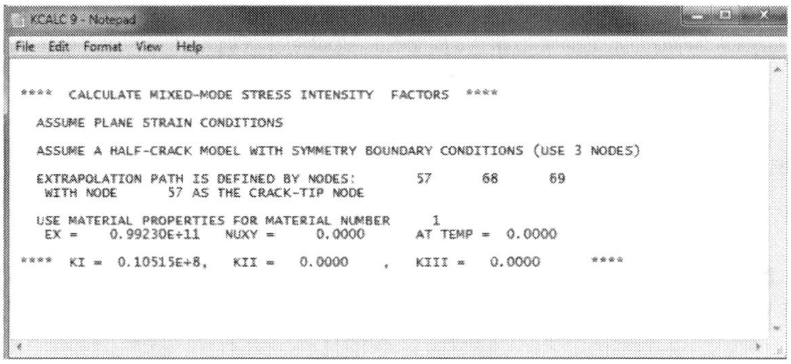

Figure 4(b). K_{Ic} of AA6061-9wt% SiC – a/W ratio = 0.45

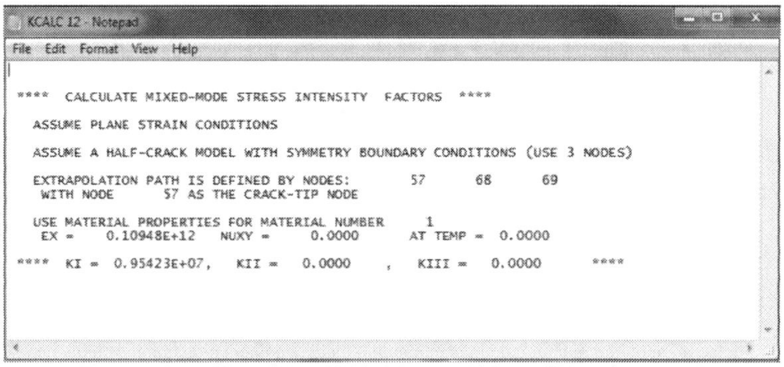

Figure 4(c). K_{Ic} of AA6061-12wt% SiC – a/W ratio = 0.45

Figure 5(a) shows the theoretical stress distribution of the plain strain and plane stress approach [1]. Figure 5(b) shows the stress distribution at the crack tip of the AA6061-6wt% SiC. The shown stress distribution resembles the Von-mises stress distribution of metallic materials [1] for the plain strain condition. This validates the simulations results. The fracture toughness at that condition is referred as the plain strain fracture toughness. As per the requirements of the plain strain fracture toughness, the model generated using finite element simulation software is satisfying the required condition. Hence the results obtained through finite element simulation are validating with the available theoretical geometrical conditions.

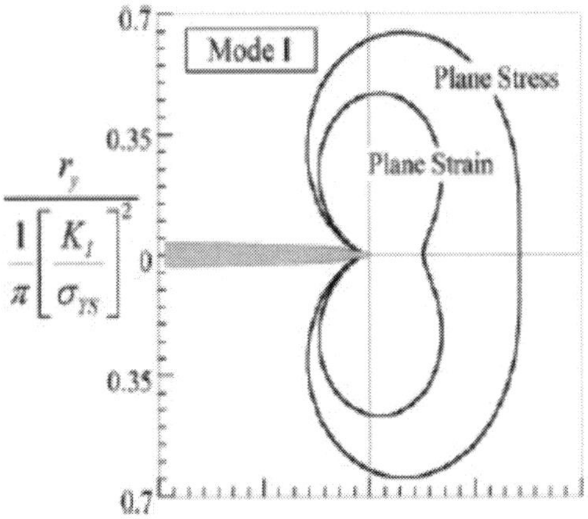

Figure 5(a). Mode I Crack-tip plastic zone shapes [1] – Theoretical approach

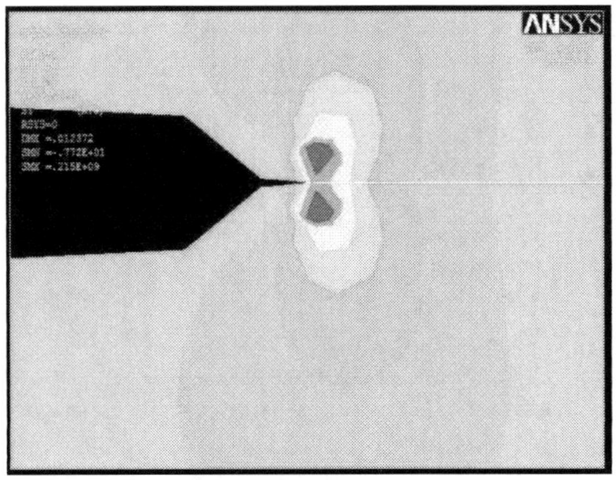

Figure 5(b). Stress around the crack tip indicating plain strain fracture toughness

Due to the improved fracture toughness of the AA6061-SiC composite, as compared to the AA6061, the composite can be the potential material utilized for the aerospace application. AA6061-SiC composite can be used as a replacement material for the use of the Helicopter rotor blade.

5. Conclusion

The finite element model of the compact tension specimen was successfully modelled using the ANSYS. The fracture toughness values obtained from the ANSYS are in close agreement with data obtained from standard K_{Ic} testing. From the comparison, it is recommended that AA6061-9%SiC composite material can be considered as a potential rotor blade material as a replacement of AA6061 in the application of main rotor blades of the helicopter. The result obtained through the finite element simulation validates with the geometrical conditions. Thus the K_{Ic} values obtained can be considered as the plain strain fracture toughness.

References

[1] T L Anderson *Fracture Mechanics-Fundamentals and Applications* 3rd Edition Taylor & Francis Group, New York 2013.
[2] Xian-Kui Zhu, James A. Joyce 2012 *Review of fracture toughness (G, K, J, CTOD, CTOA) testing and standardization*, Engineering Fracture Mechanics, Elsevier **85** 1–46.
[3] ASM Handbook, Composites, ASM International **21** 2001.
[4] Dr. A.V. Phan, 2013 Ansys Tutorial - 2-D Fracture Analysis, ANSYS Release **7.0** 1-14.
[5] Ramesh Chanwani, 2004 *3D Fracture Mechanics in ANSYS*, Zentech International Limited, UK ANSYS Conference.
[6] Jiří Běhal 2012 *Crack Growth Simulation in The Course of Industrial Equipment Life Extension*, **20th** SVSFEM ANSYS Users' Group Meeting and Conference 1-9.
[7] Mikael Amura, Lorenzo Aiello, Mario Colavita, Fabrizio De Paolis, Manuele Bernabei, 2014 *Failure of a helicopter main rotor blade*, 20th European Conference on Fracture (ECF20), Procedia Materials Science **3** 726 – 731.
[8] Dr Arjen Romeyn, 2005 *Main rotor blade failure analysis report*, Engineered system failure analysis report, November 2005.
[9] Sunrise helicopter Inc 2011 *In-flight separation of Main Rotor Blade and Collision with Terrain* Aviation Investigation Report, report no: **A11O0205**.
[10] Rolf Kieselbach, Georg Soyka, 2000 *Failure of a helicopter rotor*, Technology, Law and Insurance, T and F Online **5(3-4)** 141-146.

[11] Sylwester Klysz, Janusz Lisiecki, Marcin Kurdelski, 2010 *Material testing of the helicopter main rotor blades*, **37th** Solid Mechanics Conference.
[12] Yasmin Begum, Saleemsab Doddamani, Nagesh N 2015 *Mechanical Properties of Aluminium–Graphite Particulate Composites* NCERAME-2015 International Journal of Engineering Research & Technology (IJERT).
[13] Saleemsab Doddamani and Mohammed Kaleemulla, 2016 *Indentation Fracture Toughness of Alumnum6061-Graphite Composites*, International Journal of Fracture and Damage Mechanics **1(1)** 40-46.
[14] ASTM Standards, 2017 *Standard Test Method for Plane-Strain Fracture Toughness of Metallic Materials* ASTM International **E 399-17**.
[15] Doddamani, S., Kaleemulla, M 2017 *Fracture toughness investigations of Al6061-Graphite particulate composite using compact specimens* Frattura ed Integrità Strutturale **41** 490-497.
[16] K.K. Alaneme, A.O. Aluko 2012 *Fracture toughness (K1C) and tensile properties of as-cast and age-hardened aluminium (6063)–silicon carbide particulate composites*, Scientia Iranica A, **19 (4),** 992–996.
[17] Vikram Singh and R.C. Prasad 2004 *Tensile and fracture behavior of 6061 al-sicp metal matrix composites*, International Symposium of Research Students on Materials Science and Engineering.
[18] Saleemsab Doddamani, Yasmin Begum, Bharath K N, Rajesh A M, Mohamed Kaleemulla K, 2020, *Optimization of process parameters of fracture toughness using simulation technique considering aluminum-graphite composites*, Transactions of the Indian Institute of Metals **73(12)** 3095–3103 DOI: 10.1007/s12666-020-02113-5.
[19] Rajesh A M, Mohammed Kaleemulla, Saleemsab Doddamani 2020, *Dry sliding wear simulation of hybrid aluminum metal matrix composites*, Advanced Composites and Hybrid Materials **3(1)**120–126. https://doi.org/10.1007/s42114-020-00133-9.
[20] Saleemsab Doddamani and Mohammed Kaleemulla 2019 *Comparisons of experimental fracture toughness testing methods of Al6061-graphite particulate composites*, Journal of Failure Analysis and Prevention **19(3)** 730-737. DOI: 10.1007/s11668-019-00652-8.

Enhancement of Fracture toughness for Nano SiC reinforced AA7075 MMC's By Tensile tests and FEM Analysis

Anilkumar S. Kallimani[1], Chikkanna N[2], Suresha P[3], Hareesha G[4]

[1]Department of Mechanical Engineering, Government Engineering College, HuvinaHadagali, India

[2] Department of Aerospace Propulsion Technology, VTU, VIAT, Muddenahalli, Chickballapur, India

[3] Department of Mechanical Engineering, South East Asian College of Engineering and Technology, K R Puram Bangalore-560049, Karnataka, India,

[4]Department of Mechanical Engineering, Government Engineering College, HuvinaHadagali, India

Email: anilskallimani@gmail.com

Abstract. After the second world war, The Design and development activities in the field of Aerospace took place with a great acceleration and that resulted in to Evolution of new Aluminum alloy materials for Aircraft Structure. Recent research work reveals that particle reinforced Metal matrix composites (PAMC) of aluminum alloy reinforced with silicon carbide reinforcement have shown better mechanical and wear properties. In this work, AA7075 is chosen matrix and Nano silicon carbide(500nm) in the range of (0-8%) were added in terms of weight percentage as reinforcement. Cast MMC samples were fabricated stir casting techniques. Further the specimens were made for XRD study, and for Tensile Test. The 500nm silicon carbide content varies with the 0%, 2%, 4%, 6% &8%wt. fractions. Theoretical Stress intensity values of Compact tension specimen were compared with FEM analysis. And it was observed that higher values of toughness were obtained as the weight percentage increased. Results were compared with pure matrix material in this paper.
Keywords: 500nmSiC, AA7075, Stir Casting, CT Specimen, Fracture Toughness, FEA, ABAQUS software, XRD and EDS analysis.

1. Introduction

Metal matrix composites are obtained by popular aluminium alloy series such as 2XXX, 3XXX,6 XXX & 7XXX etc. Which are mainly used in aircraft structure design due to their high strength and weight ratio. Due to the demand in high performance of the aircrafts weight reduction methods are used in aviation sector and new materials are under search.(1-5).Aluminum alloys reinforced with Nano size of silicon carbide (MMC's) have shown greater signs in order to decrease the weight and consequently proved to be better structural components to provide strength to the structure. Vast research work has been undergoing in the area of MMC's. Because of their weight to strength ratio,

MMC's are gradually replacing the popular aluminium alloys which were used post world war in Aircraft and in other applications (6-10).

2 Experimental details

Present work describes the production of MMC's by using stir casting technique tensile test with FEA, analysis and density and porosity of AA7075 Aluminium alloy with reinforcement of silicon carbide with various %wt and compositions of AA7075 shown in below Table.

Table:1 Composition of the AA7075

Element	Wt.%	Element	Wt%
Copper	1.6	Zinc	5.5
Magnesium	2.5	Chromium	0.15
Silicon	0.4	Titanium	0.2
Iron	0.5	aluminum	Balance
Manganese	0.3		

3 Fabrications of Composites

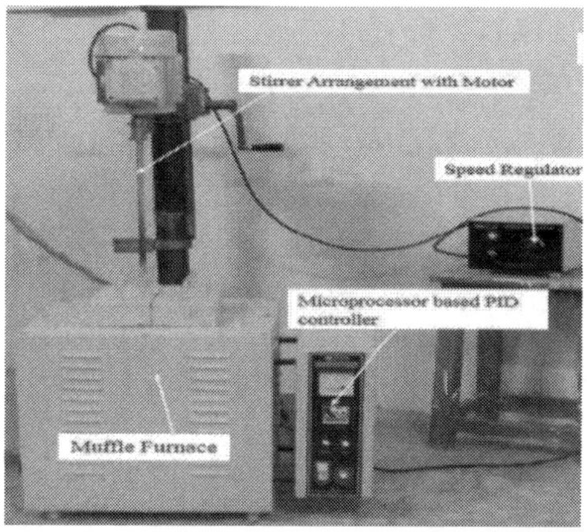

Fig: 1 Stir casting set up

The MMC's were fabricated by using the stir casting technique that setup have been shown in the fig:1. The AA 7075 is kept into the crucible furnace and supply to heat by using the electric heating source, when it is reaches to the temperature 600^0C -750^0C the AA7075 is melted in crucible than pour the weight % of SiC in the molten metal while stirs the stir by using the electrical motor. Similarly fabricated the MMC's at different weight % of SiC.

4 Experimental work

Tensile test is a Major Mechanical test to enhance the fundamental mechanical behavior of material. Tensile specimens were machined as per ASTM-E8 standards. Tests were carried out for all MMC Specimens.

IOP Conf. Series: Materials Science and Engineering 1065 (2021) 012037 doi:10.1088/1757-899X/1065/1/012037

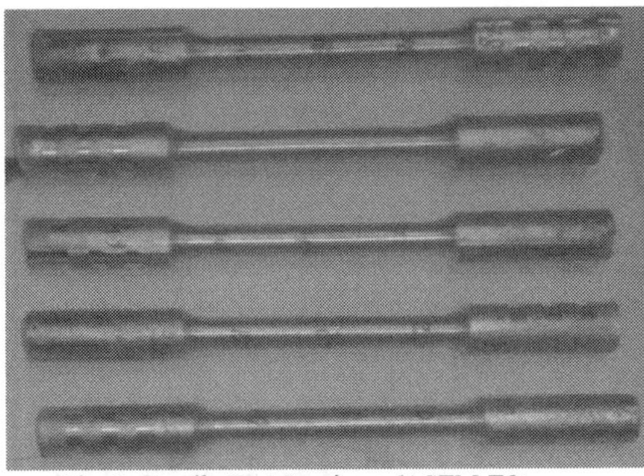

Fig. 2. Tensile Test Specimen (ASTM-E8)

Fig: 3 Universal Testing Machine

Samples for XRD

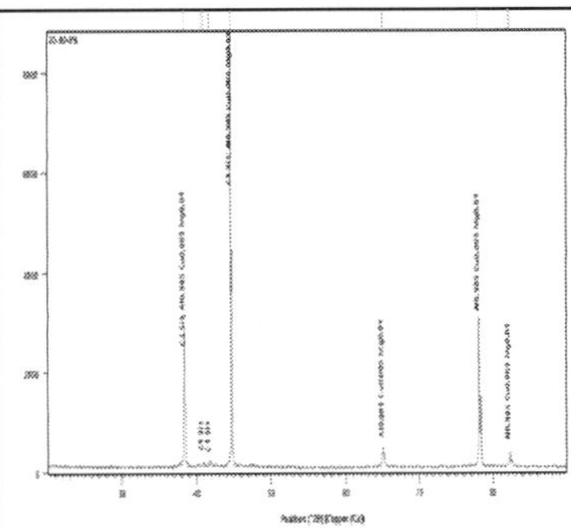

Pure AA7075

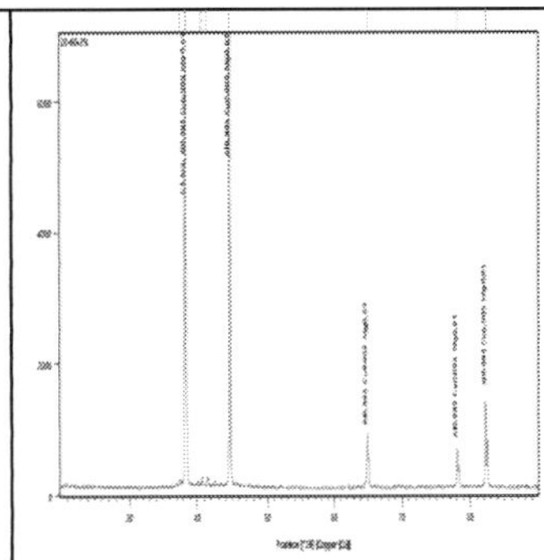

2%SiC

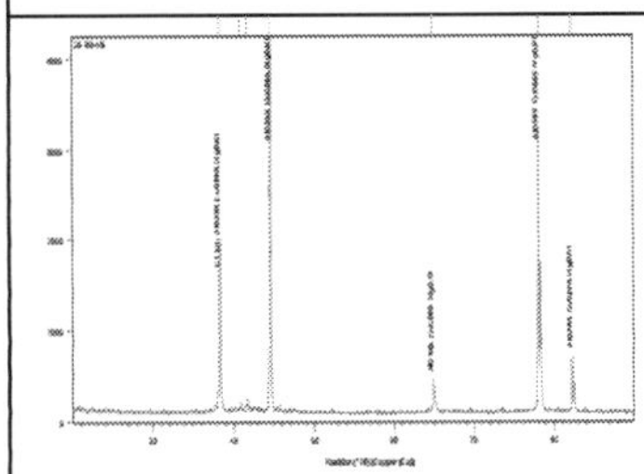

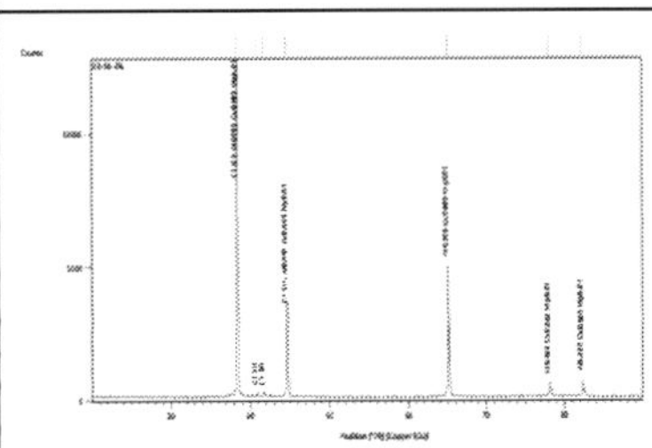

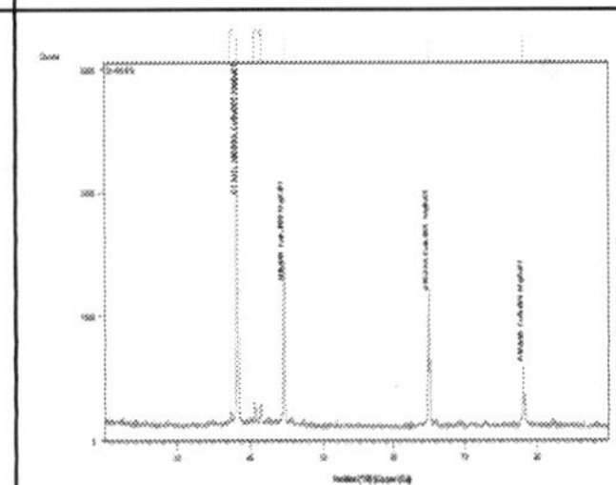

Fig 4.XRD Results

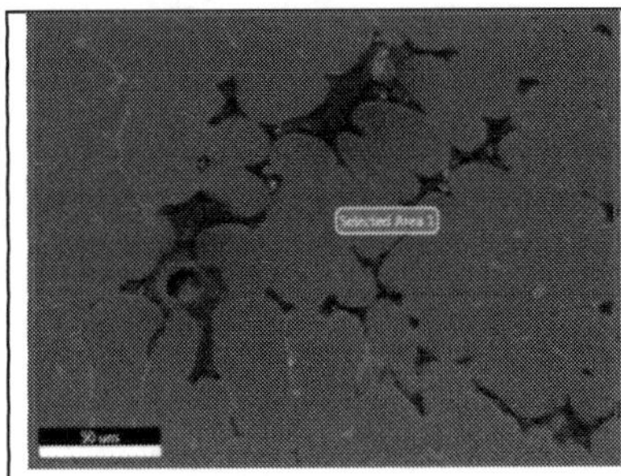

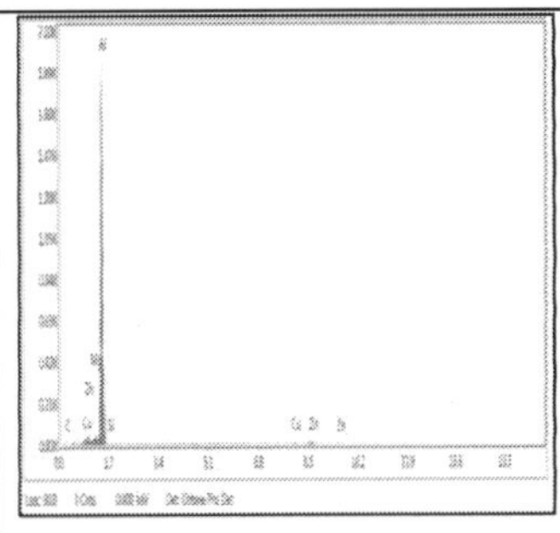

Element	Weight %	Atomic %
C K	9.71	19.95
MgK	1.79	1.82
AlK	83.02	75.95
SiK	0.39	0.34
CuK	1.5	0.58
ZnK	3.6	1.36

Fig.5. EDX analysis of MMC's

Fig.4 & 5 Shows XRD and EDX Peak patterns for the different compositions.The Peak list depicts the cubic phase of crystalline.

IOP Conf. Series: Materials Science and Engineering 1065 (2021) 012037 doi:10.1088/1757-899X/1065/1/012037

5 Tensile Test (Experimental Results)

Tensile test was carried out at a room temperature with low strain rate and the following tables shows the Tensile test results i.e Load-Deformation, Engineering Stress-strain curves, Yield strength, Ultimate Tensile strength and the final elongation for each of the specimens. were recorded. It was observed that as the weight percentage increased, higher values of Mechanical properties were obtained.

Table 2. Tensile Test Results

Initial Area mm2	67.9
Initial gauge length mm	40
Final Gauge Length mm	44.92
Yield Load KN	13.5
Yield Strength MPa	198.8
UTS Load KN	15.70
UTS MPa	231.2
% Elongation	12.3

Initial Area mm2	67.9
Initial gauge length mm	40
Final Gauge Length mm	44.92
Yield Load KN	13.5
Yield Strength MPa	198.8
UTS Load KN	15.70
UTS MPa	231.2
% Elongation	12.3

Initial Area mm2	67.9
Initial gauge length mm	40
Final Gauge Length mm	44.92
Yield Load KN	13.5
Yield Strength MPa	198.8
UTS Load KN	15.70
UTS MPa	231.2
% Elongation	12.3

Initial Area mm2	65.84
Initial gauge length mm	45
Final Gauge Length mm	49.57
Yield Strength MPa	272.67
UTS Load KN	23.01
UTS MPa	349.58
% Elongation	10.156

Initial Area mm2	65.84
Initial gauge length mm	45
Final Gauge Length mm	48.34
Yield Strength MPa	317.52
Ultimate Tensile Load KN	25.6
Ultimate Tensile Strength MPa	388.9
% Elongation	7.42

6. Youngs Modulus for MMC

Before carrying out Fracture toughness test, Elastic Modulus of each MMC samples were estimated using Rule of mixture formulae. Volume fractions of SiC were calculated using formulae

$$V_{SiC} = \frac{m_{SiC}/\rho_{SiC}}{\left[\frac{m_c - m_{SiC}}{\rho_{Al}}\right] + \frac{m_{SiC}}{\rho_{SiC}}} \qquad \textbf{[Ref. 10]}$$

$$V\_(2\%SiC) = (0.516/3.2)/([(25.801 - 0.516)/2.821] + 0.516/3.2) = 0.0176$$

$$V\_(4\%SiC) = (1.042/3.2)/([(26.066 - 1.042)/2.821] + 1.042/3.2) = 0.0354$$

$$V\_(6\%SiC) = (1.564/3.2)/([(26.075 - 1.564)/2.821] + 1.564/3.2) = 0.0532$$

$$V\_(8\%SiC) = (2.087/3.2)/([(26.094 - 2.087)/2.821] + 2.087/3.2) = 0.0711$$

Youngs modulus of MMC's is given by Kerner Equation

$$E_c = E_m\left[1 + \frac{v_p}{1 - v_p} \times \frac{15(1 - m_m)}{8 - 10m_m}\right]$$

$$E_{c2\%SiC} = 71.7\left[1 + \frac{0.0176}{1 - 0.0176} \times \frac{15(1 - 0.33)}{8 - 10(0.33)}\right] = 74.5GPa$$

$$E_{c4\%SiC} = 71.7\left[1 + \frac{0.0354}{1 - 0.0354} \times \frac{15(1 - 0.33)}{8 - 10(0.33)}\right] = 77.3GPa$$

$$E_{c6\%SiC} = 71.7\left[1 + \frac{0.0532}{1 - 0.0532} \times \frac{15(1 - 0.33)}{8 - 10(0.33)}\right] = 80.3GPa$$

$$E_{c8\%SiC} = 71.7\left[1 + \frac{0.0711}{1 - 0.0711} \times \frac{15(1 - 0.33)}{8 - 10(0.33)}\right] = 83.4GPa$$

It is observed from the calculations that modulus of elasticity increased as the weight percentage of SiC increased. These values were used to carry elastic FEM analysis for determining Fracture toughness of MMC's

7. Fracture Toughness (Theoretical)

Stress intensity factor for Compact tension specimen can be estimated by the below empirical relation.

$$K_1 = \frac{P}{B}\sqrt{\frac{\pi}{W} \times \left[16.7\left(\frac{a^{\frac{1}{2}}}{W^{\frac{1}{2}}}\right) - 104.7\left(\frac{a^{\frac{3}{2}}}{W^{\frac{3}{2}}}\right) + 369.9\left(\frac{a^{\frac{5}{2}}}{W^{\frac{5}{2}}}\right) - 573.2\left(\frac{a^{\frac{7}{2}}}{W^{\frac{7}{2}}}\right) + 360.5\left(\frac{a^{\frac{9}{2}}}{W^{\frac{9}{2}}}\right)\right]}$$

Where '***P***' is the Applied Load and '***B'*** is the thickness of the specimen ***W*** is the effective width of the specimen being the distance between the centreline of the holes and the blackface of the coupon. '***a'*** is the crack length.

Following figure shows the geometry of the Compact test specimen as per ASTM-E647-00

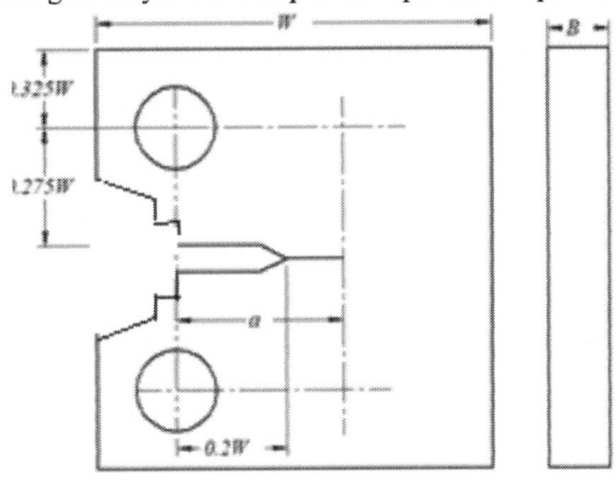

Fig.6. Compact Tension specimen Geometry

from the tensile test results Yield load was taken and following ***Kic*** Values were calculated as shown below.

%WIEGHT	*P (YIELD LOAD) N*	*KI (STRESS INTENSITY FACTOR) Mpa√m*
PURE AA7075	15700	59.87
2%SIC	18700	71.32
4% SIC	20580	78.49
6% SIC	23010	87.75
8%SIC	25600	97.63

IOP Conf. Series: Materials Science and Engineering 1065 (2021) 012037 doi:10.1088/1757-899X/1065/1/012037

8. Fracture Toughness (FEM)

Using the estimated elastic modulus and Yield loads, Elastic FEM Analysis was carried out using ABAQUS software.2D planar shell plate model was constructed in ABAQUS.

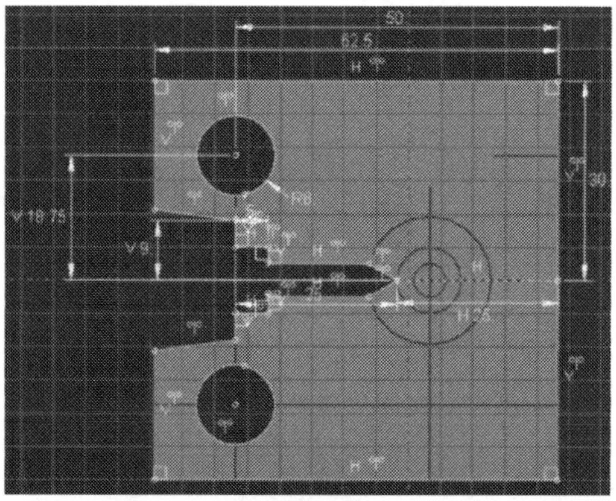

Geometry

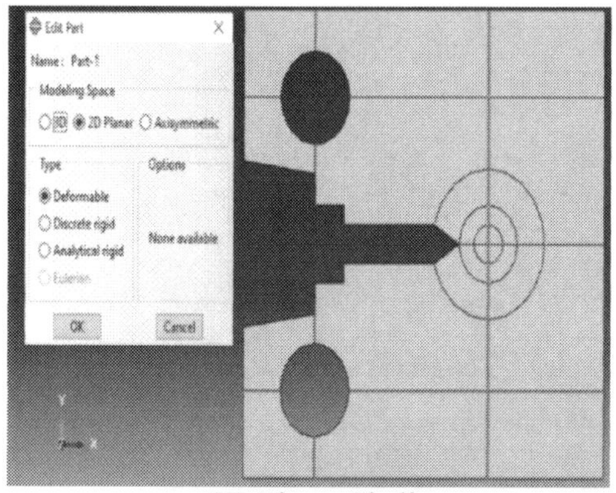

2D Planar Shell

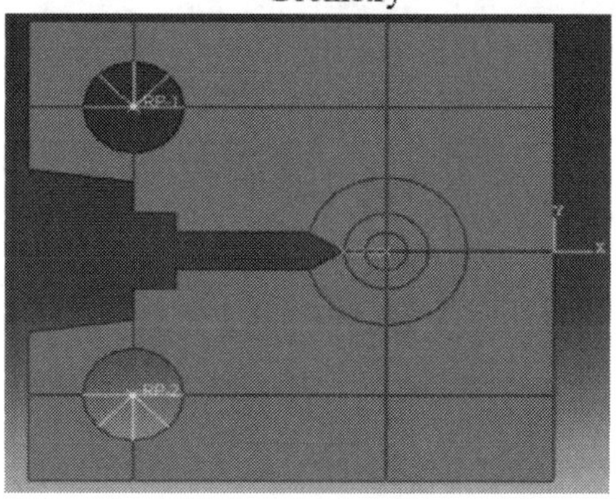

Interaction between center and inner surface of the Holes

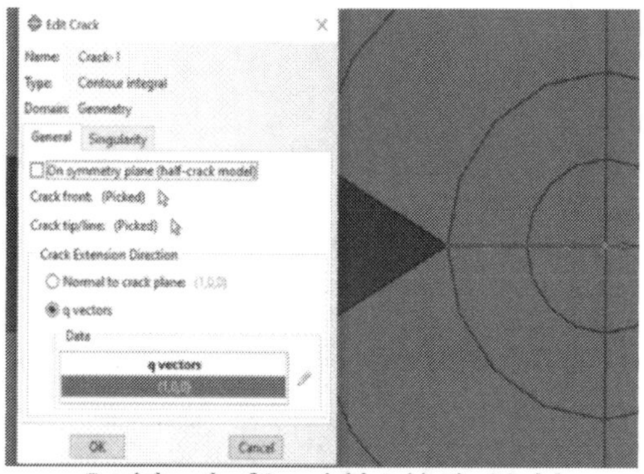

Crack length of 5mm initiated in the Model

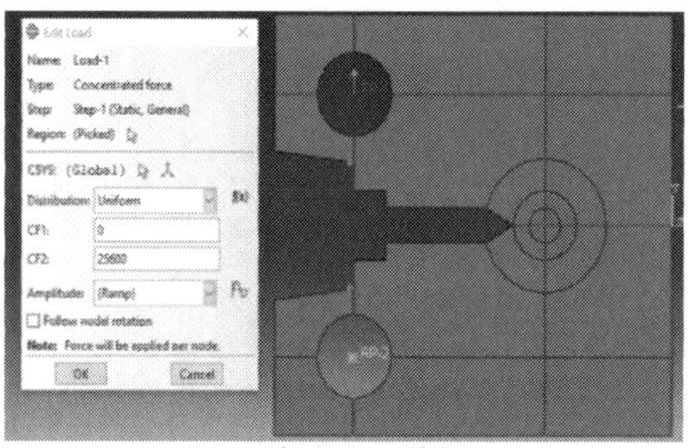

Boundary Conditions

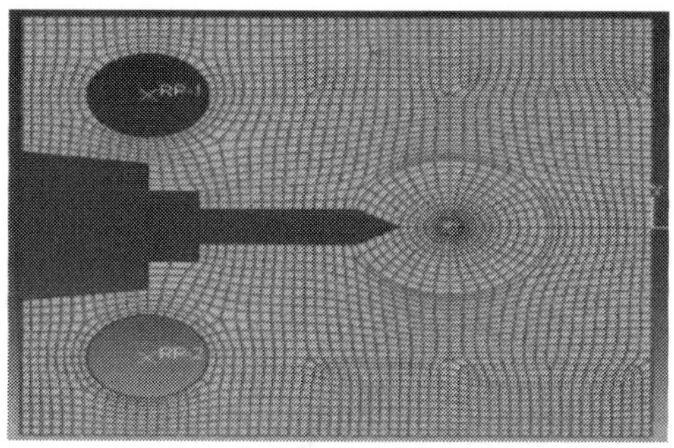

Meshed Model

Fig.7 FEM Details

Plain strain CPE4R Hexagonal elements were used to mesh the model. Total no of elements and nodes used are 3100 and 3277 respectively.

9. Results and Discussions

Stress and strains induced in the CT specimen and at the Crack tip were shown below.

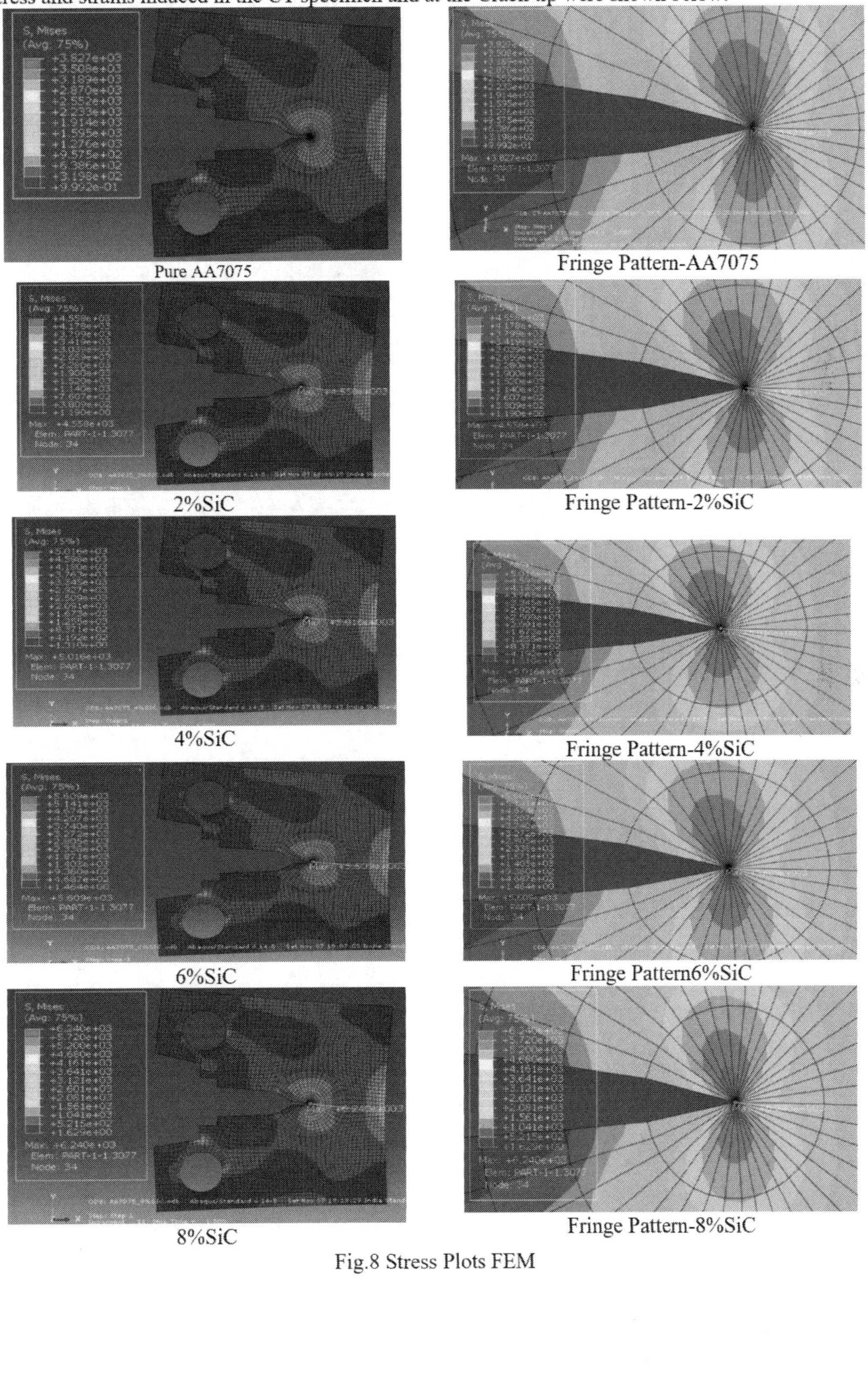

Pure AA7075 — Fringe Pattern-AA7075

2%SiC — Fringe Pattern-2%SiC

4%SiC — Fringe Pattern-4%SiC

6%SiC — Fringe Pattern6%SiC

8%SiC — Fringe Pattern-8%SiC

Fig.8 Stress Plots FEM

Table 3. Comparison of theoretical values with FEM Results

%WIEGHT	*P(YIELD LOAD) N*	*KI(Therotical) Mpa$\sqrt{m}$*	*KI(FEM) Mpa$\sqrt{m}$*
PURE AA7075	15700	59.87	56.25
2%SIC	18700	71.32	67.02
4% SIC	20580	78.49	73.76
6% SIC	23010	87.75	82.47
8%SIC	25600	97.63	91.76

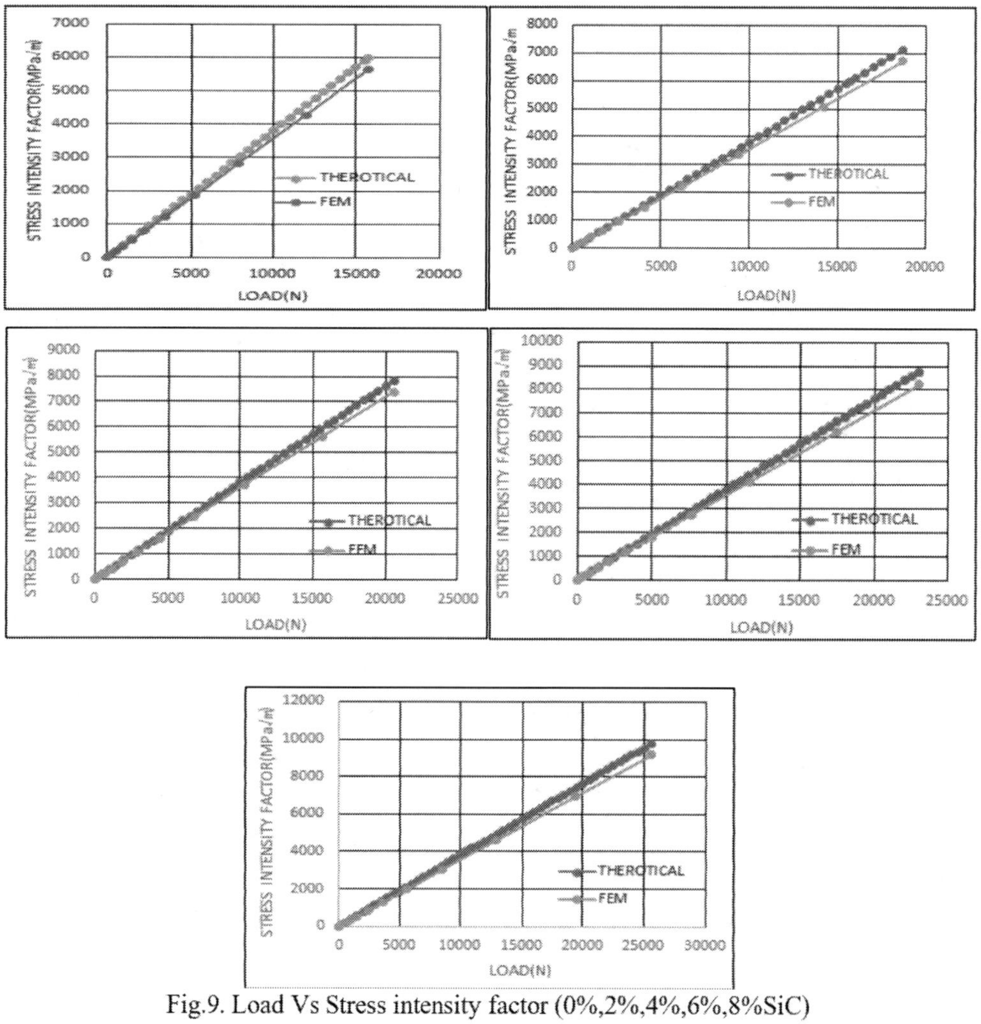

Fig.9. Load Vs Stress intensity factor (0%,2%,4%,6%,8%SiC)

10. Conclusions

In this work the MMC specimens were fabricated using stir Casting Method.
The XRD and EDX analysis done in that analysis confirmed major compositions of MMC's.
Tensile test revealed Higher values of Mechanical Properties with decrease in the percentage elongation.
Theoretical and FEM results showed Higher values of Fracture toughness for increased Weight Percentage.

References

[1] T L Anderson *Fracture Mechanics-Fundamentals and Applications* 3rd Edition Taylor & Francis Group, New York 2013.
[2] Xian-Kui Zhu, James A. Joyce 2012 *Review of fracture toughness (G, K, J, CTOD, CTOA) testing and standardization*, Engineering Fracture Mechanics, Elsevier **85** 1–46.
[3] ASM Handbook, Composites, ASM International **21** 2001.
[4] Dr. A.V. Phan, 2013 Ansys Tutorial - 2-D Fracture Analysis, ANSYS Release **7.0** 1-14.
[5] Ramesh Chanwani, 2004 *3D Fracture Mechanics in ANSYS*, Zentech International Limited, UK ANSYS Conference.
[6] Jiří Běhal 2012 *Crack Growth Simulation in The Course of Industrial Equipment Life Extension*, **20th** SVSFEM ANSYS Users' Group Meeting and Conference 1-9.
[7] Mikael Amura, Lorenzo Aiello, Mario Colavita, Fabrizio De Paolis, Manuele Bernabei, 2014 *Failure of a helicopter main rotor blade*, 20th European Conference on Fracture (ECF20), Procedia Materials Science **3** 726 – 731.
[8] Dr Arjen Romeyn, 2005 *Main rotor blade failure analysis report*, Engineered system failure analysis report, November 2005.
[9] Sunrise helicopter Inc 2011 *In-flight separation of Main Rotor Blade and Collision with Terrain* Aviation Investigation Report, report no: **A11O0205**.
[10] Rolf Kieselbach, Georg Soyka, 2000 *Failure of a helicopter rotor*, Technology, Law and Insurance, T and F Online **5(3-4)** 141-146.
[11] Sylwester Klysz, Janusz Lisiecki, Marcin Kurdelski, 2010 *Material testing of the helicopter main rotor blades*, **37th** Solid Mechanics Conference.
[12] Yasmin Begum, Saleemsab Doddamani, Nagesh N 2015 *Mechanical Properties of Aluminium–Graphite Particulate Composites* NCERAME-2015 International Journal of Engineering Research & Technology (IJERT).
[13] Saleemsab Doddamani and Mohammed Kaleemulla, 2016 *Indentation Fracture Toughness of Alumnum6061-Graphite Composites*, International Journal of Fracture and Damage Mechanics **1(1)** 40-46.
[14] ASTM Standards, 2017 *Standard Test Method for Plane-Strain Fracture Toughness of Metallic Materials* ASTM International **E 399-17**.
[15] Doddamani, S., Kaleemulla, M 2017 *Fracture toughness investigations of Al6061-Graphite particulate composite using compact specimens* Frattura ed Integrità Strutturale **41** 490-497.
[16] K.K. Alaneme, A.O. Aluko 2012 *Fracture toughness (K1C) and tensile properties of as-cast and age-hardened aluminium (6063)–silicon carbide particulate composites*, Scientia Iranica A, **19 (4),** 992–996.
[17] Vikram Singh and R.C. Prasad 2004 *Tensile and fracture behavior of 6061 al-sicp metal matrix composites*, International Symposium of Research Students on Materials Science and Engineering.
[18] Saleemsab Doddamani, Yasmin Begum, Bharath K N, Rajesh A M, Mohamed Kaleemulla K, 2020, *Optimization of process parameters of fracture toughness using simulation technique considering aluminum-graphite composites*, Transactions of the Indian Institute of Metals **73(12)** 3095–3103 DOI: 10.1007/s12666-020-02113-5.
[19] Rajesh A M, Mohammed Kaleemulla, Saleemsab Doddamani 2020, *Dry sliding wear simulation of hybrid aluminum metal matrix composites*, Advanced Composites and Hybrid Materials **3(1)**120–126. https://doi.org/10.1007/s42114-020-00133-9.
[20] Saleemsab Doddamani and Mohammed Kaleemulla 2019 *Comparisons of experimental fracture toughness testing methods of Al6061-graphite particulate composites*, Journal of Failure Analysis and Prevention **19(3)** 730-737. DOI: 10.1007/s11668-019-00652-8.

Comparative Analysis of DIP Techniques to Detect Leaf Diseases in Tomato Plant

Pushpalatha S. Nikkam[1], Archana Nandibewoor[2], Aijaz Qazi[2], Leena I. Sakri[1], Vijayendra Nargund[3], ***Aishwarya Sajjanar[2], Saloni Porwal[2], Sujit Bhosale[2], Payal Ghorpade[2]***

[1] Department of Information Science and Engineering, SDMCET, Dharwad, Karnataka, India

[2] Department of Computer Science and Engineering and, SDMCET, Dharwad, Karnataka, India

[3]Department of Plant Pathology, Agricultural University, Dharwad, Karnataka, India

geetasrikant@gamil.com

Abstract. In India agriculture is the main source of income for generating the economy. Diseases in plants are a major unavoidable problem, and hence detecting the diseases is the necessity of the day in the domain of agriculture. The main diseases found in tomato plants are viral, fungus and bacterial diseases. The detection will help improve the quantity and quality of the products with an optimum yield. In this paper a comparative analysis is carried out for the algorithms Support Vector Machine, Convolution Neural Networks, Decision tree classifier, and k-Nearest Neighbour (k-NN) with the result of 97%,97%,90% and 80% respectively.

Keywords: Tomato leaf disease detection; k-NN; Decision tree classifier; Support vector Machine; Convolution Neural Network; Early Detection; Disease classification.

I. Introduction

India is known for its agriculture revenue generation. 75% of the Indian population is dependent on agricultural area. Precautionary measurement is taken by identifying the disease in advance so that overall crop losses can be avoided. Day to day monitoring the crop is tedious time consuming process. The Tomato crop is an essential and highly spread out due to its high intake. Identifying the diseases in plants is the key to avoid losses by proper nurturing to cure them at an early state. Presently numerous pesticides are available in the market leading to less Tomato yield, such as (i) Bacterial Spot, (ii) Early Blight, (iii) Late blite, (iv) Leaf Mould,(vi) Septoria leaf spot. To overcome the pesticides control and diseases, it is essential to carry out accurate recognition of the diseases. This paper is organised by literature survey followed by methodology, results and discussion, finally conclusion.

II. Related Work

The literature survey widens the understanding of the different diseases to tomato leaves. The deep learning method uses Convolution Neural Network (CNN) with the combined feature set of Gabor transform and wavelet for classifying different tomato diseases giving an accuracy of 99% classifying on labelled dataset [1]. The CNN method with two different architecture such as AlexNet and SqueezeNet gives 95% and 94% of accuracy in detecting the tomato diseases using deep learning techniques on village dataset[2][3]. A variation of CNN model with squeezenet architecture with keras deep learning framework will produce 86.92% of accuracy over 200 classes of tomato images [4]. K means algorithm with thresholding method gives 33% and 26% on the images acquired through camera.[5]. Support Vector Machine combined with Cauchy, Laplacian Kernel and Invmult Kernel gives 100%, 98% and 78% of accuracy respectively [6]. Two serial SVM classifiers results in an accuracy of 85% in recognising the tomato leaves [7]. Supervised classification tree classifies the tomato leaf disease up to 97.3% of accuracy [8]. Artificial Neutral Network and K-means clustering over 10 different samples result and accuracy of 92.5% [9].

III. Methodology

The following algorithms are implemented to study the performance:

3.1 Convolution Neural Network
3.2 Support Vector Machine
3.3 Decision Tree

3.1 Convolution_Neural_ Network (CNN)

This algorithmic approach analyses the visual image components for detection and reorganization of the images.

ALGORITHM

1. The input images are processed by 4 CNN layers.
2. Convolution layer is the first layer where all the preprocessing like applying filter such as thresholding, edge detection, smoothing, noise is removed, many more.
3. Max Pooling and Relu is done where the matrix reduction occurs in all negative valued matrix number is removed.
4. Fully connected layer is formed and softmax function is activated.
5. Then this CNN model is trained for all the training set images.
6. The test set is used for prediction and an accuracy of 97-98% is acquired.

3.2 Support Vector Machine (SVM)

SVM Algorithm creates a line or a hyper-plane which separates the data into separate classes.

ALGORITHM

1. Generate a hyper-plane that separates the data into classes.
2. Draw a line to distinguish the different classes.
3. Checkout the nearer data points that lie closer to both the classes and the data points are called support vectors.
4. Locate the proximity between the diving plane and support vectors.
5. The distance between the data points and the separating line is a margin. The SVM algorithm is used t maximize the margin. When the margin reaches its maximum the hyper plane becomes optional.

ALGORITHM

1. Let m be the number of training data samples. Let p be an unknown point.
2. Store the training samples in an array of data points arr[]. This means each element of this array represents a tuple (x, y) and loop it.
3. Compute best distance.
4. Create a set A and let P be the smallest distance computed.

3.3 Decision Tree Algorithm

The representation of the tree tries to resolve the problem of detection and recognition of the disease.

ALGORITHM

1. Nodes correspond to attributes and leaf node corresponds to label the class.
2. Root node is the initial start of the complete dataset.
3. The computed values are stored in an excel sheet and label it.
4. Splitting the dataset set into subsets as training and test set in specified ratio. Then the features and the labels is stored in x and y train and test variables.
5. Then "CART" (Classification and Regression Tree) uses "Gini index" method to create binary splits.
6. By the use of glcmMatrix we compare the feature set least value and depending on the class value the predictions are made.

IV. Results and Discussion

(i) Convolution Neural Network

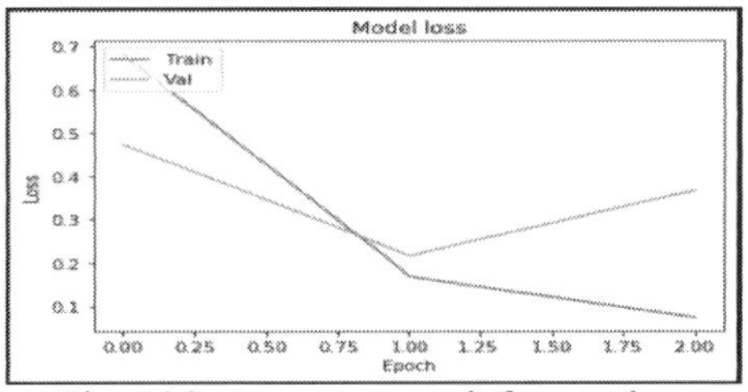

Fig. 4(a): Accuracy graph for maximum accuracy

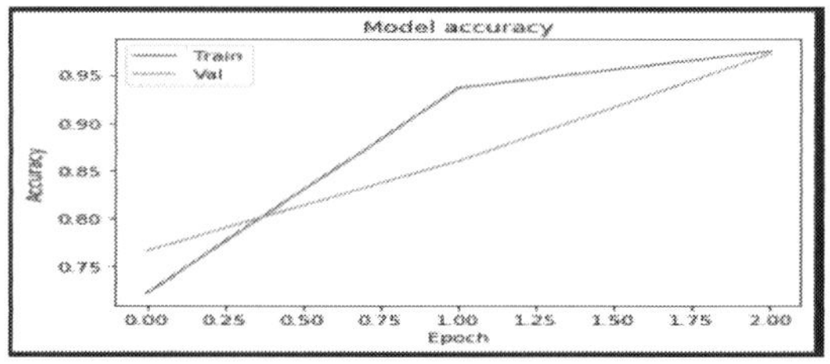

Fig.4(b): Accuracy graph for maximum loss

Fig. 4(c): Prediction of the diseases using CNN

(ii) Support Vector Machine

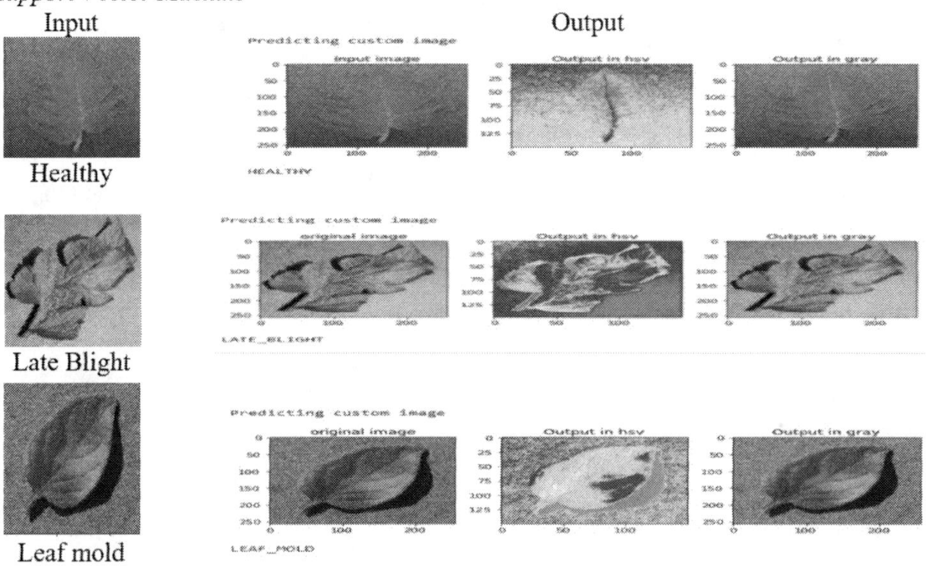

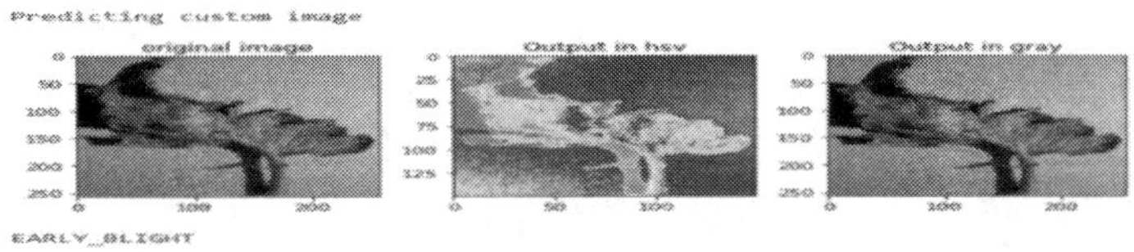

Early blight

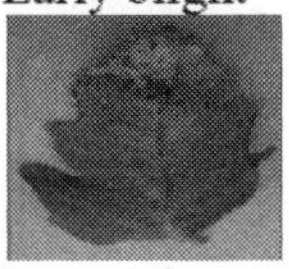

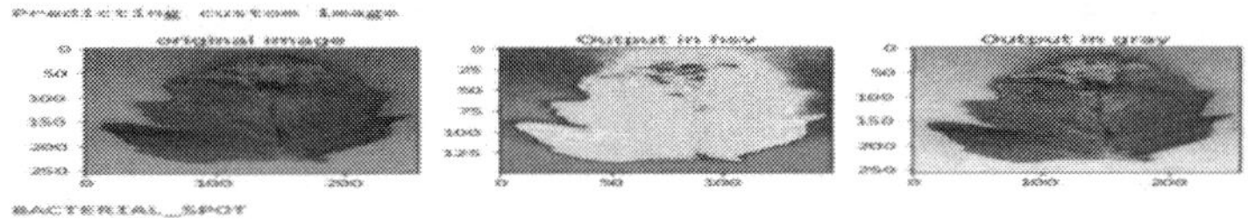

Bacterialspot

Fig. 4(d): Prediction of the diseases using SVM

```
from sklearn.metrics import classification_report, confusion_matrix
print(classification_report(Y_test, Y_predict))
```

```
              precision    recall  f1-score   support

           0       1.00      1.00      1.00        20
           1       1.00      0.89      0.94        19
           2       0.90      1.00      0.95        19
           3       1.00      0.95      0.97        19
           4       0.96      1.00      0.98        23

    accuracy                           0.97       100
   macro avg       0.97      0.97      0.97       100
weighted avg       0.97      0.97      0.97       100
```

Fig. 4(e): Prediction classification accuracy using SVM

(iii) K-Nearest Neighbour

Input Output

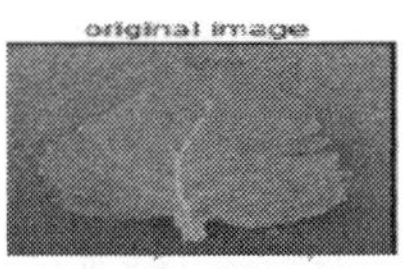

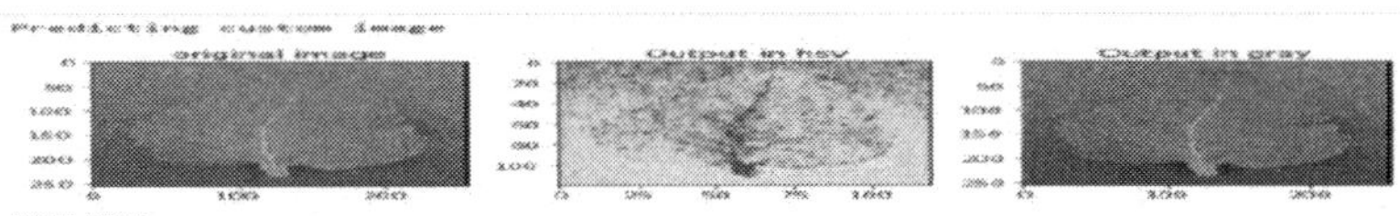

Healthy

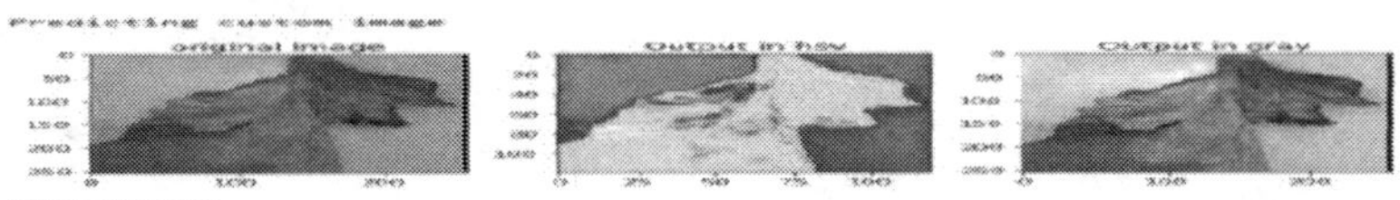

Late_Blight

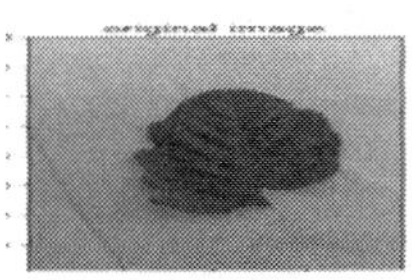

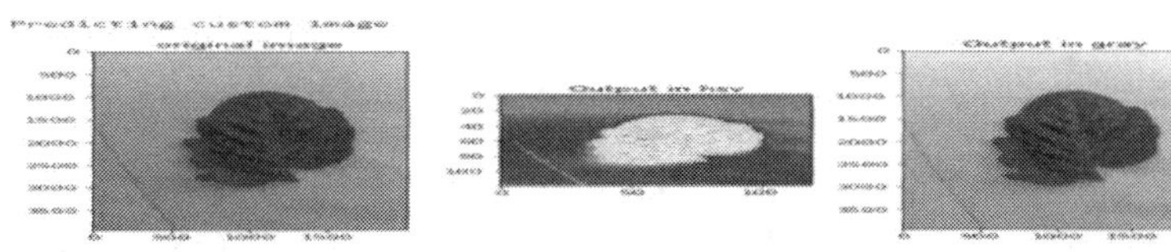

Leaf_mold

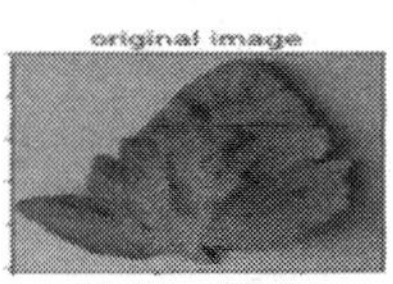

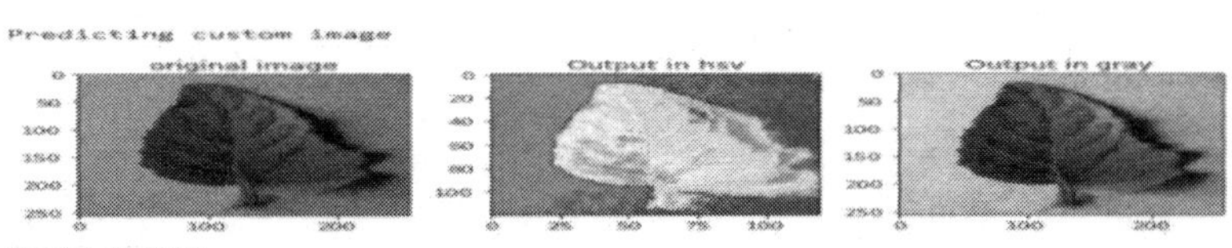

Early_blight

IOP Conf. Series: Materials Science and Engineering 1065 (2021) 012038 doi:10.1088/1757-899X/1065/1/012038

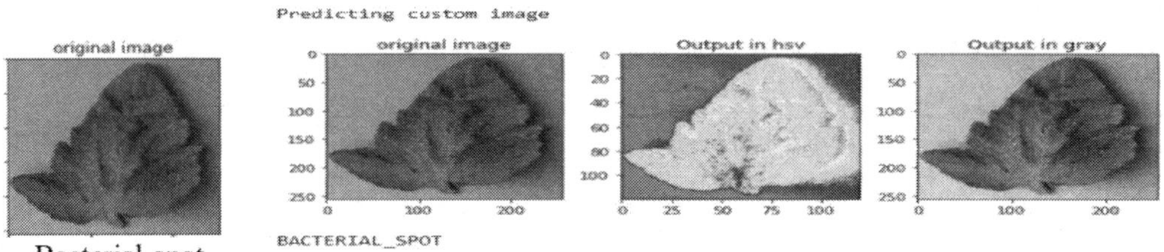

Bacterial-spot

Fig. 4(f): Prediction of the diseases using K-NN

(iv) Decision Tree

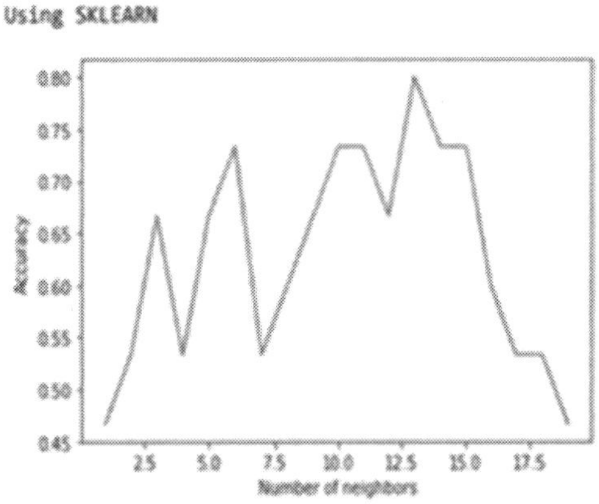

Maximum Accuracy at k=13 is 80%

4(g): Prediction classification accuracy using KNN

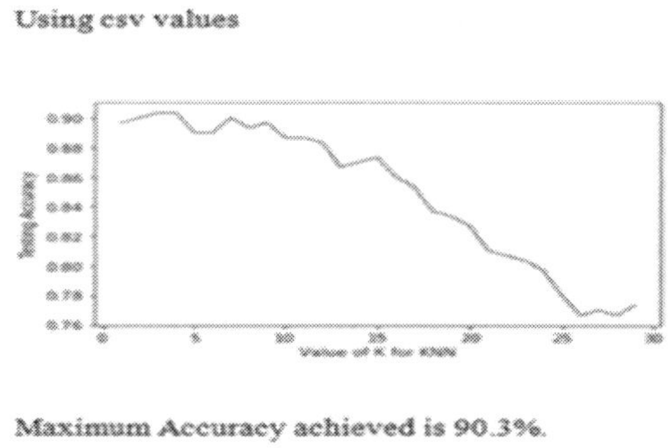

Fig.4(h): Prediction classification accuracy using KNN

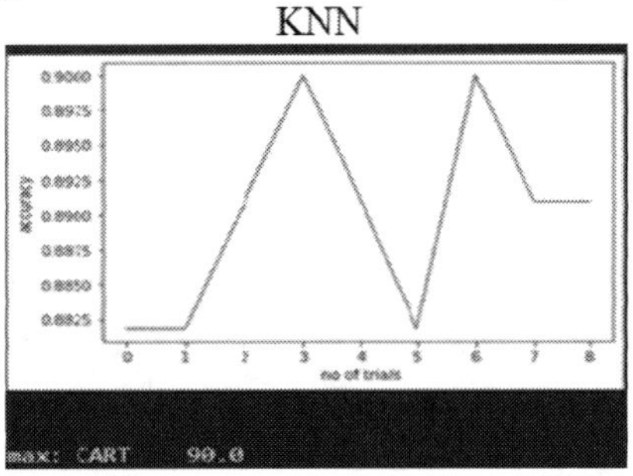

Fig. 4(i): Accuracy graph for Decision Tree

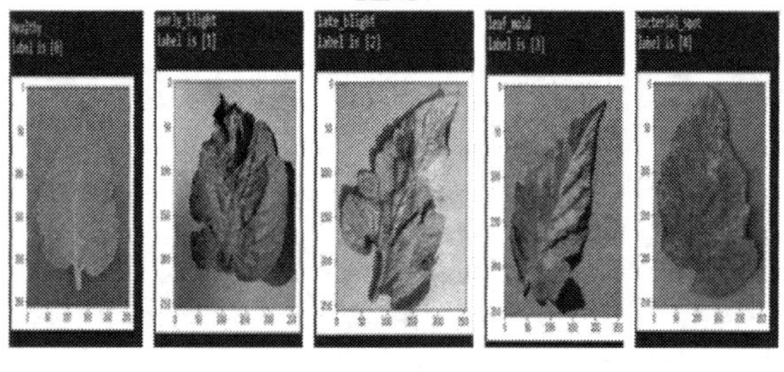

Fig.4(j): Prediction of the diseases using Decision Tree

V. Summary

A comparative analysis of four different image processing technique is carried out for disease detection using Convolution Neural Network, Support vector machine, Decision Tree and K Nearest Neighbor algorithms for leaves diseases classification. From the study it is observed that the machine trained with CNN, SVM, Decision Tree and K-NN classifier gives an accuracy up to 97%,97%,90.3% and 80% respectively.

The future scope of this work will be improving all the algorithms and train the system with detect the disease at an early stage.

References

[1] Mohammed Brahimi at. al,"Deep Learning for Tomato Diseases: Classification and Symptoms Visualization",International Journal of Applied Artificial Intelligence, VOL. 31, NO. 4, 299–315,2017

[2] HalilDurmus et al.," Disease Detection on the Leaves of the Tomato Plants by Using Deep Learning",International Journal of Innovative technology and Exploring Engineering, Vol. 9, Issue 8,June 2020.
[3]Prajwala TM et al.,"Tomato Leaf Disease Detection using Convolutional Neural Networks", Eleventh International Conference on Contemporary Computing (IC3), 2-4 August, 2018, Noida, India
[4] Akbar Hidayatuloh et al., "Identification of Tomato Plant Diseases by Leaf Image Using Squeezenet Model", International Conference on Information Technology Systems and Innovation (ICITSI) Bandung - Padang. October 22-25,2018, Indonasia.
[5] TenviMehra et al., "Maturity and Disease Detection in Tomato Using Computer Vision",Fourth International conference on Parallel, Distribute and Grid Computing(PDGC),Dec22-24,2016Waknaghat, Himachal Pradesh, India
[6] Usama Mokhtar et al.,"Tomato leaves diseases detection approach based on support vector machines",International Computer Engineering Conference (ICENCO),Dec 29-30, 2015,Egyp
[7] Youssef Es-saady et al.,"Automatic recognition of plant leaves diseases based on serial combination of two SVM classifiers ",2nd International Conference on Electrical and Information Technologies (ICEIT), May 4-7,2016, FIU Laborateries,USA
[8] H. Sabrol and K. Satish, "Tomato Plant Disease Classification in Digital Images using Classification Tree ",International Conference on Communication and Signal Processing, April 6-8, 2016, India
[9] Ch. Usha Kumari et al., "Leaf Disease Detection: Feature Extraction with K-means clustering and Classification with ANN",3rd International Conference on Computing Methodologies and Communication, March 27-29,2019, Tamil Nadu, India.

IOP Conf. Series: Materials Science and Engineering 1065 (2021) 012039 doi:10.1088/1757-899X/1065/1/012039

A Melanoma Skin Cancer Detection Using Machine Learning Technique: Support Vector Machine

Praveen Banasode*, Minal Patil, Nikhil Ammanagi

Department of Master of Computer Applications, Jain College of Engineering Belagavi-14

Affiliated to Visvesvaraya Technological University Belagavi, Karnataka state, India.

Email: [1]praveenb.jce@gmail.com

Abstract. In this paper proposed is an easy way to detect the disease and help us to know before something turns out to be serious. The aim of this work is to detect skin cancer. People can get to know what skin disease they are having and what all precaution and measures to be taken at an early stage and it will help in treating the disease successfully. The major causes of skin cancer are air pollution, UV radiation, unhealthy life style etc. The concept of machine learning will be used to determine the disease and help us to detect the result. The most commonly used classification algorithms is support vector machine (SVM). First we are analyzing the skin image then converting the images into BGR-Gray and BGR-HSV for the computer to understand and enable it to read binary codes. The result of this study will help doctors to treat disease at the initial stage and further aggravation can be avoided.

Keywords: Melanoma, Lesion, Support Vector Machine

1. Introduction

Skin cancer is a rare disease which needs to be looked into it can cause many symptoms or will have different types which can be cured or may kill. It comes to any age group and can be deadly if not treated at an early stage. Approximately one in six people develop skin cancer it's a rare disease which can affected in any way. In the medical field experts unanimously agrees that overexposure to sunlight may cause the increase of this disease.

This work is on skin cancer detection which helps us to know which type it is and what all necessary precautions should be taken. Skin cancer is the most vital thing in the medical field and we need to make the necessary arrangements to reduce this problem for the betterment of human life. Python and Machine Learning are used, which helps us to know the working and all other techniques in it. The work describes in three levels, which helps us to understand the procedure in a very precise and clear manner.

1.1 Deploying the image

The image should have a clear visibility and has to make the process in working manner. In this work some precautions are taken. The image should not be blur and low quality because if in case that is the situation then the outcome of the result will not be displayed and the accuracy won't be showing us the correct result. So, load an image and then the further steps will be processed.

1.2 Pre-processing

The pre-processing method can be used to know which image has what type of disease. Firstly the image should be in high quality which gives the result instantly and greatly help us for better accuracy. Then we have to remove the background noise to know that only the selected part of image is being used. This has to be done for every image and the process keeps on repeating until all the images are fully covered and are ready to be processed.

1.3 Classification

In this, we will know the working of it so that it can be interconnected to the next level. First, we will be having a dataset (collections of data) where we have stored a large number of images in a folder. With this, we will be analysing the images to check for its clarity and high-quality resolution and not blur. Next, we have three categories which are the types of skin cancer. In the coding part we will be doing the classification folder, depending on the type the skin cancer belongs to, it goes to that folder. After that, we resize the images of all types for the machine to understand and it creates a folder of resized images in that dataset. Now we convert the images into BGR-Gray and BGR-HSV (hue, saturation, value) for the computer to understand and read it in binary codes.

1.4 Segmentation

Next comes the segmentation where we have to take the image and only cover the part where the skin cancer is there and remove the background, the covered part will be white coloured and the background will be black. This we call segmented image or masking where the image will be analysed and gives us the proper result.

2. Literature Survey

Alaa Haddad and Shihab A. Hameed [1] proposed to find skin disease by removing noise or unwanted things analysis result can support doctor to help initial diagnosis and to know the type of disease by using the low cost solution for the skin disease detection. Amir Mirbeket et. al [2] three semi-solid phantoms are used and accurately represent the dielectric properties of normal and malignant skin tissues the penetration depth of millimetre waves malignant skin disease is calculated penetrates the human skin tissue deep enough to reach the majority of skin structure. Azadeh Noori Hoshyar et. al [3], proposed automatic detection of skin cancer, where determining and describing the different steps of processing is achieved it is well designed as a substitute of clinician in melanoma diagnosis. Enakshi Jana et. al [4] worked on three types of cancers and tested using SVM and adaboost produced best results of image classification with good accuracy which is useful in classification of normal and abnormal skin cells different types of architecture of ANN and SVM for skin cancer is used to get result with accuracy various methods are used for classification of skin cells. Haseeb Younis et. al [5] proposed different types of skin cancer and skin classification method using deep learning where it classifies the skin within 2-3 seconds pre-trained MobileNet convulsion neural network and trained over lesion dataset method used here is lightweight, fast and reliable shows a good accuracy of weightage 0.90 and 0.91 respectively. Kyamelia Roy et. al [6] proposed segmentation techniques on certain skin diseases using python language the adaptive Thresholding, edge detection, k-means clustering and morphology-based image segmentation are used to identify the skin diseases from the set. Pavels Osipovs et. al [9] proposed centralized algorithms for analysis and processing of data, thereby making sure the data is accessible and coordinates all other nodes in the system with certain limit.

3. Methodology

It helps us to know the result will have correct or somewhat similar answer to it. This algorithm is useful for the work that needs to be verified to find which is closure and which has a longer distance that can't be used in this area.

IOP Conf. Series: Materials Science and Engineering 1065 (2021) 012039 doi:10.1088/1757-899X/1065/1/012039

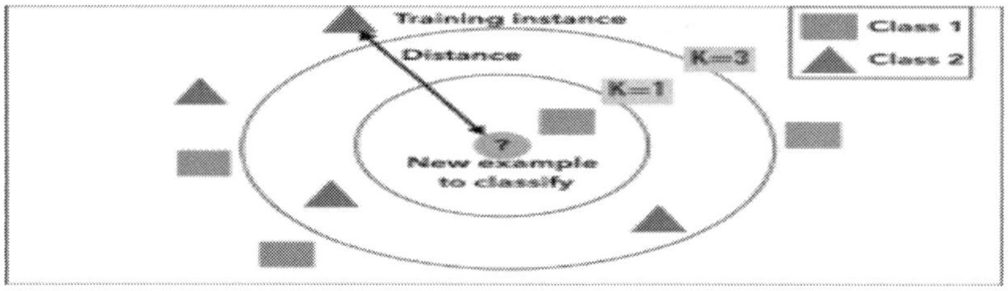

Fig 1: Space maintained from other classes
(Source: https://www.analyticsvidhya.com)

In Fig 1, class 1 shows us the amount of space it is being maintained from the other class that is displayed in the diagram. It basically helps us to know that the closest it gets to the specified area that is allocated which we see in the image that is green in colour to the class 1 which is a neighbour because it is side by side that we call it as closer member is known as KNN. And the class 2 which are further away from the target can't possibly reach the distance so this shows us the result of the diagram.

3.1 Support Vector Machine (SVM)

It is kind of an algorithm which helps us to analyse and gives the result. Let's say it has data's in that it has categorized in to two sets, one which helps us to mark the margin for the output purpose and other for the building a model to determine that which point it belongs to so this is how it works.

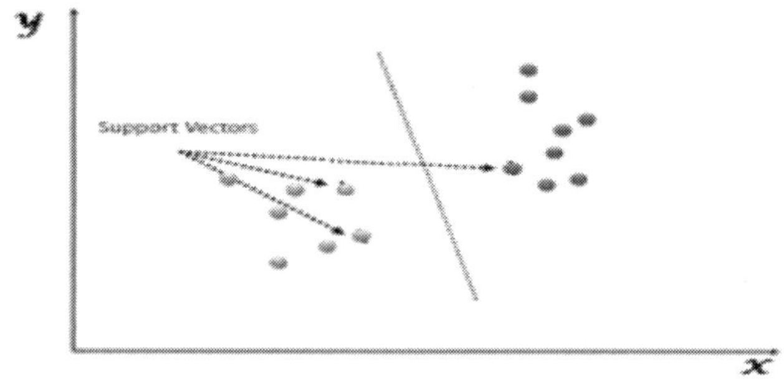

Fig 2: Co-ordinate points of supporting vector
(Source: https://www.analyticsvidhya.com)

This can be a very useful algorithm, as you can see, we have co-ordinates which helps us to separate them on what basis it is being separated will be explained. The agenda of this algorithm is that it helps to know how it works as you can see, we have a dotted line in between the green side and the red side we call it a margin which helps us to divide them and see will it be able to work or no.

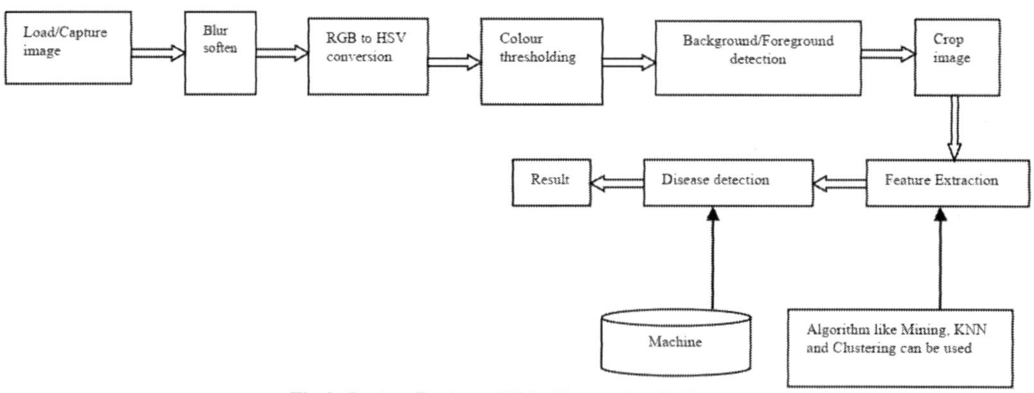

Fig 3: System Design of Skin Cancer Prediction

IOP Conf. Series: Materials Science and Engineering 1065 (2021) 012039 doi:10.1088/1757-899X/1065/1/012039

4. Results and Discussions

The dataset collected from ISIC (International Skin Imaging Collaboration) containing more 5341 images of melanoma and not melanoma, out of these images we choose 1751 images of melanoma. The training set of images is used for testing using the SVM, which classifies the affected skin area as melanoma. The feature like asymmetric behaviour, colour, and border irregularity are extracted. Finally this feature vector is given as input to SVM classifier which classifies the skin lesions [7]. Following tests are done.

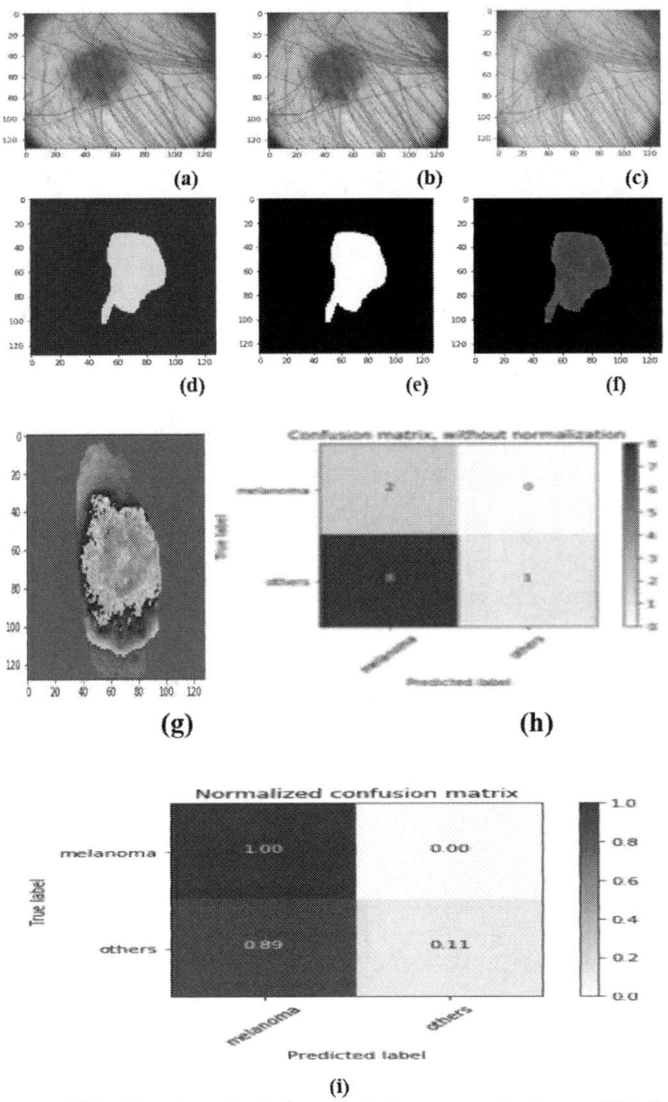

Fig 4: (a) Original Image (b) RGB to Gray Scale Image (c) Pre-processing Image (d) Labelling (e) Masking Image (f) Segmented Image (g) Image Transformation (h) Predicted Table (i) Predicted Table

a) The trained image is being deployed where it tells us the disease and the result.
b) This image is a conversion from one form to another it comes from light background.
c) It is an image which can be viewed in clear and helps us to find the spot of disease.

e) In this image, we have taken it from the original image and only taken the affected area which is white in colour so that it can be understood by anyone. This image we call it as masking.
f) This is the segmented part which helps us to know what difference does this make from masking.
h) This label tells us about the disease and how the number of images does the types have which are genuine through validation we get this model result.
g) Image transformation after SVM classification.
i) This label tells us about the accuracy of different types of diseases with the correct outcome.

By keeping other parameters in support of vector machine different tests are done to reduce the computational complexity. Following table shows the performance evaluation of SVM.

Table -1 Performance Evaluation

Sensitivity	Specificity	Accuracy
95.7%	90.2%	96.9%

The performance are evaluated by calculating Specificity and Sensitivity

$$\text{Specificity} = \frac{TrueNegative}{TrueNegative+FlasePositive} \quad \text{and} \quad \text{Sensitivity} = \frac{TruePositive}{TruePositive+FalseNegative}$$

The proposed method, using the SVM was successful. The result shows accuracy of 96.9% of detecting melanoma disease with other disease. The results explain the performance evaluation of SVM.

5. Conclusion

By using SVM algorithm we can be sure that the application which is going to be used in the coming future will be helpful for betterment of human life. Where the disease can be known at early stage or beforehand what type of disease the patient is affected with and what needs to be done before it's too late for the person. Survival rate will be more if the Melanoma is detected in time. The accuracy of detecting disease is excellent. Since machine learning plays an important role of skin cancer detection it can be a helpful factor in the medical field. By using the latest technology we can make it easier. In future it is going to be a big boon in the medical field.

References

[1] Alaa Haddad, Shihab A. Hameed et. al "Image Analysis Model For Skin Disease. 2018 Advances in Wireless and Optical Communications.
[2] Amir Mirbeik-Sabzevari "Ultra-Wideband, Stable Normal and Cancer Skin Tissue Phantoms for Millimeter-Wave Skin Cancer Imaging". IEEE publication.
[3] Azadeh Noori Hoshyar, Adel Al- Jumaily, Riza Sulaiman "Review On Automatic Early Skin Cancer Detection". IEEE publication
[4] Enakshi Jana, Dr.Ravi Subban "Research on Skin Cancer Cell Detection using Image Processing". IEEE publication.
[5] Haseeb Younis et. al "Classification of Skin Cancer Dermoscopy Images using Transfer Learning". 2018 7th International Conference on Computer and Communication Engineering (ICCCE).
[6] Kyamelia Roy et. al "Skin Disease detection based on different Segmentation Techniques". IEEE Publication.
[7] Ms. Amulya P M, Prof Mr. Jayakumar T V "A Study on Melanoma Skin Cancer Detection Techniques". International Conference on Intelligent Sustainable Systems (ICISS 2017) IEEE Xplore Compliant - Part Number: CFP17M19-ART, ISBN:978-1-5386-1959-9.
[8] Naser Alfed, Fouad Khelifi and Ahmed Bouridane "Improving a Bag of Words Approach for Skin Cancer Detection in Dermoscopic Images". CoDIT'16 - April 6-8, 2016, Malta.
[9] Pavels Osipovs, Dmitrijs Bliznuks, "Cloud Infrastructure for Skin Cancer Scalable Detection System". IEEE publication.
[10] https://www.analyticsvidhya.com/blog/2017/09/understaing-support-vector-machine-example-code/.
[11] https://www.kaggle.com/drscarlat/melanoma.

Analysis and Predicting Results of IPL T20 Matches

Praveen Banasode, Minal Patil, SupriyaVerma

Department of Master of Computer Applications, Jain College of Engineering Belagavi-14

Affiliated to Visvesvaraya Technological University Belagavi, Karnataka state, India.

Email: [1]praveenb.jce@gmail.com

Abstract. The craze of Indian Premier League (IPL) is always there in millions of people including Indian and our work is about the analysis of the data and prediction of the IPL matches. IPL Data Analysis is all about the analysing the data that is al- ready present in data set using data science, machine learning and python. This is an application design for the purpose of analysing the data by fetching the attribute from the data set and predicting the future of the match and as well as of the players .This will help in the selection of the IPL team that the team should perform good and win the match. Prediction is done for anything like which player will play well in tomorrow's match, which team will win toss and even match etc. The prediction can be done with the help of the analysis on that data set collected and by displaying proper data that is useful for the future prediction. The algorithms have given accuracy over 95%.

Keywords: Indian Premier League, IPL, Winner Prediction, Analyzation.

1. Introduction

Cricket is the biggest tournament played in all most all the countries. It is the game between two teams in which the each team has 11players the final result will be either loss or win or at the rare cases points will be shared with both teams which mean no team has lost or won. Sometimes the game is unpredictable because of that game keeps on changing each and every time.

The madness of cricket in people is like anything by looking into this, the main objective of our work is predicting the match result before the game starts based on the past statistic data that is present in the form of data set .In this the study of Indian Premier League (IPL) is done using that past 12 seasons played till the date.

The Indian Premier League (IPL) is one of the biggest T20 Cricket League in the world. It started in the year 2008, started with 8 teams initially and later added two more teams, for two year (2016 & 2017), each team plays two matches against each other the highest points earned by winning maximum matches with good net run rate will progress to the play-offs, where top 4 teams will be considered the table topper will get two chances if they lose their play-off they can play against the second eliminator and still have a chance to qualify for the final match. IPL attracts many foreign players due to its popularity and fast-paced action. Many aspiring cricketers are shaping with this league and have got a chance to play for their country because of their good performance.

The goal of work is gathering past matches data and analysing them to find out the useful information, which will help in predicting the match result. There is much other way to do prediction; it can be done by consideration of the player's performance and also the performance of the whole team.

There can be many other scenarios where it suddenly starts raining or the player get injured and not able to play. These factors will affect the prediction of the match.

2. Related Work

Deep prakash et al. [7] proposed reading the match result based on 5 features of IPL career and 5 features of International T20 career have been taken into considerations for both batsman and bowler where it gives an idea about the previous 6-8 months performance which gives a good combination of players. Sanket et al. [10] proposed dynamic winner prediction in a Twenty20 cricket match by comparing strength of playing team and also the score that the team batting first is expected to score based on the this the team can plan in setting up a good total on scoreboard and defending the score thereby increasing the chances of win percentage. Shilpi Agrawal et al. [12] proposed predicting the uncertainty of which team will IPL match based on teamwork of whole team on past data, instead of concentrating on a single player they worked on the best possible 11 in team by choosing the parameter of teamwork. Shubra singh and Parmeet kaur [13] addressed the problem of predicting the outcome of an IPL cricket also the player profilingsystem which can be a great help for the team leaders on the auction day, the auction brings a very good player into the team by bidding and making the team more strong by picking right player in terms of batting and bowling or an all-rounder. We have taken all the data from cricinfo.com from 2008 to 2019 season of total 12 IPL seasons for predicting the IPL match results based on players previous performance.

3. Proposed Model

By collecting the previous season data of each player and team, we perform training on the data by developing data analyzation. The information makes use of to calculate points of each player and also calculate the overall strength of each team based on past performance. The output is displayed with the help of various types of graphs where user can predict the match result or take decision in selecting the players for a particular match.

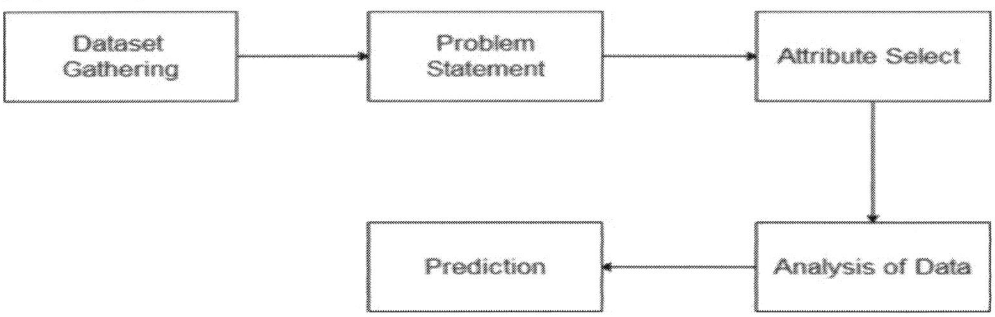

Fig. 1. Block Diagram for Proposed Model

Gathering Dataset: The first step is to collect dataset of past matches which is enough to perform the operations.

Understanding Problem Statement: Problem statement consist of the data to be displayed like, in how many matches team opted batting first after winning toss and won the match and how many choose fielding and won the match.

Attribute Selection: Base on the problem statement the data will be fetched from the.csv
File particular data is fetched.

Analysing: The raw data is converted into useful information by performing certain operations. This is done to provide information in appropriate manner.

Prediction: The future is predicated whenever that situation occurs with the help of the information the user can predict the result of the match.

There are total 756 rows and 18 columns, 756 rows imply that there were 756 IPL matches held between 2008 and 2019.

- 75% of the victorious teams that bat first won by a margin of 19 runs.
- 75% of the victorious teams that bat second won by a margin of 6 wickets.

4. Results and Discussions

This section discusses findings of our proposed model. The results are found by performing the calculations on individual players like batsman, bowler, all-rounder and fielder on different dataset. Achieved results will help the team management to decide the best playing 11 to win the match.

The Dataset used for analysis was collected from https://kaggle.com and has been scrapped from https://stats.espncricinfo.com, where data from 2008 to 2019 was available. Datasets used here are highest scores by a batsman, maximum runs scored in a particular venue, highest average score, against which bowler batsman strike rate is good and about the bowler the most dismissal kind of wickets, the bowling economy and also the Toss factor for every match played in IPL.

TABLE I: Match Attributes

Attribute	
id	dl_applied
season	matchwinner
city	result
date	matchplayer
firstteam	umpire1
secondteam	umpire2
toss	umpire3
result	Match_refree

TABLE II: Individual player's performance

Attribute	
match_id	bowler
batting_team	batsman_runs
bowling_team	extra_runs
over	total_runs
batsman	player_dismissed
non_striker	dismissal_kind

Table II consist of the twelve attribute one thousand thirty-seven instance of it. It is stored in the .csv file.

i. Run Scored By Player in All Season

batsman_run_data = bats_man_data.groupby (['season','batting_team'])
[['batsman_runs']].sum().reset_index () (1)

In above Equation batsman_run is the variable created where the season and the team is grouped (i.e. for that particular season the player played for the particular team) by using the **groupby** clause of the particular player (batsman_data) , the **sum of the run** is calculated for the every season using in-built function **sum**. Once it is done then using **reset_index ()** method the new index value is set and used.

For example, if we select any player from any team it shows his runs scored from 2008 to 2019. If the player has scored consistently in every season with good batting strike rate then that player is a X factor for the team. The following stat shows in graphical representation of a player scored in every season of IPL. This analysis helps in selecting a good batsman for the match.

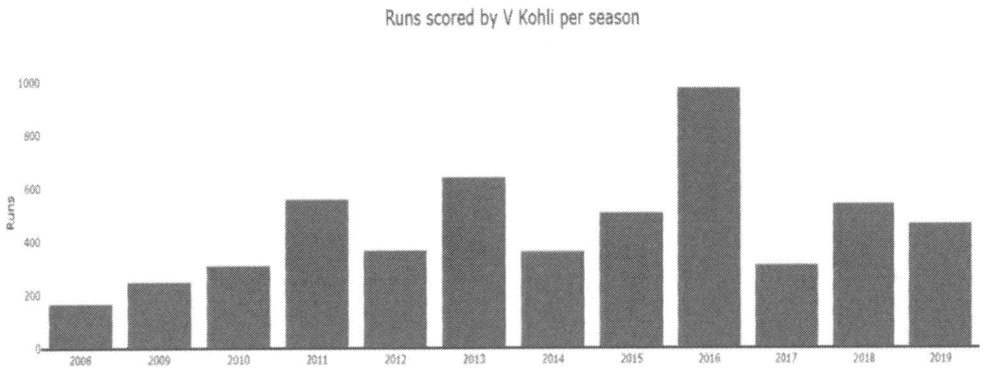

Fig. 2. Runs scored by a player in all seasons, for every team we had set a unique colour (i.e. Team Jersey Colour) in dictionary format. In the bar graph, the x-axis stores the coordinates as season and y-axis as runs. Text stores the name of the batting team

ii. Find FavouriteVenue:

venue_count = venue.groupby ('venue') [['batsman_runs']].sum ().reset_index() .sort_values('batsman_runs',ascending=False) (2)

In the above equation **venue_count** is a variable created, **venue** is the variable that stores the name of the player, the run of the player is calculated of every season and is grouped by venue using **groupby** clause then the index is reset (new value to the data set) using **reset_index (),** values of run are sorted in descending order and stored in the variable.

Every player has his favourite venue where they enjoyed playing against any team and has consistently performed well in all the matches so far they have played, may be because of the positive vibes of the city and the ground. To find out such player we did analysis of every venue.

Most runs scored by V Kohli by venue

Venue	Runs by batsman
M Chinnaswamy Stadium	2114
Feroz Shah Kotla	430
Wankhede Stadium	401
Rajiv Gandhi International Stadium, Uppal	340
Eden Gardens	307
MA Chidambaram Stadium, Chepauk	291
M.Chinnaswamy Stadium	234
Saurashtra Cricket Association Stadium	164
Maharashtra Cricket Association Stadium	156
Sawai Mansingh Stadium	131

Fig. 3. Most runs scored by a player in a particular venue (stadium). Cells display the data of the table i.e. all venues of the batsman in one column and the runs scored in an- other column.

iii. **Economy Rate of Bowler:**

bowling_data ['economy'] = bowling_data.batsman_runs / (bowling_data.ball / 6) (3)

To calculate the economy rate, formula we used is - Run Conceded/Overs bowled, using same the economy rate is calculated, bowling_data is the variable which stores the name of the bowler and it fetches the run given (run scored from that bowler ball) by the particular bowler i.e. batsman_runs .It is the divided by the total ball (of the bowler).

The economy rate of bowler plays a crucial role in selecting a good player against the strong batting opponent team. The bowler possesses skills in bowling death over and taking wickets at regular interval and restricting the team total. In our analysis we considered 50 best bowlers and we generated a scattered graph of an individual bowler.

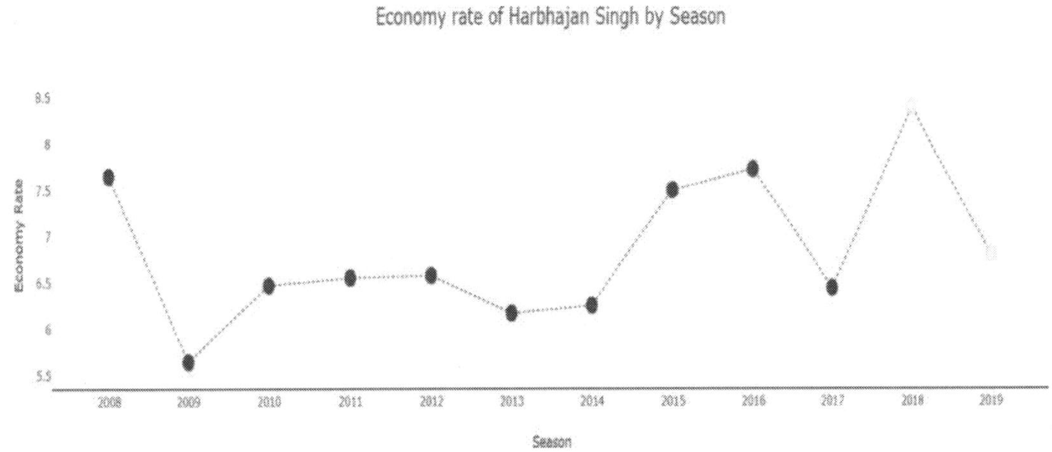

Fig. 4. Economy rate of bowler is displayed in scatter graph format, where x-axis stores the season and y-axis as economy. Text stores the name of the bowling team.

iv. Batsman Run Against Team:

runs_against_team=batsman_data.groupby ('bowling_team') [['batsman_runs']].sum(). reset_index ().sort_values ('batsman_runs', ascending=False) (4)

In the above equation runs_against_team is variable created, batsman_data is the variable that stores the name of the player, the run of the player is calculated of every season and is grouped by bowling team using groupby clause then the index is reset (new value to the data set) using reset_index (), values of run are sorted in descending order and stored in the variable.

Some batsman has performed well against few teams where they have scored consistently and contributed very well in team success, this helps the team to a pick a right player for the match.

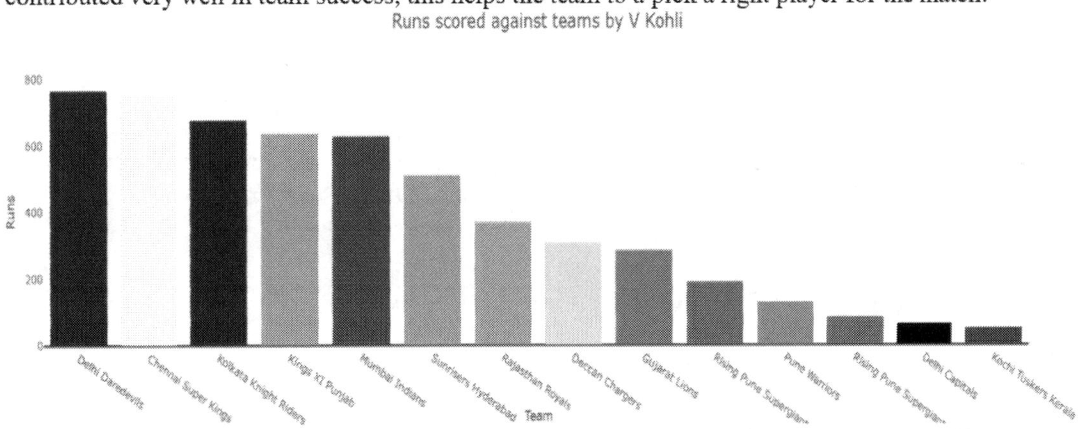

Fig. 5. Maximum runs scored by a batsman against all team, the x-axis stores the coordinates as bowling team and y-axis as runs.

v. Run Scored against Bowlers

batsman_fav_bowler = batsman_data.groupby ('bowler').agg ({'ball': len, 'batsman_runs': sum}) [['ball','batsman_runs']].reset_index()batsman_fav_bowler.sort_values ('batsman_runs', ascending=False) (5)

In equation batsman_data is the variable that stores the name of batsman and grouped by the bowler's name against which all the batsman had played, the aggregation is calculated based on total number of ball and the run scored by the batsman. The index value is reset and the sorting of the run is done of every bowler in descending order.

The Following table helps in finding a runs scored against a bowler faced a minimum 6 balls, this will help in sending a right player at crucial time to accelerate the team score by facing a minimum balls and striking at good run rate per ball.

Runs by V Kohli against bowlers (min. 6 balls)

Bowler	Runs
A Mishra	159
R Ashwin	152
UT Yadav	141
DJ Bravo	130
PP Chawla	123
JJ Bumrah	112
RA Jadeja	104
DS Kulkarni	101
SP Narine	99
TK Pathan	92

Fig.6. Run Scored against Bowlers displayed in table format, the table header as Bowler and Runs, Cells display the data of the table i.e. bowler in one column and the run scored in another column.

vi. Total Number of Wickets and Its Kind

Wickets=bowling_data.groupby(['dismissal_kind','season']).size().reset_index() (6)

In equation bowling data is the variable that stores the name of bowler and grouped by 'dismissal_kind'(the type of which the bowler took wicket like LBW, Caught etc.) group it by season , size is find i.e. number of time the wicket is taken and its kind and the index is reset to provide new index value.

The following graph shows the bowling statics of a particular bowler, which gives clear idea about the kind of dismissal of wickets.

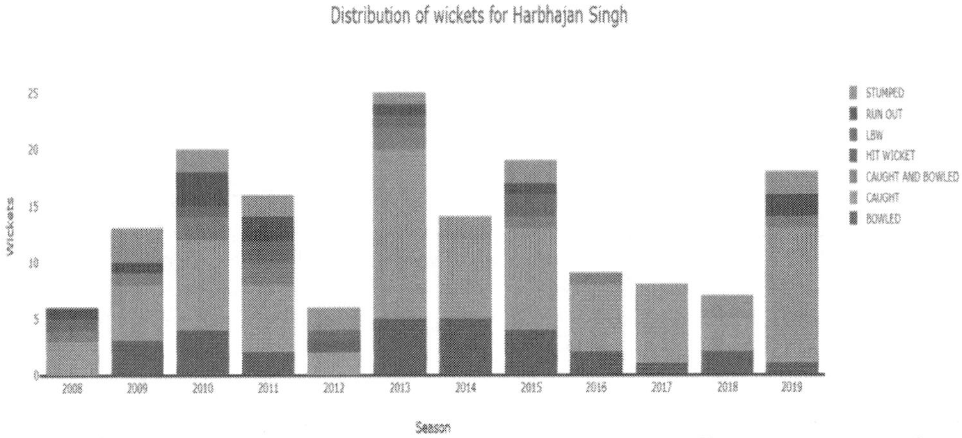

Fig.7. Total Number of Wickets and Its Kind, x-axis stores the coordinates as season and y-axis coordinate as dismissal kind.

The Toss factor plays a crucial role in deciding the match result, the analysis shows win percentage is more when the team bats first and sets up a very good total the average winning percentage is 58. And in some venues bowling first is favourable because of pitch nature where the moisture plays an important factor in swinging the ball and picking up the wickets or not allowing the batsman to score and restricting the team below average total score.

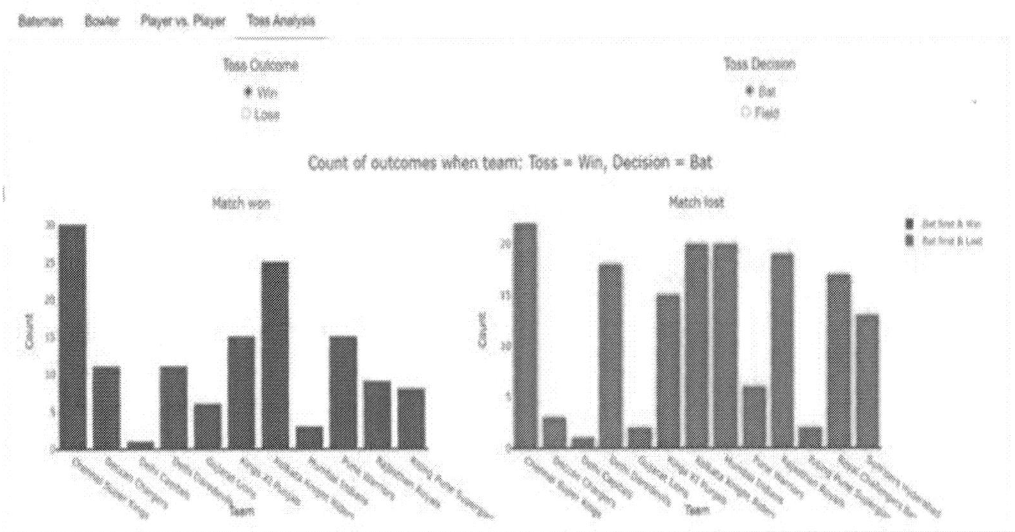

Fig.8. Toss Decision and Result

Mumbai Indians is the most successful team as they have won more than 100 plus matches and also won the most toss till 2019 followed by Chennai Super Kings and Kolkata Knight Riders. Time taken to calculate the result for all these equation is 2 seconds.

5. Conclusion

The IPL data analysis is the most famous technology in today's world. With the help of the data analysis we can find out the pros and cons of the player and work on the performance of the individual player and also the whole, it will help in taking decision. The application can be used for the selection commission to select the best player including bowler, batsman and even the fielder for the team and to perform well on the field during match. It is helpful for all type of game to work on the performance and predict the future performance of the player and team. The following methodology helps in other T20 leagues around the world like PKL,LPL,CBL,BBL,SuperSmash,T20 Blast, MSL, BPL, APL, and World League CLT20.

References

[1] Aditya Mittal, Amutha Manavalan "The IPL Model: Sports Marketing and Product Placement Sponsorship". International Journal of Humanities and Social Science Invention ISSN (Online): 2319 – 7722, ISSN (Print): 2319 – 7714 www.ijhssi.org ||Volume 6 Issue 2||February. 2017 ||PP.44-61.

[2] Amal Chaminda Kaluarachchi, Aparna S. Varde"CricAI: A classification based tool to predict the outcome in ODI cricket". Conference Paper · IEEE December2010.

[3] Amit Gupta "India and the IPL: Cricket's Globalized Empire". TheCommonwealthJournal of InternationalAffairs.

[4] Ananda Bandulasiri, "Predicting the Winner in One DayInternationalCricket". Journal of Mathematical Sciences & Mathematics Education, Vol. 3, No.1.

[5] Deep Prakash, C. Patvardhan"Data Analytics based Deep Mayo Predictor for IPL-9". International Journal of Computer Applications · October2016.

[6] Jayant Prakash, Mayank Khandelwal, and Tribikram Pradhan "Evaluation of IPL Teams and Players using Association, Correlation and Classification Rules" IEEE International Conference on Computer, Communication and Control(IC4-2015).

[7] Madan Gopal Jhawar, Vikram Pudi "Predicting the Outcome of ODI Cricket Matches: A Team Composition Based Approach". European Conference on Machine Learning and Principles and Practice of Knowledge Discovery in Databases (ECML-PKDD2016).

[8] Sasank Viswanadha, Vikram Pudi "Dynamic Winner Prediction in Twenty20 Cricket: Based on Relative Team Strengths". Conference Paper · September2017.

[9] Sachi Priyanka "Prediction of Indian Premier League-IPL 2020 using Data Mining Algorithms". International Journal for Research in Applied Science & Engineering Technology (IJRASET) ISSN: 2321-9653; IC Value: 45.98; SJ Impact Factor: 7.177. Volume 8 Issue II Feb 2020 Matches using MachineLearning".

[10] Shilpi Agrawal, Suraj Pal Singh, Jayash Kumar Sharma "Predicting Results of Indian Premier League T-20. Conference Paper · November2018.

[11] Shubhra Singh, Parmeet Kaur " IPLVisualization and Prediction Using HBase" . Information Technology and Quantitative Management(ITQM2017).

[12] https://kaggle.com

[13] https://statscricinfo.com

IOP Conf. Series: Materials Science and Engineering 1065 (2021) 012041 doi:10.1088/1757-899X/1065/1/012041

Novel Dataset Generation for Indian Brinjal Plant Using Image Data Augmentation

Balwant Gorad[1], Dr. Sirbi Kotrappa[2]

[1]Research Scholar, Computer Science and Engineering, KLE Dr. MSSCET, Belagavi, Karnataka, India – 590 008

[2]Professor, Computer Science and Engineering, KLE Dr. MSSCET, Belagavi, Karnataka, India – 590 008

goradbj@gmail.com

Abstract. Machine learning and deep learning have performed outstandingly in many computer tasks to tackle various real-world problems like disease prediction on plants etc. But unfortunately, there are many research areas like disease prediction in the agricultural crops where there is a lack of large-good quality real-world datasets. One way to solve such a problem is by using an available dataset from the internet. The problem of using an available dataset from the internet creates lots of issues. Major issues are using the dataset from different geographical locations which are deployed at other location, model overfitting due to small-sized dataset etc. The main purpose and experimentation done in this research paper are presenting different techniques to increase the size of the Indian Brinjal dataset so that deep learning models can be improved. Here data augmentation techniques to enhance the small-sized image dataset using rotation, channel shift, width shift, height shift, shear transform, brightness, scaling, uniform aspect ratio, zoom, horizontal flipping, and vertical flipping methods are used. At last, a huge high-quality training dataset of size 39,010 is generated from 350 sample images taken from the real field and 1356 high-quality images are generated to validate the model using above mentioned data augmentation techniques.

1. Introduction

Deep convolutional neural network have performed outstandingly in many computer tasks like supervised, unsupervised and semi-supervised learning. But unfortunately, many research areas are not having the good quality dataset with huge volume. One of such domain is Indian agriculture. Quality agriculture is based on collecting comprehensive information from both environment and plant. Information from environment like soil moisture, temperature, wind speed, rain fall, and humidity are to be collected and need to be analyzed. Whereas in case of plant, information like plant disease, plant growth, and insect pest is to be analyzed and monitored. Currently many researchers working in the field of machine and deep learning are using datasets which are available on internet, so satisfactory work is not happening in these areas due to limited size of dataset. To build any kind of machine learning or deep learning model, one of the major task is data acquisition; it's preprocessing and making good quality dataset ready to build classifiers. Many well-known researchers in these areas suggest having a good quality preprocessed dataset before building the model is better. If you not having good quality dataset, we may lead to less accurate models or infeasible solutions. In case of plant dataset available on internet is from different geographical locations and experiment or deployment of research models are done different geographical locations, which is leads to infeasible way due to environment factors, soil properties, types of diseases and many other factors. This research-based experimentation we have carried out for Indian Brinjal plant. This is going generate good quality preprocessed dataset with almost having size 40,336 images using various techniques like cleaning data, preparing data and manipulating data using image data augmentation and dimensionality reduction. In our research, the actual data set used in this research to generate high quality dataset in huge volume is collected manually from different farms located in western Maharashtra region. Data Augmentation consists of various techniques that enhance the size and quality of training datasets such that better deep learning models can be built using them. Finally, the outcome of this research experimentation is huge high quality Indian Brinjal dataset which can be used for building various kinds of machine or deep learning classifiers.

2. Related Work

[1] The work carried out in this paper has presented various methods for Data Augmentation, like flipping, color space, translation and noise rejection but to generate good quality dataset, only these methods are not enough. Additions to these all methods, few more methods are needed to be considered like shear transform, scaling, zoom, brightness, channel shift.

[2] The work done in this paper is based on scale & shift parameters, new images are generated from fewer images using these parameters. This researcher focused more about scale and shifting parameters only, but again only these parameters are not enough to generate good quality dataset so there is need to work more in this field.

[4] In this paper, multiple solutions to the problem of data augmentation in image classification are explored and compared. Experiments carried out here used GANs to generate images of different styles. Some questions like safety improvement and quality data generations is not yet achieved.

[5] The work carried out here by TensorFlow community is on flowers dataset. They have performed various data augmentation technique in very good way, results are good but as we need more work to be carry out in Indian agriculture using various data augmentation techniques. So still there is scope to do more work in this field.

[6] The work done in this paper is based on molecular study and enhancement of molecular dataset. They have claimed they have received good result in their work where dataset in is form of text representation of molecule. As they have carried out the work in molecular area, still scope is there, to use different techniques of augmentations in agriculture field.

3. Dataset Collection

3.1. Overview

In this research, work has been carried out on Indian Brinjal plant. We have taken training data set in the form of image. Even it is possible to have dataset in video format, in such case video frames are to be convert into multiple images.

3.2. Dataset Collection for data augmentation

Real world dataset has been taken for image data augmentation process. working with real dataset is always better while building machine or deep learning models which may be used by different researcher from same geographical locations. Along with data collection, information of different disease types on Brinjal plant have been collected from farmers and agriculture institutes/universities from the state of Maharashtra and Karnataka, India.

Total 350 images of Brinjal plant are obtained which corresponds to 7 classes. These 7 classes contain 1 healthy and 6 unhealthy classes which includes alternia leaf spot, bacterial wilt, collar rot, damping off, tobacco mosaic virus and cersocpera leaf spot.

Figure 1Real-time dataset collection

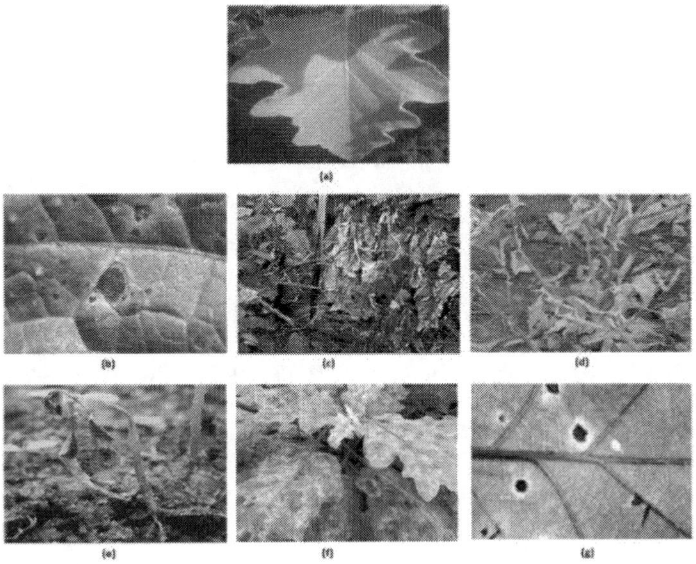

Figure 2 Six common types of brinjal diseases (a) healthy (b) alternia leaf spot, (c) bacterial wilt, (d) collar rot, (e) damping off, (f) tobacco mosaic virus and (g) cersocpera leaf spot

3.3. Overview of Samples Collected

Table 1 Disease wise number of samples used

Type of Brinjal disease	No. of samples used
Alternia leaf spot	40
Bacterial Wilt	45
Collar Rot	35
Damping Off	32
Tobacco Mosaic Virus (TMV)	55
Cersocpera leaf spot	27
Healthy samples	116
Total samples	**350**

4. Proposed Method

4.1. Overview of data augmentation

This data augmentation is very powerful tool to create more data. It modifies an image and gives back lots of unique new images, based on different operations. Operations incudes rotation, width and height shifts, brightness, shear transform, zoom, channel shift, horizontal and vertical flips, data normalization, rescale and pre-processing function. These all operations plays a vital role to generate big volume good quality dataset with little care taken during the process. The key idea of all these methods is explained in following part of this proposed method section

4.2. Uniform Aspect Ratio

First step is to ensure that, all the images present are dataset are having the same size and same aspect ratio.

4.3. Image Scaling

Once it has been ensured of have same dimensions to all images (or have some predetermined aspect ratio), it is the time to perform scaling operation on each image appropriately. We have decided to have images with width of 400 and height of 300 pixels.

4.4. Rotation

This method is used to rotate image by specified degree. If we set rotation range as 60^0 then rotation of image is done between -60^0 to $+60^0$. You can experiment part to know more about this rotation.

4.5. Width and Height Shifts

This method controls the amount of width and height shifting. It is number of floating type and its value is in between 0 and 1. This specifies upper bound by which image is randomly shifted on left or right side. In case of image is randomly shifted towards up or down.

4.6. Brightness

The brightness range indicates the range for randomly picked image brightness shift value. 0.0 indicates absolutely there is no brightness, and 1.0 indicates maximum image brightness.

4.7. Shear Transform

Shear transformation change the shape of the given image. Shear transforms and rotation is different from each other. In shear transformation, we are fixing one axis and stretching the image at a certain angle called shear angle. This is going to create a sort of 'stretch' in the given image, which we are not able to see rotation. Shear range specifies the angle of the slant in degrees.

4.8. Zoom

This random zoom technique is obtained by the zoom range argument in Keras with TensorFlow machine learning framework. A zoom with less than 1.0 magnifies the image, while a zoom with greater than 1.0 zooms out of the image.

4.9. Channel Shift

This Channel shift technique randomly shifts the channel values by using random value which is chosen from the range specified by channel shift range.

4.10. Horizontal and Vertical Flips

The horizontal flipping technique generate horizontally flipped images on a random basis whereas generates vertically flipped images.

5. Experiments

5.1. Overview of data augmentation

These experiments have been carried out using Google TensorFlow with Keras. Dataset required for these experiments synchronized with Google Drive and it is then mounted with Google collaboratory. The purpose of conducting these experiments on collaboratory is to have high end hardware configuration. During the experimentation 12.72GB RAM was available and 107.77GB disk was available.

5.2. Uniform Aspect Ratio

This technique ensures, all the images are of same size and same aspect ratio. When we have captured images from the field, it was having different uniform aspect ratio, so we have converted all images into same uniform aspect ratio.

5.3. Image Scaling

We have decided to have images with width of 400 and height of 300 pixels.

5.4. Rotation

$$\text{Rotation range} = 60^0$$

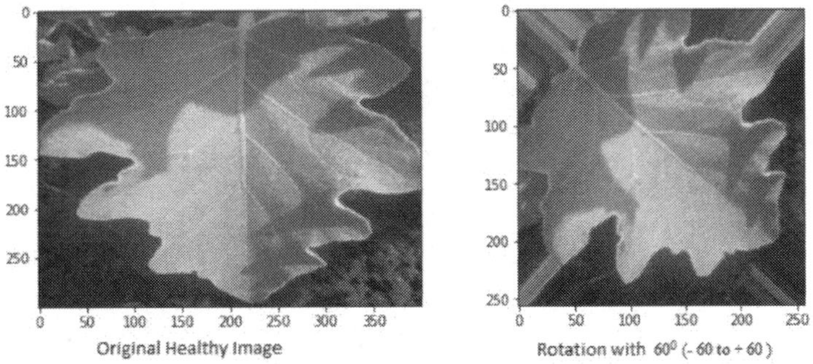

Figure 3 Augmentation using Rotation

5.5. *Width and Height Shifts*

Width shift = 30, Height shift = 10

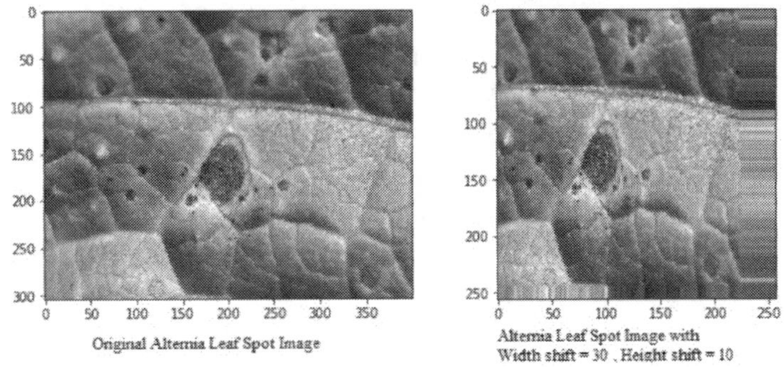

Figure 4Augmentation using Width and Height Shifts

5.6. *Brightness*

Brightness range = (0.5, 2.0)

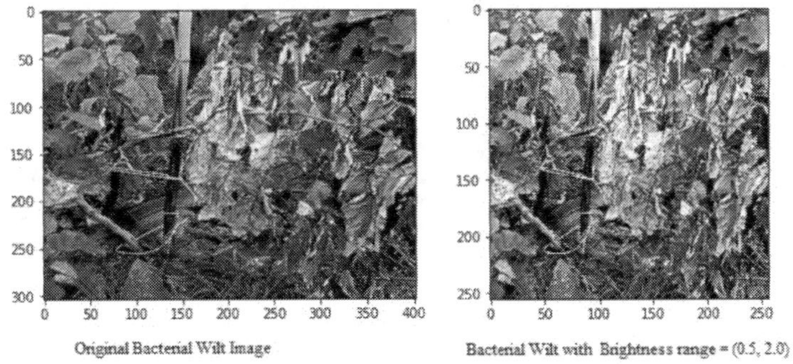

Figure 5 Augmentation using Brightness

5.7. *Shear Transform*

Shear range=40

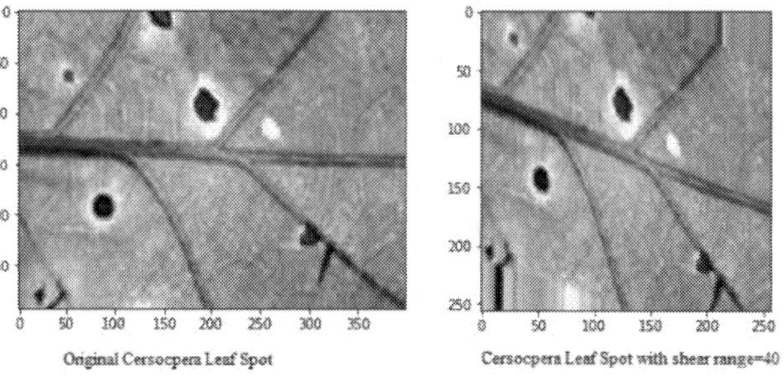

Figure 6 Augmentation using Shear Transform

5.8. Zoom

Zoom range = [0.7, 1.5]

Figure 7 Augmentation using Zoom

5.9. Channel Shift

Channel shift range=100

Figure 8 Augmentation using Channel shift

5.10. Horizontal and Vertical Flips

Horizontal flip=True

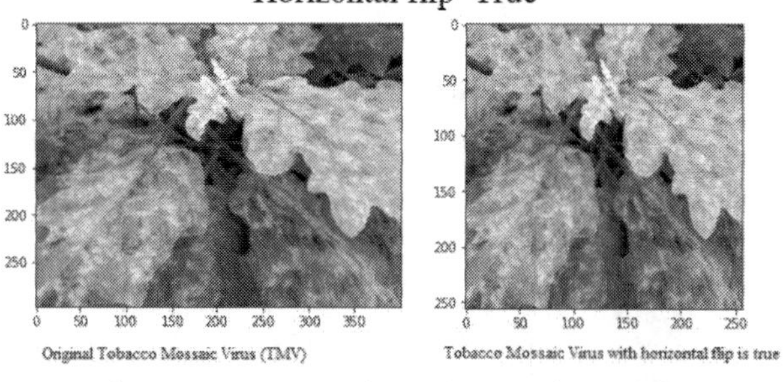

Figure 9 Augmentation using Horizontal flip

Vertical Flip = True

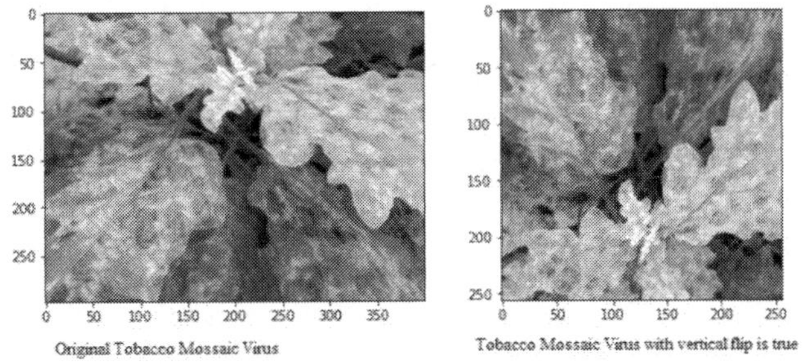

Figure 10 Augmentation using Vertical Flip

Horizontal flip = True and Vertical flip = True

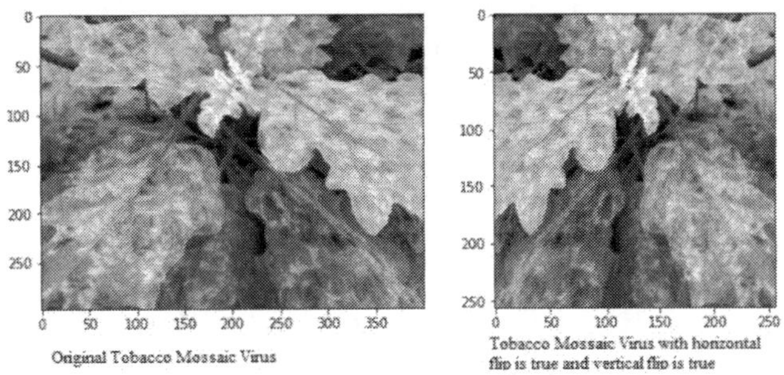

Figure 11 Using Horizontal & Vertical Flip

6. Results

Several experiments are carried out to explore effectiveness of different image data augmentation techniques. From each disease class, sample images are used to augment it. After augmenting huge sized dataset contains 40336 for both training and validation is generated.

Disease type vs. No. of augmented images for training dataset are shown below.

Table 2 Disease wise augmented images for training

Type of Brinjal disease	No. of samples used	No. of Augmented Images
Alternia leaf spot	40	1907
Bacterial Wilt	45	3136
Collar Rot	35	2621
Damping Off	32	3608
Tobacco Mosaic Virus (TMV)	55	3302
Cersocpera leaf spot	27	1835
Healthy samples	116	22601
Total	**350**	**39010**

Disease type vs. No. of augmented images for validation dataset are shown below.

Table 3 Disease wise augmented images for validation

Type of Brinjal disease	No. of samples used	No. of Augmented Images
Alternia leaf spot	40	152
Bacterial Wilt	45	252
Collar Rot	35	202
Damping Off	32	100

Tobacco Mosaic Virus(TMV)	55	104
Cersocpera leaf spot	27	99
Healthy samples	116	447
Validation Total	350	1356
Training Total		39010
Training + Validation		40336

Sample of augmented healthy dataset of size 39010 and unhealthy dataset of size 1356 is shown below.

Figure 12 Healthy augmented dataset

Figure 13 Unhealthy augmented dataset

7. Discussion

There are two types of image data augmentations, one is called data warping and second is called oversampling. Both of these types are used to improve deep learning model or image classifier. So, these augmentation techniques really help in neural network, machine learning and deep learning applications. Currently there is not any data augmentation technique available that can correct a dataset that has very less diversity. All these augmentation algorithms are performing best under the assumption that the training data and testing data are both taken from the same set of samples.

Following bar graph and line graph shows disease wise number of samples used for augmentation verses number of augmented images for training the classifier.

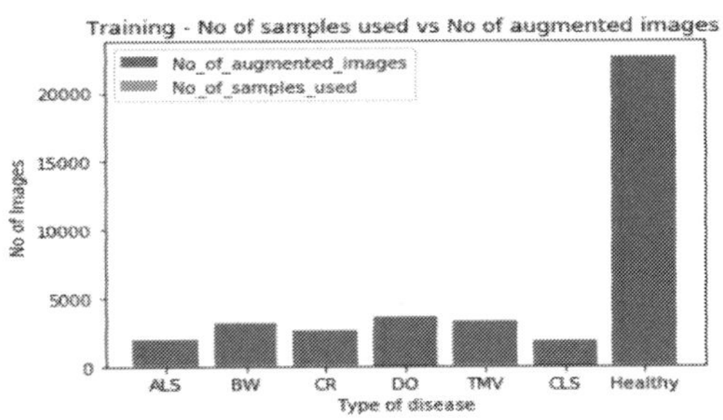

Figure 14 Samples vs. Augmentation bar graph for train dataset

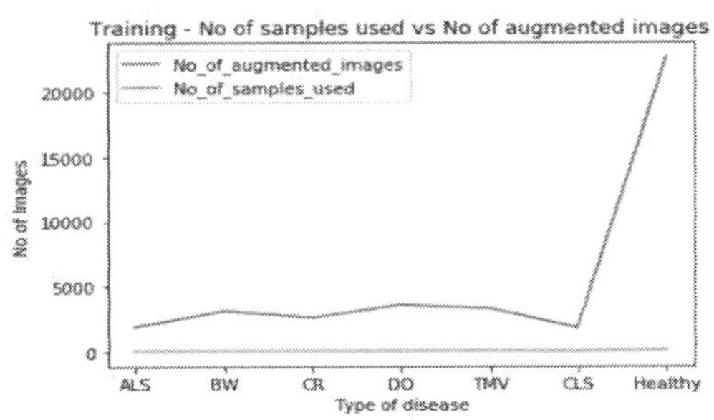

Figure 15 Samples vs. Augmentation line graph for train dataset

Following bar graph and line graph shows disease wise number of samples used for augmentation verses number of augmented images for validating the classifier.

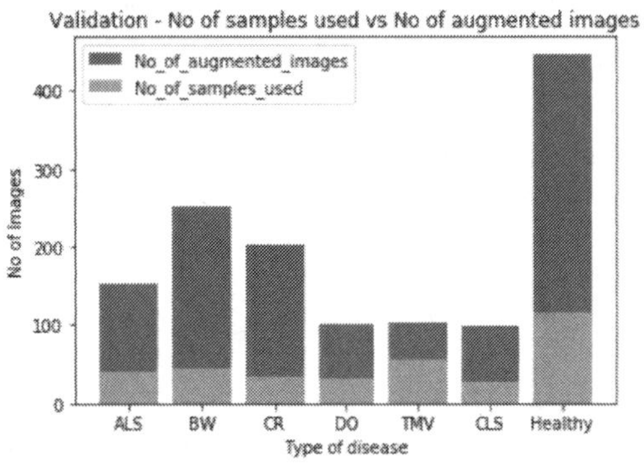

Figure 166 Samples vs. Augmentation bar graph for validation dataset

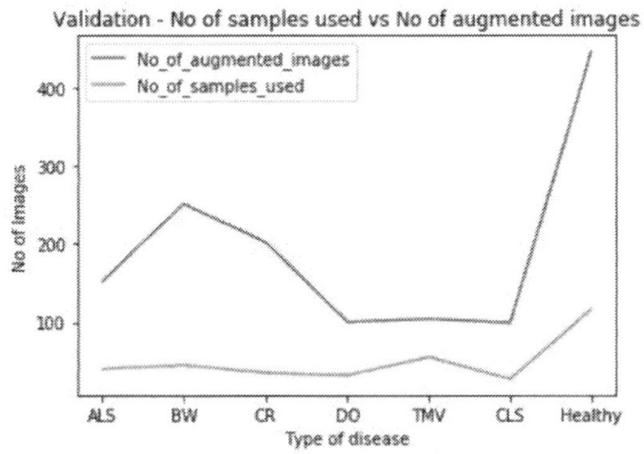

Figure 17 Samples vs. Augmentation line graph for validation dataset

8. FUTURE WORK

Once quality dataset with enough volume is ready by performing all above data augmentation techniques, such dataset cab be used as part of future work to train the classifier. Such classifiers are used to make classification in many applications in the field of agriculture, healthcare, automotive industries and many more. Even one more step can be done is that, after building and testing classifiers, hyper parameter tuning can be done which improves performance of models. Apart from techniques discussed above, some more techniques like normalization, scaling and preprocessing functions can be used to generate a greater number of images as a part of future work.

9. CONCLUSION

Experimentation done in this research presents series of data augmentation techniques which gives solution to overfitting in machine and deep learning models due to limited size of dataset. In many areas of machine learning, deep learning and big data, this technique gives tremendous benefits to the researchers. This image data augmentation technique is much useful for generating better quality datasets. In coming future, this technique will be used in many applications of emerging areas. Data augmentation is not going to reduce complete noise from input dataset but overall, this is one of recommended technique to create better quality dataset. The major problem of deep learning and machine learning models is overfitting and with the help of these data augmentation technique, the problem of model overfitting can be defeated.

Keywords

Data Augmentation, Machine Learning, Deep Learning, Deep Convolutional Neural Network, Big Data, Image Data Augmentation, Data Science, Image Classification, Machine Learning Classifier, Indian Agriculture, Neural Network

Acknowledgment

We would like to thank the KLE Dr. MSSCET, Belagavi, Karnataka; Visvesvaraya Technological University, Belagavi, Karnataka and Shriram Agriculture College, Mahatma Phule Agriculture University, Solapur, Maharashtra for helping us to carry out this research. Additionally, Opinions, findings, conclusions, or recommendations in this paper are solely of the authors' and do not reflect the views of the above-mentioned organizations.

References

[1] Connor Shorten and Taghi M. Khoshgoftaar, "A survey on Image Data Augmentation for Deep Learning", Journal of Big Data, Sprinjer, Shorten and Khoshgoftaar J Big Data (2019)

[2] Atsuhiro Noguchi1 and Tatsuya Harada, "Image Generation From Small Datasets via Batch Statistics Adaptation", The University of Tokyo, 2RIKEN, arXiv:1904.01774v4 [cs.CV] 23 Oct 2019
[3] Mr. Balwant J Gorad, Dr. S. Kotrappa, "Comparative Study and Review for Development of Disease Prediction System for Indian Crops", IOSR Journal of Computer Engineering (IOSR-JCE), e-ISSN: 2278-0661,p-ISSN: 2278-8727, Volume 21, Issue 2, Ser. I (Mar - Apr 2019), PP 68-73
[4] Jason Wang, Luis Perez, "The Effectiveness of Data Augmentation in Image Classification using Deep Learning", Standford University, Google, 2018
[5] Deep mnist for experts https://www.tensorflow.org/get_started/ mnist/pros Accessed: 2017-05-19.
[6] E. Jannik Bjerrum. SMILES Enumeration as Data Augmentation for Neural Network Modeling of Molecules. ArXive-prints, Mar. 2017.
[7] Mr. B J Gorad, Dr. S. Kotrappa, Dr. Jibi Abraham, "Development of Intelligent Monitoring IoT Solution for Indian Agriculture Fields", NCRET, SITCOE, India
[8] Y. Kubo, G. Tucker, and S. Wiesler. Compacting Neural Network Classifiers via Dropout Training. ArXiv e-prints, Nov. 2016.
[9] X. Liang, Z. Hu, H. Zhang, C. Gan, and E. P. Xing. Recurrent topic-transition GAN for visual paragraph generation. CoRR, abs/1703.07022, 2017.
[10] Y. Ma and D. Klabjan. Convergence analysis of batch normalization for deep neural nets. CoRR, abs/1705.08011, 2017
[11] M. Marchesi. Megapixel Size Image Creation using Generative Adversarial Networks. ArXiv e-prints, May 2017. 2[14] C. N. Vasconcelos and B. N. Vasconcelos. Increasing deep learning melanoma classification by classical and expert knowledge based image transforms. CoRR, abs/1702.07025, 2017
[12] B. Wang and D. Klabjan. Regularization for unsupervised deep neural nets. CoRR, abs/1608.04426, 2016
[13] S. C. Wong, A. Gatt, V. Stamatescu, and M. D. McDonnell. Understanding data augmentation for classification: when to warp? CoRR, abs/1609.08764, 2016
[14] S. Xiang and H. Li. On the Effects of Batch and Weight Normalization in Generative Adversarial Networks. ArXive-prints, Apr. 2017.
[15] Krizhevsky A, Sutskever I, Hinton GE. ImageNet classification with deep convolutional neural networks. Adv Neural Inf Process Syst. 2012;25:1106–14.

IOP Conf. Series: Materials Science and Engineering 1065 (2021) 012042 doi:10.1088/1757-899X/1065/1/012042

Passenger data analysis of Titanic using machine learning approach in the context of chances of surviving the disaster

Md Arfinul Haque[1], Shivaprasad G[2,4] and Guruprasad G[3]

[1]PG Student, Department of CSE, Manipal Institute of Technology, Manipal Academy of Higher Education, Manipal, India

[2]Department of CSE, Manipal Institute of Technology, Manipal Academy of Higher Education, Manipal, India

[3]Department of CSE, Yenopaya Institute of Technology, Mangalore, India

[4]shiva.prasad@manipal.edu

Abstract. Titanic disaster occurred about 100 years back but still it attracts the researchers to understand and study that how some passengers survived and others perished. In this work, the characteristics of the passengers will be identified and the relationship of survival chance from the disaster is found. Feature engineering techniques will be performed, where the alphabetic values will be changed to numeric values, the family size will be calculated. Also, we will extract the title from the name, and deck label from ticket number. Classification is done using Decision tree machine learning classification algorithm using two classes which are survived and not survived. R programming has been used for its implementation. Clustering is performed using KMeans machine learning algorithm. Its implementation has been done using Python programming.

1. Introduction

One the most infamous Shipwrecks in history is the sinking of the RMS Titanic. It happened on April 15, 1912. When an iceberg collided with the titanic, it sank and 1502 passengers died from the total of 2224 passengers as well as the crew people got killed. The International community got shocked by this sensational tragedy and because of this, now there are better safety regulations for the ships.

Based on data set obtained from Kaggle, the purpose is about survival prediction of the titanic [4]. We will split the historical data set that we obtained from kaggle into two groups – train set and test set. They are the two different files in csv format. The train data will be an input for the training model. This data will be used to build the model for the generations of predictions for the test data. There can be many flaws in the train set because of which training of the model on the basis of such data is not possible or can leads to an incorrect output. It is possible that in the csv file of the train set, there will be few blank cells where no values is mentioned. For an example, if the age of a person is not mentioned then this a flaw in the train data set. Hence, it has to be filled with appropriately calculated age value before training phase. This is feature engineering that we do to make the train data set perfect for giving it as an input for training the model. The titles for detecting the crew members and other professional among the passengers also has to be grabbed. Such titles can be like Dona, Lady, Capt, Sir, Doc etc.

Male and female will be identified using those titles of the names. The titles belonging to a female and male professionals will help to classify the female and male people, then all such titles which were in huge number will be shorted down into titles like Master, Miss, Mr, Mrs and Rare Title. The surname from the passenger names has to be grabbed for detecting the members in the family. This can be done by breaking the passenger names into some meaningful variables. The unique surnames will be grabbed to analyze the families. After applying further such feature engineering on the train data, the training model will be applied using classification algorithm. Decision tree machine learning algorithm will be used for predicting the survival status of each passenger in the test data set. Decision tree is used to train the machine about the survival of the people in titanic. Two classes will be used for predicting the survival status [3]. At every level in the decision tree the probability of the survival will be identified. Every node in the decision tree can be identified using the node number and at every node number the number of observations can be observed. The information at every node in the tree will be explained using the parameters like predicted class, expected loss, P(node), class counts, probabilities. These parameters will be used to identify the information of people whose family size is zero or was travelling alone. For family, at every node the information will be explained using complexity parameter, predicted class, expected loss, P(node), class counts, probabilities, left son and right son. Further the primary splits and surrogate splits will be displayed using age, passenger class and family size. Overall summary of decision tree classification will be examined using conditional probability, number of splits, relative error and expected error. Accuracy of the predicted test data will be calculated with respect to the survived information available in the train data set.

Similarly, there are modifications that needs to be performed on training data set. The prediction that whether or not the titanic passengers survived the sinking will be found for each passenger in the test set. Different machine learning techniques will be applied on the data set available. By the results obtained from those techniques we will be able to understand the statistics more clearly. Algorithms used includes feature engineering, decision tree visualization, random forest etc. [1].

2. Problem Definition

We have to perform the complete analysis about what kind or sort of titanic passengers likely to survive in the titanic disaster. The type of passengers include those who travelled alone and passengers with their family. Passengers type can be analyzed by family size, family size has to be categorized into three categories like singleton, small and large family. Family size as 1 will belong to singleton, from 2 to 4 family size as small family and family size greater than 4 as a large family. Depending on three family categories, the survival rate is small or large can be identified. It has been illustrated in the Fig.1 for understanding the family size categorization. This will help in training the data about survival and then same can be predicted in test data.

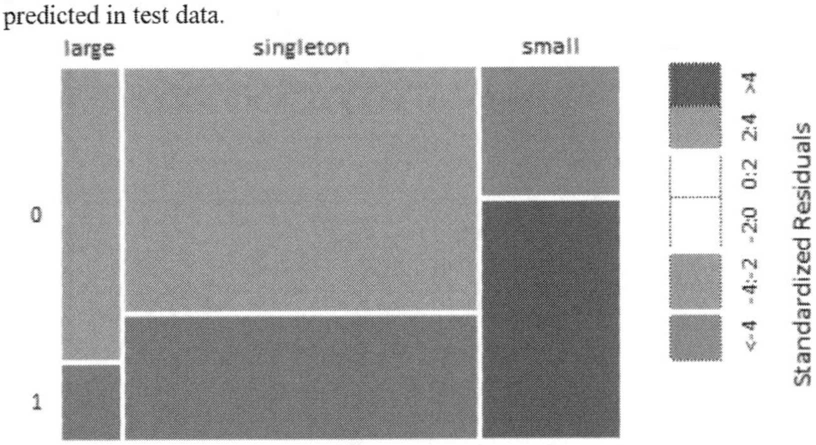

Fig. 1. Family Size Survival

IOP Conf. Series: Materials Science and Engineering 1065 (2021) 012042 doi:10.1088/1757-899X/1065/1/012042

Similarly, the crew members also have to be classified by their working departments. Finally, the output will be the survival status as 1 and not survived as 0. Survival status has to be understood using classification algorithm and using also using a clustering algorithm. For classification and clustering, decision tree and KMeans algorithm has to be implemented. Accuracy has to be checked for the machine learning algorithms implemented [5]. Expected accuracy is greater than 80 percent but for future work more than the resulted accuracy rate has to be implemented. This is possible by modifying and feature engineering the train data more accurately. Therefore, for this, we will be applying the tools of machine learning for prediction of passengers survived the tragedy. In the decision tree, there will be 2 classes i.e. male and female then depending on the survival status 0 and 1, the decision tree has to be grown. For the survival status of male and female respectively, further survival rate will be calculated. Same way the decision tree will grow depth wise. Each node of decision tree will indicate either 1 or 0 which are the status of survived and not survived passengers.

3. Objective

In the test data, all the attributes of train data will be mentioned except the survived column. So, the objective is to get the survival status of each passenger in the test data. The same expected result is illustrated in the Fig.2 and also the accuracy value in this prediction when compared with the survived status of train data set.

```
> my_prediction <- predict(my_tree, test_new, type = "class")
> head(my_prediction)
892 893 894 895 896 897
  0   0   0   0   0   0
Levels: 0 1
> my_prediction
 892  893  894  895  896  897  898  899  900  901  902  903  904  905  906
   0    0    0    0    0    0    0    0    1    0    0    0    1    0    1
 907  908  909  910  911  912  913  914  915  916  917  918  919  920  921
   1    0    0    1    0    0    0    1    0    1    0    1    0    0    0
 922  923  924  925  926  927  928  929  930  931  932  933  934  935  936
   0    0    0    0    0    0    1    0    0    0    0    0    0    1    1
 937  938  939  940  941  942  943  944  945  946  947  948  949  950  951
   0    0    0    1    0    0    0    1    1    0    0    0    0    0    1
 952  953  954  955  956  957  958  959  960  961  962  963  964  965  966
   0    0    0    0    0    1    1    0    0    1    0    0    0    0    1
 967  968  969  970  971  972  973  974  975  976  977  978  979  980  981
   0    0    1    0    0    1    0    1    0    0    0    1    1    1    1
 982  983  984  985  986  987  988  989  990  991  992  993  994  995  996
   0    0    1    0    1    0    1    0    0    0    1    0    0    0    1
 997  998  999 1000 1001 1002 1003 1004 1005 1006 1007 1008 1009 1010 1011
   0    0    0    0    0    0    1    1    1    1    0    0    1    1    1
1012 1013 1014 1015 1016 1017 1018 1019 1020 1021 1022 1023 1024 1025 1026
   1    0    1    0    0    1    0    1    0    0    0    0    0    0    0
1027 1028 1029 1030 1031 1032 1033 1034 1035 1036 1037 1038 1039 1040 1041
   0    0    0    0    0    0    1    0    0    0    0    0    0    0    0
1042 1043 1044 1045 1046 1047 1048 1049 1050 1051 1052 1053 1054 1055 1056
   1    0    0    0    0    0    1    0    0    1    1    0    1    0    0
```

Fig. 2. Survival status in test data

In order to get the expected result, there is a need to apply 3 techniques which are feature engineering technique, classification algorithm and clustering algorithm.

Feature Engineering [7][8][9][10][11] is required for modifying the train data set because there can be some empty spaces, title for some passengers will be missing, department for some crew members may not be mentioned. Unique surname has to be extracted then calculation of family size will be required for analyzing the families through unique surnames. Visualization of relationship between the family size and survival is required. Hence for all such errors this technique has be applied before applying the machine learning techniques for training the model. The best way for doing this using R programming for data extraction and analyzing using graphs.

Classification Technique [12][13] is used to determine that for which class or category a given observation will belong to. There are many algorithms related to classification but for this work, the decision tree algorithm is employed. All nodes in the tree will indicate either the passenger survived or not survived using 0 and 1 respectively. At the first level, the relation between the child nodes and the root node will indicate the gender of passenger. Further, the relation will be based on age, Passenger class and family size. The implementation will be done using the library called "rpart" in R programming. Accuracy for the resulted survival status in the test data has to be calculated by comparing with the survival status of passengers in train data.

Clustering Technique [14][15] is used to group a set of observations in such a way that observation of similar types belongs to a group. There can be many groups defined while clustering. This technique has also to be implemented for examining passenger's survival in titanic disaster. This has been illustrated using KMeans algorithm in Python as illustrated in the Fig. 2.

4. Background

This research work is part of machine learning and data mining domain. A computer can get an ability to learn without being explicitly programmed by machine learning. Algorithms with machine learning techniques can be constructed in such a way that a computer can learn from and also make predictions on data available. Computational statistics is closely related or we can say overlapped in machine learning. Computers can do predictions on data available by machine learning techniques. Even machine learning is very closely related to mathematical optimization by which many methods, theory and applications gets developed.

5. Implementation

For implementation of the task to be performed, R and Python has been used. There are different R libraries involved which can be understood by the table (Table 1).

Feature engineering technique where Passenger names has to be broken down into some meaningful variables which will be used for predictions or for creating new variables. As passenger title is mentioned in passenger name, we can use a new variable as surname to represent the families. After that all the unique surnames of all the passengers in the train data has to be grabbed to analyze the families. The family size will be calculated using this result of unique name extraction. The family size calculation will be done using the formula as:

$$\text{Family Size} = \text{Parents Children} + \text{Siblings Spouses} + 1$$

Further, the missing value of passengers age has to be calculated for filling such missed age values of passengers. After applying all such feature engineering methods, while analyzing the data, it can be noticed that the death of passengers who travelled alone was more compared to the survival of singleton passengers. As illustrated in Fig.3, death was less than survival for passengers whose family size was 2. Similar condition was observed with the passengers with family size of 3 and 4. This indicates that survival status for small sized family was more compared to singleton and large family size. Further, after analyzing the data, the training model will be generated.

Table 1. Different R Libraries Involved

Purpose	R Library Involved
Visualization	library('ggplot2')
	library('ggthemes')
	library('scales')
Data manipulation	library('dplyr')
Imputation	library('mice')
Decision tree	library(rpart)

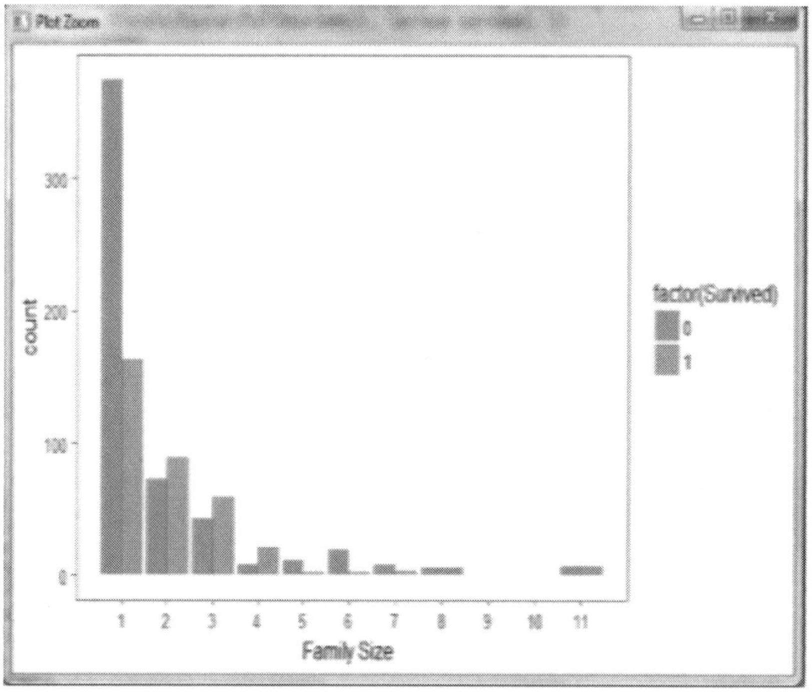

Fig. 3. Family size & Survival

Classification technique will be applied for determining the class or category of a given observation to which it will belongs to. All nodes in the tree will indicate either the passenger survived or not survived using 0 and 1 respectively. As illustrated in the Fig. 4, at the first level the relation between the child nodes and the root node will indicate the gender of passenger. Further, the relation will be based on age, Passenger class and family size. The implementation will be done using the library called "rpart" in R programming. For every node, the survival status has to be displayed. The splits are categorized into primary and surrogate split. Each split is based on sex, Passenger class, family size, and age. Splits will appear only if a passenger has travelled with family. At every node, the number of observations will be displayed.

Two classes have been used to classify the survival i.e- survived and not survived. After generating the tree using 'rpart' in R language, we can notice that totally 435 nodes will be generated [2]. Also level wise error is decreasing or accuracy is increasing. At first level, the accuracy is zero as error is 100 percent. The summary of the information from the decision tree has been shown in Fig. 5 where columns indicate the number of splits, relative error and expected error at each level. Conditional probability at each level will get reduced from root level to the leaf. Conditional level has been taken as 0.0001 while implementing. This indicates that, when the conditional probability of a node will reach this, then further node generation will get stopped. The expected output for the survival status for each passenger has been displayed in the Fig. 2 by 1 and 0 which are indicating survived and not survived passengers in the titanic disaster.

```
Console
 $ family_size: num  1 2 1 1 3 1 1 3 1 3 ...
> my_tree <- rpart(Survived ~ Age + Sex + Pclass  + family_size, data = train_new,
 method = "class", control=rpart.control(cp=0.0001))
>
> my_tree
n= 891

node), split, n, loss, yval, (yprob)
      * denotes terminal node

  1) root 891 342 0 (0.61616162 0.38383838)
    2) Sex=male 577 109 0 (0.81109185 0.18890815)
      4) Age>=6.5 553  93 0 (0.83182640 0.16817360)
        8) Pclass>=1.5 433  50 0 (0.88452656 0.11547344) *
        9) Pclass< 1.5 120  43 0 (0.64166667 0.35833333)
         18) Age>=36.5 81  22 0 (0.72839506 0.27160494)
           36) Age>=53 22   3 0 (0.86363636 0.13636364) *
           37) Age< 53 59  19 0 (0.67796610 0.32203390)
             74) Age< 47.5 46  11 0 (0.76086957 0.23913043) *
             75) Age>=47.5 13   5 1 (0.38461538 0.61538462) *
         19) Age< 36.5 39  18 1 (0.46153846 0.53846154)
           38) Age< 33.5 29  13 0 (0.55172414 0.44827586)
             76) Age>=28.5 8   2 0 (0.75000000 0.25000000) *
             77) Age< 28.5 21  10 1 (0.47619048 0.52380952)
              154) Age< 24.5 10   3 0 (0.70000000 0.30000000) *
              155) Age>=24.5 11   3 1 (0.27272727 0.72727273) *
           39) Age>=33.5 10   2 1 (0.20000000 0.80000000) *
      5) Age< 6.5 24   8 1 (0.33333333 0.66666667)
       10) family_size>=4.5 9   1 0 (0.88888889 0.11111111) *
       11) family_size< 4.5 15   0 1 (0.00000000 1.00000000) *
    3) Sex=female 314  81 1 (0.25796178 0.74203822)
      6) Pclass>=2.5 144  72 0 (0.50000000 0.50000000)
       12) family_size>=4.5 27   3 0 (0.88888889 0.11111111) *
       13) family_size< 4.5 117  48 1 (0.41025641 0.58974359)
```

Fig. 4. Classification

```
> summary(my_tree)
Call:
rpart(formula = Survived ~ Age + Sex + Pclass + family_size,
    data = train_new, method = "class", control = rpart.control(cp = 1e-04))
  n= 891

           CP nsplit rel error    xerror       xstd
1 0.444444444      0 1.0000000 1.0000000 0.04244576
2 0.030701754      1 0.5555556 0.5555556 0.03574957
3 0.023391813      3 0.4941520 0.5116959 0.03467453
4 0.020467836      4 0.4707602 0.5087719 0.03459945
5 0.005847953      6 0.4298246 0.4766082 0.03374384
6 0.004385965     11 0.3976608 0.4853801 0.03398272
7 0.002923977     13 0.3888889 0.4853801 0.03398272
8 0.000100000     17 0.3771930 0.4824561 0.03390356

Variable importance
        Sex         Age      Pclass family_size
         49          19          18          13

Node number 1: 891 observations,    complexity param=0.4444444
  predicted class=0  expected loss=0.3838384  P(node) =1
    class counts:   549   342
   probabilities: 0.616 0.384
  left son=2 (577 obs) right son=3 (314 obs)
  Primary splits:
      Sex         splits as  RL,          improve=124.42630, (0 missing)
      Pclass      < 2.5       to the right, improve= 43.78183, (0 missing)
      family_size < 1.5       to the left,  improve= 17.43059, (0 missing)
      Age         < 6.5       to the right, improve= 10.05326, (0 missing)
```

Fig. 5. Summary of data

Decision tree can be generated using the prp function in R programming, where the type of decision has been selected as type 4. The decision tree helps to understand the survival status in a very user friendly way. At each level, the visualization is easier compared to reading decision tree summary. There is another way to generate more informative decision tree using "fancyRpartPlot" function in R programming. In this type of decision tree, the decision nodes, chance nodes and leaf nodes can be identified in different colors.

Overall accuracy of trained model for getting the survival status of passengers in the titanic disaster s 85.52 % as displayed in the Fig.6, which is more than that of the expected accuracy in our objective. Further in future work, we will be try to improve the accuracy of classification.

```
> class.pred <- table(predict(my_tree, type="class"), train_new$Survived)
> #error
> #error rate below:
> 1-sum(diag(class.pred))/sum(class.pred)
[1] 0.1447811
> # accuraccy below:
> sum(diag(class.pred))/sum(class.pred)
[1] 0.8552189
> library(my_tree)
```

Fig. 6. Classification accuracy

Clustering technique is used to group a set of observations in such a way that observations of similar types belong to a group [6]. There can be many groups defined while clustering. Clustering implementation has been performed using Python programming. Installation of numpy, pandas and scipy packages are required for its implementation in Python. KMeans algorithm will be implemented using KMeans library in sklearn. This will generate the survival predictions of all the passengers in the test data.

6. Conclusion

On training data set, modification has been performed to fill the missing values and other related errors for fulfilling the feature engineering technique. Then, the machine learning algorithms are applied for classification and clustering techniques. It can be concluded that, passengers travelled with small family size whose family size was from 2 to 4 survived more compared to death. The death expectation is more with the passengers who travelled alone and who travelled with large family size. Then, the classification with 2 classes has been performed using decision tree where survival got analyzed at every level. Also, the accuracy rate observed is 85%. Survival status was also analyzed using clustering techniques. Further as future work, Feature engineering has to be performed more accurately to gain more knowledge from the titanic data. Also, other classification and clustering algorithm can be applied to improve the accuracy value. The exploratory data analysis has been carried out on Titanic passenger data set in this work. The practical connotation will be extending or applying these algorithms on other real-world datasets like natural disaster data and covid-19 data set for further analysis and inference.

References

[1] Cicoria S, Sherlock J, Muniswamaiah M and Clarke L 2014 *Proc. of Student-Faculty Research Day (New York)* Classification of titanic passenger data and chances of surviving the disaster p 1-6

[2] Cortes C, Vapnik V 1995 *Support-vector Networks* (Machine Learning Vol 20) p 273–297

[3] Russell S J, Norvig P 2003 *Artificial Intelligence: A Modern Approach (*Prentice Hall) 2nd Ed

[4] Finlay S 2014 *Predictive Analytics, Data Mining and Big Data: Myths, Misconceptions and Methods* (New York: Palgrave Macmillan) 1st ed

[5] James G, Witten D, Hastie T and Tibshirani R 2013 *An Introduction to Statistical Learning with Applications in R* (New York: Springer) 1st ed
[6] Andrew Ng CS229 Notes 2012 (Stanford University)
[7] Jonathan W, Charlotta L and Renato U 2020 Automated machine learning: Review of the state-of-the-art and opportunities for healthcare (Artificial Intelligence in Medicine Vol 104 Elsevier)
[8] Guyon I, Elisseeff A 2006 *An Introduction to Feature Extraction* (Feature Extraction. Studies in Fuzziness and Soft Computing Vol 207 Springer, Berlin, Heidelberg)
[9] Heaton J *An empirical analysis of feature engineering for predictive modeling* 2016 Proc. SoutheastCon (Norfolk VA) pp. 1-6
[10] Khalid S, Khalil T and Nasreen S 2014 *A survey of feature selection and feature extraction techniques in machine learning* Proc. Int. Conf. on Science and Information (London) pp. 372 78
[11] Aparna U R., Paul S 2016 *Feature selection and extraction in data mining* Online Int. Conf. on Green Engineering and Technologies (Coimbatore) pp. 1-3
[12] Manal B, Liping Z 2019 *A review of machine learning algorithms for identification and classification of non-functional requirements* (Expert Systems with Applications: X Vol.1)
[13] Neeraj N, Shikha A and Sanjay S 2015 *Recent Advancement in Machine Learning Based Internet Traffic Classification* (Procedia Computer Science Vol. 60) pp. 784-791
[14] Youguo L, Haiyan W 2012 *A Clustering Method Based on K-Means Algorithm* (Physics Procedia, Vol. 25) pp. 1104-09
[15] Ahmed O, Fatima-Zahra B, Ayoub A L and Samir B 2018 *Big Data technologies: A survey* (Journal of King Saud University - Computer and Information Sciences Vol. 30 Iss. 4) pp 431-48

Performance Investigation of Single-Phase Capacitor Start – Capacitor Run Induction Motor Drive

Mr. Antheesh R [1], Mr. Subrahmanya Bhat [2] and Dr. Rajkiran Ballal I [3]

[1] Assistant Professor, Department of Electrical & Electronics Engineering, SDM Institute of Technology, Ujire. Karnataka, India

[2] Assistant Professor, Department of Electrical & Electronics Engineering, SDM Institute of Technology, Ujire. Karnataka, India

[3] Former Principal, Mangalore Marine College and Technology, Kuppepadavu, Mangalore. Karnataka, India

Abstract: The capacitor-start capacitor-run single-phase induction motors are analyzed by means of double-revolving field theory. This paper presents a practical approach for obtaining equivalent circuit parameters of a single-phase induction motor drive. The mathematical modelling of the machine is done by considering both the rotating fields. The input parameters of a single-phase capacitor-start, capacitor-run induction motor are investigated through simulation and experimental results under normal power frequency operation. The performance of the above single-phase induction motor is analyzed for different load conditions.

Keywords— Single Phase Induction Motor, Double revolving field theory

1. Introduction

An induction motor is extensively used for domestic, commercial, industrial purposes. The single-phase induction motor has two stator windings main winding and the auxiliary winding or starting winding with a laminated iron core arranged which are perpendicularly displaced [1]. When the stator is energized by a single-phase supply, a varying magnetic field is generated. Even though the field winding is stationary, for every half cycle of input the field reverses its polarity. Hence, the resultant alternating flux cannot rotate the stationary rotor [2]. The working of induction motor may be described by Double Revolving Field Theory [3]. Mathematically, this field can be divided into two components, which rotate in opposite directions [4].The resultant flux which is generated due to these opposite fields will be equal to Zero hence the torque is Zero. The direction of rotation of the rotor remains same, hence motor quickly accelerates and reaches speed slightly less than synchronous speed.

2. Equivalent Circuit Diagram

The equivalent circuit of a single-phase induction motor is shown in Figure 1. The stator winding carrying the supply current produces a time varying magnetic field that is stationary in space. The equivalent circuit assumes two imaginary rotor windings based on double revolving field theory. One winding produces a

IOP Conf. Series: Materials Science and Engineering 1065 (2021) 012043 doi:10.1088/1757-899X/1065/1/012043

magnetic flux in the direction of rotating field with a slip 's' while the other imaginary rotor winding produces a magnetic flux opposite to rotating field with the slip as (2-s). From each imaginary rotor windings, a forward torque and reverse torque will be produced. The net torque will be the difference value of power and torque may be obtained by using the respective equations [6].

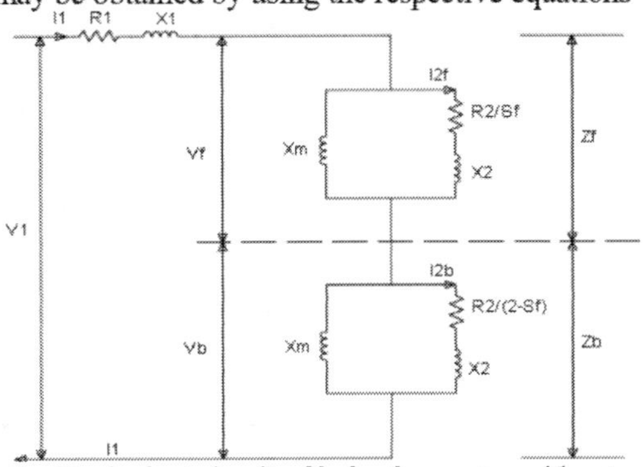

Figure.1 Equivalent circuit of induction motor without core loss

The Slip for Backward Flux is ,

$$S_b = \frac{N_S - (-N)}{N_S} = \frac{N_S + N}{N_S}$$

$$S_b = 2 - S$$

The impedance of the stator is: $Z_1 = R_1 + jX_1$

The impedance of the rotor is given as:

$$Z_f = \frac{jX_m\left(\frac{R_2}{S_f} + jX_2\right)}{\frac{R_f}{S_f} + j(X_2 + X_m)} = R_2 + jX_f$$

$$Z_b = \frac{jX_m\left(\frac{R_2}{2 - S_f} + jX_2\right)}{\frac{R_f}{(2 - S_f)} + j(X_2 + X_m)} = R_b + jX_b$$

$$Z_{input} = Z_1 + \frac{Z_f + Z_b}{2}$$

The input current: $I_1 = \frac{V_1}{Z_{input}}$

The input power of the motor is: $P_{input} = V_1 \cdot I_1 \cdot cos\varphi$

The power converted into mechanical is:

$$P_{conv} = (1 - S_f)(P_{agf} - P_{agb}) = \frac{(1 - S_f)I_1^2}{2}(R_f - R_b)$$

Where R_f and R_b are real parts of Z_f and Z_b

The shaft output power is: $P_{output} = P_{conv} - P_{rot}$

Torque motor: $T_{shaft} = \frac{P_{output} \cdot 60}{2 \cdot \pi \cdot N_S}$

The motor efficiency in percentage is given as:

$$Efficiency = \frac{P_{output}}{P_{input}} \times 100$$

The forward rotor parameters:

IOP Conf. Series: Materials Science and Engineering 1065 (2021) 012043 doi:10.1088/1757-899X/1065/1/012043

$$V_f = I_1 \cdot Z_f$$

$$I_{2f} = \frac{V_f}{\frac{R_2}{S_f} + \frac{jX_2}{2}} \; (Amps)$$

$$P_f = I_{2f}^2 \cdot \frac{R_2}{S_f} \; (Watts)$$

$$T_f = \frac{P_{output} \cdot 60}{2 \cdot \pi \cdot N_S} \; (N - m)$$

The backward rotor parameters:

$$V_b = I_1 \cdot Z_b$$

$$I_{2b} = \frac{V_b}{\frac{R_2}{2 - S_f} + \frac{jX_2}{2}} \; (Amps)$$

$$P_b = I_{2f}^2 \cdot \frac{R_2}{2 - S_f} \; (Watts)$$

$$T_b = \frac{P_b \cdot 60}{2 \cdot \pi \cdot N_S} \; (N - m)$$

Resultant torque: $T = T_f - T_b$

The power input of the motor is $: P = P_f - P_b$

3. Modelling of single phase induction motor

As the resistances and inductances of the starting and running windings of the capacitor start-run induction motor is asymmetrical, to obtain the mathematical model of the motor, it is required to transform all the variables to the stationary reference frame (d-q) fixed to the stator. The Machine model is shown in figure 2.

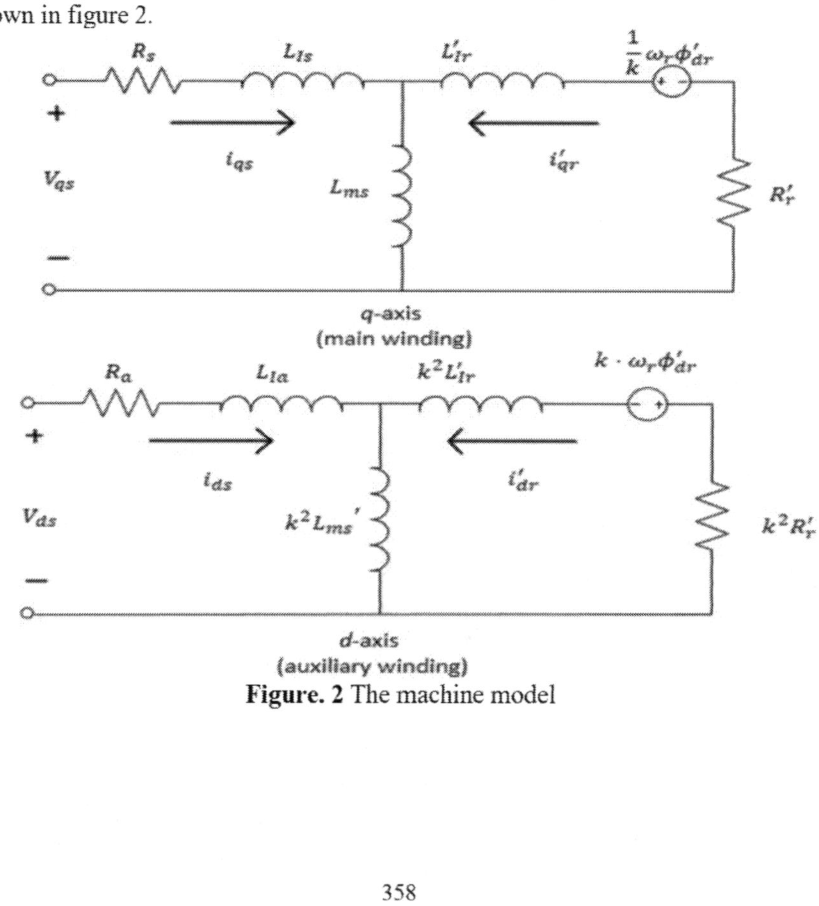

Figure. 2 The machine model

4. Experimental Results

The experiments were conducted on the single phase induction motor drive with the following specifications. The drive consists of the single phase AC source, induction motor and mechanical load. The Specifications of Single Phase Induction Motor is shown in Table 1.

Table 1. Specification of Motor

Specification of Single Phase Induction Motor	
Voltage	230V
Current	7 A
Speed	1440 r p m
Power	1.5hp
Frequency	50Hz

To obtain circuit parameters, No load test, Blocked Rotor Test and stator resistance measurement test were conducted on the motor. Based on the test results, mentioned in the table-2, the parameters of equivalent circuit were obtained as shown in figure 3.

Table 2. Test Results on Single Phase Induction Motor

No Load Test			**Blocked Rotor Test**		
V_o (V)	**I_o (A)**	**W_o (W)**	**V_{sc} (V)**	**I_{sc} (A)**	**W_{sc} (W)**
220	**2.25**	**157.5**	**78**	**7**	**300**

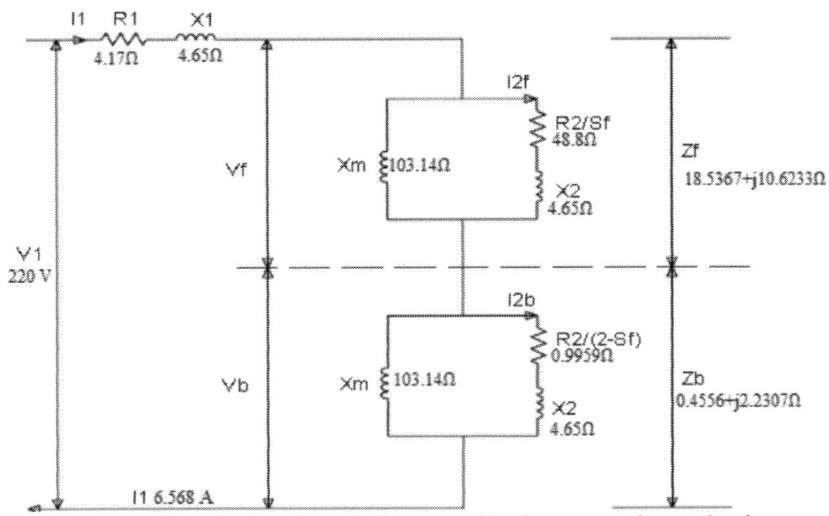

Figure.3 Equivalent Circuit diagram with above experimental values

The equivalent circuit parameters are found by carrying the no load and blocked rotor tests. The obtained parameters of motor are utilized to create experimental machine model in Matlab Simulink to equalize the machine parameters both in experiment and simulation. Comparison of the experimental and simulation results under different loading conditions were obtained for the above mentioned drive motor.

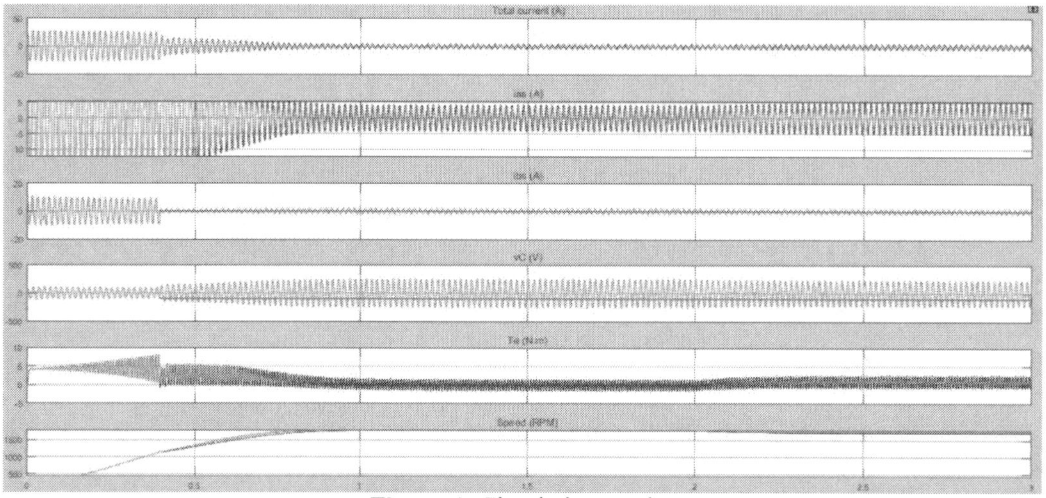

Figure 4. Simulation results

The simulation of drive circuit has been carried out to obtain the variations in the load current, torque and speed of the motor. The figure 4. Shows the simulated results under steady state operating conditions.

In order to validate the above results the Performance Characteristics under different loading conditions were experimentally obtained for the above mentioned drive motor by conducting direct load test. The torque versus slip and load current versus efficiency characteristics were obtained. The figure 4 shows the performance characteristics experimentally observed.

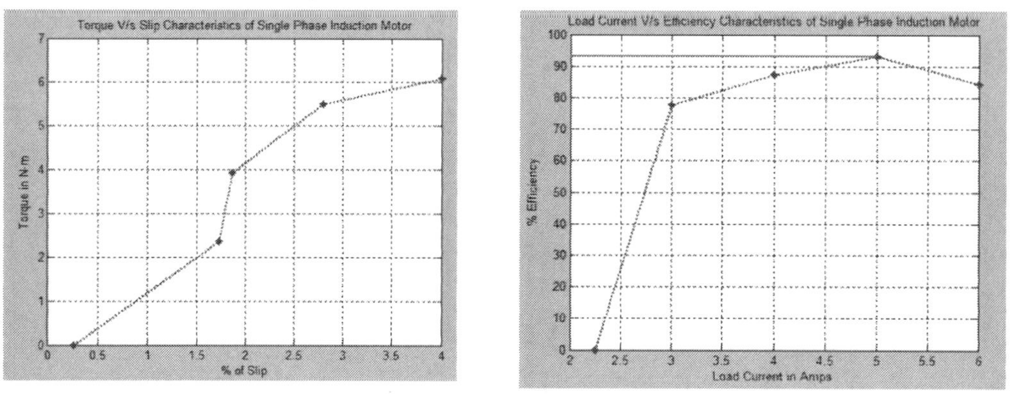

Figure.5. (a)*: Torque v/s Slip Characteristics* (b) Load *Current v/s Efficiency characteristics*

Conclusion

This paper presents practical method of obtaining the equivalent circuit of a single- phase induction motor drive. The performance parameters of a single-phase induction motor were investigated through simulation and experimental results under power frequency operation, and mathematical modeling of the machine is developed. The simulation and experimental results shows that as the load current increases the motor operates at better efficiency.

References:

[1] Manuel Guerreiro, D F and Armando C, *A Phasor Speed Control of a Single- or Two-Phase Induction Motor*, Proceeding of the International Conference on Electrical Machines, 1,2008.

[2]. Vinamra K G and Yogesh C, *Modeling & Simulation of PWM Controlled Cyclo converter Fed Split Phase Induction Motor*, International Journal of Advanced Research in Electrical, Electronics and Instrumentation Engineering,1(3), 2012

[3] Animesh K, Nihar R R, Pradip K S, Rajarshi M and Gautam K P, *Dynamic Modelling Of Single-Phase Permanent Capacitor Induction Motor and Study of Non-Linear Phenomenon*, International Journal Of Computational Engineering Research (ijceronline.com). 2(5), 2012, 1690-1694.
[4] Mohd A I and Gurmeet S, *A Review on Influence of Rotor Geometry on the Performance of Single-Phase Capacitor-Run Induction Motor*, International Journal of Advanced Research in Electrical, Electronics and Instrumentation Engineering 3(6), 2014, 10216-224.
[5] Yahaya Asizehi ENESI, *Performance Characteristics and Double Revolving Theory of Single Phase Induction Motor*, Leonardo Journal of Sciences, ISSN 1583-0233, Issue 23, July-December 2013, Page number 41-45
[6] McGraw-H, *Revolving Theory of Single Phase Induction Motor*, McGraw-Hill Companies, 2003, p. 467-469.

CLOUD DATA SECURITY MODEL USING MODIFIED DECOY TECHNIQUE IN FOG COMPUTING FOR E-HEALTHCARE

Dr. P. Maragathavalli[1], S. Atchaya[2], N. Kaliyaperumal[2] and S. Saranya[2]

[1]Assistant Professor, Department of Information Technology, Pondicherry Engineering College, East Coast Road, Pillaichavadi, Puducherry, [1]marapriya@pec.edu

[2]Student Member, Department of Information Technology, Pondicherry Engineering College, East Coast Road, Pillaichavadi, Puducherry.

[2]atchaya5899@pec.edu

Abstract In recent years, the cloud provides the facility of storing the possible files to a remote database which can be retrieved on demand. Healthcare cloud infrastructure is used to store the medical records which in turn help in managing and tracking the records of the user. The primary care providers use Electronic Medical Record (EMR) for diagnosing and evaluating medical data. Big data analytics is an important technology in enormous business areas which includes banking, social media, machine sensor data and medical welfare. Healthcare cloud provides various storage facilities such as Google drive, dropbox business and one drive in which security issues like data theft attacks to be considered as the serious security breaches of the cloud. In this paper, a modified decoy technique is proposed in order to ensure more security for the user's Medical Big Data (MBD). The tri-party authentication key uses key agreement protocol which is modelled by bilinear pairing cryptography. The proposed system detects the attacker and sends the information such as IP address, access time and date to the user. It also provides the facility to block the account from further access. Using the triple-des algorithm in bilinear pairing the parameters like throughput is increased and computational complexity is decreased.

Keywords: *Remote Database; Security and Privacy; Medical Big Data Analytics; Electronic Medical Record; Key Agreement Protocol.*

1. INTRODUCTION

Hybrid cloud for healthcare data sharing and accessing facilities is provided to store the medical records of the patients. It also helps in managing the users record and tracking the records of them, even if the patient moves across any part of the world [3]. It has several security issues such as privacy and security, policy issues, transparency, data protection, and licensing. To achieve a secure healthcare cloud fog computing and cloud is used in our proposed methodology in the users Medical Big Data (MBD)[2].

This paper sections are ordered as follows: The literature survey is described in section II. The proposed methodology with design, coding and implementation in section III. Section IV describes the input and output specification. Implementation environments are described in section V, performance parameters in section VI and the conclusion of this paper is concluded section VII.

2. LITERATURE SURVEY

Medical big data analysis: preserving security and privacy with hybrid cloud technology [1] has provided various mechanisms and algorithms for the security of the cloud. Analysing the several techniques it provided the alternative security model using bilinear pairing protocol. In this proposed model the parameters like computational complexity is considered in which in the result it decreases by 20%. The limitation in the proposed work is time consumption which is high.

Lightweight and privacy and medical service access for healthcare cloud [4] which uses Attribute Based Signature (ABS) technology to overcome the above limitations. By providing the stronger security model it focuses on the time consumption which in the result reduces the response time by 0.25s. The proposed system has some limitations which it is suitable for small scale medical access.

Improving security and privacy attribute-based data sharing in cloud computing technology [5] has provided various mechanisms and algorithms for the security of the cloud. Analysing the several techniques it provided the alternative security model using Ciphertext policy- Attribute based encryption (CP-ABE). In this proposed model the parameters like computational overhead is considered in which in the result it decreases by 40%. The proposed work provides the weak security model.

A novel smart healthcare design, simulation and implementation using healthcare using 4.0 processes [6] has provided various mechanisms and algorithms for the security of the cloud. Analysing the several techniques it provided the alternative security model using Blockchain which uses the existing dataset. In this proposed model the parameters like time consumption is considered in which in the result it reduces by 1s. The proposed work has the limitation in adding the additional features to the system.

Secure verifiable database supporting efficient dynamic operations in cloud computing [7] has provided various mechanisms and algorithms for the security of the cloud. Analysing the several techniques it provided the alternative security model using BLS signature which uses the existing dataset. In this proposed model the parameters like time throughput is considered in which in the result the throughput is increased by 20%. The proposed work has the limitation in which the pairing is not efficient.

In the above paragraphs, the literature survey is done from an existing system which has some limitations in time consumption, pairing, weaker security model and those limitations are considered in our project.

3. PROPOSED METHODOLOGY

Decoy technique helps in generating the decoy files or the duplicate medical records and detects unauthorized users with the help of user profiling. It also helps in securing users' medical records by confusing the attacker with a decoy gallery.

3.1 Design

The proposed system focuses to keep Original Medical Big Data (OMBD) secret by user profiling the attacker is detected and Decoy Medical Big Data (DMBD) is used as a honeypot. DMBD data is secured by pre-existing values and an efficient tri-party authentication key is proposed.

In this method initially both the user and attacker will be accessing the DMBD and they are considered as the unauthorized users. This DMBD is placed in the fog computing layer near user profiling and it contains the fake Medical Big Data (MBD) which makes the attacker believe that they are accessing the OMBD. When the user gets to know that it is not his profile he will be going to the next process of verification and after passing some security challenges he will be accessing the OMBD. By user profiling the attacker is detected and SMS is sent to the user that his account is being accessed and it contains information like access time, date and IP address.

Now if the user wants to upload his medical records for this the tri-party needs to be communicated. The secret key is generated from the Private Key Generator (PKG) and it is received by the tri-party. Now each party authenticates with another party to communicate with others secretly. Then the medical records are encrypted using the triple-des algorithm. These encrypted data are stored in the cloud and again decrypted for viewing purposes using this algorithm.

Thus the two level security is proposed firstly DMBD as the honeypot and the other by keeping the encrypted medical record in the cloud secretly Fig.1 shows the proposed system.

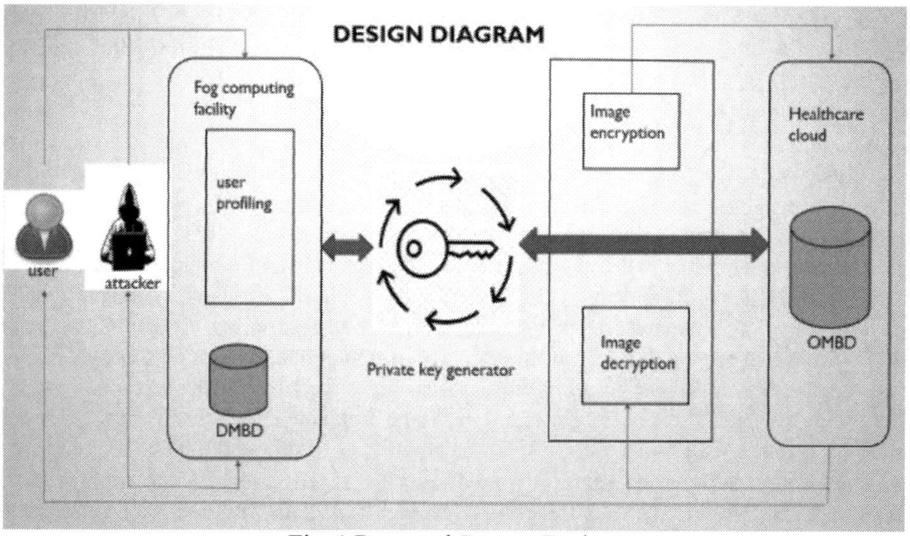

Fig. 1 Proposed System Design

3.2 Coding and Implementation

User profiling and decoy technique detects abnormal patterns such as login count, download data, security failure questions.

The pseudo-code for user profiling and decoy technique is given below

Step 1: Calculate the user login count, cntLoginFail increase.

Step 2: Check the downloaded datas, cntDataDownload increase.

Step 3: Count the security failure question, cntSecQueFail increase.

Step 4: IF cntLoginFail > 3 && cntDataDownload > 3 && cntSecQueFail > 3 && Key matches. If the above condition is passed then the user ***is an attacker*** *or a normal user.*

Step 5: Send the information to the user.

Step 6: Block the attacker.

3.3 Encryption

When the user uploads the medical record, the encryption process is done. The medical record is encrypted by asymmetric encryption technique Triple DES with the secret key that is generated by the private key generator (PKG). Now the original record is converted into the cipher image. This encrypted cipher image is kept securely in the cloud storage with the help of a cloud service provider which is shown in Fig. 2.

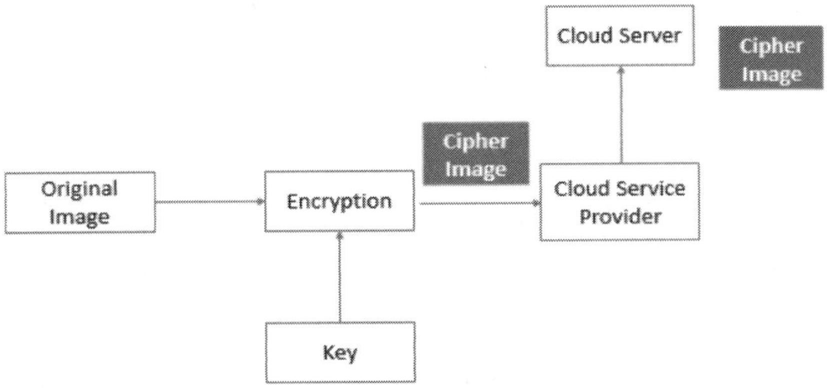

Fig. 2 Encryption Process

3.3.1 Changed pseudo-code for encryption

cipherimage64bits = Encrypt(64bitimagelst,i)
cipherimage64bits = Decrypt(64bitimagelst,j)
cipherimage64bits = Encrypt(64bitimagelst,k)
cipherimage64bits = Inverse(Finalpermutation(cipherimage64bits))
for cimage in cipherimage
64bits:cipherimage = cipherimage + cimage
Output: cipherimage

3.4 Decryption

Decryption is the process of converting encrypted data into the plain text. The cipher image stored in the cloud service provider is decrypted into the original data with the secret key which is generated by the private key generator and the original data is available to the user is shown in Fig. 3.

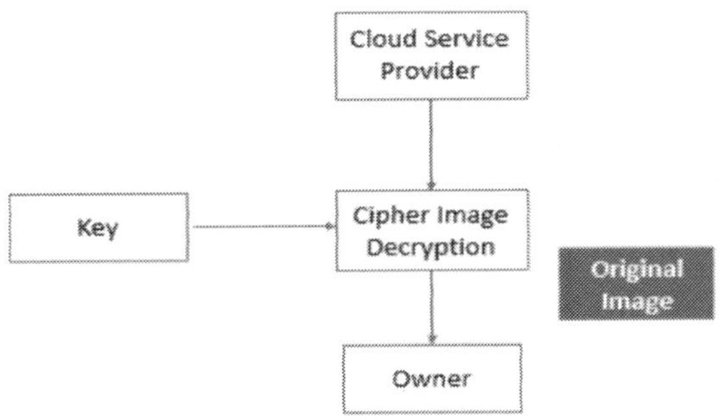

Fig. 3 Decryption Process

3.4.1 Modified pseudo-code for decryption

cipherimage64bits = Decrypt(64bitimagelst,i)
cipherimage64bits = Encrypt(64bitimagelst,j)
cipherimage64bits = Decrypt(64bitimagelst,k)
image64bits = Inverse(FinalPermutation(image64bits))
for image in image
64bits: imager=imager + image
Output: imager

4. INPUT – OUTPUT SPECIFICATION

4.1 Input specification

- Input data – plain medical image

- Format for each input data – cipher text (after encryption)

4.2 Output specification

- Output data – plain medical image (after decryption)
- Format for each output data – cipher text

The input-output specification is shown in Fig. 4.

Fig. 4 Input and Output Data Processing

5. IMPLEMENTATION

5.1 Data set description

The Multimedia Dataset of 128MB with 150 entries is used as an input to this project. The multimedia dataset is converted into cipher text which is used as input data for this project. The output of our project is also a cipher text.

URL: https://www.kaggle.com/kmader/siim-medical-images

5.2 Hardware Requirements

Processor: 64 bit, 1.5 GHZ minimum per Core
RAM: 2GB minimum
Hard disk: 10GB

5.3 Software requirements

Real time SMS provider: MYSMSMANTRA
Front end: Python 3.8
Back end: SQL Server

5.4 Implemented code for encryption using Triple DES

IOP Conf. Series: Materials Science and Engineering 1065 (2021) 012044 doi:10.1088/1757-899X/1065/1/012044

3des.py - C:\Users\Hp\Desktop\New folder\3des.py (3.8.2)

File Edit Format Run Options Window Help

```
##Encryption
lst=[]
with open('Capture.png','rb') as f:
    jpgdata=f.read()
for fl in jpgdata:
    lst.append (fl)
    f.close()
    data = jpgdata
k=triple_des("PROJECTFILEKEY12",CBC,"\0\0\0\0\0\0\0\0",pad=None,padmode= PAD_PKCS5)
print ("please wait for a moment")
jpgdata=k.encrypt (data)
with open('Encrypted.png','wb') as f:
    f.write(jpgdata)
print("Encryption complete.")
##Decryption
lst=[]
with open('Encrypted.png','rb') as f:
    jpgdata=f.read()
for fl in jpgdata:
    lst.append(fl)
    f.close()
    data = jpgdata
k =triple_des ("PROJECTFILEKEYTR", CBC,"\0\0\0\0\0\0\0\0", pad=None, padmode=PAD_PKCS5 )
print("please wait for a moment")
jpgdata=k.decrypt (data)
with open ('Encrypted1.png','wb') as f:
    f.write(jpgdata)
print("Decryption complete.")
##Encryption
lst=[]
with open('Encrypted1.png','rb') as f:
    jpgdata=f.read()
for fl in jpgdata:
    lst.append (fl)
    f.close()
    data = jpgdata
k=triple_des("PROJECTFILEKEY14",CBC,"\0\0\0\0\0\0\0\0",pad=None,padmode= PAD_PKCS5)
print ("please wait for a moment")
jpgdata=k.encrypt (data)
with open('Encrypted2.png','wb') as f:
    f.write(jpgdata)
print("Encryption complete.")
```

6. PERFORMANCE PARAMETERS

Based on the following comparisons, the triple des algorithm gives a better throughput, computational complexity and time consumption as a result triple-des is the best algorithm which is used in our proposed system.

6.1 Throughput

The throughput is done for the encryption and decryption and proposed system is compared with the existing system. Throughput is estimated by the average number of data blocks that transfer successfully to the time taken which is measured by bits per second (bps).

$$\textbf{Throughput} = \frac{\textbf{Number of data blocks successfully transferred}}{\textbf{Time taken}}$$

Throughput is increased by 20% since the data transfer in encryption by using the triple des algorithm provides efficiency in throughput whereas encryption by blowfish increases by 10% only.

6.2 Computational complexity

Computational complexity is the amount of time taken by the process. It is estimated by the number of bilinear pairing cryptography which is shown in below equation,

$$\boldsymbol{B(n) = O(n\text{^}2)}$$

Where B(n) is a bilinear pairing function and O(n^2) is a complexity analysis function. Complexity time taken for bilinear pairing is reduced by using secret key and tri party authenticate key when compared to the existing system it is reduced by 10%.

6.3 Time consumption

The time consumption is the process of time required to complete the key generation.

$$\boldsymbol{time\ consumption = \frac{Amount\ of\ time\ taken\ for\ individual\ key\ size}{Total\ rate\ of\ the\ time\ consumption}}$$

Time consumption is calculated with respect to the key size generated in bilinear pairing cryptography and time consumption is reduced by 25% by using triple des algorithm than blowfish algorithm.

7. CONCLUSION

A cloud data security model is developed for protecting the patient's information in the EHR system by fog computing in which two levels of security have been provided. By user profiling the attacker can be detected in that decoy gallery as the honey pot. The Original Medical Big Data (OMBD) is maintained secretly in cloud storage and Decoy Medical Big Data (DMBD) in fog computing layer. When a user wants to access the OMBD needs verification. Thus, by storing the OMBD as the hidden gallery, OMBD becomes secure. For communication between OMBD and DMBD, a Private Key Generator (PKG) is used. By using the proposed methodology, the throughput can be increased by 20% as well as the computational complexity and time can also be reduced significantly than the existing techniques.

8. REFERENCES

[1] Shanmugapriya. E and Kavitha.R. "Medical big data analysis: preserving security and privacy with hybrid cloud technology", Soft Computing, Springer 23, 2585–2596 (2019). https://doi.org/10.1007/s00500-019-03857-z.

[2] M. Joshi, K. Joshi and T. Finin, "Attribute Based Encryption for Secure Access to Cloud Based EHR Systems," 2018 IEEE 11th International Conference on Cloud Computing (CLOUD), San Francisco, CA, 2018, pp. 932-935, doi: 10.1109/CLOUD.2018.00139.

[3] M. Du, K. Wang, X. Liu, S. Guo and Y. Zhang, "A Differential Privacy-Based Query Model for Sustainable Fog Data Centers," in IEEE Transactions on Sustainable Computing, vol. 4, no. 2, pp. 145-155, 1 April-June 2019, doi: 10.1109/TSUSC.2017.2715038.

[4] J. Liu, H. Tang, R. Sun, X. Du and M. Guizani, "Lightweight and Privacy-Preserving Medical Services Access for Healthcare Cloud," in IEEE Transaction on Dependable and secure computing, vol. 7, pp. 106951-106961, 2019, doi: 10.1109/ACCESS.2019.2931917.

[5] L. Zhang, Y. Cui and Y. Mu, "Improving Security and Privacy Attribute Based Data Sharing in Cloud Computing," in IEEE Systems Journal, vol. 14, no. 1, pp. 387-397, March 2020, doi: 10.1109/JSYST.2019.2911391.

[6] Kumar, R. Krishnamurthi, A. Nayyar, K. Sharma, V. Grover and E. Hossain, "A Novel Smart Healthcare Design, Simulation, and Implementation Using Healthcare 4.0 Processes," in IEEE Transaction on Dependable and secure computing, vol. 8, pp. 118433-118471, 2020, doi: 10.1109/ACCESS.2020.3004790.

[7] J. Shen, D. Liu, M. Z. A. Bhuiyan, J. Shen, X. Sun and A. Castiglione, "Secure Verifiable Database Supporting Efficient Dynamic Operations in Cloud Computing," in IEEE Transactions on Emerging Topics in Computing, vol. 8, no. 2, pp. 280-290, 1 April-June 2020, doi: 10.1109/TETC.2017.2776402.

[8] Abouelmehdi. K, Beni-Hessane. A and Khaloufi. H, "Big healthcare data: preserving security and privacy." Big Data 5, 1 (2019). https://doi.org/10.1186/s40537-017-0110-7

[9] Mithun Mukherjee; Mohamed Amine Ferrag; Leandros Maglaras; Abdelouahid Derhab; Mohammad Aazam, "Security and Privacy Issues and Solutions for Fog," in Fog and Fogonomics: Challenges and Practices of Fog Computing, Communication, Networking, Strategy, and Economics, Wiley, 2020, pp. 353-374, doi: 10.1002/9781119501121.ch14.

[10] Claudio A. Ardagna, Rasool Asal, Ernesto Damiani, "From Security to Assurance in the Cloud: A Survey", Publication: ACM Computing Surveys, July 2019, Article No.: 2, https://doi.org/10.1145/2767005.

[11] Lahar Singh Nishad, Akriti, Sarvesh Kumar and Sumitra Beniwal, "Security, Privacy Issues and challenges In Cloud Computing: A Survey" ICTCS '16: Proceedings of the Second International Conference on Information and Communication Technology for Competitive Strategies, March 2019, Article No.: 47, Pages 1–7, https://doi.org/10.1145/2905055.2905253.

IOP Conf. Series: Materials Science and Engineering 1065 (2021) 012045 doi:10.1088/1757-899X/1065/1/012045

Incorporating LSTM Method on Modified Deep Learning Technique for Sensor Based Human Activity Recognition System

Dr. P. Maragathavalli[1], Ms. R.Vivitha[2], Ms. M. Saranya[2] and Mr. Uppu Dinesh[2]

[1]Assistant Professor, Department of Information Technology Engineering, Pondicherry Engineering College, Kalapet, Pondicherry, marapriya@pec.edu

[2]Student Member, Department of Information Technology Engineering, Pondicherry Engineering College, Kalapet, Pondicherry,

vivitha.2707@pec.edu

Abstract. In today's scenario, mobile phones or smartphones are rapidly becoming the essential communication device in people's lives. Human Activity Recognition has become the popular field of research in different fields such as medical care, monitoring and education. The sensors in smartphones enable us to use them for diverse applications. Healthcare is a major domain where human activity recognition is used. In this paper, a human activity recognition system has been developed that can detect six activities of daily living (ADL) along with human fall. Human fall occurs due to an accident that can cause serious injuries which may lead to major medical issues when the issue is not addressed properly. The proposed system uses a variant of deep learning technique to detect human activities and human fall. The accuracy is significantly increased by nearly 4% when compared with previous results.

Keywords: Activities of Daily Life; Mobile sensors; Accelerometer; Gyroscope; Monitoring; Healthcare; Medicare

1. Introduction

In recent times, smartphones have become an indispensable part in the life of humans. The daily routine of an individual can be captured by using smartphones. Most widely used sensors are an accelerometer and gyroscope. The in-built sensors in smartphones enable us to capture the individual's movement. This information can be utilised to recognise several day-to-day routine activities like standing, walking, laying, sitting, walking upstairs and walking downstairs [2]. The same data is also utilized for detecting human fall. Daily life data can be used to study the individual's way of life and their order of routine bodily activities. By monitoring daily activities any difference from routine activities can be easily identified, as lack of physical activity causes most of the chronic diseases and detecting human falls can also help in bringing timely help to the people when they are in need. HAR has numerous applications in the various fields such as elderly care, monitoring the patient, education, military, and rehabilitations [2][3][8].

IOP Conf. Series: Materials Science and Engineering 1065 (2021) 012045 doi:10.1088/1757-899X/1065/1/012045

Fall detection systems [3] are systems that help in monitoring human activity and immediately alert at a time of catastrophe. It differentiates between the Activity of daily living (ADL) and fall events (forewarning during its occurrence) are the main objectives of the fall detection systems. These systems aid to send immediate notification to medical entities, caretakers, family members to ensure that timely care can be given.

This paper proposes an enhanced human activity recognition model for fall detection using modified Bidirectional Recurrent Neural Network (BRNN) incorporated with Long Short-Term Memory (LSTM), a Deep Learning (DL) approach. First of all the data for the proposed system is collected from the smartphone sensors such as the accelerometer and gyroscope. Secondly, the data is then preprocessed and segmented. Then the system is trained and tested for six human activities and fall. The Mobifall dataset is used which contains sensor data of 66 subjects performing four different falls and twelve different ADLs. Finally, metrics like accuracy, f-measure, precision and confusion matrix are used to measure the performance of the system[2][3][8].

The organization of the paper is as follows: Section 2 presents existing and a few of the correlated works. Section 3 provides the outline of the modified BRNN-LSTM system proposal. Experimentation and evaluation of the model is presented in Section 4. Section 5 bestows the outcome. Section 6 presents the conclusion with works for the future direction.

2. Background and Related Work

The basic steps involved in human activity recognition system comprises the following stages/steps: gathering of data, pre-processing of data, Feature Extraction, categorizing of activities and analysis of results, as shown in Figure 1.

Various sensors are used to gather the input data. The next step is pre-processing which helps in elimination of noise and deals with erring values which can be done by applying data cleansing methods. Cleansed data is then segmented into windows. Segmentation of windows is done using many ways like sliding windows, event-based windows and energy windows. When pre-processing and segmentation is completed, feature extraction is done for increasing the efficiency of classification algorithms. To accomplish this task, deep learning methods are used [2][3]. The features extracted are used to categorize the activities using various classification algorithms and metrics such as accuracy, F1-score, precision and confusion matrix are used to assess the performance of the system [1][2][3].

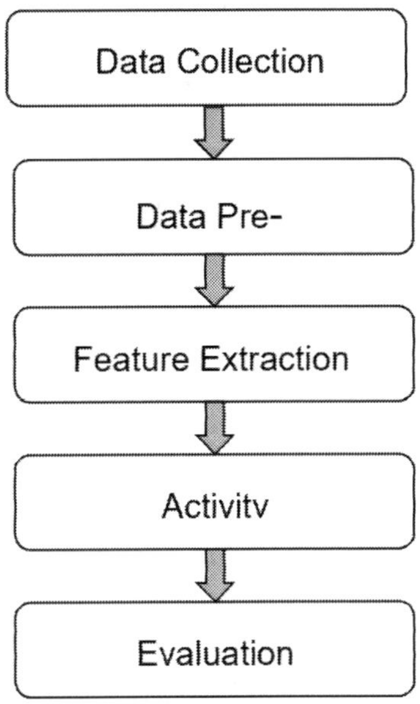

Fig. 1. Stages involved in Human Activity Recognition

The ability for automatic feature extraction is one of the main reasons for the popularity of deep learning methods in recent times. In the existing works, Human Activity Recognition (HAR) is done by various Deep Learning (DL) Techniques namely CNN[1][3], RNN, RNN-LSTM[2][4] as shown in Table 1. The main issue in the existing work is the issue of high computation power that makes real-time predictions arduous as most systems fail to handle it.

Human Activity Recognition is a time series problem such that the sequence of sensor readings are in time T. LSTM[2] cells can catch relationships in time dependent data without having to mix the timesteps together as a 1D convolutional neural network (CNN)[3] would do. At present, as "big data" emerges, LSTM architecture can offer better outcomes and in various unrealized applications. The hope for potential improvement still remains, though in the past years, researchers have worked to make great progress in the field of HAR. The present state is predicted by the LSTM cells from the information of the former state.

Apparently, we can observe that in real-world human trajectories are ceaseless. Therefore, the memory cells may not aptly obtain the required information by running in a single direction. In the proposed system, a novel method using bidirectional recurrent neural networks combined with long short-term memory cells is produced to assure the uprightness of the data fed to the different layers of neural networks and the accuracy of the existing model has been improved.

Table 1. Comparison of existing deep learning techniques

Sl. No.	NAME OF THE JOURNAL, YEAR	TITLE OF THE PAPER	TECHNIQUES USED	METRICS	ADVANTAGES AND DISADVANTAGES
1.	XXVI Brazilian Congress on Biomedical Engineering, Springer 2019.	Human Activity Recognition Based on Convolutional Neural Network	Convolutional Neural Network	Accuracy, precision	- Lower computational cost. - Information loss.
2.	International Conference on Computational Intelligence and Data Science (ICCIDS), Elsevier 2019.	A Lightweight Deep Learning Model for Human Activity Recognition on Edge Devices	Deep learning approach using Recurrent Neural Network and Long Short Term Memory	Accuracy, precision, recall, F1-score	- Reduces communication latency, cost and network traffic. - Does not support multi-sensor data.
3.	Robotics and Automation Letter, IEEE 2019.	HMFP-BRNN: Real Time Hand Motion Filtering and Prediction via Deep Bidirectional RNN	Deep bidirectional recurrent neural network approach	Accuracy, F1-score	- Produces precise outputs. - Overfitting of data occurs.
4.	Sensors Journal, 2019.	Deep Learning to Predict Falls in Older Adults Based on Daily-Life Trunk Accelerometry	Deep learning architecture using Convolutional Neural Network and Long Short Term Memory	F1-score	- Proven to learn sequential data. - Information loss.
5.	Sensors Journal, IEEE Access 2019.	Advanced Sensing and Human Activity Recognition in Early Intervention and Rehabilitation of Elderly People	Supervised machine learning approach	Accuracy	- Provides accurate results, - More sensors are required.

3. Illustration of the Proposed Model

The proposed system is evolved using BRNN and LSTM. It has two hidden layers and 30 neurons. Each module in the system is explained as follows.

3.1. Recurrent Neural Network (RNN)

RNN algorithm is introduced for solving problems that involve continuous data. As HAR is a time series problem, the input is in sequential form. It comprises three layers - i/p, hidden & o/p in which hidden layers include multiple nodes. Figure 2 shows the visual representation of a RNN node. The present hidden state value h_t and the output value o_t is generated by the generating function function of each hidden node t is calculated using formula:

$$h_t = \varepsilon(w_{hh}h_{t-1} + w_{ih}x_t + b_h) \quad (1)$$

$$y_t = \varepsilon(w_{ho}h_t + b_o) \quad (2)$$

where, x_t = new input state, h_{t-1} = old hidden state, h_t = new hidden value,
w_{hh} =hidden-hidden state weight, w_{ih} = input-hidden state weight, w_{ho} = hidden-output state weight, b_h = hidden state bias, b_o = output state bias, o_t = new output state, ε = activation function. ε is called an activation function.

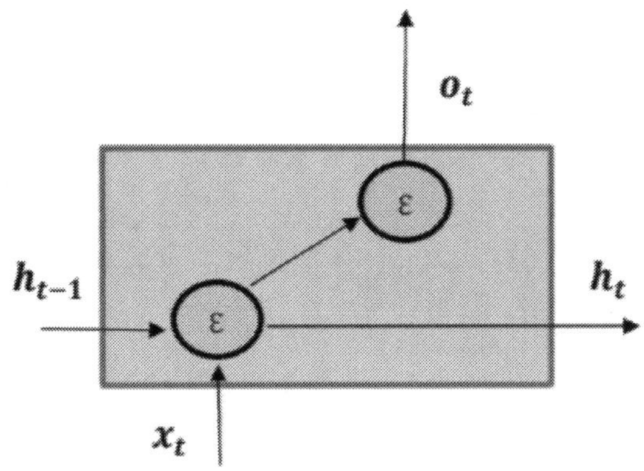

Fig. 2. Sample Node of Recurrent Neural Network (RNN)

3.2. Long Short Term Memory (LSTM)

The main drawback of using RNN is it's inability to handle exploding and vanishing gradients which decreases the performance of the network model in wide-range of temporal dependencies between inputs and human activities. Use of LSTM enables us to overcome the drawback of RNN, by using LSTM memory cells instead of RNN nodes. LSTM memory cells are shown in Figure 3. Each memory cell's characteristics are administered by the gates in LSTM cells. The activation functions of gates control the state of each cell. Input gate, forget gate and output gates are the gates to which the input values are fed. Activation function is represented by the symbol ε.

The following equations show each timestep(t) calculation of hidden state value.

$$i_t = \varepsilon_i(w_{xi}x_t + w_{hi}h_{t-1} + w_{ci}c_{t-1} + b_i) \quad (3)$$

$$f_t = \varepsilon_f(w_{xf}x_t + w_{hf}h_{t-1} + w_{cf}c_{t-1} + b_f) \quad (4)$$

$$c_t = f_t c_{t-1} + i_t\varepsilon_c(w_{xc}x_t + w_{hc}h_{t-1} + b_c) \quad (5)$$

$$o_t = \varepsilon_o(w_{xo}x_t + w_{ho}h_{t-1} + w_{co}c_t + b_o) \quad (6)$$

$$h_t = o_t\varepsilon_h(c_t) \quad (7)$$

W_{xi},w_{hi},w_{ci},w_{xf},w_{hf},w_{cf},w_{xc},w_{hc},w_{xo}, and w_{co}, are weights (w_{xi} = input to input weight, w_{hi} = hidden input weight), b_i,b_o,b_c, and b_f bias weights.

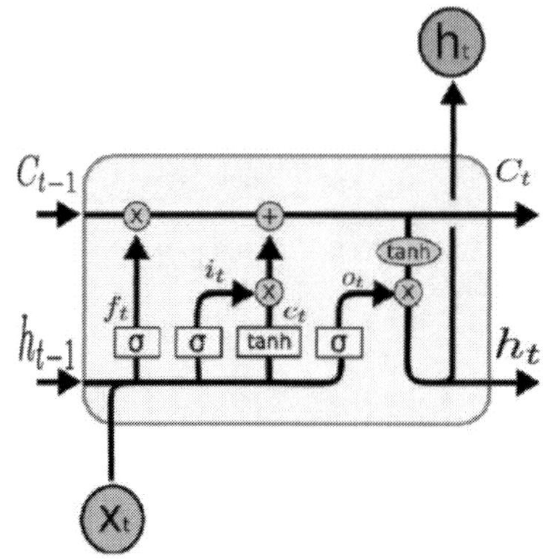

Fig. 3. Long-Short Term Memory (LSTM) Cell

3.3. Proposed modified BRNN-LSTM Model

Accelerometer and gyroscope readings are splitted into fixed size window T with 128 timesteps. The neural network input consists of reading (y_1, y_2, y_3,........,y_{T-1}, y_T) obtained at time T, where y_T is the reading obtained at any time instance t. The readings from the windows that are segmented are then given to the modified BRNN-LSTM network. The output is combined to single final output from different states using Rectified Linear Unit (ReLU).

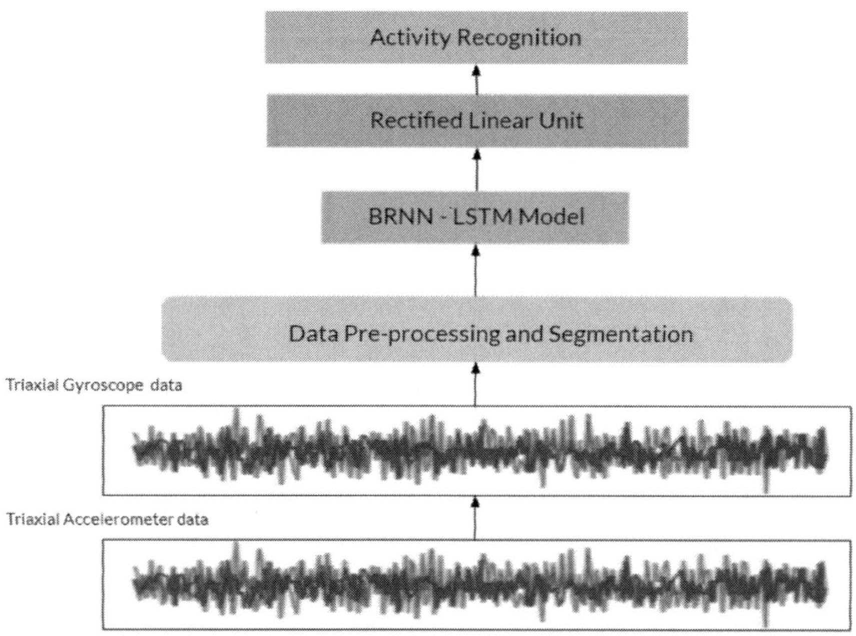

Fig. 4. Proposed System Design

In the proposed system design shown in Figure 4, the accelerometer and gyroscope data are given as input. The raw data of the sensors are preprocessed by a high pass butterworth filter which removes noise and also enhances the fine details and highlights the edges. After preprocessing, data

IOP Conf. Series: Materials Science and Engineering 1065 (2021) 012045 doi:10.1088/1757-899X/1065/1/012045

segmentation takes place by using a sliding window approach. Now, the preprocessed and segmented data is sent to the BRNN-LSTM layers.

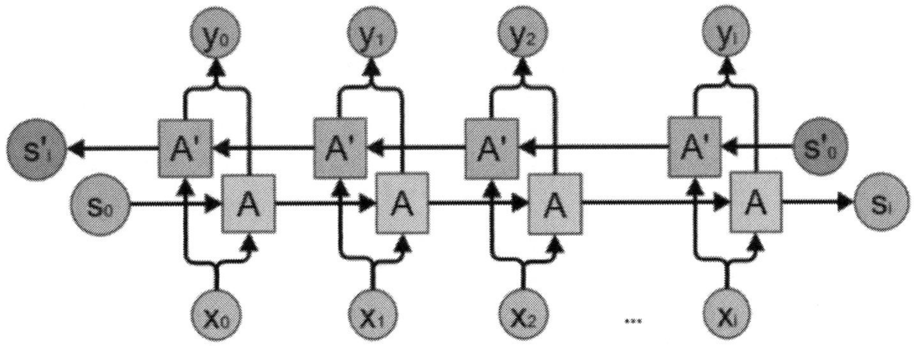

Fig. 5. Bidirectional Recurrent Neural Network-Long Short Term Memory (BRNN-LSTM) Diagram

```
def lstm_network(feature_matrix,conf_net):
#get input for the LSTM
#Stack two LSTM layers, both layer has same shape
with tf.variable_scope('forward_pass'):
    lstm_layers_forward = rnn.MultiRNNCell([rnn.BasicLSTMCell(conf_net.hidden_unit) for _ in range
        (conf_net.layer_size)])
with tf.variable_scope('backward_pass'):
    lstm_layers_backward = rnn.MultiRNNCell([rnn.BasicLSTMCell(conf_net.hidden_unit) for _ in range
        (conf_net.layer_size)])
#get LSTM outputs
outputs,_,_= rnn.static_bidirectional_rnn(lstm_layers,lstm_layers_backward,inputs,dtype=tf.float32)
# Get last time step's output feature for a "many-to-one" style classifier.
lstm_last_output = outputs[-1]
```

Fig. 6. Pseudo code for the modified BRNN-LSTM method

Improvement in the proposed system shows that not only the former information is considered to produce the present result but the upcoming information is also considered as shown in Figure 5. The fundamental approach of BRNN is to separate the neurons of a traditional RNN into two directions, one for forward states (positive time direction), and the other one for backward states (negative time direction). This helps in accurate prediction of human activities especially sitting and standing as those two activities have less accuracy in the existing methods.

4. Experiment and Evaluation

The enhanced human activity recognition system for fall detection is executed using python 3.5 and tensorflow 1.7.

4.1. Dataset Description

Data from the accelerometer and gyroscope sensors of a smartphone were recorded. Mobifall is an open-source dataset that consists of data gathered from a smartphone when participants are performing activities of different types and a range of falls. The data set comprises four types of falls, twelve distinct ADLs collected from overall 66 subjects. The tabulated description of the activities recorded are shown in Figure 7 shown below. The ADLs are divided into three sections such as mentioned in Figure 7. Data set is taken from URL https://bmi.hmu.gr/the-mobifall-and-mobiact-datasets-2.

Dataset	Simple Movements	Standard Normal Life Movements	Sporting Activities	Falls
MobiFall & MobiAct	- Sitting on a chair - Stepping in a car - Stepping out of a car - Standing	- Normal walking - Going downstairs - Going upstairs	- Jogging - Jumping	- Forwards (use of hands to dampen fall) - Forwards (first impact on knees) - Sideward bending legs - Backward (while trying to sit down)

Fig. 7. ADLs and falls executed by the experimental subjects

The x, y, z coordinates differ for different activities. When observing the plot for walking it shows maximum spikes along z direction whereas maximum spikes can be observed in x and y direction in case of jogging. The activation function ε defines the output of a neuron in terms of the input. From *MobiFall* data, 70% is used for training in the BRNN-LSTM layer and the remaining 30% for testing.

4.2. Performance Metrics

The performance of the system is analyzed using three widely used metrics such as accuracy, precision and f-measure are given below:

Table 2. Performance metrics

Metrics	Formula
Accuracy: Number of correct predictions over total number of predictions.	$Accuracy = \frac{TP+TN}{TP+TN+FP+FN}$ where,TP = true positives, TN = true negatives, FP = false positives, FN = false negatives.
Precision: Number of actual true predictions over total true predictions.	$Precision = \frac{1}{C}\left(\sum_{C=1}^{C} \frac{tp_c}{tp_c+fp_c}\right)$ where, C = total no.of classes, tp_c = true positives for class c, fp_c = false positive for class c.
F-measure: Harmonic mean of precision and recall.	$f\text{-}measure = \sum_{C=1}^{C} 2\left(\frac{n_c}{N}\right) * \frac{precision_c * recall_c}{precision_c + recall_c}$ where, N = total no.of inputs, n_c = no.of inputs in class c, $precision_c$ = precision value for certain class c, $recall_c$ = recall value for certain class c.

5. Conclusion

An Enhanced HAR system using modified BRNN-LSTM for fall detection is developed in this paper. The system is able to predict six different activities and also detect falls successfully. The developed system is expected to produce higher accuracy than many existing machine learning and deep learning methods by extracting the most representative features from the data. Especially, it will be able to differentiate between sitting and standing which many existing models have not shown better results.

Future directions can be included with more complex activities for recognition and also be extended to support multi-sensor data.

6. References

[1] Coelho Y., Rangel L., dos Santos F., Frizera-Neto A., Bastos-Filho T. (2019), "Human Activity Recognition Based on Convolutional Neural Network". In: Costa-Felix R., Machado J., Alvarenga A. (eds) XXVI Brazilian Congress on Biomedical Engineering. IFMBE Proceedings, Vol. 70/2. Springer, Singapore. https://doi.org/10.1007/978-981-13-2517-5_38

[2] Preeti Agarwal, Mansaf Alam (2019), "A Lightweight Deep Learning Model for Human Activity Recognition on Edge Devices". *International Conference on Computational Intelligence and Data Science (ICCIDS)*, Procedia Computer Science 167, pp.2364-2373, Elsevier, pp. 2364-2373.
[3] Casilari E., Lora-Rivera R., García-Lagos F. (2019), "A Wearable Fall Detection System Using Deep Learning". In: Wotawa F., Friedrich G., Pill I., Koitz-Hristov R., Ali M. (eds) Advances and Trends in Artificial Intelligence. From Theory to Practice. IEA/AIE 2019. Lecture Notes in Computer Science, vol 11606. Springer, Cham. https://doi.org/10.1007/978-3-030-22999-3_39
[4] Pienaar, S. W., & Malekian,R.(2019), "Human Activity Recognition using LSTM-RNN Deep Neural Network Architecture". arXiv preprint ArXiv: 1905:00599.
[5] Ignatov, A. (2018), "Real-time human activity recognition from accelerometer data using Convolutional Neural Networks". Applied Soft Computing, 62, pp. 915-922.
[6] Johan Wannenburg, et. al. (2018), "Wireless Capacitive-Based ECG Sensing for Feature Extraction and Mobile Health Monitoring", *IEEE Sensors Journal*, Vol. 18, No. 14, pp. 6023-6032.
[7] S. Shahtalebi, S. F. Atashzar, R. V. Patel and A. Mohammadi (2019), "HMFP-DBRNN: Real-Time Hand Motion Filtering and Prediction via Deep Bidirectional RNN," in *IEEE Robotics and Automation Letters*, Vol. 4, No. 2, pp. 1061-1068. doi: 10.1109/LRA.2019.2894005.
[8] Ahmed Nait Aicha, Gwenn Englebienne, Kimberley S.Van Schooten, Mirjam Pijnappels and Ben Krose (2019), "Deep Learning to Predict Falls in Older Adults Based on Daily-Life Trunk Accelerometry", in *IEEE Sensors Journal*, 18(5), 1654. doi: 10.3390/218051654.
[9] J. C. Dogan and M. S. Hossain (2019), "A Novel Two-Step Fall Detection Method Using Smartphone Sensors", *2019 IEEE International Conference on Smart Computing (SMARTCOMP)*, Washington DC, USA, pp. 434-438. doi: 10.1109/SMARTCOMP.2019.00083.
[10] M. M. Hassan, A. Gumaei, G. Aloi, G. Fortino and M. Zhou (2019), "A Smartphone-Enabled Fall Detection Framework for Elderly People in Connected Home Healthcare," in *IEEE Network*, Vol. 33, No. 6, pp. 58-63. doi: 10.1109/MNET.001.1900100.
[11] J. Lee and H. Tseng (2019), "Development of an Enhanced Threshold-Based Fall Detection System Using Smartphones With Built-In Accelerometers," *IEEE Sensors Journal*, vol. 19, no. 18, pp. 8293-8302. doi: 10.1109/JSEN.2019.2918690.
[12] Li,H., and Trocan, M. (2018), "Deep Learning of Smartphone Sensor Data for Personal Health Assistance", Microelectronics Journal.
[13] "Mobifall Dataset" [Online]. Available: https://bmi.hmu.gr/the-mobifall-and-mobiact-datasets-2

IOP Conf. Series: Materials Science and Engineering 1065 (2021) 012046 doi:10.1088/1757-899X/1065/1/012046

Experimental Investigations of Load Characteristics of Induction Motor Drive

Mr. Ajithanjaya Kumar[1], Mr. Subrahmanya Bhat[2] and Dr. Rajkiran Ballal [3]

[1]Assistant Professor, Department of Electrical & Electronics Engineering, St Joseph college of Engineering, Vamanjoor, Mangalore. ajithk@sjec.ac.in,

[2]Assistant Professor, Department of Electrical & Electronics Engineering, SDM Institute of Technology, Ujire. manyavenur@sdmit.in ,

[3]Former Principal, Mangalore Marine College and Technology, Kuppepadav, Mangalore. ballalrk@yahoo.co.in .

Abstract. The three phase induction motors are most widely used for electric drive applications due to its simple, rugged construction and easy maintenance. Induction motors are considered as workhorse of the industry. This paper presents an experimental investigation of load characteristics along with equivalent circuit parameters of three-phase induction motor drive. The dynamic mathematical model of the machine under different reference frames is analysed with the parameters of the motor obtained through experiments. The equivalent circuit model under steady state is developed; performance of the induction motor is determined form the developed equivalent circuit and the results obtained are verified using the dynamic model under different reference frames. A 5 hp induction motor is used for the experimental verification.

1. Introduction

The induction machine is an important class of machine, widely used in industry. Almost 85% of the industrial motors are the induction motors. Almost 50% of the industrial power requirements are used by induction motors. Torque is developed due to inducement of current in the rotor, which is at asynchronous speed, hence induction motors are known as asynchronous motors. Due to ease in speed control, dc motors were widely used in industry for constant speed applications. Wide range of speed control is also possible in induction motors due to developments in power electronics. The Induction motors have the advantages such as cheap, robust and low maintenance over other classes of motor, they are used for wide applications in industries ranging from fractional kW applications to MW applications [1].

By knowing the parameters of the induction motor, operating performance of the induction motor such as efficiency, torque, current and power factor could be easily determined. The field efficiency of the induction motor may be determined by various ways as from name plate details, slip method, equivalent circuit method, shaft torque method and segregated loss method. The equivalent circuit method is the one of the standard approach of deterring the machine performance parameters as specified by IEEE Standard 112-2004 and IEC Standard 60034-2-1[3]. With the equivalent circuit method, the performance of the machine is easily predicted by knowing the impedance of the machine.

One of the easiest methods of deterring the equivalent circuit parameters is by performing Blocked rotor test and No load test [2]. The induction motor may be considered as a rotating transformer having stationary primary winding and a rotating secondary winding. As like transformers, the circuit parameters are easily obtained by performing the non loading tests as Blocked rotor test and No load test. Induction motor is also represented by the equivalent circuit as per the IEEE recommendation [7]. The Thevenin circuit model developed from the equivalent circuit parameters is useful in studying the performance of the machine in steady state where all electrical transients during load changes and variations in stator frequency are neglected.

Iron loss, friction and windage losses are neglected in the dynamic mathematical models. The dynamic mathematical models are used for the transient and steady state behavior of the machines. The three phase machine is transformed into two phase machine by adopting the power invariance concept. In this paper, performance of the machine is evaluated using stator reference frame, rotor reference and synchronously rotating reference frame models and steady state performance evaluated from equivalent circuit model developed for a 5hp motor is validated. In the dynamic model of the machine with all reference frames, the assumptions made are air gap is uniform, stator and rotor windings are balanced, mmf distribution and inductance versus rotor position are sinusoidal. Further saturation and parameter variations are neglected. To achieve fast and precise control of speed and torque, field orientation control can be used. Decoupling the stator current vector into torque producing and flux producing components helps in decoupled control of flux linkage and electromagnetic torque.

2. Tests on Induction Motor for discovering motor parameters

The parameters of the equivalent circuit model is determined by perfuming non loading tests such as No load test and Blocked rotor test. The stator resistance is determined by using volt-amp method. Name plate details of the squirrel cage induction motor used for conducting tests is given in Table 1.

Table 1. Name plate details

Rated Voltage (Volts)	Rated Current (Amps)	Power Rating (HP)	No Load Speed (RPM)	Number of Poles (P)	Supply Frequency (HZ)	Type of Stator Connection
415	7.5	5	1498	4	50	Delta

2.1 Stator Resistance Measurement

With rotor at stationary, dc voltage is applied to one of the phase winding of the stator. The dc resistance is calculated from the readings of voltmeter and ammeter. The ac resistance is obtained by multiplying dc resistance by a factor of 1.25. The effect temperature variation is neglected.

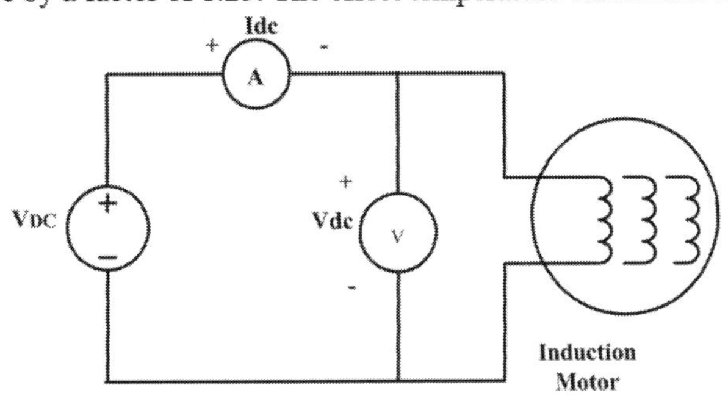

Fig 1. DC resistance measurement of stator winding

Table 2. Stator resistance measurement

Sl No	Voltage(Volts)	Current(Amps)	DC Resistance(Ohm)	Average Resistance(Ohm)	Stator AC Resistance(Ohm)
1	6.82	1.5	4.55		
2	9.34	2	4.67	4.57	5.71
3	11.21	2.5	4.484		

2.2 No Load Test

With rated voltage and frequency applied, motor is run with no load. The voltage, current and power input are measured. Since motor is with no load, power input accounts for the stator copper loss and stator iron loss. From the measurements, no load power factor, magnetizing current and inductance, core loss resistance is estimated.

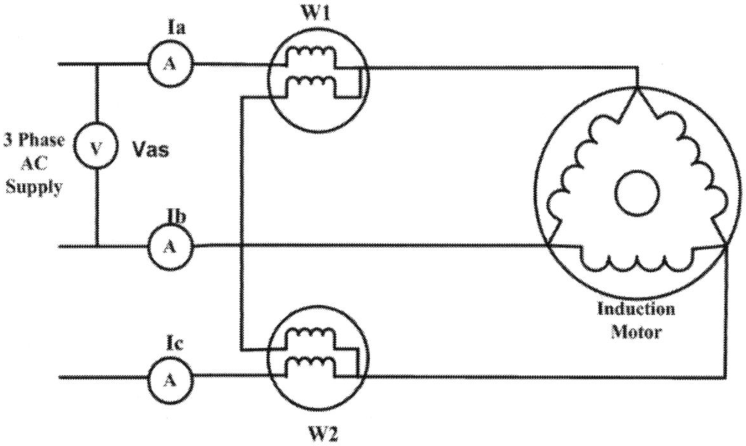

Fig 2. No load test circuit

Table 3. No Load Test observations

Voltage(V_{as})	Ia	Ib	Ic	No Load Current(Io)	W1(watts)	W2(watts)	Power Input(Poc)	Speed(rpm)
415	4.15	4.10	4.10	4.12	-672	1080	408	1498

2.3 Blocked Rotor Test

Rotor is locked to keep it in standstill and reduced voltage is applied so that rated current flows through stator. The voltage, current and power inputs are measured. From the measurements equivalent circuit is developed which helps to determine rotor resistance and total leakage reactance.

Table 4. Blocked Rotor Test observations

Voltage (V_{SC})	Current (I_{SC})	W_1 (Watts)	W_2(Watts)	Power Input (Psc) (Watts)
82	7.5	-11	510	499

3. Equivalent Circuit Model from No Load and Blocked Rotor tests

From the no load and blocked rotor test measurements, the equivalent circuit model (IEEE model) is developed.

3.1 From no load test:

No load current /phase = $\frac{4.12}{\sqrt{3}}$ = 2.38A

No load power factor cosΦo = $\frac{P_{oc}}{\sqrt{3}\,V_L I_L}$ =0.1377

No load impedance/phase Zo = $\frac{V_L}{I_{phase}}$ =174.47Ω

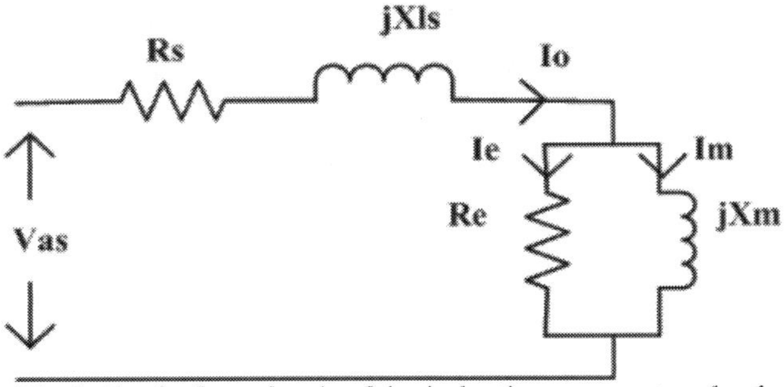

Fig 3. Equivalent circuit of the induction motor at no load

Magnetizing current = Im = I_{ophase} sinΦo=2.36A
Core loss branch current Ic= I_{ophase} cosΦo=0.33A
With neglecting stator impedances, Magnetizing inductance = Lm = $\frac{V_{as}}{2\pi f_s I_m}$ = 0.5597H

Core loss resistance Rc=$\frac{V_{as}}{I_c}$ = 1267.17 Ω

3.2 From Blocked Rotor Test

Short circuit current/phase = $\frac{I_{sc}}{\sqrt{3}}$ = 4.42 A

Power factor cosΦsc=$\frac{Psc}{\sqrt{3}VscIsc}$=0.5219

Locked rotor impedance =Zsc=$\frac{Vsc}{Isc}$=18.795 Ω

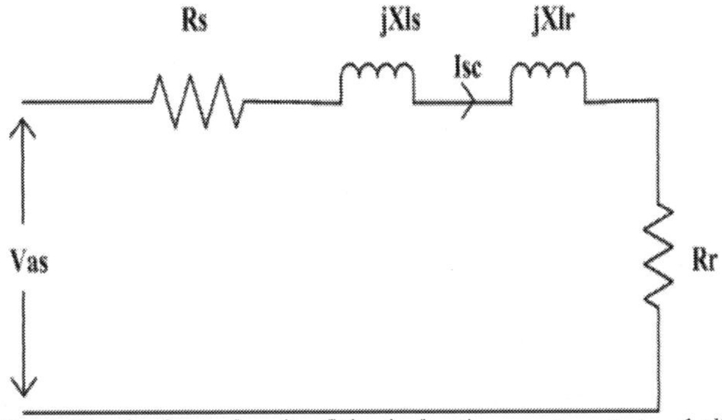

Fig 4. Equivalent circuit of the induction motor at standstill

Stator referred rotor resistance per phase=Rr = Zsc cosΦsc-Rs= 4.11 Ω
Total leakage reactance = Xeq = Zsc sinΦsc= 16.03 Ω
Stator leakage reactance Xls = rotor leakage reactance Xlr = $\frac{Xeq}{2}$ = 8.015 Ω
Stator leakage inductance Ls = rotor leakage inductance Lr = $\frac{Xls}{2\pi f}$ = 25.512 mH
The stator impedances, core loss resistance and magnetizing inductance are recalculated by considering stator impedance as shown in Fig.3
Magnetizing branch voltage = Vm= Vas-(Rs+jXls)Io
Magnetizing reactance = $\frac{Vm}{Im}$ = 169 Ω
Core loss resistance = $\frac{Vm}{Ic}$=1227.14 Ω

3.3 Equivalent circuit model and Thevenin equivalent of induction motor circuit model

From the parameters obtained by the no load test and blocked rotor test, equivalent circuit of the induction motor is shown in Fig 5.

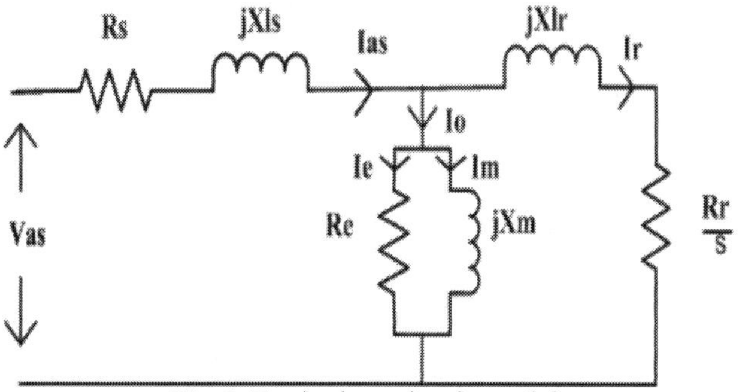

Fig 5. Equivalent circuit model

In Fig 5,. Re represents the core loss, friction and windage loss. These losses are separated out by no load and blocked rotor tests. By subtracting these losses from gross mechanical power output, exact circuit (IEE circuit) model is developed as in Fig 6.
The Thevenin's equivalent circuit model of induction motor is developed by finding the Thevenin's voltage and Thevenin's impedance to the left of the rotor impedances as

$$|V_{TH}| = \left|Vas\ \frac{jXls}{Rs+j(Xls+Xm)}\right| = 396\ \text{V} \tag{1}$$

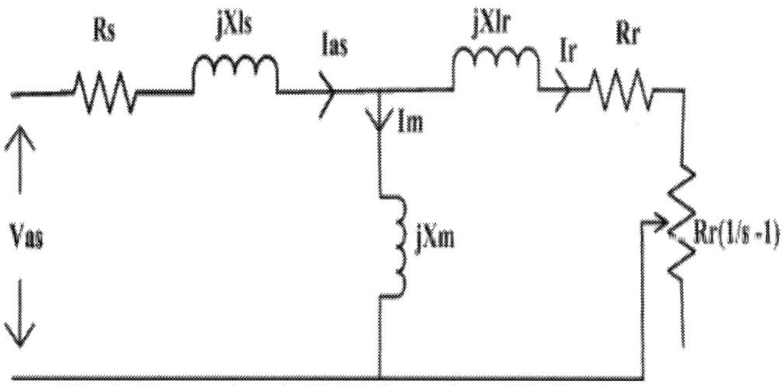

Fig 6. Exact circuit (IEEE) model of induction motor

$$Z_{TH} = \frac{jxm\ (Rs+jXls)}{Rs+j(Xls+Xm)} = R_{TH}+jX_{TH} = 5.19+j7.819 \tag{2}$$

From the parameters obtained from equations (1) and (2), the Thevenin's equivalent circuit model of an induction motor is shown in Fig 7.
From the Thevinin;s equivalent circuit, slip for maximum electromagnetic torque is obtained as

$$\frac{Rr}{s_{max,T}} = \sqrt{R_{TH}{}^2 + (X_{TH} + X_{lr})^2} \tag{3}$$

Substituting the obtained values in equation (3), the slip for maximum electromagnetic torque obtained as $S_{max,T} = 0.2466$

$Z_{Total} = 21.85+j15.834\ \Omega$

$|I_2| = \frac{|V_{TH}|}{|Z_{Total}|} = 14.67$ A

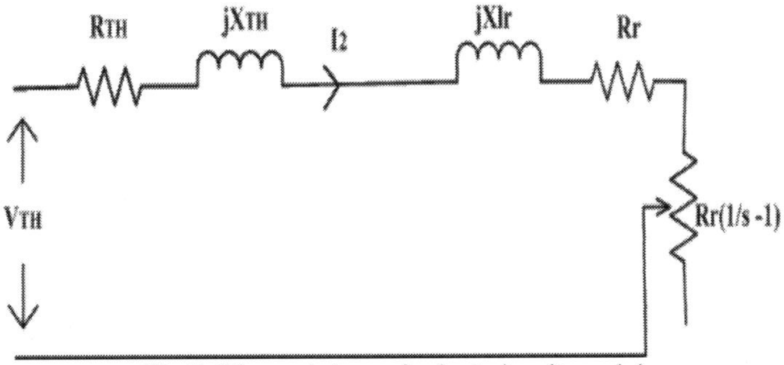

Fig 7. Thevenin's equivalent circuit model

The maximum electromagnetic torque = Temax = $\frac{3I_2^2 R_2}{S_{maxT}\ \omega_s}\frac{P}{2}$ = 68.5 N-m

Where ω_s is the synchronous speed in electrical radian/sec = 314.15 rad/sec

4. Analysis of induction motor performance by the generalized dynamic models

The performance of the induction motor is analyzed by the generalized dynamic models in synchronously rotating reference frame, rotor reference frame and stator reference frame. The performance parameter is determined for the slip $S_{max,T}$ of 0.2466. In dynamic model, three phase machine is transformed to two phase machine. Stator and rotor voltages and currents are represented in d-q axes.

The machine parameters obtained from no load and blocked rotor teat given in Table 5 is used in the analysis of the machine by dynamic model with different reference frames.

Table 5. The Induction Motor parameters

Rs (ohm)	Rr (ohm)	Ls = Lls+Lm Henry	Lr=Llr+Lm Henry	Lm Henry
5.7	4.11	0.5634	0.5634	0.5379

The model of the induction machine in arbitrary reference frame is given as

$$\begin{bmatrix} V_{qs}^c \\ V_{ds}^c \\ V_{qr}^c \\ V_{dr}^c \end{bmatrix} = \begin{bmatrix} Rs+LsP & \omega_c Ls & LmP & \omega_c Lm \\ -\omega_c Ls & Rs+LsP & -\omega_c Lm & LmP \\ LmP & (\omega_c-\omega_r)Lm & Rr+LrP & (\omega_c-\omega_r)L_r \\ -(\omega_c-\omega_r)Lm & LmP & -(\omega_c-\omega_r)L_r & Rr+LrP \end{bmatrix} \begin{bmatrix} i_{qs}^c \\ i_{ds}^c \\ i_{qr}^c \\ i_{dr}^c \end{bmatrix} \quad (4)$$

Where ω_c is the speed of arbitrary reference frame in electrical radian/sec

ω_r is the speed of rotor in electrical radian/sec

P is the differential operator d/dt

The subscripts s and r refers to stator and rotor quantities respectively.

4.1 Synchronously rotating reference frame model

As the reference frame rotates at the synchronous speed, $\omega_c = \omega_s$ (5)

$\theta c = \theta_s = \omega_s$ t (6)

$V_{as}=V_m \sin \omega_s t = 586.89 \sin \omega_s t$

$V_{bs} = V_m \sin(\omega_s t-2\pi/3) = 586.89 \sin(\omega_s t-2\pi/3)$

$V_{cs} = V_m \sin(\omega_s t+2\pi/3) = 586.89 \sin(\omega_s t+2\pi/3)$

$$\begin{bmatrix} V_{qs}^e \\ V_{ds}^e \\ V_o \end{bmatrix} = [T_{abc}^e] \begin{bmatrix} V_{as} \\ V_{bs} \\ V_{cs} \end{bmatrix} \quad (7)$$

$[T_{abc}^e]$ Is the transformation matrix from abc to dqo variables

IOP Conf. Series: Materials Science and Engineering 1065 (2021) 012046 doi:10.1088/1757-899X/1065/1/012046

$$[T^e_{abc}] = \frac{2}{3}\begin{bmatrix} cos\theta_s & \cos(\theta_s - \frac{2\pi}{3}) & \cos(\theta_s + \frac{2\pi}{3}) \\ sin\ \theta_s & \sin(\theta_s - \frac{2\pi}{3}) & \sin(\theta_s + \frac{2\pi}{3}) \\ \frac{1}{2} & \frac{1}{2} & \frac{1}{2} \end{bmatrix} \quad (8)$$

Substituting for V_{as}, V_{bs}, V_{cs} and equation (8) in equation (7)

$$\begin{bmatrix} V^e_{qs} \\ V^e_{ds} \\ V_o \end{bmatrix} = \begin{bmatrix} 0 \\ V_m \\ 0 \end{bmatrix} = \begin{bmatrix} 0 \\ 586.89 \\ 0 \end{bmatrix} \quad (9)$$

Obtained d and q axes stator voltages are dc quantities, hence responses are also dc quantities. Thus system becomes linear.

$$Pi^e_{qs} = Pi^e_{ds} = Pi^e_{qr} = Pi^e_{dr} = 0 \quad (10)$$

Substituting equations (5), (6), (9) and (10) in equation (4)

$$\begin{bmatrix} 0 \\ V_m \\ 0 \\ 0 \end{bmatrix} = \begin{bmatrix} Rs & \omega_s Ls & 0 & \omega_s Lm \\ -\omega_s Ls & Rs & -\omega_s Lm & 0 \\ 0 & \omega_{sl} Lm & Rr & \omega_{sl} Lr \\ -\omega_{sl} Lm & 0 & -\omega_{sl} Lr & Rr \end{bmatrix} \begin{bmatrix} i^e_{qs} \\ i^e_{ds} \\ i^e_{qr} \\ i^e_{dr} \end{bmatrix}$$

Which is of the form V=ZI where Z is the impedance matrix.

$\omega_{sl} = \omega_s - \omega_r$ = slip speed = 77.47 radian/sec

$$Z = \begin{bmatrix} 5.7 & 176.99 & 0 & 168.98 \\ -176.99 & 5.7 & -168.98 & 0 \\ 0 & 41.67 & 4.11 & -43.64 \\ -41.67 & 0 & -43.64 & 4.11 \end{bmatrix}$$

Currents are obtained as $I = Z^{-1} V = \begin{bmatrix} i^e_{qs} \\ i^e_{ds} \\ i^e_{qr} \\ i^e_{dr} \end{bmatrix} = \begin{bmatrix} -13.86 \\ 16.88 \\ 11.62 \\ -17.20 \end{bmatrix}$

Electromagnetic Torque $= T_e = \frac{3}{2}\frac{P}{2} L_m (i^e_{qs} i^e_{dr} - i^e_{ds} i^e_{qr})$ N-m where P is the number of poles.

Te = 68.82 N-m

Actual phase stator currents $i_{abc} = [T^e_{abc}]^{-1} i^e_{qdo}$

$$\begin{bmatrix} i_{as} \\ i_{bs} \\ i_{cs} \end{bmatrix} = \begin{bmatrix} cos\omega_s t & sin\omega_s t & 1 \\ cos(\omega_s t - \frac{2\pi}{3}) & sin(\omega_s t - \frac{2\pi}{3}) & 1 \\ cos(\omega_s t + \frac{2\pi}{3}) & sin(\omega_s t + \frac{2\pi}{3}) & 1 \end{bmatrix} \begin{bmatrix} i^e_{qs} \\ i^e_{ds} \\ 0 \end{bmatrix}$$

$$\begin{bmatrix} i_{as} \\ i_{bs} \\ i_{cs} \end{bmatrix} = \begin{bmatrix} 21.84 \sin(\omega_s t - 39.46°) \\ 21.84 \sin(\omega_s t - 159.28°) \\ 21.84 \sin(\omega_s t + 80.59°) \end{bmatrix}$$

4.2 Rotor reference frame model

The speed of the rotor reference frame is the speed of the rotor. $\omega_c = \omega_r$ (11)

Angular position $\theta c = \theta r$ (12)

IOP Conf. Series: Materials Science and Engineering 1065 (2021) 012046 doi:10.1088/1757-899X/1065/1/012046

q-d axis stator voltages in rotor reference frame are $\begin{bmatrix} V_{qs}^r \\ V_{ds}^r \\ V_o \end{bmatrix} = [T_{abc}^r] \begin{bmatrix} V_{as} \\ V_{bs} \\ V_{cs} \end{bmatrix}$ (13)

The transformation matrix from abc to dqo variables $[T_{abc}^r] = \frac{2}{3} \begin{bmatrix} \cos\theta_r & \cos(\theta_r - \frac{2\pi}{3}) & \cos(\theta_r + \frac{2\pi}{3}) \\ \sin\theta_r & \sin(\theta_r - \frac{2\pi}{3}) & \sin(\theta_r + \frac{2\pi}{3}) \\ \frac{1}{2} & \frac{1}{2} & \frac{1}{2} \end{bmatrix}$

Form equation (xiii), $\begin{bmatrix} V_{qs}^r \\ V_{ds}^r \\ V_o \end{bmatrix} = \begin{bmatrix} V_m \sin\omega_{sl}t \\ V_m \cos\omega_{sl}t \\ 0 \end{bmatrix} = \begin{bmatrix} -jV_m \\ V_m \\ 0 \end{bmatrix} = \begin{bmatrix} -j586.89 \\ 586.89 \\ 0 \end{bmatrix}$ (14)

As the stator voltages appear at slip frequency in rotor reference frames, hence currents are also at slip frequency in steady state. Thus P = jω_{sl} (15)

Substituting equations (11), (12),(14) and (15) in in equation (4)

$$\begin{bmatrix} -jV_m \\ V_m \\ 0 \\ 0 \end{bmatrix} = \begin{bmatrix} Rs + j\omega_{sl}L_s & \omega_r Ls & j\omega_{sl}L_m & \omega_r Lm \\ -\omega_r Ls & Rs + j\omega_{sl}L_s & -\omega_r Lm & j\omega_{sl}L_m \\ j\omega_{sl}L_m & 0 & Rr + j\omega_{sl}Lr & 0 \\ 0 & j\omega_{sl}L_m & 0 & Rr + j\omega_{sl}Lr \end{bmatrix} \begin{bmatrix} i_{qs}^e \\ i_{ds}^e \\ i_{qr}^e \\ i_{dr}^e \end{bmatrix}$$

$$Z = \begin{bmatrix} 5.7 + j43.65 & 133.34 & j41.67 & 127.31 \\ -133.34 & 5.7 + j43.65 & -127.31 & j41.67 \\ j41.67 & 0 & 4.11 + j43.65 & 0 \\ 0 & j41.67 & 0 & 4.11 + j43.65 \end{bmatrix}$$

Currents are obtained as I = Z^{-1} V = $\begin{bmatrix} i_{qs}^r \\ i_{ds}^r \\ i_{qr}^r \\ i_{dr}^r \end{bmatrix} = \begin{bmatrix} 21.83\angle -129.47° \\ 21.83\angle -39.47° \\ 20.74\angle 55.90° \\ 20.74\angle 145.90° \end{bmatrix}$

Actual phase stator currents $i_{abc} = [T_{abc}^r]^{-1} i_{qdo}^r$

$$\begin{bmatrix} i_{as} \\ i_{bs} \\ i_{cs} \end{bmatrix} = \begin{bmatrix} cos\omega_r t & sin\omega_r t & 1 \\ cos(\omega_r t - \frac{2\pi}{3}) & sin(\omega_r t - \frac{2\pi}{3}) & 1 \\ cos(\omega_r t + \frac{2\pi}{3}) & sin(\omega_r t + \frac{2\pi}{3}) & 1 \end{bmatrix} \begin{bmatrix} i_{qs}^r \\ i_{ds}^r \\ 0 \end{bmatrix} = \begin{bmatrix} 21.83 \sin(\omega_s t - 39.46°) \\ 21.83 \sin(\omega_s t - 159.27°) \\ 21.83 \sin(\omega_s t + 80.47°) \end{bmatrix}$$

Electromagnetic Torque =Te = $\frac{3}{2}\frac{P}{2} L_m(i_{qs}^r \overline{i_{dr}^r} - i_{ds}^r \overline{i_{qr}^r})$ N-m = 68.82 N-m

4.3 Stator reference frame model

As the reference frame is stator, ω_e=0

$$\begin{bmatrix} V_{qs}^s \\ V_{ds}^s \\ V_o \end{bmatrix} = [T_{abc}^s] \begin{bmatrix} V_{as} \\ V_{bs} \\ V_{cs} \end{bmatrix} = \frac{2}{3} \begin{bmatrix} 1 & -\frac{1}{2} & -\frac{1}{2} \\ 0 & -\frac{\sqrt{3}}{2} & \frac{\sqrt{3}}{2} \\ \frac{1}{2} & \frac{1}{2} & \frac{1}{2} \end{bmatrix} \begin{bmatrix} V_{as} \\ V_{bs} \\ V_{cs} \end{bmatrix} = \begin{bmatrix} V_m \sin\omega_s t \\ V_m \cos\omega_s t \\ 0 \end{bmatrix}$$

Since the stator voltages appear at synchronous frequency in the reference frame, currents are also at synchronous frequency at steady state. Thus P= jω_s in the arbitrary reference frame model. Thus the model of induction motor is

$$\begin{bmatrix} V_{qs}^{c} \\ V_{ds}^{c} \\ V_{qr}^{c} \\ V_{dr}^{c} \end{bmatrix} = \begin{bmatrix} Rs + LsP & 0 & LmP & 0 \\ 0 & Rs + LsP & 0 & LmP \\ LmP & -\omega_r Lm & Rr + LrP & -\omega_r L_r \\ \omega_r Lm & LmP & \omega_r L_r & Rr + LrP \end{bmatrix} \begin{bmatrix} i_{qs}^{c} \\ i_{ds}^{c} \\ i_{qr}^{c} \\ i_{dr}^{c} \end{bmatrix}$$

$$Z = \begin{bmatrix} 5.7 + j176.99 & 0 & j168.98 & 0 \\ 0 & 5.7 + j176.99 & 0 & j168.98 \\ j168.98 & -127.31 & 4.11 + j176.99 & -133.34 \\ 127.31 & j168.98 & 133.34 & 4.11 + j176.99 \end{bmatrix}$$

$$I = Z^{-1} V = \begin{bmatrix} i_{qs}^{s} \\ i_{ds}^{s} \\ i_{qr}^{s} \\ i_{dr}^{s} \end{bmatrix} = \begin{bmatrix} 21.82\angle -39.47^\circ \\ 21.83\angle 50.54^\circ \\ 20.74\angle 145.90^\circ \\ 20.75\angle -124.07^\circ \end{bmatrix}$$

Actual phase stator currents $i_{abc} = [T_{abc}^{r}]^{-1} i_{qdo}^{r}$

$$\begin{bmatrix} i_{as} \\ i_{bs} \\ i_{cs} \end{bmatrix} = \begin{bmatrix} 1 & 0 & 1 \\ -\frac{1}{2} & -\frac{\sqrt{3}}{2} & 1 \\ -\frac{1}{2} & \frac{\sqrt{3}}{2} & 1 \end{bmatrix} \begin{bmatrix} i_{qs}^{s} \\ i_{ds}^{s} \\ 0 \end{bmatrix} = \begin{bmatrix} 21.83 \sin(\omega_s t - 39.46^\circ) \\ 21.83 \sin(\omega_s t - 159.44^\circ) \\ 21.83 \sin(\omega_s t + 80.52^\circ) \end{bmatrix}$$

$$Te = \frac{3}{2}\frac{P}{2} L_m (i_{qs}^{s} \overline{i_{dr}^{s}} - i_{ds}^{s} \overline{i_{qr}^{s}}) \text{ N-m} = 68.82 \text{ N-m}$$

5. Conclusion

The equivalent circuit model of the induction motor is the easiest method of determining the machine parameters. In the laboratory, the machine parameters are determined by dc resistance measurement, no load test and blocked rotor tests. From the developed Thevenin's equivalent circuit model, the maximum electromagnetic torque and the slip at which this occurs is determined. The dynamic model of the machine in all the three reference frames is analyzed. The results obtained from steady state model are validated with dynamic models. The sinusoidal signals are transformed to dc signals in the synchronous reference frame. Decoupling the torque and flux producing components helps in decoupled control of flux linkage and electromagnetic torque, hence control of the induction motor similar to separately excited dc motor is possible. In cases of stator reference frame and rotor reference frame, stator and rotor q-d axes currents are complex and displaced from each other by 90°. Rotor reference frame model is useful when the power and switching elements are to be controlled on rotor side. In case of stator reference frame, input variables are well defined, stator and rotor d-q axes voltages are determined through simple algebraic equations. This model is well suitable for control applications of stator-controlled inductions motors as applicable to inverter controlled and phase controlled induction motor drives. Further the performance of the machine can be analysed by considering the effect of variation in stator and rotor winding temperature, iron and core loss.

References

[1] Diptarshi Bhowmick and Suparna Kar Chowdhury 2018 Parameter and Loss Estimation of Three Phase Induction Motor from Dynamic Model using H - G Diagram and Particle Swarm Optimization *2018 IEEE 8th Power India International Conference (PIICON)* (Kurukshetra, India), DOI: 10.1109/POWERI.2018.8704353

[2] Phumiphak T and Chat-uthai C 2002 Estimation of induction motor parameters based on field test coupled with genetic algorithm International Conference on Power System Technology(Kunming: China) IEEE *Xplore* Print ISBN:0-7803-7459-2

[3] Ayasun S and Nwankpa C O 2005 Induction motor tests using MATLAB/Simulink and their integration into undergraduate electric machinery courses IEEE Transactions on Education (Volume: 48, Issue: 1, Feb. 2005) Page(s): 37 - 46

[4] Aminu M, Ainah P K, Abana M and Abu U A 2018 Identification of induction Machine Parameters using Only No Load test Measurement Nigerian Journal of Technology (NIJOTECH) Vol. 37, No. 3, July 2018, pp. 742 – 748
[5] France O. Akpojedje, Ese M. Okah and Yussuf O. Abu 2016 A Comparative Analysis of Three Phase Induction Motor Performance Evaluation International Journal of Engineering and Techniques - Volume 2 Issue 3, May – June 2016 ISSN: 2395-1303 Page 64
[6] Sunil Sehra, Gautam K K and Vijay Bhuria 2012 Performance Evaluation of Three Phase Induction Motor Based on No Load and Blocked Rotor Test Using MATLAB International Journal of Science, Environment and Technology, Vol. 1, No 5, pp. 541 – 547
[7] Pandey K K and Zope P H 2013 Estimating Parameters of a Three-Phase induction motor using Matlab/Simulink International Journal of Scientific & Engineering Research, Volume 4, Issue 12, ISSN 2229-5518
[8] IEEE Standard Test Procedure for Polyphase Induction Motors and Generators Publisher :IEEE Print: ISBN 0-7381-3977-7 SH95211 PDF: ISBN 0-7381-3978-5 SS95211
[9] Tripti Rai and Prashanth Debre 2016 Generalized modeling model of three phase induction motor Proc. International Conference on Energy Efficient Technologies for Sustainability (ICEETS) (Nagercoil: India) Publisher: IEEE Print ISBN:978-1-5090-1534-4
[10] Sonakshi Gupta and Dr. Sulochana Wadhwani 2014 Dynamic Modeling of Induction Motor Using Rotor Rotating Reference Frame International Journal of Advanced Research in Electrical, Electronics and Instrumentation Engineering Vol. 3, Issue 6, June 2014 ISSN (Print) : 2320 – 3765 ISSN (Online): 2278 – 8875
[11] Krishnan R 2013 *Electric Motor Drives Modelling, Analysis and Control* (Delhi PHI) Chapter 5 pp 193-217

IOP Conf. Series: Materials Science and Engineering 1065 (2021) 012047 doi:10.1088/1757-899X/1065/1/012047

Comparison between CMOS full adder and PTL full adder

Amitha.M[1], Deepa[2]

[1]Master of Technology in VLSI Design and Embedded Systems, Dept. of Electronics and Instrumentation Engineering, Siddaganga Institute of Technology, Tumakuru.

[2]Assistant Professor, Dept. of Electronics and Instrumentation Engineering, Siddaganga Institute of Technology, Tumakuru

Email.id: rmdeepa@yahoo.com

Abstract: In electronic industry the level of integration is an important aspect as makes the electronic device simpler and more reliable. The device density increases with the better level of integration. Power dissipation, area occupied and delay are some of the important factors that need to be considered. These parameters play a vital role in manufacturing portable electronic gadgets. Designing of full adder using conventional CMOS design and PTL has been shown in this paper. A comparison is made between the two designs with respect to power dissipation, delay and area (number of transistors). Mentor Graphics Tool is used in design and simulation of the full adder. The PTL full adder has a smaller number of transistors and lower power dissipation compared to CMOS full adder. With all the comparison made it will be concluded that the PTL full adder performance is better than the CMOS full adder.

Key words: Power dissipation, delay, area, PTL, mentor graphics, full adder.

1. Introduction

Full adder is a type of adder that adds three inputs and gives two outputs. Out of three, two will be the present inputs and the third input will be the carry from the previous stage. Full adder is shown in Fig 1, where 'A' and 'B' are the actual inputs, 'C' is the carry from the previous operation. SUM and CARRY OUT are the two outputs. The Truth Table shown in Table 1 gives an idea about outputs for various input combinations. For any arithmetic operations like multiplication, subtraction, division, DSP designs adders are the fundamental blocks. Many binary adders are formed using full adders. Hence, if any enhancements have to be made to improve the performance it can be made at the root i.e., adders. This in turn helps in bettering the performance of the electronics circuits which follow adder. The low power VLSI design is of great importance due to portable electronic products. Lowering the threshold voltage and the input voltages are some of the effective ways to reduce the power consumption. The adders are designed in various techniques and they all are concentrated on reducing the area, power dissipation and delay [1]. The area and complexity can be reduced by decreasing the transistor count. The speed depends on transistor size, delay in the critical path and parasitic capacitance [4]. Although all the designs perform the similar function, each design has its own advantages and disadvantages. With a combination of one or more logic styles better result can be got.

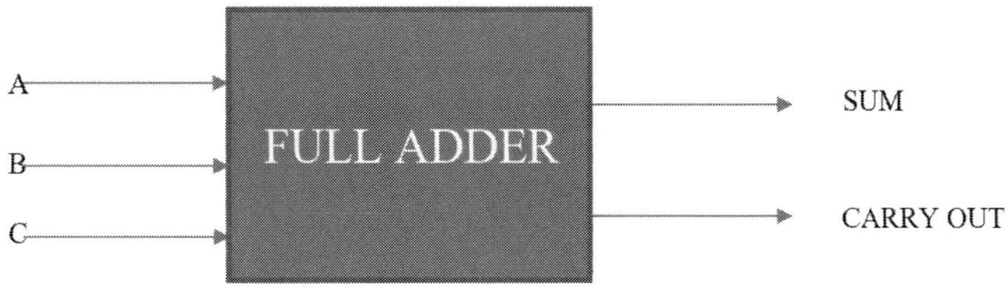

Fig 1: Block diagram of full adder

Table 1: Full adder truth table

A	B	C	SUM	CARRY OUT
0	0	0	0	0
0	0	1	1	0
0	1	0	1	0
0	1	1	0	1
1	0	0	1	0
1	0	1	0	1
1	1	0	0	1
1	1	1	1	1

SUM = A^B^C = [A+B+C].[~CARRY OUT]+ABC --- (1) CARRY OUT= AB+BC+CA = AB+C [A+B] --- (2)

1.1. Mentor Graphics

Mentor graphics is a well-known technology provider in Electronics Design and Automation. This has built software and hardware solutions that help companies to make their quality electronic products at a higher production rate and also at lower cost. Mentor Graphics is used mainly to develop electronic devices that are fully customized [3]. One small flaw in the chip can ruin the entire system hence it is important to eliminate them at the initial stages. Mentor Graphics helps one to prepare the details related to design such as initial schematics, layout and physical design of the device; also verification of the circuit can be done before it is fabricated. At each stage, the design can be verified by simulating it and check whether the circuit is functioning properly.

Here is the 1bit full adder designed using both Complementary Metal Oxide Semiconductor (CMOS) design and also using Pass Transistor Logic (PTL). Both the designs are drawn using Mentor Graphics. The paper is organized in sections. The section 2 is about Conventional CMOS design. The section 3 is about the PTL full adder. In section 4 a comparison is made between both the designs.

2. Conventional CMOS design

Conventional CMOS design is based on normal complementary metal oxide semiconductor design with P-MOS acting as pull-up and N-MOS as pull-down transistor. This design requires 28 transistors, thus the silicon area and the complexity is more for larger designs. These are more immune to noise. The design and the simulation result for the 28T full adder is as shown in Fig 2. The schematics are drawn as per the equation (1) and (2). It is easy to analyse the conventional CMOS full adder by using the principle of mirror. In Fig 2(a) it can be seen that the NMOS network is identical to that of the PMOS network. This fully symmetric design reduces the complexity of the

IOP Conf. Series: Materials Science and Engineering 1065 (2021) 012047 doi:10.1088/1757-899X/1065/1/012047

layout design as it makes layout design, placement and routing easier. The main disadvantage of this configuration is that it has slower speed as it has to drive heavier load. Requirement of high input capacitance and buffer are also some of the disadvantages. High robustness against voltage scaling and transistor sizing are the advantages of CMOS full adder design.

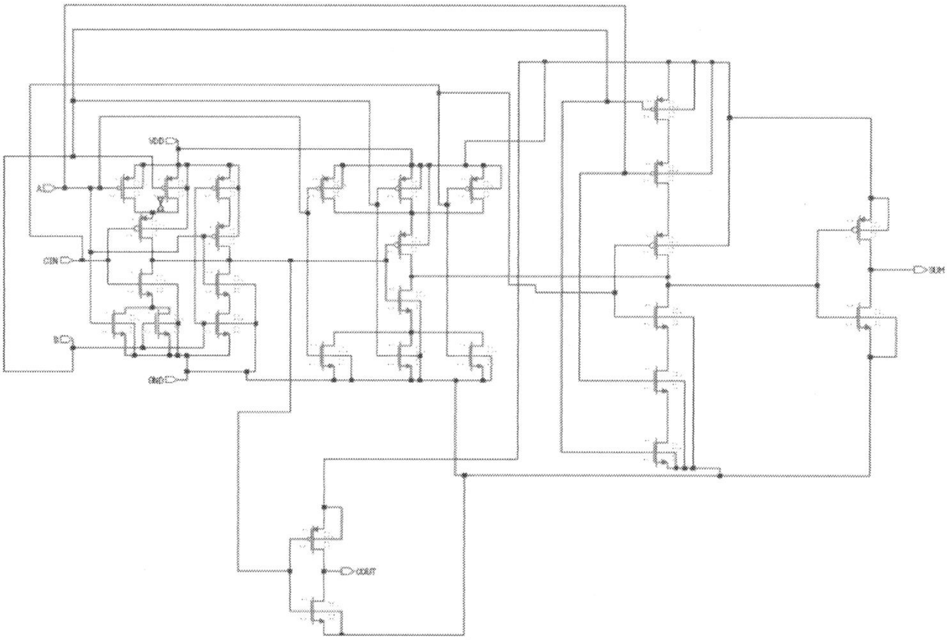

Fig 2(a): Schematic of conventional CMOS full adder

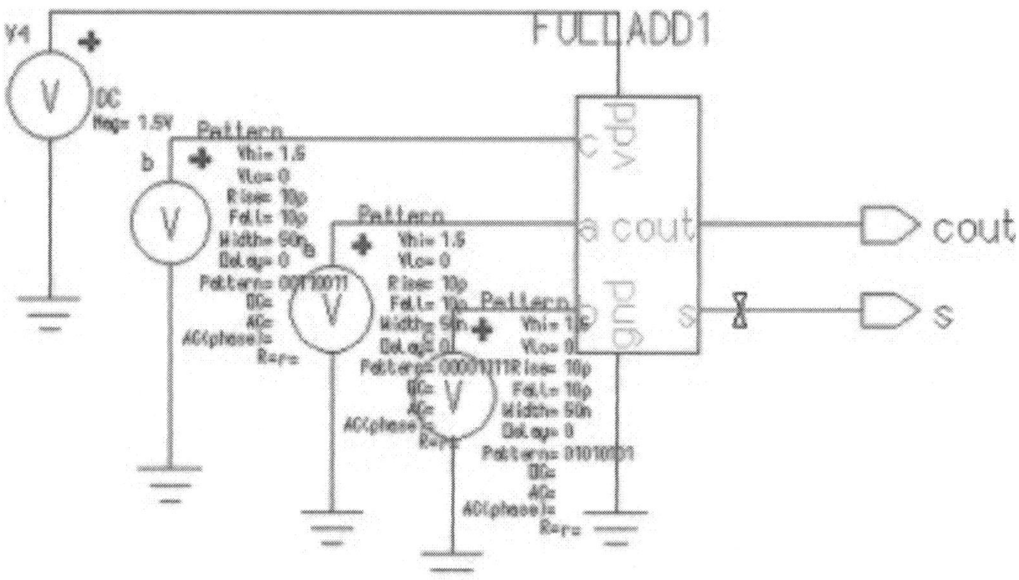

Fig 2(b): Block diagram of Conventional CMOS Full adder

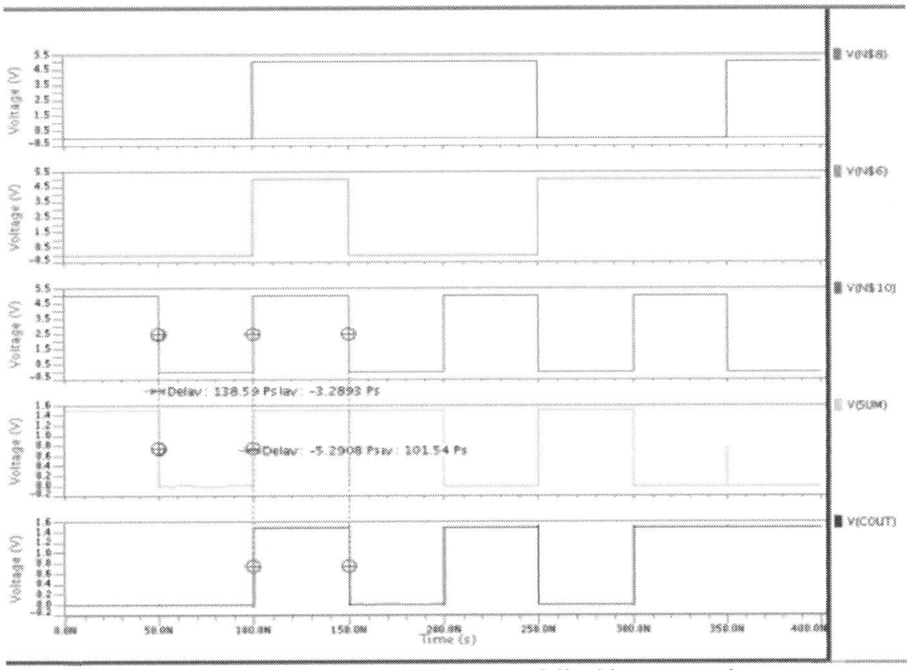

Fig 2(c): Waveforms showing inputs to full adder, sum and carry

Delay for sum=(τPHL+τPLH)/2 = (-3.289+138.59)ps/2 =70.93psec
Delay for carry=(τPHL+τPLH)/2 = (-5.291+101.54)ps/2 =53.4155psec Power desipation=6.372 n watts
CMOS full adder has been designed, test schematic is drawn, and delay of 70.93psec for sum and 53.4155psec for carry and power dissipation of 6.372n watts has been obtained.

3. PTL Full Adder

The power dissipation is one of the main factors in VLSI circuits. The 1 bit CMOS 2T full adder was not efficient to build larger circuits with high performance as the complexity would increase. Hence, Pass Transistor Logic (PTL) adders were designed. This Logic helps us eliminate redundant transistors. The reduction in the transistor count in turn reduces the power dissipation of a device and at the same time maintains a proper throughput [2].In 1 bit and 8 bit full adder, including of NMOS based PTL increases the performance as there will be more transistors in less area and hence proves itself to be an effective power handling logic.

Pass transistor Logic comes along with the advantages such as [5] high speed, low power dissipation and low interconnection effect. High speed results in lesser delay and the reason for this is small node capacitances. Lower power dissipation is due to less number of transistors. During the logic low the charge flowing through the transistor is drained. These charges are reused and given as input to control the transistor gates. This shows that there is no direct way to ground. Hence, the power dissipation is almost zero. The small area of circuit results in low interconnection effect. There are some disadvantages of PTL design techniques such as 1) threshold drop across the pass transistor 2) at the regenerative inverter the input voltage value is not Vdd. There will be multiple drops at the output as the NMOS can pass '0' signal correctly but can never pass '1' accurately. The threshold loss is always encountered. The below Fig 3 shows schematic of PTL full adder, its block diagram and waveforms related to input and output.

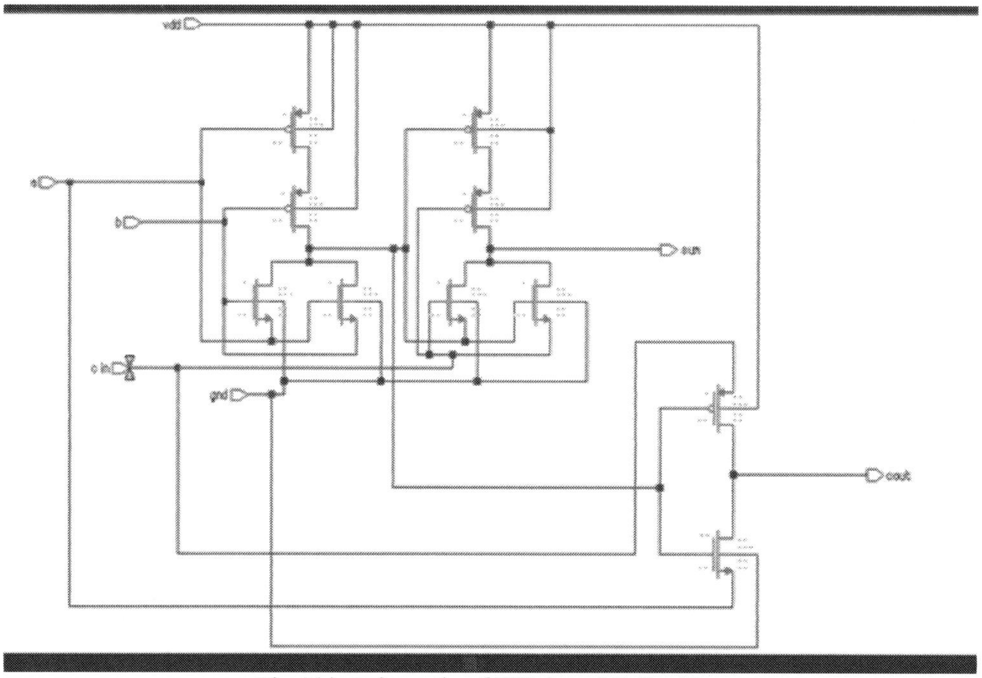

Fig 3(a): Schematic of PTL based Full adder

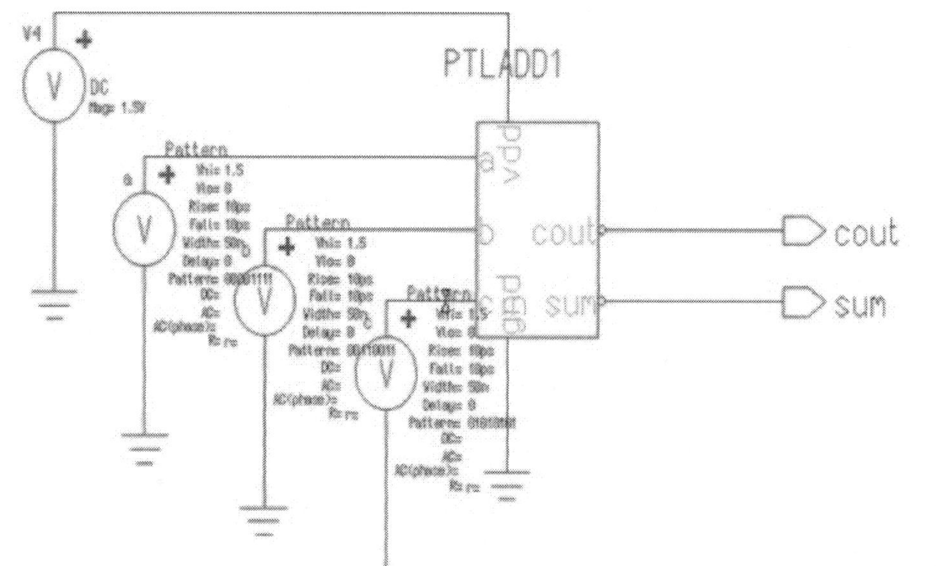

Fig 3(b): Block diagram of PTL full adder showing inputs and outputs

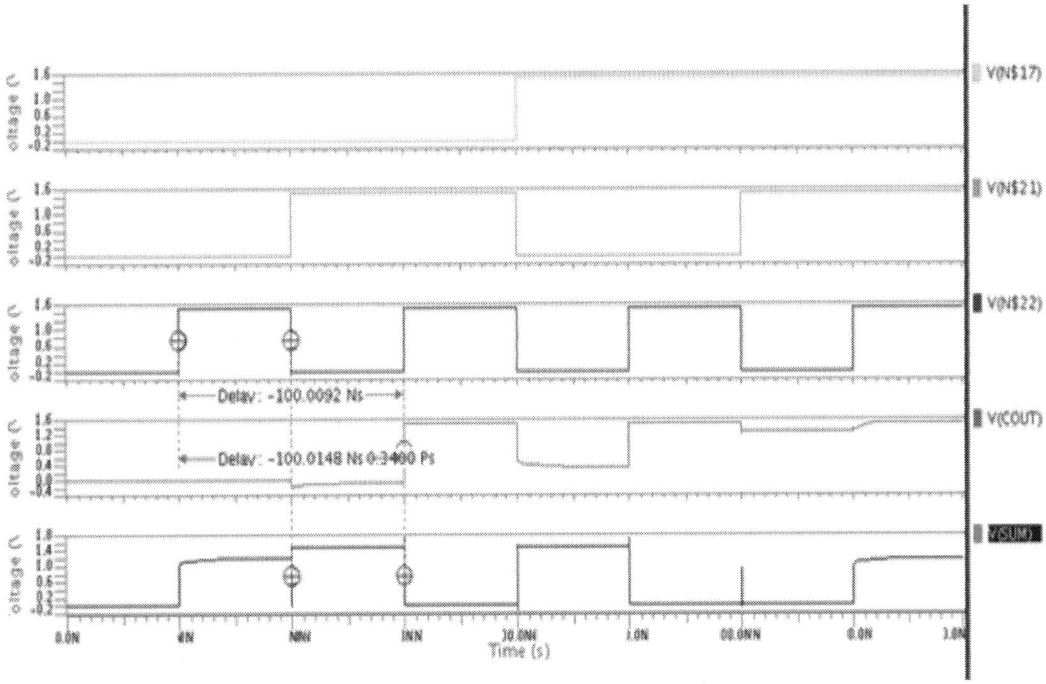

Fig 3(c): Waveforms of inputs, sum and carry

Delay=-100.0148ns
Power dissipation: 1.381nW
Advantage: Complexity reduces due to less number of transistors and delay reduces.
Disadvantage: NO full swing in the output voltages and leakage power is high.

4. Result and Conclusion

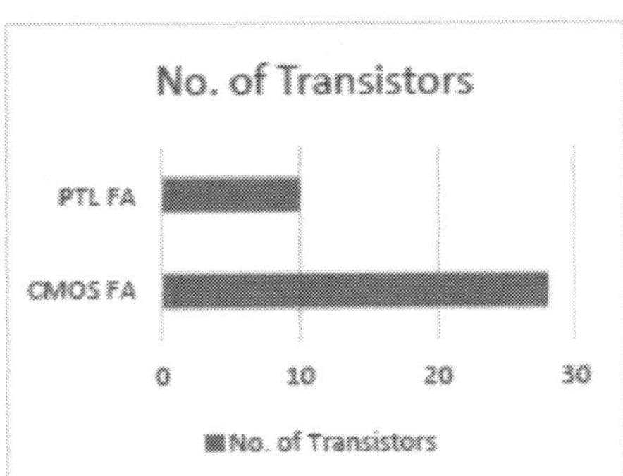

(a)

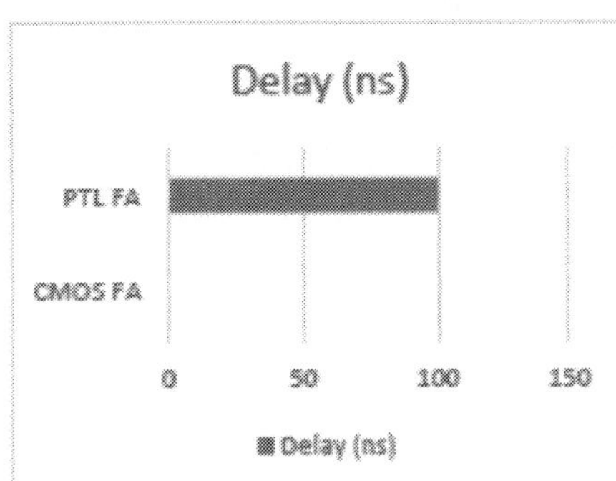

(b)

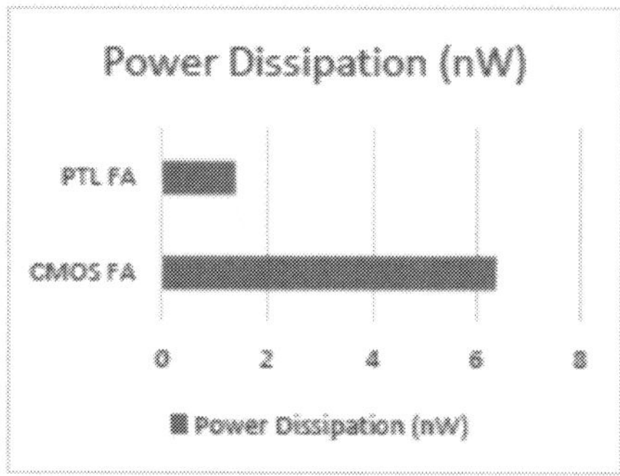

(c)

Fig 4: Comparison among PTL FA and CMOS FA. (a)No. of Transistors is the factor for comparison. (b)Delay is taken as the factor for comparison (c)Power Dissipation is taken as the factor for comparison

Table 2: Comparison between CMOS FA and PTL FA

Parameter	**Conventional CMOS FA**	**PTL FA**
No. of transistors	28	10
Delay	0.0793 ns	100.11 ns
Power Dissipation	6.372nW	1.381nW

Fig 4 has different graphs drawn for comparison between CMOS FA and PTL FA. Table 2 summarizes the practical values got by running the simulation of both the adders. By the above observations it can be concluded that the PTL full adder needs less area, less delay and also less power dissipation. Whereas the CMOS full adder requires many transistors hence occupies more area. More delay is encountered and also more power dissipation. Hence PTL full adder is far more advantageous than CMOS full adder.

References

[1] Mariano Aguirre-Hernandez, Monico Linares-Aranda, "CMOS Full-Adders for Energy Efficient Arithmetic Applications", IEEE Transactions on Very Large Scale Integration (VLSI) Systems, 2011

[2] Hamed Naseri;Somayeh Timarchi, "Low-Power and Fast Full Adder by Exploring New XOR and XNOR Gates", IEEE Transactions on Very Large Scale Integration (VLSI) Systems, 2018

[3] T.K. Schneider, A.J. Schwab;J.H. Aylor, "A full custom VLSI design methodology using Mentor Graphics design software in an educational environs", Sixth Annual IEEE International ASIC Conference and Exhibit

[4] Mehedi Hasan, Md. Jobayer Hossein, Mainul Hossain;Hasan U. Zaman, Sharnali Islam, "Design of a Scalable Low-Power 1-Bit Hybrid Full Adder for Fast Computation", IEEE Transactions on Circuits and Systems II: Express Briefs, 2020

[5] Jyoti Kandpal;Abhishek Tomar, Mayur Agarwal;K. K. Sharma, "High-Speed Hybrid-Logic Full Adder Using High-Performance 10-T XOR–XNOR Cell", IEEE Transactions on Very Large Scale Integration (VLSI) Systems, 2020

An implementation of a device to assist the visually Impaired / blind people for easy navigation through bus

Dr. V. Siddeswara Prasad [1], Sagar R [2]

[1]Professor & Dean (Students Welfare), Dept. of E&IE, Siddaganga institute of technology, Tumkur, India

[2] PG Scholar, Dept. of E&IE, Siddaganga institute of technology, Tumkur, India E-mail:

sagar1947rater@gmail.com

Abstract. Talking signs, control stick, echolocations are generally helpful in exploring the visually impaired people to arrive at their goal, but this can't be used for longer distance of travelling in the traffic .So a wireless sensor network is proposed to help the blind or visually impaired person to navigate easily through the bus in the traffic safely .The proposed system has two modules one is in the hand of blind and other inside the bus .Both the modules has a ZigBee transceiver. The bus arrival informati on is given from the ZigBee in the bus module to the ZigBee of the blind module. When the blind tells his destination through the microphone it is then recognized by the voice recognition software in the blind module .The input is then analyzed by the microcontroller and if it is matched with the stored data base the controller guides the blind to take the right bus to reach his/her desired destination otherwise he has to wait for other bus .The buzzer and LED is connected with the device inside the bus to alert the driver when the blind module starts communication with the device present in the bus. When the bus reached the destination, it is announced through earphone of the blind module with the help of voice synthesizer. Along with this the buzzer is ma de and LED glows to provide the much more alert to the blind to get down in his/her desired destination. So, this project not only helps the blind but also for the elder or senior people for their independent navigation.

Keywords— Universal Asynchronous Receiver Transmitter, Node Microcontroller unit

1. INTRODUCTION

1.1 HISTORY OFBLINDNESS

In the beginning primitive times the blind or visually challenged people were more reddishly treated. During the early civilization, it is described that blind child were deserted and make them to die without food and many of them as a food for the wild animals and also used them in many in many cruel activities like prostitution, begging and so on. But at the middle ages of civilized society starts to care them by building Alms houses for the blind. The first "School for the Blind" was established in France in 1784.Louis Braille invented the tactile reading and writing systems. So, it has been revised and there are several competing tactile methods were introduced in the further days. Congress passed an act called The American with Disabilities Act in 1990.In the upcoming days an International Braille and Technology Centre was established by the National Federation of the Blind to secure and evaluate the blind related developing technologies which were introduced to the blind .If anyone want to publish thebook they need to send the electronic copy to the federation before getting the copy right for his work. In 2004 the president signed this legislation which was passed by the congress. By selling Louis Braille coins the federation collect the four million dollars and invest them for the Braille literacy projects.

More than a quarter of the world's population 2.2 billion people are having blindness or visual impairment. According to the first World Vision Report released by the World Health Organization (WHO) on October 8,2019 one billion of cases were prevented. There was praise for India in the report for its National program for control of Blindness (NPCB). The report illustrates that in 2016-17 6.5 million people in India were provided with cataract surgery achieving a cataract surgical rate of over 6,000 per million population and that time 32 million children were screened in their schools. And approximately 750,000 spectacles were distributed. Due to these concerned attempts there was drastic decrease in the visual impairments rate from 1.1 percent in 2001-02 to 0.45 percent during the years 2015-19 in India.

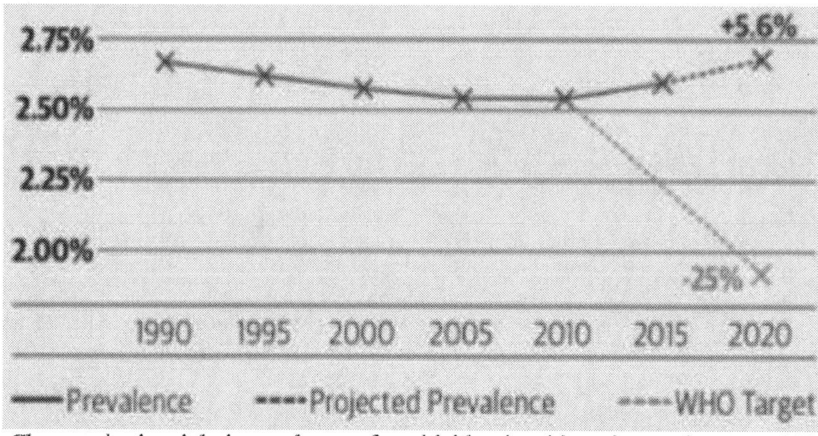

Figure 1. Changes in the global prevalence of avoidable visual impairment between 1990 and 2015 and projections for 2020

Due to the digital revolution around the world now a days there are many innovative ideas are developed in order to help the blind people. Some of them are listed as follows

1. Any text reading through Ring like device
2. A device with Touch screen capable of creating Figures and Braille
3. 3D printing Technology
4. Smart glass for blind
5. The "APP STORE "for the blind

2. HARDWARE IMPLEMENTATIONS

This chapter deals with the description of various hardware components used in this project. The hardware components required is

1 Raspberry pi
2 ZigBee Transceiver
3 Node MCU

2.1 Raspberry pi 3B+ module

In the proposed system raspberry pi 3b+ module is used as a source to connect microphone, earphone to assist the blind during his navigation. Raspberry pi resembles like a credit card sized computer which

IOP Conf. Series: Materials Science and Engineering 1065 (2021) 012048 doi:10.1088/1757-899X/1065/1/012048

is designed computer which is designed by the United Kingdom in order to teach the students regarding computer science. Basic modules of Raspberry pi have 256 megabytes of RAM it has been evolved and upgraded to model B and model B+ to have 512 MB to 1GB of RAM. The model B+ uses micro SD card for booting and storage as it is not having inbuilt hard disk in it.Because if its Quad core 1.2GHz Broadcom BCM2837 64bit CPU, 1GB RAM and less power consumption capacity when compared with other Raspberry pi modules a Raspberry pi 3B+ module is used in this project. It also includes inbuilt BCM43438 wireless LAN and Bluetooth Low Energy (BLE). 4 USB 2 ports helps to connect various required components to the board.

2.2. ZIGBEE TB 387 2. 4 GHZ

In the proposed system a ZigBee module TB 387 2.4GHz ZigBee module is used for communication between the blind device ZigBee module and the bus device Zigbee module as it has high range of operation when compared with the B luetooth low energy module i. e. up to 500 -meter data transfer range and it consumes less power for its operation. The TB 387 is based on the 2. 4GHz frequency band wireless transparent data transmission module. This module supports AT commands, baud rate, I D number, frequency settings, version information. When the device is in AT command mode serial port is used to set the device parameters and if it is in data transfer mode then the data and its data frame number, row count can be transmitted and retransmi t ted at the reliable range with the higher data rate of 250kbps. In the proposed system both the ZigBee modules are paired with the same baud rate of 9600 and using the Universal Asynchronous Receiver Transmitter (UART)protocol the serial communication is m ade between the two modules.

2.3. NODE MCU ESP8266

The Node MCU is an open source firmware and development kit that helps to build an IoT product required within a few Lua script lines. This module includes firmware that runs on the ESP8266 Wi-Fi SoC from Espressif system and hardware which is based on the ESP-12 module Node MCU is used to store real-time data and controls the home appliances using blynk App.Node MCU can be interfaced with Arduino, Raspberry pi, AVR development board etc. In the proposed project Node MCU has Station (STA) mode using which it can connect to existing wi-fi network and can act as HTTP server with assigning IP address using the network. Node MCU gets IP from Wi-Fi router to which it is connected. With this IP address, in this project using Telegram app that it can act as an HTTP server to which any wi-fi device can connect.

3. SYSTEM SOFTWARE

3.1. Raspbian OS

In the proposed system the Blind device uses the raspberry pi with the Raspbian OS using which the device is controlled for various operations. It has web browser, menu bar, a file manager and there is no shortage for desktop shortcut for pre-installed applications. Raspberry pi is an unofficial port of of Debian Wheezy armhf which optimizes the code and compile it to run in it. Due to this performance of the device is increased which helps them to use in several; applications and in many floating-point arithmetic operations. By using the upgraded instructions of the ARMv6 CPU in raspberry pi helps other applications to gain performance. Raspberry pi is primarily introduced and invented by team work of Mike Thompson (mpthompson) and Peter Green (plugwash) and other raspberry pi community members. It supports varies programing languages C, C++, Java, Python etc. In the proposed system python version
3.7.3 is used for programing the blind module.

3.1.1. Speech Recognition in Python using Google Speech API

Speech recognition play an important role now a days for many applications like home automation, artificial intelligence and in several security requirements etc. in the proposed system the google speech recognition api is used to recognize the speech through the microphone connected to the raspberry pi using python version 3.7.3 programing language. this is done by installing the speech recognition library and importing it to the required code to recognize the input which is given by the blind through the external microphone.

3.1.2. Convert Text to Speech in Python

There are different APIs are available to convert text to speech in python. But in that gTTS (google Text to Speech) API is better because of its easy tool usage and installation process. Using this tool, a text written can be converted into a speech and saved in the mp3 file format. It can be used any time by importing it into a program when the user required it.

The gTTS API supports many languages including English, Hindi, Tamil, French, German and many more. Using this tool, the speed of the converted speech (audio) can be adjusted as slow or fast according to the user requirements

3.2. Arduino IDE

Arduino IDE software tool is open source which helps to develop many wireless network IoT related projects. It contains a text editor for writing code, a message area, a text console and a toolbar. Programs can be coded and uploaded to test board which is connected to the PC and so that the after uploading completes the light blinks in the board and the results are shown in the serial monitor. The Node MCU is programmed with the Arduino IDE software tool using C programing language by installing the inbuilt library of esp8266 shown in the library list.

4. METHODOLOGY

4.1. BLOCK DIAGRAM OF THE PROPOSED SYSTEM

A Wireless sensor network is designed to help the blind or visually impaired person to easily navigate through the bus. The proposed system has two modules one is in the hand of blind and other inside the bus. Both the modules have a ZigBee transceiver the bus arrival information is given from the ZigBee in the bus module to the ZigBee of the blind module. When the blind tells his destination through the microphone it is then recognized by the voice recognition software in the blind module. The input is then analyzed by the microcontroller and if it is matched with the stored data base the controller guides the blind to take the right bus to reach his desired destination otherwise he has to wait for other bus. when the bus reached the destination it is announced through earphone of the blind module with the help of voice synthesizer. So, this project not only helps the blind but also for the elder or senior people for their independent navigation.

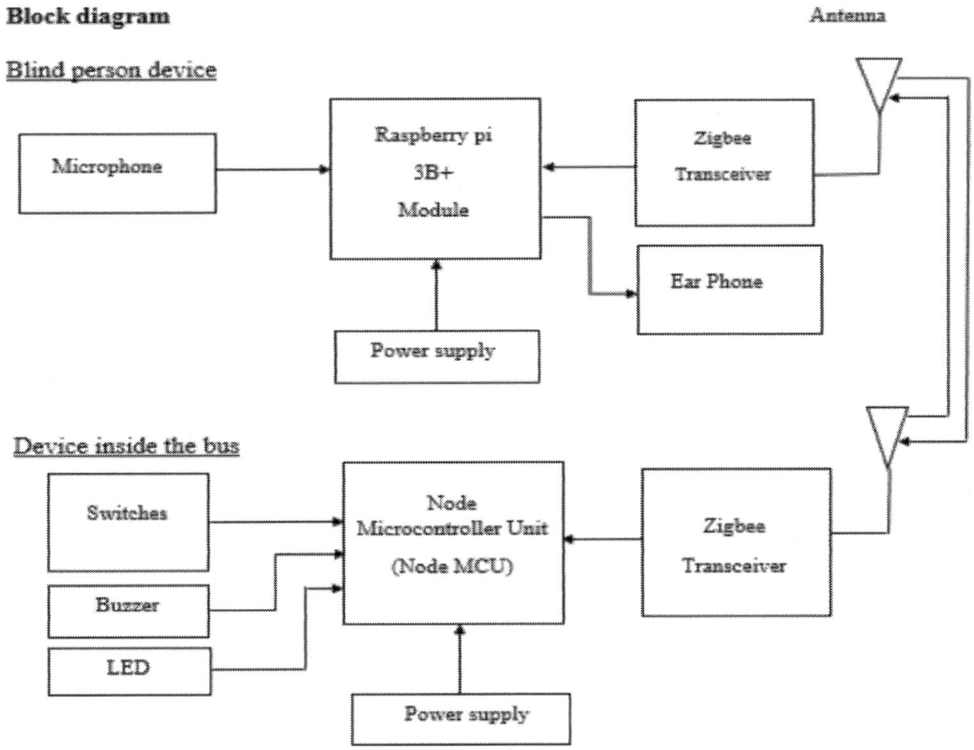

Figure 2. The proposed system to assist the visually impaired/blind people for easy navigation through bus

Implementation and working

Stage 1: Acquisition of bus arrival information
The ZIGBEE TB 387 2.4 GHZ is used at both the blind and the device inside the bus. The bus arrival information is transmitted through the signal from the ZigBee in the bus module to the ZigBee in the blind module within 500 -meter range.
Stage 2: Intake of the destination to be travelled by the blind
The audio input of the required destination from the blind is taken through the microphone attached to the blind module.
Stage 3: Reception of destination information by bus module
The voice input from the blind is then converted into the required text format with the help of python 4.0 voice recognition software installed in the raspberry pi 3b+ module and sends that to a Node Microcontroller Unit (Node MCU) which is present in the bus.
Stage 4: Bus information Processing
The controllers compare the translated text with the bus data base if it matched then only the bus information is sent to the blind module otherwise the blind has to wait for the next bus to reach his /her destination. If the matched is found the blind gets into the bus and alert is given to the driver with the help of Buzzer and the LED.
Stage 5: Audio output for blind interaction
The received signal is then decoded and with the help of voice play back software package the blind is guided through the earphone attached to the blind module to reach his destination safely.

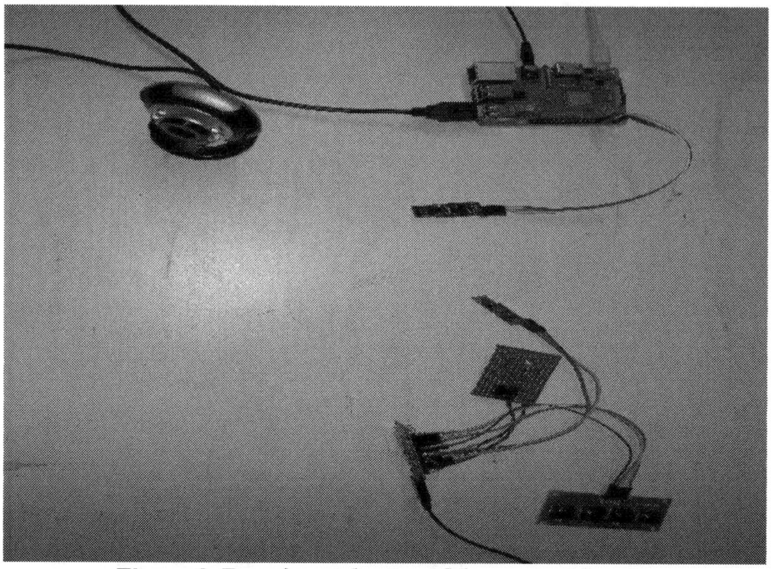

Figure 3. Experimental setup of the proposed system

4.2. Interfacing diagram of the proposed system

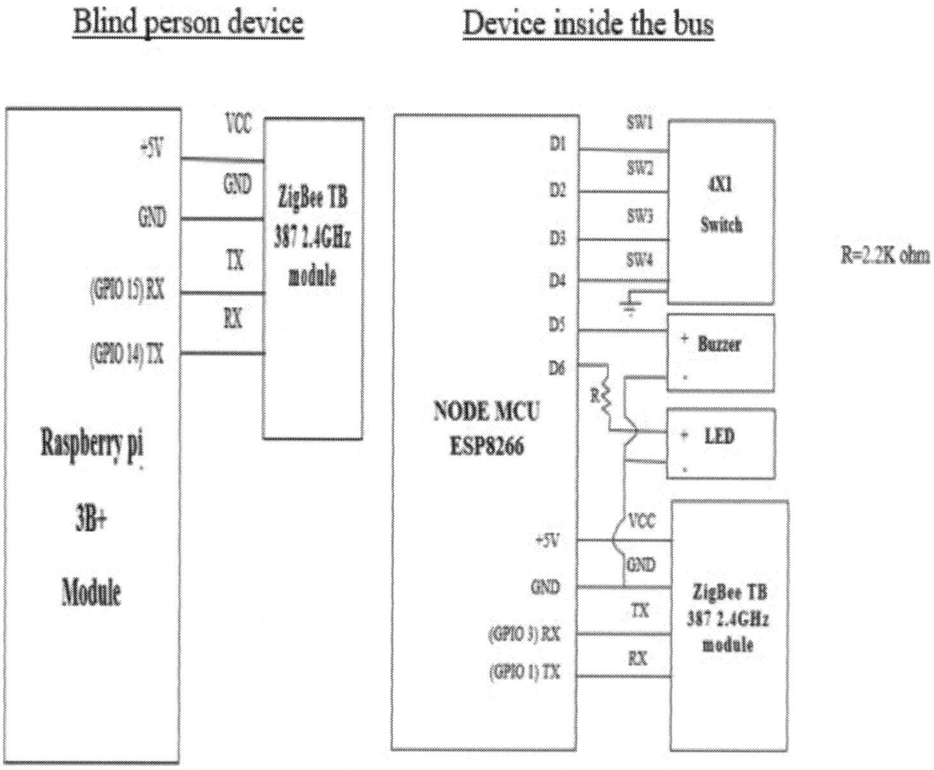

Figure 4. Interfacing diagram of the proposed system

From the interfacing diagram Raspberry pi and Node MCU is connected with their ZigBee module. The Tx and RX pin of ZigBee is connected to the GPIO 15 and GPIO 14(RX and TX) pins of

Raspberry pi 3b+ module. Likewise, ZigBee TX and RX of bus module is connected to the Node MCU RX and TX i.e. GPIO 3 and GPIO 1 pins. Here 4X1 switches is designed to guide the blind about his /her destination reached information by pressing the switch button by the driver when the destination reached. These 4x1 switch is connected to digital GPIO pins of Node MCU from D1 to D4. Buzzer is connected to D5 and LED is connected to D6 through a 2.2K ohm resister.

All these stages of implementation are explained in the flow chart shown in fig 5.

4.3. Flow Chart

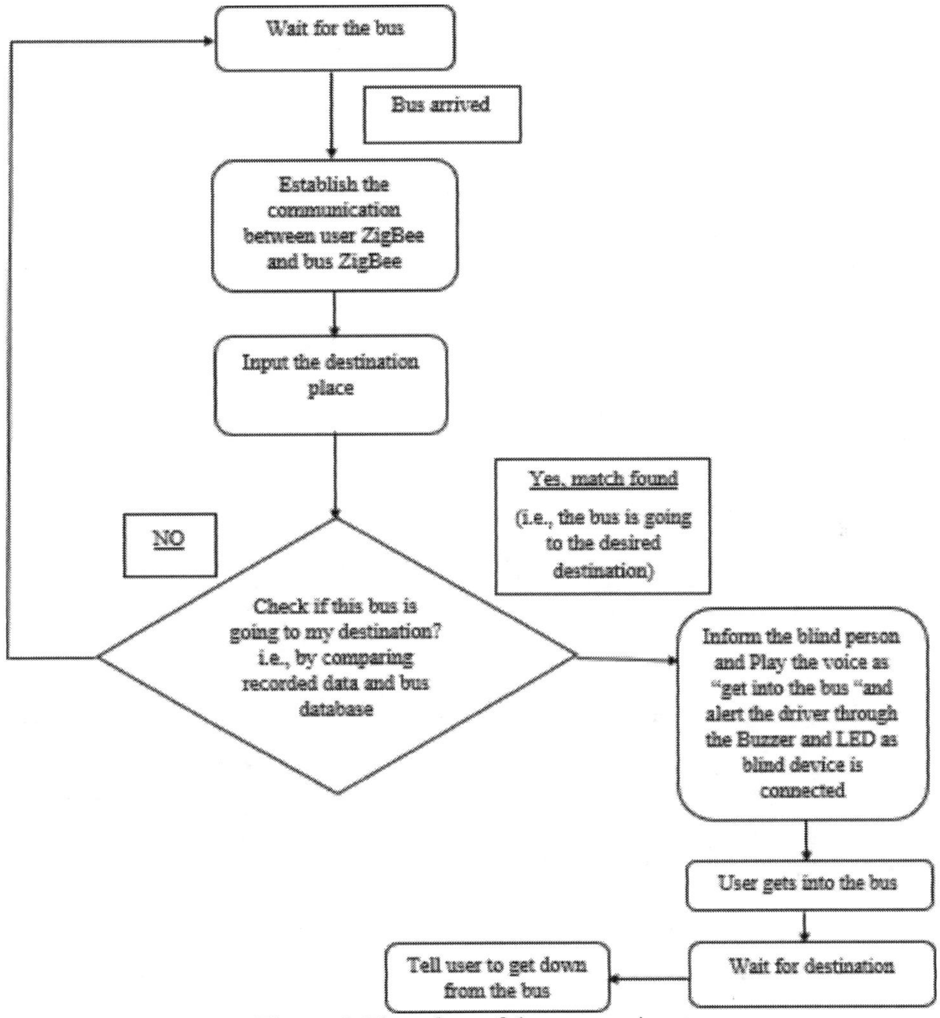

Figure 5. Flow chart of the proposed system

5. EXPERIMENTAL RESULTS

The Proposed systems is successfully tested and results are verified. In this section four test are conducted and their results are found successfully.

5.1. Test 1: If desired destination of Blind is Vijayanagar (Destination 1)

IOP Conf. Series: Materials Science and Engineering 1065 (2021) 012048 doi:10.1088/1757-899X/1065/1/012048

The bus arrival information is listened to the blind through ear phone which is connected to the blind module. This is shown in the Figure 6.

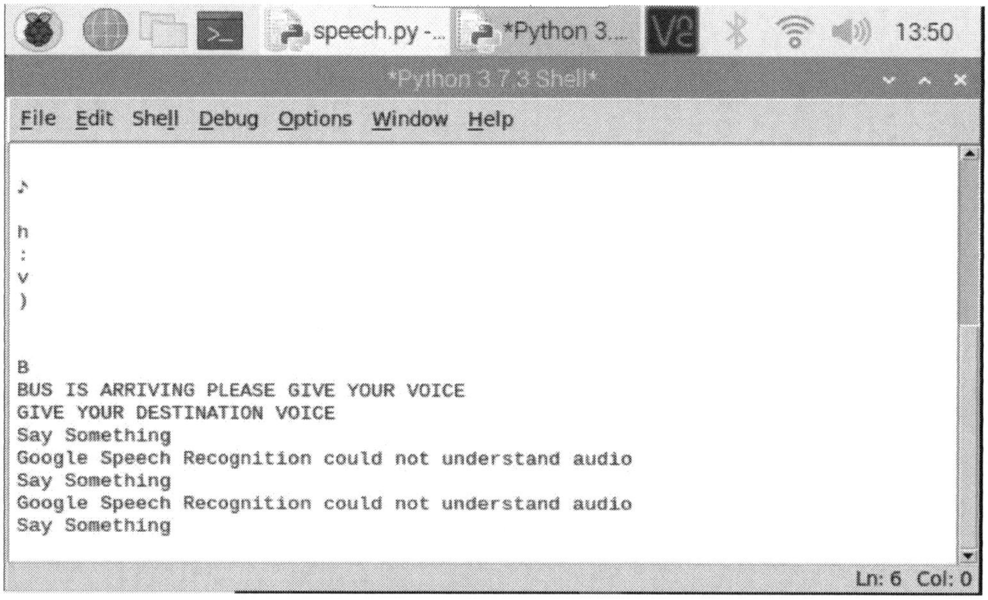

Figure 6. Bus arriving information listen from the blind module

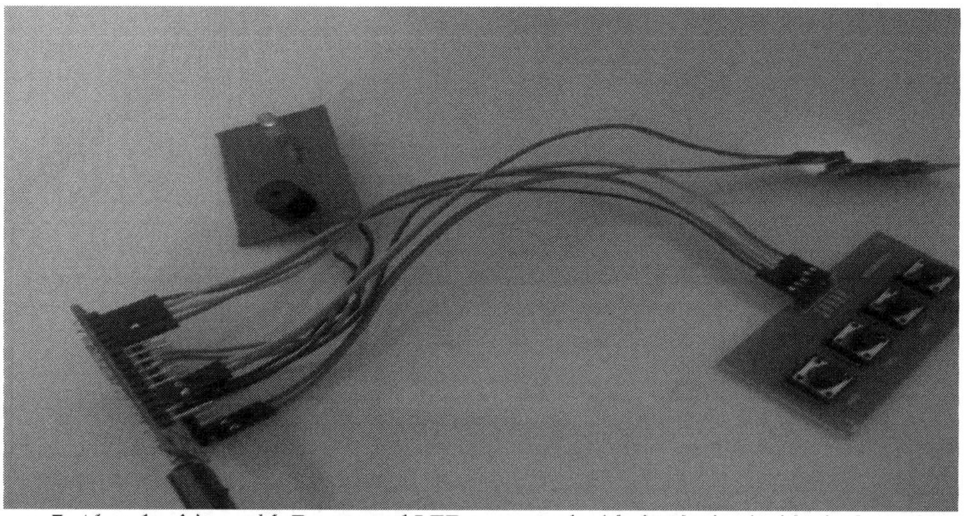

Figure 7. Alert the driver with Buzzer and LED connected with the device inside the bus when the blind module is communicated with the device inside the bus

After the bus arrived the blind need to say his/her desired destination through the microphone, here the desired destination of the blind is Vijayanagar. The system ignores the unmatched location as the location is not present in the stored data base of the bus module.

It is then recognised by the python 3.7 voice recognition software i.e. Google speech recognition API software and in controller it is matched with stored database Vijayanagar.

After the match is found the alert for the driver is made through Buzzer and LED connected to the controller so that driver can know some blind device is connected as show in Figure

Blind can take the right bus parked Infront of him and all the information regarding his desired destination place is announced through the earphone.

Table 1. Assignment of 4x1 switch to each destination location

4X1 Switch	Destination
SW1	Destination 1(Vijayanagar)
SW2	Destination 2(Majestic)
SW3	Destination 3(Yeshwanthpur)
SW4	Destination 4(Tumkur)

Table 2. Operation of switches which is controlled by driver

SW1	SW2	SW3	SW4	Destination
0	1	1	1	Destination 1 reached
1	0	1	1	Destination 2 reached
1	1	0	1	Destination 3 reached
1	1	1	0	Destination 4 reached

Note: 4X1 pull up switches is used in which '0' indicates that the switch is pressed i.e., destination reached information and '1' indicates that Next stop to alert the blind to reach his/her destination.

Example: When driver knows that next stop is blind destination location i.e. destination location 2 then he presses next stop switch SW1 to alert the blind and when the bus reached the desired destination location 2 then he presses the SW2 that is assigned for the required destination location 2 of blind.

When the bus reached the destination, the driver presses the SW1 assigned for the destination1, i.e. Vijayanagar connected with the controller announces that desired destination is reached through the earphone connected with the blind module and also Buzzer is made and LED blinks. The buzzer and LED are used for to provide an additional alert for the blind along with the output heard through earphone when bus reaches his/her desired location or destination All these results are shown in Figure .8 and Figure .9

IOP Conf. Series: Materials Science and Engineering 1065 (2021) 012048 doi:10.1088/1757-899X/1065/1/012048

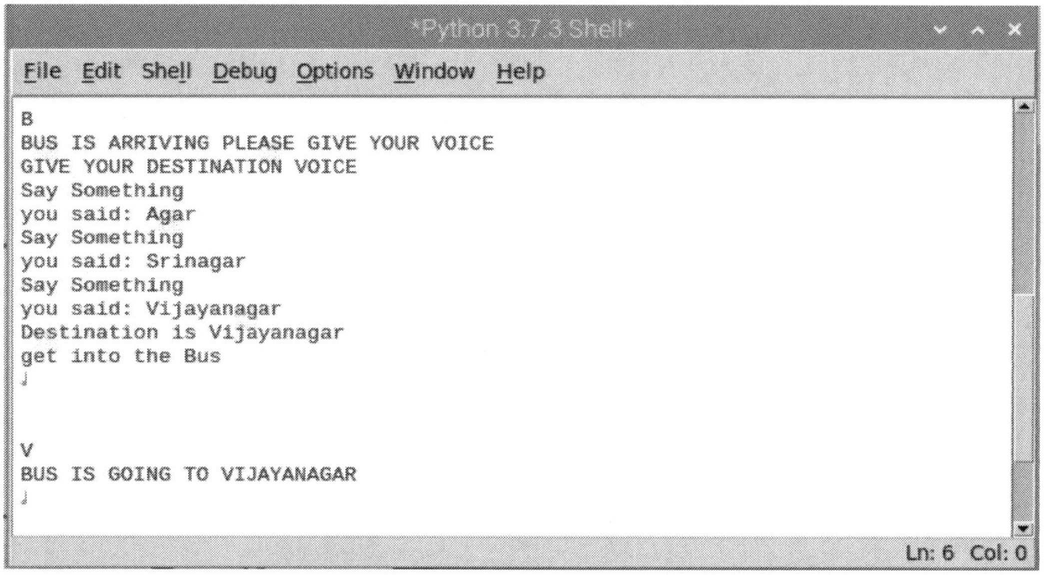

Figure 8. Comparing the recorded data with the database stored in bus and if match is found alert the blind to get into the bus

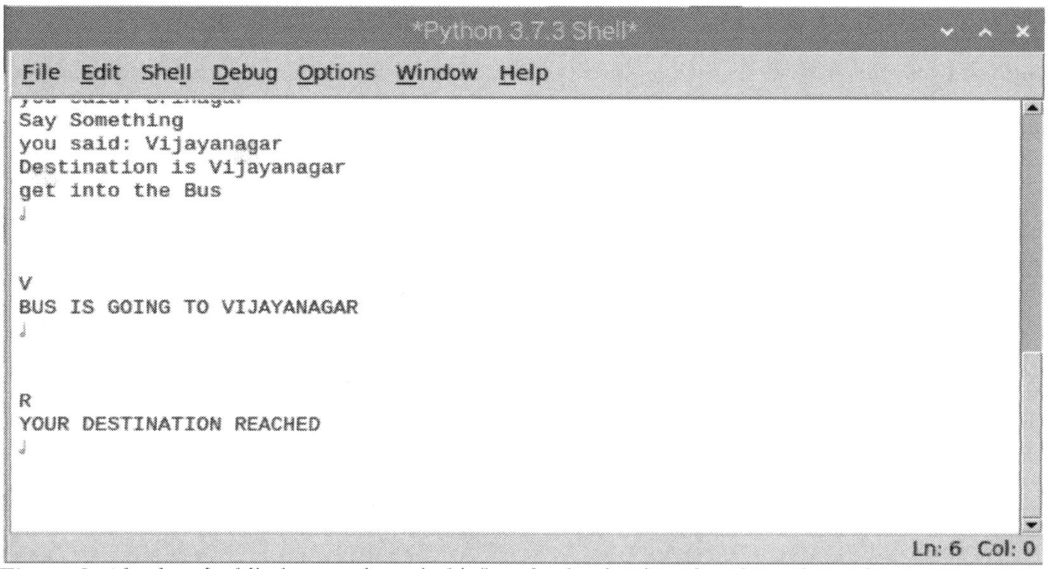

Figure 9. Alerting the blind to get down in his/her destination location through earphone when driver presses the particular 4x1 switch assigned for each destination location i.e. Destination1(Vijayanagar)

5.2. Test 2: If the desired destination of Blind person is Majestic (Destination 2)

When the blind person wishes to travel Majestic then the device listens his requirement and guide them to reach their destination safely. The test is done and the results are found which is shown in Figure 10 and Figure 11 and hence conclude that the device is successfully tested and verified.

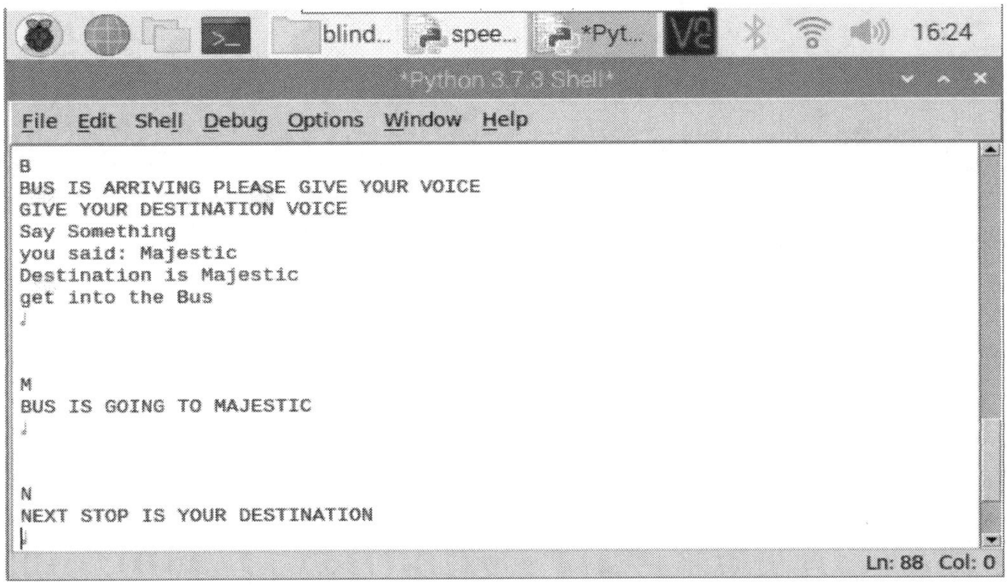

Figure 10. Comparing the recorded data with the database stored in bus and if match is found alert the blind to get into the bus

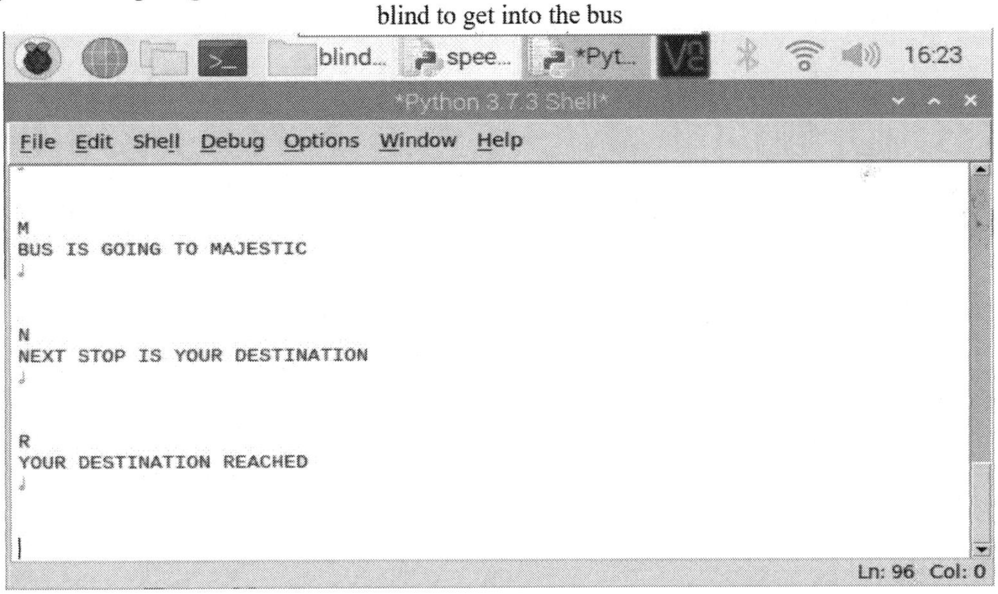

Figure 11. Alerting the blind to get down in his/her destination location when the driver presses the particular 4x1 switch assigned for each destination location i.e., Destination 2(Majestic).

5.3. Test 3: If the desired destination of Blind person is Yeshwanthpur (Destination 3)

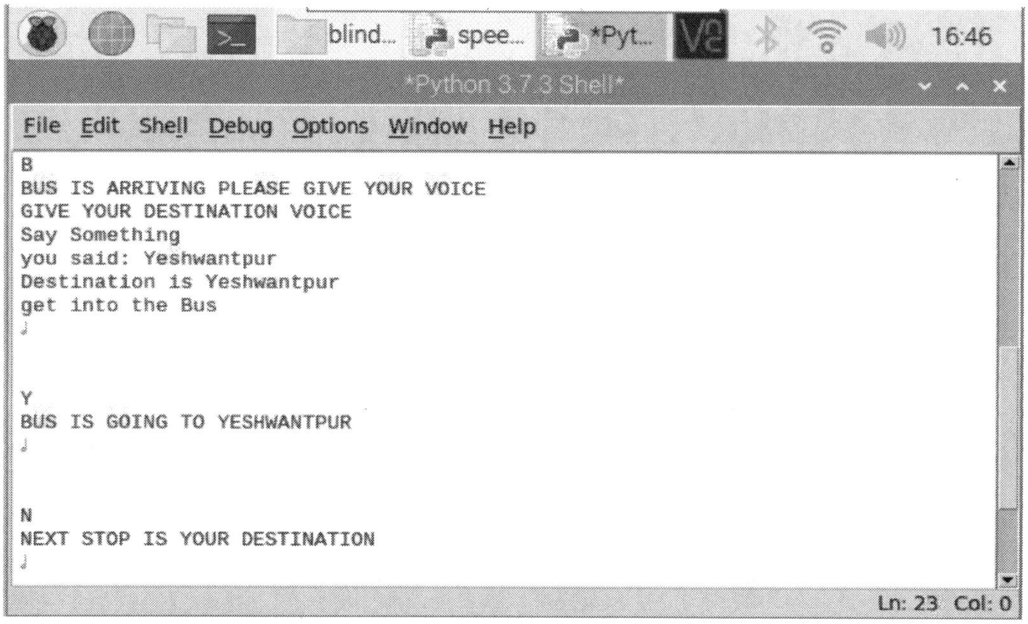

Figure 12. Comparing the recorded data with the database stored in bus and if match is found alert the blind to get into the bus

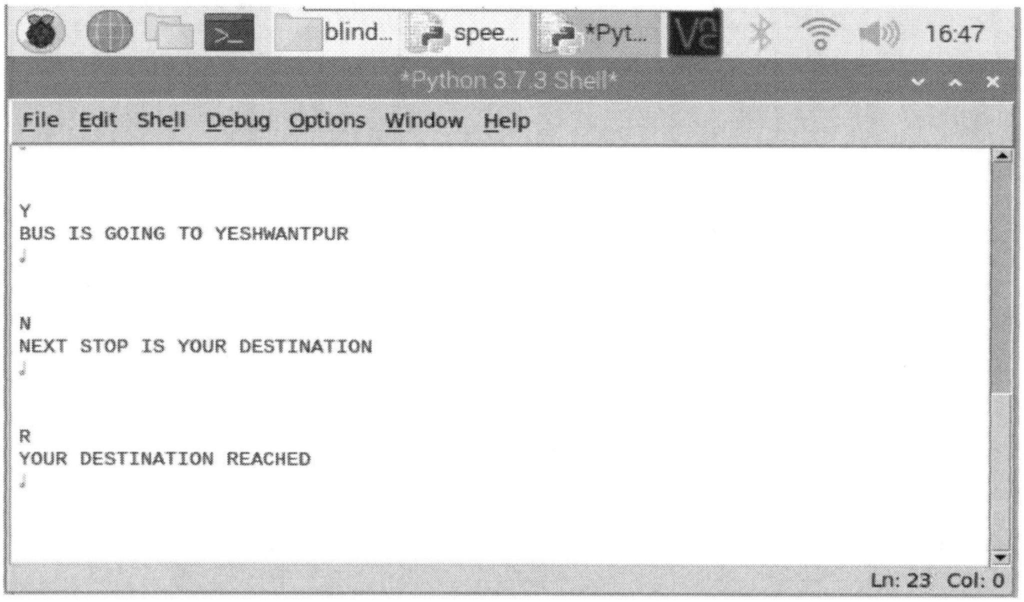

Figure 13. Alerting the blind to get down in his/her destination location when the driver presses the particular 4x1 switch assigned for each destination location i.e. Destination 3(Yeshwanthpur)

5.4. Test 4: If the desired destination of Blind person is Tumkur (Destination 4)

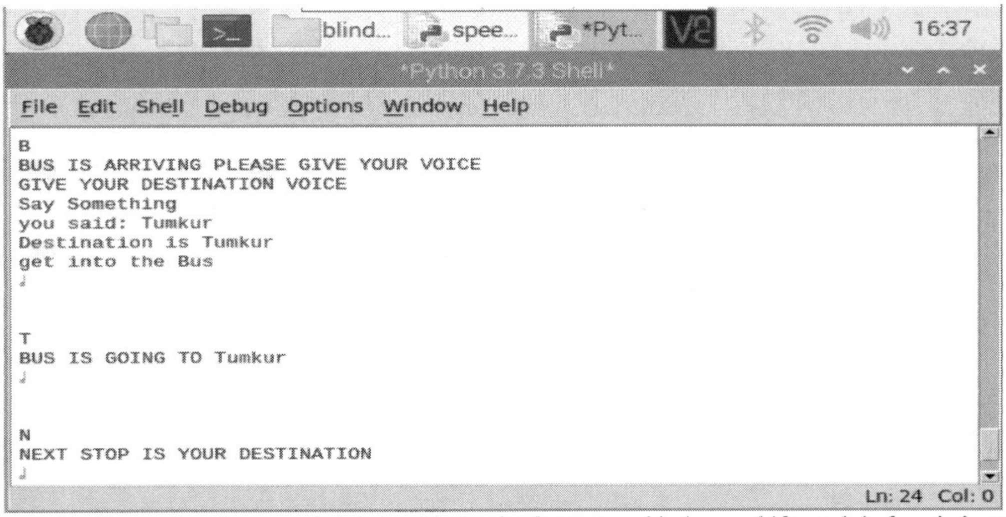

Figure 14. Comparing the recorded data with the database stored in bus and if match is found alert the blind to get into the bus

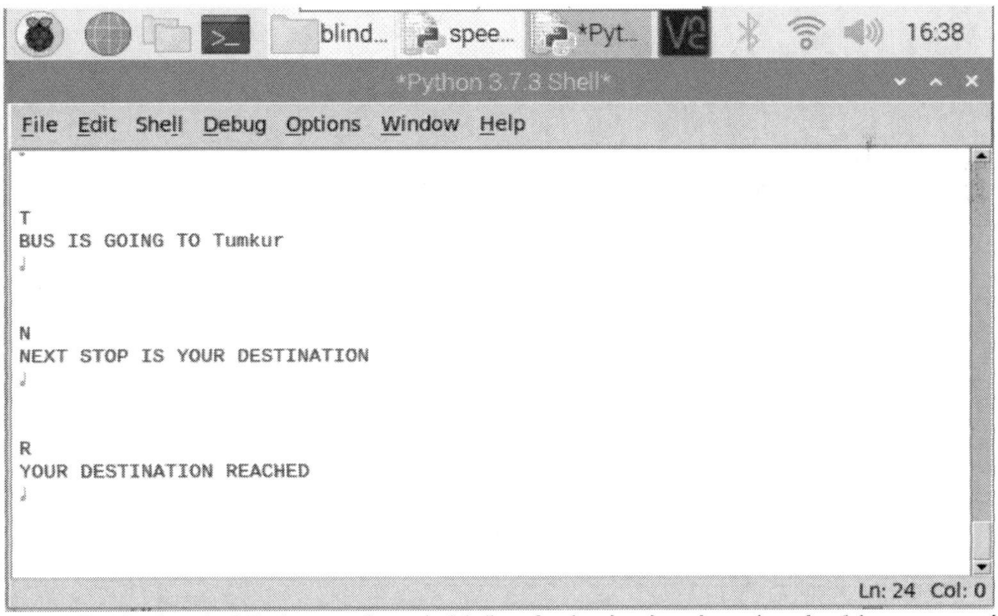

Figure 15. Alerting the blind to get down in his/her destination location when the driver presses the particular 4x1 switch assigned for each destination location i.e. Destination 4(Tumkur)

With all these tests and their corresponding results, the system is successfully verified for different destination locations

5.5. Error Analysis

If by chance driver presses the different switch rather than pressing the particular destination switch will misguide the blind to reach his/her desired destination location. Here the destination 4 is the drop location of the blind so the driver has to press the SW4 but he presses the other switch rather than pressing the SW4 so it may lead to misguide the blind. The results are shown in the Figure 16.

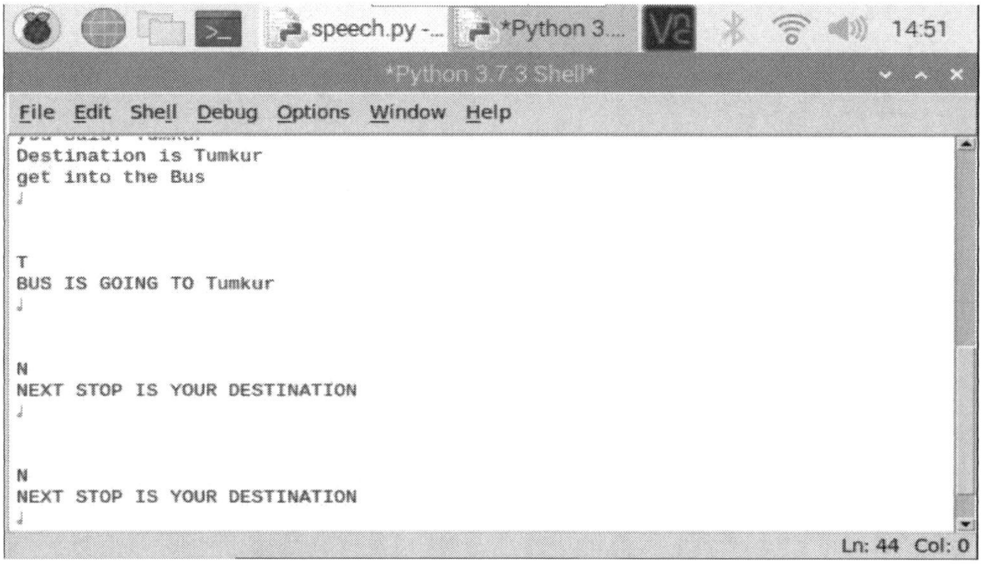

Figure 16. Error analysis of the proposed system

5.6. ADVANTAGES OF PROJECT

1. Safety concerns for blinds
2. Cost effective and consumes less power
3. Automatic operation
4. Continuously quick monitor
5. High alert system

5.7. DISADVANTAGES OF PROJECT

1. Difficult to illiterates.
2. If network fails, then it is not possible to intimate.
3. Network oriented system.
4. If by chance driver presses the different switch rather than pressing the particular destination switch will misguide the blind to reach his/her desired destination location.
5. As the destination locations increases the number of switches increases which increases the complexity of the system
6. The driver has a chance to misguide the blind or visually impaired person by unknowingly pressing the different switch rather than pressing the particular assigned destination switch while the bus is in the desired destination of blind person.

6. CONCLUSION

In this project a pair of devices is built in order to help the blind and impaired people to make their navigation to his/her destination location through bus easy with the safety concerns. A low power operating with continuously quick monitor and high alert system is designed to fulfill the traveling needs of blind people. With the help of the proposed module a visually impaired person can navigate to his/her desired destination without any problems while travelling. The module developed is not only helps the blind but also for the senior citizens of the society for their safe navigation.

7. FUTURE ENHANCEMENT

This system can further be improved by using GSM with this module in order to help the blind if in case he/she met with any threat or emergency during travelling .This can be done by sending an alert message, live location of him to his /her relatives and also these information is sent to the nearest police station

Because of its ability, performance and security applications the proposed system can be adopted for office security, colleges, hospitals and also in parking system. The gps can replace the switches to assist the blind to get down in his/her required destination location and also the system can be built in a offline mode by using offline maps like Navit offline mapping tool instead of switches as well as the offline speech recognition software.

REFERENCES

[1] Liang-Bi Chen, Jian-Ping Su, et al 2019, *"An Implementation of an Intelligent Assistance System for Visually Impaired/Blind People"*, IEEE Consumer Electronics Magazine, Vol. **7**, no. 3, pp. 12-20.

[2] Jinqiang Bai, Shiguo Lian, et al 2018," *Virtual-Blind-Road Following Based Wearable Navigation Device for Blind People"*, IEEE Transactions on Consumer Electronics, Vol:**64**.

[3] A. Aladrén; G. López-Nicolás, Luis Puig et al, 2016, *"Navigation Assistance for the Visually Impaired Using RGB-D Sensor with Range Expansion"*, IEEE Systems Journal, Vol :**10**.

[4] Farooq Shaikh,Vishal Kuvar et al,2017, *"Ultrasonic sound-based navigation and assistive system for visually impaired with real time location tracking and Panic button"* 2017 2nd International Conference on Communication and Electronics Systems (ICCES).

[5] He Zhang, Cang Ye 2017, *"An Indoor Wayfinding System Based on Geometric Features Aided Graph SLAM for the Visually Impaired"*, IEEE Transactions on Neural Systems and Rehabilitation Engineering, Vol: **25**, Journal Article.

[6] Jinqiang Bai, Shiguo Lian et al,2017, *"Smart guiding glasses for visually impaired people in indoor environment"*, IEEE Transactions on Consumer Electronics, Vol: **63**, Journal Article.

[7] Mun-Cheon Kang, Sung-Ho Chae *et al*, 2017, *"An enhanced obstacle avoidance method for the visually impaired using deformable grid"*, IEEE Transactions on Consumer Electronics, Year: 2017, Vol: **63**, Journal Article.

[8] Moritz Fischer, Michael Renzler et al, 2019 *"Development of a Smart Bed Insert for Detection of Incontinence and Occupation in Elder Care"*, IEEE Access, Vol: **7**, Journal Article.

[9] Moritz Fischer, Michael Renzler et al, 2019, *"Development of a Smart Bed Insert for Detection of Incontinence and Occupation in Elder Care"*, IEEE Access, Vol: **7**, Journal Article.

[10] Jorge Adorno, Yueng DeLaHoz et al 2016, *"Smartphone-based Floor Detection in Unstructured and Structured Environments"*, IEEE Access, IEEE International conference.

[11] Shripad Bhatlawande, Manjunatha Mahadevappa Jayanta Mukherjee et al 2014, *"Design, Development, and Clinical Evaluation of the Electronic Mobility Cane for Vision Rehabilitation"*, IEEE Transactions on Neural Systems and Rehabilitation Engineering, Vol: **22**, Journal Article.

IOP Conf. Series: Materials Science and Engineering 1065 (2021) 012049 doi:10.1088/1757-899X/1065/1/012049

An ensemble classification based approach for breast cancer prediction

Vijayalakshmi G V Mahesh[1], Mohan Kumar M[2]

[1]Department of Electronics and Communication Engineering, BMS Institute of Technology and management, Bangalore, India

[2]Department of Electronics and Communication Engineering, Yenepoya Institute of Technology, Mangalore, India

E-mail: vijayalakshmi@bmsit.in, mohankumar@yit.edu.in

Abstract. Breast cancer is the second major reason for deaths in women. Early detection of the breast cancer and receiving the appropriate treatment can reduce the death rates as survival becomes hard in the higher stages of the tumor growth. Application of machine learning in healthcare play a key role in aiding the clinical experts to detect the disease at a early stage and perform precise assessment. This paper proposes a pattern recognition methodology that uses breast cancer biomarkers as the attributes and ensemble classification approach for accurately detecting the presence of cancer. The proposed method was evaluated for the samples of the breast cancer Coimbra dataset by fusing the decisions of naive Baye's , radial basis function neural network and linear discriminant analysis classifiers based on majority voting rule. The experimental results demonstrated the enhanced performance of the system with fusion of classifiers as compared to the single classifiers.

1. Introduction

An uncontrolled way of cell division leads to cancer and these cells often invade into other healthy tissues. Breast cancer is the most common and is the leading cause of death in women. It can be invasive or non invasive: invasive cancer spreads from the ducts or glands to other parts of the breast while noninvasive cancer does not spread from the original tissue. Based on this, the following types of breast cancer are identified (i) Ductal carinoma in situ(DCIS):It is non invasive where the cancer cells are confined only to the ducts of the breast, (ii) Lobular carinoma in situ(LCIS): A non invasive condition where the cancer cells are limited to the milk producing glands of the tissue, (iii) Invasive ductal carcinoma(IDC): This is invasive type of cancer. The cancer cells develop in the milk ducts of the breast and then invade to the nearby tissues and organs. IDC is the most common type of breast cancer and (iv) Invasive lobular carinoma(ILC): This type of cancer begins in the lobules of the breast and spreads to the surrounding tissue.

Based on the size of the tumor and the extent of its spread into nearby tissues and organs the doctors recognize the different stages of cancer as stage 1, stage 2, stage 3 and stage 4 breast cancer. Stage 1 is a primary tumor with its size around 2cms while stage 4 is advanced where the tumor can have any size and the cancer cells have spread to nearby tissues and organs. Early detection of the cancer i.e., in the initial stages improves possibility of survival. If the cancer reaches stage 4, the chances of survival comes down. Thus early detection of cancer is very crucial and regular checks and screening can aid in detecting the symptoms earlier. Clinical experts diagnose the cancer after routine screening. The use of biomarkers and imaging tests helps the clinicians to identify the breast cancer. The most widely used image screening methods are Mammogram, ultrasound and Magnetic resonance imaging(MRI). Further researchers are trying to find new breast imaging methods to improve the detection and treatment of the cancer and have come up with Scintimammagraphy(molecular breast imaging), Positron emission mammography(PEM), Electrical impedance imaging(EIT) and Elastography

Though imaging methods play an important role in decision making followed by treatment. They require expertise to deliver significant and accurate information . For example, extraction of feature descriptors from images using mathematical techniques to detect and differentiate the tumor and non tumorous region. Also the reliability of the detection method needs to be assessed based on (i) Accuracy, (ii) Precision, (iii) Correlation to the disease and (iv) Usability. Biomarkers[1], on the other hand are gaining lot of importance in accurately detecting and diagnosing the cancer as a result of advancements in bioinformatics tools. Recognition of significant biochemical changes in cancer has facilitated towards advances in detection and its treatment. They can be relied upon for early detection and accurate evaluation. Breast and prostate cancers are diagnosed comparatively earlier as compared to other cancers. Additionally, the cancer biomarkers are significant in detecting the disease for asymptomatic persons at the early stage. Ideally a portion of a protein or a protein in the blood or urine samples is considered as a marker for tumor detection. Thus this paper proposes a pattern recognition system that use biomarkers as features or descriptors combined with machine learning algorithms to effectively identify the presence or absence of cancer. Accordingly, this is a classification task where the classifier predicts the class label of the sample. The performance of the model is later quantitatively evaluated using the metrics derived from the confusion matrix.

The rest of the paper is organized as follows, Section 2 presents a comprehension of related work. Section 3 describes the methodology of the works. The experimental results and discussion are provided in section 4. Finally, section 5 concludes the paper.

2. Related work

Biomarkers in cancer are the characteristics that are measured and they describe the pathological state of an living being. Cancer biomarkers are categorized into predictive, diagnostic and prognostic based on how they are handled. The first cancer biomarker was recognized from the blood sample of a patient with colon cancer [2]. Later different biomarkers were developed to identify other cancers. The standards and measurements of breast cancer biomarkers are provided in [3] and a overview is found in[2].Further, the advancements in technology and data analytics is seen to solve various health related issues to improve the quality of health care and provide effective treatment. Recent years have witnessed a biggest transformation of Artificial intelligence in health care domain powered by machine learning algorithms that can assist the clinicians/experts in diagnosing the disease and build up effectual solutions. Thus the acquired dataset is fed to the algorithms that can discover the patterns and anomalies to make right decisions about the disease diagnosis.

Many researchers have put forward their ideas in using cancer biomarkers with machine learning algorithms to improve the precision of detection and analysis. Li and Chen[4] investigated the adaption of cancer attributes in detecting the disease which can reduce the probability of death rate. They used Wisconsin Breast cancer and Breast Cancer Coimbra datasets with Decision tree(DT), Random forest(RF), Support Vector Machine(SVM), Neural Network(NN) and Logistics

Regression(LR) Classifiers and compared their performances . The experimental study showed that the Random forest algorithm provided better results in terms of prediction accuracy, F -score and Area Under the Curve(AUC) as compared to other classifiers. Farah Sardouk et al.,[5] studied the predictors of Breast Cancer Coimbra dataset to estimate the presence or absence of the cancer. They created six machine models using Adaboost, Regression , Random Forest, Jrip, RBFNN and J48 classifiers in WEKA and MATLAB environments. The models were assesses using True positive rate, False positive rate, precision, recall, F1-score MCC and AUC. The results proved that the machine learning algorithms were efficient in predicting the outcome. Miguel Patricio et al., [6] at university of Coimbra collected clinical attributes from 166 individuals based on anthropometric data. The measured data or features were then fed to logistic regression, Support vector machine and random forest machine learning algorithms to evaluate the probability of using the anthropometric data as biomarkers for cancer prediction. which were given to the developed a model that can be used as a biomarker. The evaluation of the classifier models for the metrics sensitivity, AUC and specificity indicated the better performance of SVM for the following predictors: Glucose, Resistin, BMI and age. Thus they can be efficient biomarkers of breast cancer.

Jin Yue et al.,[7] carried out statistical analysis, spectral analysis and least squares fitting analysis on the Breast Cancer Coimbra dataset to check the interclass variation. The analysis was able to clearly indicate the significant distinction between the two classes: cancerous and non cancerous. Yolanda D. Austria et al.,[8] investigated 11 classification algorithms and their variants : LR, k Nearest Neighbor(kNN), SVM, DT, RF, Gradient boosting and naive Baye's for predicting the breast cancer on the Coimbra dataset. The performance of the models were evaluated using accuracy and prediction time. The evaluation displayed an higher accuracy of 74.14% with Gradient boosting classifier and non linear SVM was fast in predicting the label without any delay. The work also identified the anthropometric data body mass index(BMI) as the principal predictor. Ratula Ray et al.,[9] explored the use of DT, kNN, RF and Guassian naive Baye's machine learning algorithms in aiding the pathologists to make precise decision about the presence of the disease considering both image and numeric(biomarkers) databases. Authors used breast histopathology images from Kaggle and breast Coimbra dataset for the evaluation of classifier models. The analysis indicated the better performance of random forest classifier which provided precision , recall and F1 score of 71%, 71% and 70% respectively for the numeric dataset. Similarly it presented precision , recall and F1 score of 91%, 92% and 92% respectively. The study also implies the importance of resistin and BMI biomarkers that influence the decision process.

From the evaluation of the contributions it is found that majority of the methods used various machine learning algorithms independently to improve the success rate of cancer detection. This paper presents a model that makes use of a ensemble classification where the classifiers are fused at the decision level to enhance the performance of the system.

3. Methodology

The methodology of the proposed system to predict the presence of breast cancer is shown in Figure1.The frame work includes (i) Feature extraction (ii) Classifier/Machine learning model creation (iii) Classification.

The extraction of suitable features or descriptors are essential in deciding the accuracy of the system. The anthropometric data measured from the input samples or the biomarkers form the features. These features are given to different classifiers for models creation which can predict the class label of the given sample. To enhance the presentation of the system, the decisions of the individual classifiers are combined based on decision level fusion strategy. The created models are evaluated to validate their performance. The details of the process is described in subsequent sub sections.

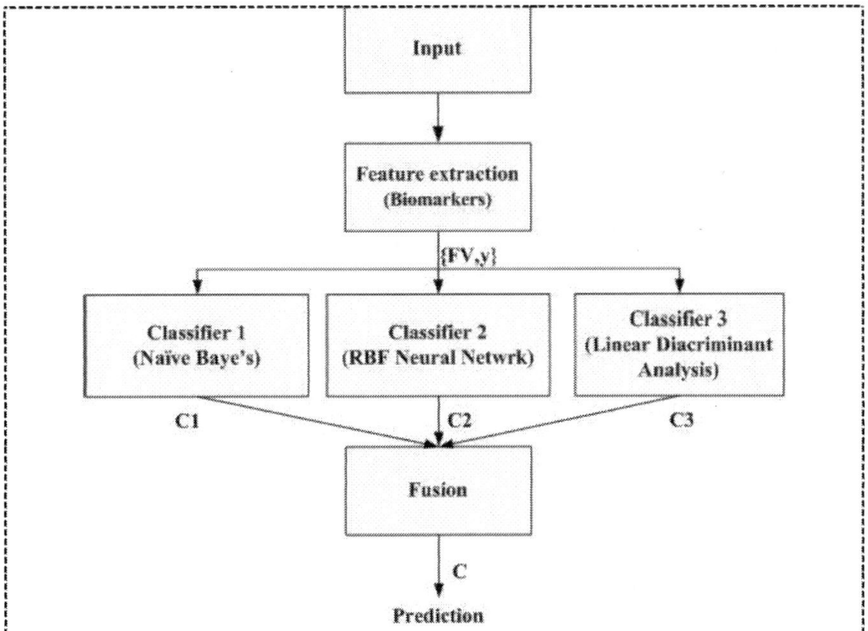

Figure 1. Methodology of the proposed system

3.1 Dataset and Feature extraction

The work considered Breast Cancer Coimbra dataset[6] collected from 116 persons. From each participant nine clinical features were measured as displayed in Table 1.

Table 1. Anthropometric data(Features)

Feature	Description	Unit
A	Age	Years
BMI	Body Mass Index	Kg/m2
G	Glucose	Mg/dL
I	Insulin	µU/mL
HOMA	Homeostasis Model Assessment index	--
L	Leptin	Ng/mL
Ad	Adiponectin	µg/mL
R	Resistin	Ng/mL

These clinical measures are the anthropometric data collected during procedural blood analysis. Out of 116 input samples, 64 samples are with breast cancer and the remaining 52 are healthy samples. An observation of the features of dataset as illustrated in Figure 2 identifies the distinction between the cancerous and non-cancerous samples that clearly indicate that they are potential biomarkers for the detection of the pathology .

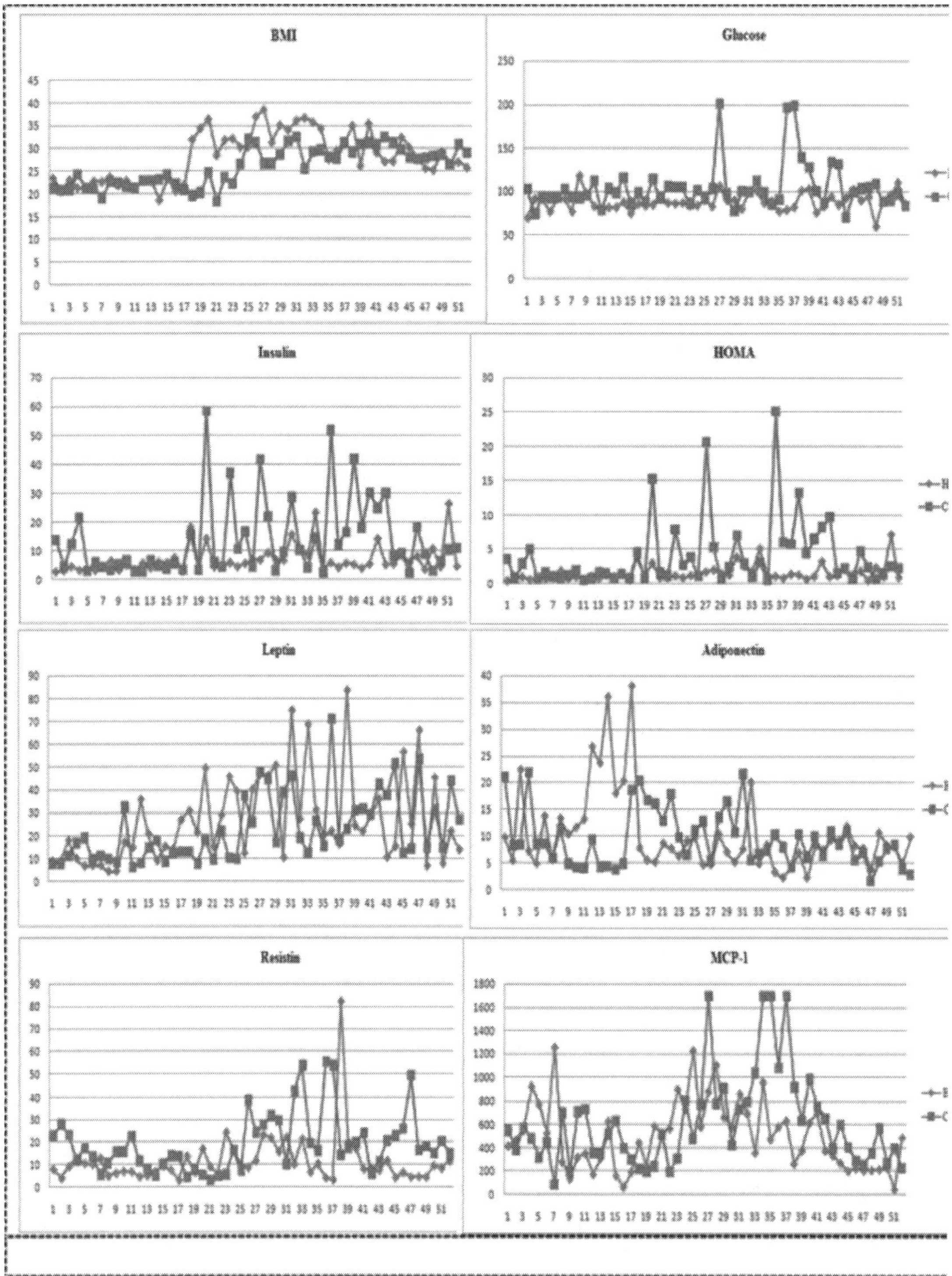

Figure 2. Features of Healthy and Cancerous samples

The features are combined to form a feature vector as shown,

$$FV=[\ A, BMI,\ G, I, HOMA, L,\ Ad, R, MCP\text{-}1] \qquad (1)$$

which is provided to the classifiers for training and testing. To validate the proposed system, the dataset with its features was divided into two non overlapping training and testing sets with a ratio of 75:25. The training set is labeled using the class labels y={'1' , '0'}, where '1' indicates healthy and '0' indicates Cancerous. The labeled training set {FV,y} is passed to the classifier for creating the model. The model created is then tested using the samples from the testing set to predict its outcome C ϵ y.

3.2Machine Learning Algorithms

The classifier predicts the outcome of an unseen sample 'Q' as a function of labeled training set and the learning parameters of the algorithm .

$$C=f(FV_Q, \{[FV,y]: \text{training parameters}\}$$

where FV is the feature vector of the training samples and FV_Q is the feature vector of the query/unseen sample.
The proposed system used three classifiers: Naive Baye's(NB), Radial Basis Function Neural network(RBFNN) and Linear Discriminant Analysis(LDA).

3.2.1 Naive Baye's classifier

It is a simple probabilistic classifier which is based on Baye's theorom[10]. It requires fewer number of training samples for prediction. Baye's rule can be formulated as

$$P(y/X)=\frac{P(X/y).P(y)}{P(X)} \quad (2)$$

where X signify the features and y is the label representing the category of the input sample. In the context of the work presented here, the above equation can be rewritten by expanding the feature set as,

$$P(C/f_1,f_2,\ldots\ldots,f_9) = \frac{P(f_1/C).P(f_2/C).P(f_3/C)\ldots\ldots P(f_9/C)\,P(C)}{P(f_1)P(f_2)\ldots\ldots P(f_9)} \quad (3)$$

where $(f_1, f_2, \ldots\ldots f_9)$=FV

For all the elements of the dataset, the denominator is constant, thus we get

$$P(C/f_1,f_2,\ldots\ldots,f_9) \text{ proportional } P(C)\prod_{i=1}^{9} P(f_i/C) \quad (4)$$

Here C can be one of the class label belonging to y. Thus the outcome C is found based on maximum probability, i.e.,

$$C=\text{argmax}_C\, P(C)\prod_{i=1}^{9} P(f_i/C) \quad (5)$$

The classifier finds its applications in biometrics(face recognition), healthcare, weather prediction, security, emotion recognition, sentiment analysis etc.

3.2.2 Radial Basis Function Neural network(RBFNN)

RBFNN[11] is one of the variants of feed forward neural network that uses radial basis functions as the activation function in the hidden layer. The network converges faster to reach the performance goal. The NN architecture has three layers: Input layer, Hidden layer and Output layer. The number of neurons in the input layer correspond to the dimension of the feature vector and the neurons in output layer correspond to the number of class labels. Each layer is associated with its activation function. The activation functions in the hidden layer are implemented as Guassian functions.

The labeled training set {FV,y} is provided to the network for training . During training the parameters such as the network weights, RBF activation function centres and their distributions are tuned reach the performance goal: Mean squared error(MSE)

For a given input FV, the output at a node of output layer is computed as,

$$C_i = \sum_{j=0}^{J} w_{ij}\, \varphi_j\,(\|FV - M_j\|) \quad (6)$$

Where, FV is the input to the network/Feature vector

w_{ij} is the weight from hidden neuron j to output neuron i

$\|.\|$ L_2 / Euclidean norm

φ_j is the RBF in the hidden neuron j with center M_j.

3.2.3 Linear Discriminant Analysis(LDA)

LDA classifier proposed by Fisher [12]is based on the concept of reducing the within(intra) class variation and maximizing the between(inter) class variation. This classifier finds class separation by minimizing the spread within class and maximizing the distance between each class. Thus it is based on the statistical measures mean and variance. For the given dataset and class labels, LDA works by projecting it on to hyperplane to find the scalar, of all the possible lines, the line that gives maximum class separability is chosen.

The best projection vector is selected by maximizing the Fisher's criteria given by,

$$J(W) = \frac{W^T S_B W}{W^T S_W W} \quad (7)$$

here J(W) indicates the measure of difference between means of the classes

S_B represents the *within class* scatter matrix S_W represents *between class* scatter matrix W is the projection matrix.

Maximizing J(W) yields which is Fisher's Linear Discriminant

$$W = \text{argmax}(J(W)) = S_W^{-1}(\mu 1 - \mu 2), \quad (8)$$

where, (μ1- μ2) is the distance between the projected means for two classes.

4. Results and Discussion

The experiment was conducted to identify the presence of the breast cancer for the given sample based on fusion of classifiers. The proposed method is validated with hold out method where the dataset with 116 samples was separated into training and testing sets with the ratio of 3:1. Thus the training set has 87 samples and testing set 29 samples.

As a first step, all the training samples were passed to feature extractor that provided the feature vector FV with 9 dimensions. Each feature of the FV has different ranges, so the data is normalized using Z- score normalization that gives normalized FV with zero mean and standard deviation of one as shown

$$FV_n = \frac{FV - \mu}{\sigma} \quad (9)$$

Later the FV_n was labeled to form training data set $Z_i = \{FV_n, y\}_{i=1}^{87}$. The training data set was accepted by the Naive Baye's(NB), Radial Basis Function Neural network(RBFNN) and Linear Discriminant Analysis(LDA) classifiers for creating the models. During training, the learnable parameters of the classifiers were modified based on the performance function. The RBFNN was optimized for a spread constant of 3.5 and for NB classifier the prior probabilities was set to [0.2 0.2]. The trained models were tested with the testing data set $Z_{iQ} = \{FV_n\}_{i=1}^{29}$, where Q represents the Query or unseen sample.

The results of the individual classifiers were observed and analyzed using the parameters Receiver Operating Characteristics(ROC), Classification accuracy(CA), Precision, Recall(Sensitivity), Specificity and F1 score which are computed using the elements of confusion matrix[13] as shown,

$$\text{Classification accuracy(CA)} = \frac{(TP+TN)}{(TP+TN+FP+FN)} \tag{10}$$

$$\text{Precision} = \frac{(TP)}{(TP+FP)} \tag{11}$$

$$\text{Recall(Sensitivity)} = \frac{(TP)}{(TP+FN)} \tag{12}$$

$$\text{Specificity} = \frac{(TN)}{(TN+FP)} \tag{13}$$

$$\text{F1 score} = 2 * \frac{(\text{Precion} * \text{Recall})}{(\text{Precision} + \text{recall})} \tag{14}$$

where TP - Number of true positives FN- Number of false negatives FP- Number of false positives TN- Number of true negatives

The ROC obtained for all the classifiers is displayed in Figure 3 and other metrics are shown in Table 2.

The analysis of ROC and the performance measures indicate the poor presentation of the classifiers when used individually. The highest CA, precision and specificity was obtained from RBFNN that provided 69%, 70% and 81.25%respectively.

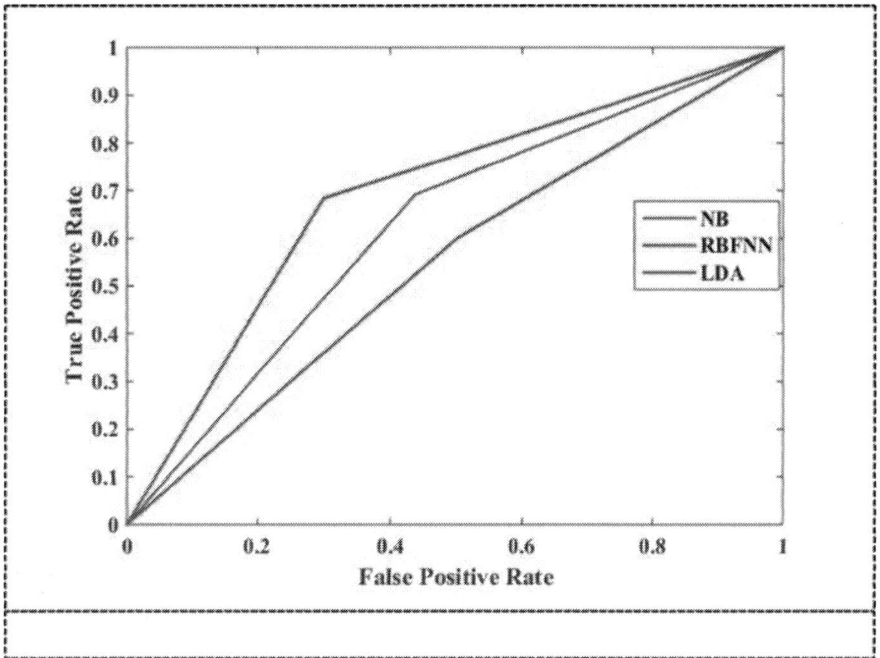

Figure 3. ROC obtained from the testing set

Table 2. Classifier results

Classifier	NB	RBFNN	LDA	Ensemble classification
CA(%)	62.06	69	58.17	75.86
Precision(%)	56.25	70	53.84	75
Recall(%)	69.23	53.84	53.84	69.23
Specificity(%)	56.25	81.25	62.5	81.25
F1-score	0.6206	0.6086	0.5383	0.7199

To improve the performance of the system, the decision of the individual classifiers can be fused[14] which is applied in the proposed frame work. A review of the literature finds different fusion strategies based on max, min, sum, product, mean, median and majority vote rules. This work utilizes binary classifiers with class label vector y={'1', '0'}. Let C_1, C_2 and C_3 be the outputs of NB, RBFNN and LDA classifiers such that Ci ϵ y. The outputs are fused as shown,

$$C= F(C1, C2, C3) \quad (15)$$

here F signifies the fusion rule

Now based on the fusion, we arrive at the final decision by opting for majority voting rule which requires odd number of single classifiers. This is the simple and a sensitive approach that assigns the sample with a most frequent class labels. Finally to demonstrate the significance of ensemble classification, the final decision was analyzed using the performance metrics and the results are tabulated in Table 1. The results show an 75% precision with ensemble classification that represents a low false positive rate . The analysis of F1-score which is the harmonic mean of precision and recall is advantageous in our work as the dataset considered has uneven class distribution(52 healthy and 64 cancerous samples). The range of F1-score is [0 1]. Thus a value of 0.7199 is better as compared to individual classifier results.

It is observed from the results that, the ensemble classification provided the results greater than the average of the single classifiers indicating a enhanced performance of the system subjected to CA, precision, recall, specificity and F1-score.

5. Conclusion

This paper proposes a pattern recognition system with ensemble of classifiers for classifying a sample into two classes viz presence of breast cancer and healthy. The work utilized anthropometric data /biomarkers as the features for data representation . These features or descriptors were provided to the NB, RBFNN and LDA machine learning algorithms for classifier model creation and evaluation. The experimental study on breast cancer Coimbra dataset with the proposed framework revealed the efficacy of the biomarkers as better features. Further the analysis of results obtained from fusing the decisions of single classifiers indicated improved performance measures in terms of classification accuracy, precision, recall, specificity and F1-score. This justified the deployment of ensemble classifiers in better discrimination of classes leading to enhanced performance of the system

6. References

[1] Kim SH, Hoffmann U, Borggrefe M, Akin I, Behnes M. Advantages and limitations of current biomarker research: from experimental research to clinical application. Current Pharmaceutical Biotechnology. 2017 May 1;18(6):445-55.

[2] Chatterjee, Sabarni K., and Bruce R. Zetter. "Cancer biomarkers: knowing the present and predicting the future." (2005): 37-50.

[3] Cole KD, He HJ, Wang L. Breast cancer biomarker measurements and standards. PROTEOMICS–Clinical Applications. 2013 Jan;7(1-2):17-29.

[4] Li Y, Chen Z. Performance evaluation of machine learning methods for breast cancer prediction. Appl Comput Math. 2018 Oct 18;7(4):212-6.

[5] Sardouk F, Duru AD, Bayat O. Classification of breast cancer using data mining. American Scientific Research Journal for Engineering, Technology, and Sciences (ASRJETS). 2019 Jan 4;51(1):38-46.
[6] Patrício M, Pereira J, Crisóstomo J, Matafome P, Gomes M, Seiça R, Caramelo F. Using Resistin, glucose, age and BMI to predict the presence of breast cancer. BMC cancer. 2018 Dec 1;18(1):29
[7] Yue J, Zhao N, Liu L. Prediction and Monitoring Method for Breast Cancer: A Case Study for Data from the University Hospital Centre of Coimbra. Cancer Management and Research. 2020;12:1887.
[8] Austria YD, Jay-ar PL, Maria Jr LB, Goh JE, Goh ML, Vicente HN. Comparison of Machine Learning Algorithms in Breast Cancer Prediction using the Coimbra Dataset. cancer. 2019;7:10.
[9] Ray R, Abdullah AA, Mallick DK, Dash SR. Classification of Benign and Malignant Breast Cancer using Supervised Machine Learning Algorithms Based on Image and Numeric Datasets. InJournal of Physics: Conference Series 2019 Nov (Vol. 1372, No. 1, p. 012062). IOP Publishing.
[10] Rish I. An empirical study of the naive Bayes classifier. InIJCAI 2001 workshop on empirical methods in artificial intelligence 2001 Aug 4 (Vol. 3, No. 22, pp. 41-46).
[11] Mahesh VG, Raj AN. Invariant face recognition using Zernike moments combined with feed forward neural network. International Journal of Biometrics. 2015;7(3):286-307.
[12] Bouzalmat A, Kharroubi J, Zarghili A. Comparative study of PCA, ICA, LDA using SVM classifier. Journal of Emerging Technologies in Web Intelligence. 2014 Feb 1;6(1):64-8.
[13] Fawcett T. ROC graphs: Notes and practical considerations for researchers. Machine learning. 2004 Mar 16;31(1):1-38.
[14] Mi A, Wang L, Qi J. A multiple classifier fusion algorithm using weighted decision templates. Scientific Programming. 2016 Jan 1;2016.

Optimization of Nano Capacitive Pressure Sensor for Medical Applications

Kirankumar B. Balavalad*, PraveenKumar B. Balavalad & Somanath Pidashetti***
*Electronics & Communication Engineering Department, Basaveshwar Engineering College (A), Bagalkot- 587103, Karnataka, India. kiranb4004@gmail.com
**Automobile Engineering Department, Basaveshwar Polytechnic College (A), Bagalkot-587103, Karnataka, India, praveenmech410@gmail.com

Abstract.: Nano/NEMS sensors have emerged from MEMS technology. Nano sensors have got lot of applications in medical and health care applications. Nano sensors have lot of attributes which make them very unique and special, such as: small size, low mass, high sensitivity and low cost of production. In this paper we present optimization of nano capacitive pressure sensor dimensions for medical applications. Capacitive pressure sensors are well suited for low pressure sensing applications. Typically, in medical field involve low pressure levels ranging from few Pascals to kilo Pascals. DoE is used for the optimization of sensor dimensions. The sensor is optimized to operate over the pressure range up to 100 kPa. Over this pressure range the sensor dimensions are optimized to get better response in terms of capacitance. The proposed optimization illustrates the dimensions best suited for the design of nano capacitive pressure sensor for operational range up to 100 kPa.

Keywords: Optimization, nano sensor, nano capacitive pressure sensor, DoE, ANOVA.

1. Introduction

NEMS/Nano sensors are now being widely explored in the medical field. Medical field need more and more tiny sensors and actuators which can work efficiently and reliably. Medical field is one, where always we need to deal with extremely low-pressure levels (usually ranging from 0 to 1bar). Capacitive pressure sensors have proven to be most suitable for low pressure sensing applications. Currently lots of research activities are going on to drag micron sized sensors to nano sized. Nano sized sensors have got many advantages compared to the conventional MEMS (micron sized) sensors. NEMS is integration of electronics and mechanical components on the nano scale. Using NEMS technology researchers have developed devices like sensors, actuators, gears, accelerometers, cantilevers, tweezers and many other nano devices. Different properties of the NEMS-based devices, which make them unique, are low mass, high electrical strength, high mechanical resonance frequencies, potentially large quantum mechanical effects and a high surface-to-volume ratio useful for surface-based sensing mechanisms. The global NEMS market is expected to reach $108.88 million by 2022. NEMS based sensors are well suited for medical applications as they are highly sensitive and can be easily mounted in/on the human body. NEMS devices are being developed for different health care applications such as, blood pressure sensing, intercranial pressure sensing, drug delivery systems, blood sugar detection systems and various other medical emergency detection systems. Capacitive pressure sensors provide better sensitivity and are temperature invariant [1-3]. These sensors make use of parallel plate transduction principle, were applied pressure alters the capacitance between two plates as the gap between the two plates decreases. [4, 5]. Capacitive sensors consume less power and are invariant of temperature but response will be nonlinear [6-10]. Capacitive sensors are compatible with most mechanical structures. Most of the time theory of thin plates is used in the design considerations of the capacitive pressure sensors [11]. Paper [11], describes the use of theory of thin plates in the design of sensors. The paper provides the simulation of all available transduction mechanisms in the pressure sensing domain. A critical review of the capacitive pressure sensors is given [12], considering special types of capacitive sensors like slotted, touch mode etc. Many papers in the literature describe the design and development of MEMS based capacitive pressure sensors for medical applications. Diaphragm shapes typically used are square, circular, rectangular, amongst all square diaphragm provides proper stress distribution in the diaphragm. A proper analysis of different shapes of the diaphragm are presented in [13].

Paper [14], shows the mathematical formulation and simulation of elliptical diaphragm. There are few papers that describe the optimization of MEMS based capacitive pressure sensor for medical applications. [15] presents, a highly sensitive MEMS capacitive pressure sensor for biomedical applications. The sensor has been designed to measure pressures in the range of 0 to 60 mmhg. In paper [16], authors have optimized MEMS based capacitive pressure sensor for intraocular pressure measurement used for monitoring glaucoma. Authors [16], suggest that the diaphragm in capacitive pressure sensor is the important factors that affect the sensor's performance. Taguchi and Two-Level Factorial approach are used by authors in [16] for optimization of the diaphragm thickness, slot width, and slot length.

II. Design of Nano Capacitive Pressure Sensor

The Nano sensor is basically made of two electrode i.e Top and Bottom electrode. The working of this sensor is dependents upon the top plate because its moveable and bottom plate is always fixed. whenever you can apply the external pressure on the top electrode it will move to words the bottom electrode than the distance between two electrodes will decreases hence capacitance will increase

.

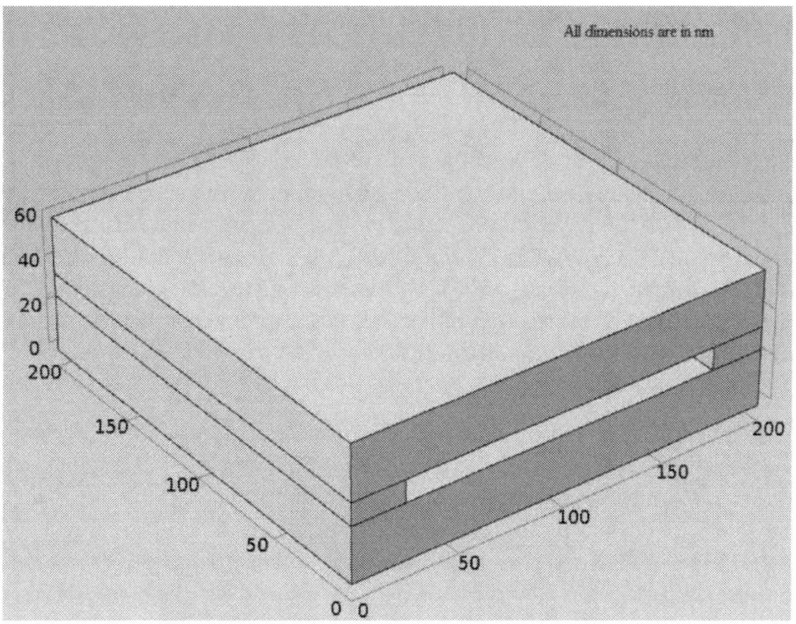

Fig. 1. Model of the Nano Capacitive Pressure

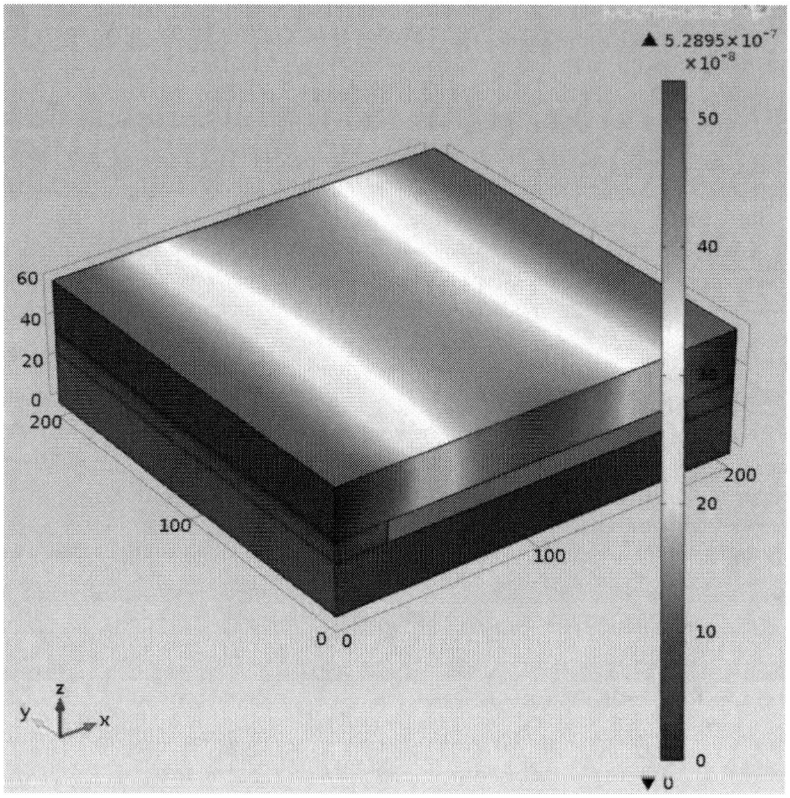

Fig. 2. Simulation model of the Nano Pressure Sensor

III. Optimization of Nano Capacitive Pressure Sensor

DoE based optimization of sensor dimensions is proposed in this paper [16, 17]. In this section the optimization of Sensor for medical applications is considered. A square diaphragm is designed in the proposed work. The sensor dimensions like length/width, thickness and the gap between the plates is optimized for the better performance of sensor at nano level. Input to the device is external pressure, the performance of the sensor mainly depends on the diaphragm dimensions and on the gap between the two electrodes. In this work the focus in on finding the effect of individual sensor dimensions on the performance of the sensor. DoE was designed and performed on sensor dimensions using predefined combinations with L9 orthogonal array. ANOVA (Analysis of Means) is also applied on the sensor, which gives a better insight of effect analysis of individual sensor dimensions. The factors affecting the sensor performance are capacitive sensor dimensions viz., 1. Length/Width (A). 2. Thickness of diaphragm (B). 3. Plate gap (C).

The above-mentioned factors are considered here for effect analysis and hence find the optimum combination. Each factor is assigned with the three levels. The literature reveals that the lower bound on the length/width of the sensor diaphragm is 200 nm and upper bound is 700 nm. The levels defined for thickness are 25 nm, 50 nm and 100 nm. The levels for the electrode gap are followed by considering the fabrication feasibility. The levels for gap considered are 10 nm, 20 nm and 30 nm. Table 1, presents the factors and their levels assigned.

Table 1. Factors and their levels

Factors	Level 1	Level 2	Level 3
Length/Width	200 nm	500 nm	700 nm
Thickness	25 nm	50 nm	75 nm
Gap	10 nm	20 nm	30 nm

Simulation for the above parameters/factors were carried out using L9 orthogonal array. In this array 1, 2, 3 represent the levels of respective factors. In the array, for any pair of columns, all combinations of factor levels occur and they occur equal number of times, hence the array is orthogonal. In Table 2, the factors combinations and the responses are tabulated i.e., output capacitance of the sensor. Simulation was conducted for five repeated times and an average value is taken for say experiment i, where i =1, 2, 3, … 9 different experiments/simulation combinations.

Table 2. Experiment as per L9 orthogonal array

Expt. No	Column Number and Factor Assigned			Output Capacitance (fF) at 100kPa pressure
	1. Length/ Width (nm)	**2. Thickness (nm)**	**3. Gap between the electrodes (nm)**	
1.	200	25	10	$1.45983e^{-16}$
2.	200	50	20	$6.18834e^{-17}$
3.	200	100	30	$4.14376e^{-14}$
4.	500	25	20	$9.93238e^{-16}$
5.	500	50	30	$1.45983e^{-16}$
6.	500	100	10	$2.0372e^{-16}$
7.	700	25	30	$1.97167e^{-15}$
8.	700	50	10	$9.41765e^{-16}$
9.	700	100	20	$4.34157e^{-16}$

From the response data in the Table 3, the overall mean is computed, as

$$m = \frac{1}{9}\sum_{i=1}^{9} C_i = \frac{1}{9}[C_1 + C_2 + \cdots + C_9] = 5.14 \times 10^{-15}$$

3.1. Effect of Factors at Different Levels

This section presents the effect of the Length/width, Thickness and Gap factors at different levels. All the mean values and effect values are in nf.

$$m_{A3} = \frac{1}{3}[C_7 + C_8 + C_9] = 1.115 \times 10^{-15}$$

The effect of Length/width (A) at level 2 is given by the result of experiment number 4, 5, 6.

$$m_{A2} = \frac{1}{3}[C_4 + C_5 + C_6] = 4.47 \times 10^{-16}$$

The effect of Length /width (A) at level 1 is given by the result of experiment number 4, 5, 6.

$$m_{A1} = \frac{1}{3}[C_1 + C_2 + C_3] = 1.38 \times 10^{-14}$$

The effect of length /width (A) at level A3 is given by, $(m_{A3}\text{-}m) = -4.025\times10^{-15}$ The effect of length/width (A) at level A2 is given by, $(m_{A2}\text{-}m) = -4.69\times10^{-15}$. The effect of length /width (A) at level A1 is given by, $(m_{A1}\text{-}m) = 8.66\times 10^{-15}$.

3.1.2. Effect of Thickness (B) at Different Levels

The effect of Thickness (B) at level 3 is given by the result of experiment number 3, 6, 9.

$$m_{B3} = \frac{1}{3}[C_3 + C_6 + C_9] = 1.48 \times 10^{-14}$$

The effect of Thickness (B) at level 2 is given by the result of experiment number 2, 5, 8.

$$m_{B2} = \frac{1}{3}[C_2 + C_5 + C_8] = 1.14 \times 10^{-15}$$

The effect of Thickness (B) at level 1 is given by the result of experiment number 1, 4, 7.

$$m_{B1} = \frac{1}{3}[C_1 + C_4 + C_7] = 3.51 \times 10^{-15}$$

The effect of Thickness (B) at level B3 is given by, $(m_{B3}\text{-}m) = 9.66\times 10^{-15}$.
The effect of Thickness (B) at level B2 is given by, $(m_{B2}\text{-}m) = -4\times 10^{-15}$.
The effect of Thickness (B) at level B1 is given by, $(m_{B1}\text{-}m) = -1.63\times 10^{-15}$.

3.1.3. Effect of Gap(C) at Different Levels

The effect of Gap (C) at level 3 is given by the result of experiment number 3, 5, 7.

$$m_{C3} = \frac{1}{3}[C_3 + C_5 + C_7] = 1.45 \times 10^{-14}$$

The effect of Gap (C) at level 2 is given by the result of experiment number 2, 4, 9.

$$m_{C2} = \frac{1}{3}[C_2 + C_4 + C_9] = 4.96 \times 10^{-16}$$

The effect of Gap (C) at level 1 is given by the result of experiment number 1, 6, 8.

$$m_{C1} = \frac{1}{3}[C_1 + C_6 + C_8] = 2.60 \times 10^{-16}$$

The effect of Gap (C) at level C3 is given by, $(m_{C3}\text{-}m) = 9.36\times 10^{-15}$.
The effect of Gap (C) at level C2 is given by, $(m_{C2}\text{-}m) = -4.644\times 10^{-15}$.
The effect of Gap (C) at level C1 is given by, $(m_{C1}\text{-}m) = -4.88\times 10^{-15}$.
now we can find the optimum combinations of factors. The optimum combinations are indicated where the contribution of the level to deviate away from the mean. Table 5, presents the optimum levels indicated with /* mark.
Table 3. Optimum Combinations. The means plot describes thickness has got least effect, so it can be neglected while considering the combinations. But considering all factor effects the optimum combinations to have enhancement in output voltage the combinations need to be preferred are A3B1C3 or A3B1C2

Table. 3. Optimum Combinations

Factors	Level 1	Level 2	Level 3
A. Length/ Width	$1.38 \times 10^{-14*}$	4.47×10^{-16}	1.115×10^{-15}
B. Thickness	3.51×10^{-15}	1.14×10^{-15}	$1.48 \times 10^{-14*}$
C. Gap	2.60×10^{-16}	4.96×10^{-16}	$1.45 \times 10^{-14*}$

The objective of the Taguchi methods is to predict optimum condition/combinations. The optimum predicted conditions as mentioned in the Table 3. A3B1C3. Using the additive model, the value of C under optimum combinations as,

$$C = m + (m_{A1} - m) + (m_{B3} - m) + (m_{C3} - m) = 4.824 \times 10^{-14}$$

A verification simulation is conducted after determining the optimum conditions and predicting the response with these combinations. The result of the verification simulation is compared with the predicted value of C. Predicted value of C is 4.824×10^{-14} and the verification simulation gave the response for C as $4.14376e^{-14}$. So this actually concludes that this model is adequate in describing C in terms of the variables defined.

IV. ANOVA: Analysis of Means Approach to Find the Effect of Capacitor Dimension of Sensor Response

The effect of different factors of capacitor and an optimized levels and combinations were defined using means plot in the previous section. But the better insight of relative effect of different factors can be obtained by decomposition of variance, known as Analysis of Variance (ANOVA). ANOVA also provides the variance for the factor effects and variance of the prediction error. It involves three decompositions viz., grand sum of squares, sum of squares due to mean and total sum of squares. Where Total sum of squares = (Grand sum of squares – sum of squares due to mean) The model of experimentation is same as defined in Section II. All factor and their levels defined in above section i.e., in DoE are followed in this section too. Same responses i.e., the output capacitance for input pressure of 100 kPa is considered for ANOVA.

ANOVA:

1. Grand Sum of Squares (GSS) is given by

$$GSS = \sum_{i=1}^{9} C_i^2 = 1.716 \times 10^{-27}$$

2.Sum of Squares due to Mean (GSM) is given by

GSM = (number of experiments)× m^2.

$$GSM = 9 \times 5.14 \times 10^{-15} = 2.377 \times 10^{-28}$$

3.The Total Sum of Squares (TSS) is given by

$$TSS = \sum_{i=1}^{9}[C_i - m]^2 = 7.32 \times 10^{-17}$$

4. TSS=GSS-GSM

$$TSS = 1.716 \times 10^{-27} - 2.377 \times 10^{-28} = 1.4783 \times 10^{-27}$$

5. Sum of Squares due to individual factors

1) Sum of squares due to Factor A.

$$\text{Length/Width} = [3(m_{A1} - m)^2 + 3(m_{A2} - m)^2 + 3(m_{A3} - m)^2] = 3.386 \times 10^{-28}$$

2) Sum of Squares due to Factor B

$$\text{Thickness} = [3(m_{B1} - m)^2 + 3(m_{B2} - m)^2 + 3(m_{B3} - m)^2] = 3.349 \times 10^{-28}$$

3) Sum of Squares due to Factor C.

$$\text{Gap} = [3(m_{C1} - m)^2 + 3(m_{C2} - m)^2 + 3(m_{C3} - m)^2] = 3.989 \times 10^{-28}$$

Based on the above findings, ANOVA table is prepared, and is shown in Table 4.

Table. 4. ANOVA Table

Factors	**DoF**	**Sum of Squares**	**Mean Squares**	**F Ratio**
A. Length	2	316.87	334.612	3216.91
B. Thickness	2	392.31	336.173	3231.91
C. Gap	2	746.24	373.122	3587.13
Error	1	0.10	0.104	0.104
Total	7	1455.53		
(Error)	4			

From ANOVA Table 5, factor A is responsible for (316.87/1455.53) = 21.77% percent of variation of Capacitance. Similarly All factor effects are tabulated in table7.

Table 5. Effect of individual factors on the output of sensor

Factors	**Percentage effect of factors for variation in output voltage in '%'**
Length	21.77
Thickness	26.94
Gp of the sensor	51.26

V. Conclusion

The paper presents a very unique and novel technique of finding the individual effect of piezoresistor dimensions on the output of the sensor & the optimum combinations of levels required to enhance the capacitance of the nano sensor. Design of Experiment & ANOVA methods are used to achieve the purpose. The results & observations suggest the best combination of levels to optimize the response of the sensor is $A_1B_3C_3$ i.e, 200nmX75nmX30nm. The predicted response after optimization (using DoE) was 4.824×10^{-14} f and the validation simulation using the optimum combinations shows the response value of 4.143×10^{-14} f. And ANOVA method describes that the effect of gap of the sensor has highest effect on the output variations of the sensor. Length has 21.77% effect next is thickness with 26.94% effect and highest is the effect of gap 51.26%. Therefore, to enhance the output response of the sensor, gap between the electrodes has to be carefully addressed.

REFERENCES

[1] W. P. Eaton, J. H. Smith, Micromachined pressure sensors: review and recent development,*Smart Materials and Structures*, Vol. 6, 1997, pp. 530-539.

[2] K. N. Bhat, M. M. Nayak, MEMS Pressure Sensor-An overview of challenges in 'Technology and Packaging', *Journal of ISSS*, Vol. 2, No. 1, March 2013, pp. 39- 71.

[3] T. Supriya, Dr. V. Jeyalakshmmi, Survey on Pressure Sensors in the Previous Decades, *International Journal of Emerging Technology and Advanced Engineering, ISO 9001:2008*, Vol. 3, Issue 7, July 2013.

[4] C. S. Sander, J. W. Knutti, J. D. Meindle, A monolithic capacitive pressure sensor with pulse period output, *IEEE Transactions on Electron Devices*, 1980, pp. 927-930.

[5] Gitesh Mishra, Neha Paras, Arti Arora, P. J. George, Simulation of MEMS Capacitive Pressure Sensor Using Comsol Multhysics, *International Journal of Applied Engineering Research*, Vol. 7, No. 11, 2012.

[6] Tamas Karpati, Andrea Edit Pap, Sandor Kulinyi, Prototype MEMS Capacitive Pressure Sensor Design and Manufacturing, *Electrical Engineering and Computer Science*, 57, 1, 2013, pp. 3-7.

[7] Sensata Technologies, Pressure Transducers TechnicalNote 1, Understanding And Specifying Capacitive Pressure Transducers, Mitch Berkson, Original, Dave Field, Update, 2007.

[8] Ko, W.H.; Wang, Q. "Touch mode capacitive pressure sensors for industrial applications", IEEE International Conference, Micro Elec. Mech. Syst. (MEMS) 1997, 284-289.

[9] Chapter 6, Experimental Methods for Engineers, 6th edition, J.P. Holman, 1994, PRESSURE MEASUREMENTS, www.mech.uq.edu.au/courses/metr3100/ST5_Press_Meas.pdf.

[10] Y. Hezarjaribi, Golestan University, Gorgan, Iran, M. N. Hamidon, A. R. Bahadorimehr University Putra Malaysia, 43400 UPM Serdang, Selangor, Malaysia , S. H. Keshmiri, University of Ferdowsi, Mashhad, Iran "Capacitive pressure sensor technology and applications", ICSE 2008 Proceedings. 2008, Johor Bahru, Malaysia.

[11] S. Timoshenko, S. Woinowsky-Krigger, Theory of Plates and Shells, Mc. Graw-Hill, 1959, pp. 13, 105, 202.

[12] Kirankumar B. Balavalad, B. G. Sheeparamatti, "A Critical Review on MEMS Capacitive Pressure Sensors", Sensors & Transducers journal, IFSA publication, ISSN: 2306-8515, e-ISSN 1726-5479 Volume 187, Issue 4, pp-120-128, April 2015.

[13] Rohit Goyal, "Performance Optimization of Silicon Capacitive Pressure Sensor", Advanced Research in Electrical and Electronic Engineering Print ISSN: 2349-5804; Online ISSN: 2349-5812 Volume 2, Number 4 January-March (2015) pp. 316-320 © Krishi Sanskriti Publications http://www.krishisanskriti.org/areee.html

[14] Rishabh Bhooshan Mishra and S. Santosh Kumar 2019 J. Phys.: Conf. Ser. 1240 012068.

[15] M. Shahiri-Tabarestani, B. A. Ganji and R. Sabbaghi-Nadooshan, "Design and simulation of high sensitive capacitive pressure sensor with slotted diaphragm," 2012 International Conference on Biomedical Engineering (ICoBE), Penang, 2012, pp. 484-489. doi: 10.1109/ICoBE.2012.6179064

[16] Muslihah Ali, Abdullah C. W. Noorakma, Norliana Yusof, W. N. F. MohamaN. Soin, S. F. Wan Muhamad Hatta "Optimization of MEMS Intraocular Capacitive Pressure Sensor" IEEE-ICSE2016 Proc. 2016, Kuala Lumpur, Malaysia.

[17] Madhav Padke, Quality Engineering Using Robust Design, Prentice Hall, 1989. https://www.ee.iitb.ac.in/~apte/CV_PRA_TAGUCHIINTRO.ht

AUTHOR INDEX